Precoding and Signal Shaping for Digital Transmission

Precoding and Signal Shaping for Digital Transmission

Robert F. H. Fischer

IEEE

The Institute of Electrical and Electronics Engineers, Inc., New York

WILEY-INTERSCIENCE

A JOHN WILEY & SONS, INC., PUBLICATION

Published simultaneously in Canada.

Library of Congress Cataloging-in-Publication Data is available.

Fischer, Robert, F. H.
Precoding and Signal Shaping for Digital Transmission/
p. cm.
Includes bibliographical references and index.
ISBN 0-471-22410-3 (cloth: alk. paper)
ISBN 13: 978-0-471-22410-5

10 9 8 7 6 5 4 3 2 1

Contents

Preface

This book is the outcome of my research and teaching activities in the field of fast digital communication, especially applied to the subscriber lines network, over the last ten years. It is primarily intended as a textbook for graduate students in electrical engineering, specializing in communications. However, it may also serve as a reference book for the practicing engineer. The reader is expected to have a background in engineering and to be familiar with the theory of signals and systems—the basics of communications, especially digital pulse-amplitude-modulated transmission, are presumed.

The scope of this book is to explain in detail the fundamentals of digital transmission over linear distorting channels. These channels—called intersymbol-interference channels—disperse transmitted pulses and produce long-sustained echos. After having reviewed classical equalization techniques, we especially focus on the applications of precoding. Using such techniques, channels are preequalized at the transmitter side rather than equalized at the receiver. The advantages of such strategies are highlightened, and it is shown how this can be done under a number of additional constraints. Furthermore, signal shaping algorithms are discussed, which can be applied to generate a wide range of desired properties of the transmitted or received signal in digital transmission. Typically, the most interesting property is low average transmit power. Combining both techniques, very powerful and flexible schemes can be established. Over recent years, such schemes have attracted more and more interest and are now part of a number of standards in the field of digital transmission systems.

I wish to thank everyone who supported me during the preparation of this book. In particular, I am deeply indebted to my academic teacher Prof. Dr. Johannes Huber for giving me the opportunity to work in his group, for his encouragement, his valuable advice in writing this book, and for the freedom he gave me to complete my work. The present book is strongly influenced by him and his courses that I had the chance to attend.

Many thanks to all proofreaders for their diligent review, helpful comments, and suggestions. Especially, I would like to acknowledge Dr. Stefan Müller-Weinfurter for his detailed counsel on earlier versions of the manuscript, and Prof. Dr. Johann Weinrichter at the Technical University of Vienna for his support. Many thanks also to Lutz Lampe and Christoph Windpassinger for their critical reading. All remaining inadequateness and errors are not due to their fault, but because of the ignorance or unwillingness of the author.

Finally, I express thanks to all colleagues for the pleasant and companionable atmosphere at the Lehrstuhl für Informationsübertragung, and the entire Telecommunications Laboratory at the University of Erlangen-Nürnberg.

ROBERT F. H. FISCHER

Erlangen, Germany
May 2002

1

Introduction

Reliable digital transmission is the basis of what is commonly called the "information age." Especially the boom of the Internet and its tremendous growth are boosting the ubiquity of digital information. Text, graphics, video, and sound are certainly the most visible examples. Hence, high-speed access to the global networks is one of the key issues that have to be solved. Meanwhile not only business sites are interested in fast access, but also private households increasingly desire to become connected, yearning for ever-increasing data rates.

Of all the network access technologies currently under discussion, the *digital subscriber lines (DSL)* technique is probably the most promising one. The copper subscriber lines, which were installed over the last decades, were only used for the *plain old telephone system (POTS)* or at most for *integrated services digital network (ISDN)* services. But dial-up (voice-band) modems with data rates well below 50 kbits/s are only able to whet the appetite for Internet access.

During the 1980s it was realized that this medium can support data rates up to some megabits per second for a very high percentage of subscribers. Owing to its high degree of penetration, the use of copper lines for digital transmission can build an easy-to-install and cost-efficient bridge from today's analog telephone service to the very high-speed fiber-based communications in the future. Hence, copper is probably the most appealing candidate to solve the "last-mile problem," i.e., bridging the distance from the central office to the customer's premises.

Initiated by early research activities and prototype systems in Europe, at the end of the 1980s an the beginning of the 1990s, broad research activities began which led to what is now commonly denoted as digital subscriber lines. Meanwhile a whole family of philosophies and techniques are being designed or are already in practical use. The first instance to be mentioned is *high-rate digital subscriber lines (HDSL)*, which provide 2.048 Mbits/s (E1 rate in Europe) or 1.544 Mbits/s (DS1 rate in North America) in both directions, typically using two wire pairs. HDSL can be seen as the successor of ISDN primary rate access. Contrary to HDSL, which is basically intended for commercial applications, *asymmetric digital subscriber lines (ADSL)* are aimed at private usage. Over a single line, ADSL offers up to 6 Mbits/s from the central office to the subscriber and a reverse channel with some hundred kbits/s—hence the term asymmetric. Interestingly, ADSL can coexist with POTS or ISDN on the same line. Standardization activities are presently under way for *single-pair digital subscriber lines (SDSL)* (sometimes also called symmetric DSL), which will support 2.312 Mbits/s in both directions while occupying only a single line. Finally, *very high-rate digital subscriber lines (VDSL)* have to be mentioned. If only some hundred meters, instead of kilometers, have to be bridged, the copper line can carry up to 50 Mbits/s or even more.

The purpose of this book is to explain in detail the fundamentals of digital transmission over channels which disperse the transmitted pulse and produce long-sustained echos. We show how to equalize such channels under a number of additional constraints. Thereby, we focus on *precoding techniques*, which do preequalization at the transmitter side, and which in fact enable the use of channel coding. Moreover, *signal shaping* is discussed, which provides further gains, and which can be applied to generate a wide range of desired properties for transmit or received signal. Combining both strategies, very powerful and flexible schemes can be established.

Even though most examples are chosen from the DSL world, the concepts of equalization and shaping are applicable to all scenarios where digital transmission over distorting channels takes place. Examples are power-line communication with its demanding transmission medium, or even mobile communications, where the time-varying channel is rather challenging.

We expect the reader to have an engineering background and to be familiar with the theory of signals and systems, both for the continuous-time and discrete-time case. This also includes knowledge of random processes and their description in the time and frequency domains. Also, the basics of communications, especially digital pulse-amplitude-modulated transmission, are assumed.

1.1 THE STRUCTURE OF THE BOOK

Figure 1.1 depcits the organization of this book.

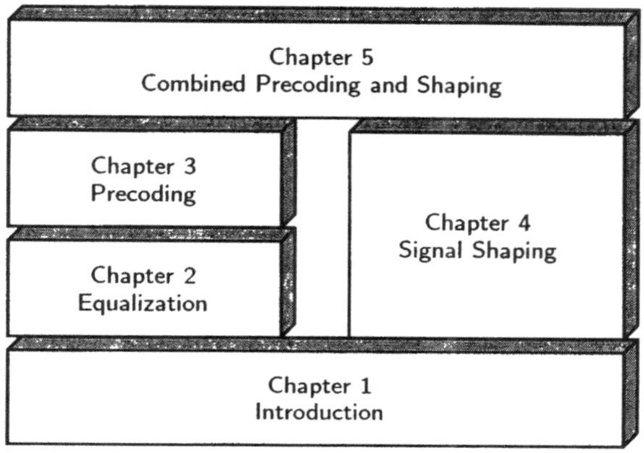

Fig. 1.1 Organization of the book.

Following this introduction, the topics of the four chapters are as follows:

Chapter 2: ***Digital Communications via Linear, Distorting Channels***

The fundamentals of digital communications over linear, distorting channels are discussed. After the problem description, linear equalization techniques are discussed. The optimal receiver is derived and the achievable signal-to-noise ratio is evaluated.

The performance can be improved via noise prediction. This leads to the concept of decision-feedback equalization, which is discussed and analyzed in detail.

After a summary on discrete-time end-to-end descriptions of the transmission and equalization schemes, further performance improvement by maximum-likelihood sequence estimation is explained briefly.

Chapter 3: ***Precoding Schemes***

This chapter is devoted to precoding schemes. First, Tomlinson-Harashima precoding is introduced and analyzed. Various aspects such as the compatibility with coded modulation and signal shaping are discussed. Then flexible precoding, an alternative scheme, is addressed. Combined coding and precoding is a topic of special interest.

Both precoding schemes are compared and the differences and dualities are illustrated via numerical simulations.

Finite-word-length implementation, in particular that of Tomlinson-Harashima precoding, is regarded. Thereby, a new, nonrecursive precoding structure is proposed.

Finally, some interesting information-theoretical aspects on precoding are given.

Chapter 4: ***Signal Shaping***

In this chapter, signal shaping, i.e., the generation of signals with least average power, is discussed. By using the signal points nonequiprobable, a power reduction is possible without sacrificing performance. The differences and similarities between shaping and source or channel coding are studied. Then, performance bounds on shaping are derived.

Two shaping schemes are explained in detail: shell mapping and trellis shaping. The shaping algorithms are motivated and their performance is covered by numerical simulations.

In the context of trellis shaping, the control of the power spectral density is studied as an example for general shaping aims.

The chapter closes with the optimization of the signal-point spacing rather than resorting to nonequiprobable signaling.

Chapter 5: ***Combined Precoding and Signal Shaping***

Combined precoding and signal shaping is addressed. In addition to preequalization of the intersymbol-interference channel, the transmit signal should have least average power.

In particular, the combination of Tomlinson-Harashima precoding and trellis shaping, called trellis precoding, is studied. Then, shaping without scrambling is presented, which avoids the disadvantages of trellis precoding and, without changing the receiver, can directly replace Tomlinson-Harashima precoding.

Besides average transmit power, further signal parameters may be controlled by shaping. Specifically, a restriction of the dynamic range at the receiver side and a reduction of the peak-to-average power ratio of the continuous-time transmit signal, are considered.

After a geometrical interpretation of combined precoding and shaping schemes is given, the duality of precoding/shaping to source coding of sources with memory is briefly discussed.

Appendices:

Appendix A summarizes the ***Wirtinger Calculus***, which is a handy tool for optimization problems depending on one or more complex-valued variables.

The ***Parameters of the Numerical Simulations*** given in this book are summarized in Appendix B.

In Appendix C, an ***Introduction to Lattices***, which are a powerful concept when dealing with precoding and signal shaping, is given.

The ***Calculation of Shell Frequency Distribution*** in shell-mapping-based transmission schemes is illustrated in Appendix D.

Appendix E generalizes precoding schemes and explains briefly ***Precoding for MIMO Channels***.

Finally, in Appendix F a ***List of Symbols, Variables, and Acronyms*** is given.

Note, the bibliography is given individually at the end of each chapter.

1.2 NOTATION AND DEFINITIONS

1.2.1 Signals and Systems

Continuous-time signals are denoted by lowercase letters and are functions of the continuous-time variable $t \in \mathbb{R}$ (in seconds), e.g., $s(t)$. Without further notice, all signals are allowed to be complex-valued, i.e., represent real signals in the equivalent complex baseband. By sampling a continuous-time signal—taking $s[k] \triangleq s(kT)$—where T is the sampling period, we obtain a sequence of samples $s[k]$, numbered by the discrete-time index $k \in \mathbb{Z}$ written in square brackets. If the whole sequence is regarded, we denote it as $\langle s[k] \rangle$.

The *Fourier transform* of a time-domain signal $x(t)$ is displayed as a function of the frequency $f \in \mathbb{R}$ (in hertz) and denoted by the corresponding capital letter. Transform and its inverse, respectively, are defined as

$$X(f) = \mathcal{F}\{x(t)\} \triangleq \int_{-\infty}^{\infty} x(t)\, \mathrm{e}^{-\mathrm{j}2\pi f t}\, \mathrm{d}t\,, \tag{1.2.1a}$$

$$x(t) = \mathcal{F}^{-1}\{X(f)\} \triangleq \int_{-\infty}^{\infty} X(f)\, \mathrm{e}^{\mathrm{j}2\pi f t}\, \mathrm{d}f\,. \tag{1.2.1b}$$

The correspondence between the time-domain signal $x(t)$ and its Fourier transform $X(f)$ is denoted briefly as

$$x(t) \circ\!\!-\!\!\bullet\, X(f)\,. \tag{1.2.2}$$

The z-*transform* of the sequence $\langle x[k] \rangle$ and its inverse are given as

$$X(z) = \mathcal{Z}\{x[k]\} \triangleq \sum_k x[k]\, z^{-k}\,, \tag{1.2.3a}$$

$$x[k] = \mathcal{Z}^{-1}\{X(z)\} \triangleq \frac{1}{2\pi \mathrm{j}} \oint X(z)\, z^{k-1}\, \mathrm{d}z\,, \tag{1.2.3b}$$

for which we use the short denomination $x[k] \circ\!\!-\!\!\bullet\, X(z)$, too.

Regarding the Fourier pair (1.2.2), the spectrum $X^{(\mathrm{d})}(\mathrm{e}^{\mathrm{j}2\pi f T})$ of the sampled signal $x^{(\mathrm{d})}[k] \triangleq x(kT)$ and that of the continuous-time signal $x(t)$ are related by

$$X^{(\mathrm{d})}(\mathrm{e}^{\mathrm{j}2\pi f T}) = \frac{1}{T} \sum_\mu X(f - \tfrac{\mu}{T})\,. \tag{1.2.4}$$

Because the spectrum is given by the z-transform, evaluated on the unit circle, we use the denomination $\mathrm{e}^{\mathrm{j}2\pi f T}$ as its argument. Moreover, this emphasizes the periodicity of the spectrum over frequency.

1.2.2 Stochastic Processes

In communications, due to the random nature of information, all signals are members of a *stochastic process*. It is noteworthy that we do not use different notations when dealing with the process or a single sample function/sequence thereof. *Expectation* is done across the ensemble of functions belonging to the *stochastic process* and denoted by $\mathsf{E}\{\cdot\}$.

Autocorrelation and *cross-correlation sequences* of (wide-sense) stationary processes shall be defined as follows

$$\phi_{xx}[\kappa] = \mathsf{E}\{x[k+\kappa] \cdot x^*[k]\} , \qquad (1.2.5a)$$

$$\phi_{xy}[\kappa] = \mathsf{E}\{x[k+\kappa] \cdot y^*[k]\} . \qquad (1.2.5b)$$

The respective quantities for continuous-time processes are defined accordingly.

The power spectral density corresponding to the autocorrelation sequence $\phi_{xx}[\kappa]$ of a stationary process is denoted as $\Phi(\mathrm{e}^{\mathrm{j}2\pi fT})$, and both quantities are related by the z-transform

$$\Phi(\mathrm{e}^{\mathrm{j}2\pi fT}) = \mathcal{Z}\{\phi_{xx}[\kappa]\} . \qquad (1.2.6)$$

When dealing with cyclostationary processes (e.g., the transmit signal in pulse amplitude modulation), the average power spectral density is regarded.

Finally, $\Pr\{\cdot\}$ stands for the *probability* of an event. If the random variable x is distributed continuously rather than discrete, its distribution is characterized by the *probability density function (pdf)* $\mathsf{f}_x(x)$. In case of random variables x, conditioned on the event y, we give the conditional pdf $\mathsf{f}_x(x|y)$.

1.2.3 Equivalent Complex Baseband Signals

It is convenient to represent (real-valued) bandpass signals[1] by its corresponding *equivalent complex baseband signal*, sometimes also called *equivalent low-pass signal* or *complex envelope* [Fra69, Tre71, Pro01].

Let $x_{\mathrm{HF}}(t)$ be a real-valued (high-frequency) signal and $X_{\mathrm{HF}}(f)$ its Fourier transform, i.e., $x_{\mathrm{HF}}(t) \circ\!\!-\!\!\bullet X_{\mathrm{HF}}(f)$. The equivalent complex baseband signal $x(t)$ corresponding to $x_{\mathrm{HF}}(t)$ is obtained by first going to one-sided spectra, i.e., generating the *analytic signal* to $x_{\mathrm{HF}}(t)$ [Pap77], and then shifting the spectrum by the frequency f_0, such that the relevant components are located around the origin and appears as a low-pass signal. Usually, when regarding carrier-modulated transmission, the transformation frequency f_0 is chosen equal to the carrier frequency. Mathematically, we have

$$x(t) \triangleq \frac{1}{\sqrt{2}}\left(x_{\mathrm{HF}}(t) + \mathrm{j}\,\mathcal{H}\left\{x_{\mathrm{HF}}(t)\right\}\right) \cdot \mathrm{e}^{-\mathrm{j}2\pi f_0 t} , \qquad (1.2.7a)$$

where $\mathcal{H}\{\cdot\}$ denotes *Hilbert transform* [Pap77]. Conversely, given the complex baseband representation $x(t)$, the corresponding real-valued signal is obtained as

$$x_{\mathrm{HF}}(t) = \sqrt{2}\,\mathrm{Re}\left\{x(t) \cdot \mathrm{e}^{+\mathrm{j}2\pi f_0 t}\right\} . \qquad (1.2.7b)$$

[1]To be precise, the only requirement for the application of equivalent complex baseband representations is that the signals are real-valued, and hence one half of the spectrum is redundant.

Note, normalization in (1.2.7) is chosen such that the original signal and its equivalent complex baseband representation have the same energy, i.e.,

$$\int_{-\infty}^{\infty} x_{\mathrm{HF}}^2(t)\, \mathrm{d}t = \int_{-\infty}^{\infty} |x(t)|^2\, \mathrm{d}t \tag{1.2.8}$$

holds [Tre71, Hub92, Hub93].

Regarding (1.2.7), the spectra of $x_{\mathrm{HF}}(t)$ and $x(t)$, respectively, are related to each other by

$$X(f) = \frac{1}{\sqrt{2}}\left(1 + \operatorname{sgn}(f + f_0)\right) \cdot X_{\mathrm{HF}}(f + f_0)\,, \tag{1.2.9a}$$

where $\operatorname{sgn}(x) \triangleq x/|x|$ is the sign function, and by

$$X_{\mathrm{HF}}(f) = \frac{\sqrt{2}}{2}\left(X(f - f_0) + X^*(-(f + f_0))\right)\,. \tag{1.2.9b}$$

If $h_{\mathrm{HF}}(t)$ denotes the impulse response of a linear, time-invariant system and $x_{\mathrm{HF}}(t)$ and $y_{\mathrm{HF}}(t)$ are its in- and output signal, respectively, we have the relation $y_{\mathrm{HF}}(t) = x_{\mathrm{HF}}(t) * h_{\mathrm{HF}}(t)$ ($*$: convolution). In order to hold the desirable relation $y(t) = x(t) * h(t)$ in the equivalent baseband, the impulse response of a system has to be transformed according to

$$h_{\mathrm{HF}}(t) = 2\,\mathrm{Re}\left\{h(t) \cdot \mathrm{e}^{+\mathrm{j}2\pi f_0 t}\right\}\,, \tag{1.2.10}$$

where $h(t)$ is the complex impulse response corresponding to $h_{\mathrm{HF}}(t)$ [Tre71, Pro01].

Finally, regarding the definitions of equivalent complex signals (1.2.7) and that of autocorrelations (1.2.5), correlation functions are transformed according to

$$\phi_{x_{\mathrm{HF}}x_{\mathrm{HF}}}(\tau) = \mathrm{Re}\left\{\phi_{xx}(\tau) \cdot \mathrm{e}^{+\mathrm{j}2\pi f_0 \tau}\right\}\,. \tag{1.2.11}$$

The respective power spectral densities are then related by

$$\Phi_{xx}(f) = \left(1 + \operatorname{sgn}(f + f_0)\right) \cdot \Phi_{x_{\mathrm{HF}}x_{\mathrm{HF}}}(f + f_0)\,. \tag{1.2.12}$$

In particular, real white Gaussian noise with power spectral density $\Phi_{x_{\mathrm{HF}}x_{\mathrm{HF}}}(f) = \frac{N_0}{2}$, $\forall f$, results in an equivalent complex Gaussian process with power spectral density $\Phi_{xx}(f) = N_0$, $f > -f_0$, and zero else. When filtering equivalent complex signals, the frequency components for $f \leq -f_0$ are irrelevant by definition. Hence it is convenient to define the power spectral density of white, complex-valued Gaussian noise in the equivalent complex domain simply to be equal to N_0 for all frequencies.

1.2.4 Miscellaneous

Vectors and *matrices* are denoted by bold-faced letters. Usually, vectors are written with lowercase letters, whereas uppercase letters stand for matrices.

A shadowed letter is used for the special *sets of numbers*. In particular, the set of natural numbers (including zero) is denoted by $\mathbb{N}$, the set of integers by $\mathbb{Z}$, the set of real numbers by $\mathbb{R}$, and the set of complex numbers is abbreviated by $\mathbb{C}$.

REFERENCES

[Fra69] L. E. Franks. *Signal Theory*. Prentice-Hall, Inc., Englewood Cliffs, NJ, 1969.

[Hub92] J. Huber. *Trelliscodierung*. Springer Verlag, Berlin, Heidelberg, 1992. (In German.)

[Hub93] J. Huber. *Signal- und Systemteoretische Grundlagen zur Vorlesung Nachrichtenübertragung*. Skriptum, Lehrstuhl für Nachrichtentechnik II, Universität Erlangen–Nürnberg, Erlangen, Germany, 1993. (In German.)

[Pap77] A. Papoulis. *Signal Analysis*. McGraw-Hill, New York, 1977.

[Pro01] J. G. Proakis. *Digital Communications*. McGraw-Hill, New York, 4th edition, 2001.

[Tre71] H. L. van Trees. *Detection, Estimation, and Modulation Theory—Part III: Radar-Sonar Signal Processing and Gaussian Signals in Noise*. John Wiley & Sons, Inc., New York, 1971.

2

Digital Communications via Linear, Distorting Channels

Over the last decades, digital communications has become one of the basic technologies for our modern life. Only when using digital transmission, information can be transported with moderate power consumption, high flexibility, and especially over long distances, with much higher reliability than by using traditional analog modulation. Thus, the communication world has been going digital.

When regarding digital transmission, we have to consider two dominant impairments. First, the signal is corrupted by (usually additive) noise, which can be thermal noise of the receiver front-end or crosstalk caused by other users transmitting in the same frequency band. Second, the transmission media is *dispersive.* It can be described as a *linear* system with some specific transfer function, where attenuation and phase vary over frequency. This property causes different frequency components to be affected differently—the signal is distorted—which in turn broadens the transmitted pulses in the time domain. As a consequence, successively transmitted symbols may interfere with one another, a phenomenon called *intersymbol interference (ISI)*. Depending on the application, ISI can affect hundreds of succeeding symbols as, e.g., in digital subscriber lines.

The ISI introduced by the linear, distorting channel calls for some kind of *equalization* at the receiver. Unfortunately, equalization of the amplitude distortion also enhances the channel noise. Thus, the receiver has to regard both the linear distortions and the noise, when trying to compensate for, or at least mitigate, the ISI.

The aim of this chapter is to give an overview on topics, characterized by the following questions:

- Given a certain criterion of optimality and some additional restrictions, what is the best choice for the receiver input filter?

and

- How can the transmitted data be recovered appropriately from the sequence produced by the receive filter?

We start from simple linear equalization known from basic system theory and then successively develop more elaborate receiver concepts. In each case the basic characteristics are enlightened and the achievable performance is given and compared to an ISI-free channel.

2.1 FUNDAMENTALS AND PROBLEM DESCRIPTION

The most important and widely used digital modulation techniques are *linear* and *memoryless*. In particular, we focus on digital *pulse amplitude modulation (PAM)*, where the continuous-time transmit signal $s(t)$ is given by the convolution of a discrete-time sequence $\langle a[k] \rangle$ of information symbols and a pulse shape $g_T(t)$ (see, e.g., [Pro01, Bla90, And99] or any textbook on digital communications)[1]

$$s(t) = \sum_{k} a[k] g_{\mathrm{T}}(t - kT) \ . \tag{2.1.1}$$

For baseband transmission $s(t)$ has to be real, whereas if passband, i.e., modulated transmission, is regarded, $s(t)$ is complex-valued, given as the equivalent complex baseband signal (e.g., [Fra69, Tre71, Hub92b, Hub93b, Pro01]). The discrete-time index $k \in \mathbb{Z}$ numbers the symbols, which are spaced by T seconds, the duration of the modulation interval, and t is continuous time measured in second (s).

The *information* or *data symbols* $a[k]$ are taken from a finite set $\mathcal{A}$, the *signal set* or *signal constellation* with cardinality $M = |\mathcal{A}|$. Depending on the choice of the signal constellation $\mathcal{A}$, different families of PAM are possible. Restricting $\mathcal{A}$ to solely comprise uniformly spaced points on the real line, we arrive at *amplitude-shift keying (ASK)*. Sometimes PAM is used synonymously for ASK. If the points constitute a (regular) two-dimensional grid in the complex plane, the transmission scheme is called *quadrature amplitude modulation (QAM)*, and selecting the points uniformly spaced on the unit circle results in *phase-shift keying (PSK)*.

First, we assume transmission without channel coding and equiprobable data symbols $a[k]$. Moreover, if the number of signal points is a power of two, say $M = 2^{R_{\mathrm{m}}}$, then the binary data stream to be transmitted is simply partitioned into blocks of R_{m} information bits, and each block is *mapped* onto one of the $2^{R_{\mathrm{m}}}$ possible

[1] A sum $\sum_k(\cdot)$, where the limits are not explicitly given, abbreviates $\sum_{k=-\infty}^{\infty}(\cdot)$.

symbols. Mapping is done memoryless, independent of preceding or succeeding blocks. The number $R_{\rm m}$ is also called the rate of the modulation. If M is not a power of two, mapping can be done based on larger blocks of binary data, generating blocks of data symbols. This approach is sometimes called *multidimensional mapping*. For details on mapping strategies see Chapter 4; for the moment it is sufficient to think of a simple symbol-by-symbol mapping of binary data. Generation of the transmit signal is illustrated in Figure 2.1.

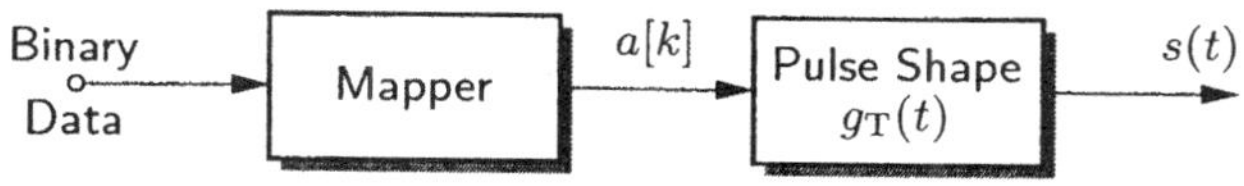

Fig. 2.1 Generation of the PAM transmit signal.

Subsequently, we always assume an ulindependent, identically distributed (i.i.d.) data sequence $\langle a[k]\rangle$ with zero mean value. Thus, the *autocorrelation sequence* is given by ($\mathsf{E}\{\cdot\}$: expectation)

$$\phi_{aa}[\kappa] \triangleq \mathsf{E}\left\{a[k+\kappa]a^*[k]\right\} = \begin{cases} \mathsf{E}\left\{|a[k]|^2\right\} \triangleq \sigma_a^2 , & \kappa = 0 \\ 0 , & \text{else} \end{cases} . \tag{2.1.2}$$

This implies that the *power spectral density (PSD)* of the data sequence is white, i.e., constant, with value σ_a^2.

The (possibly complex-valued) pulse shape $g_{\rm T}(t)$ constitutes the second part of the transmit signal generation. Because the discrete-time sequence $\langle a[k]\rangle$ has a periodic, and thus infinitely broad spectrum with respect to continuous time, it has to be filtered to achieve spectral efficiency. Because of the white data sequence, the average PSD of the transmit signal $s(t)$ is proportional to $|G_{\rm T}(f)|^2$, where $G_{\rm T}(f) \triangleq \mathcal{F}\{g_{\rm T}(t)\}$ is the Fourier transform of the pulse shape $g_{\rm T}(t)$.

Obviously, the natural unit of the continuous-time signals, e.g., the transmit signal $s(t)$, is volts (V). Alternatively, one may think of ampere (A), or volts per meter (V/m), or any other suitable physical value. Here, we always implicitly normalize all signals (by 1 V), and thus only treat dimensionless signals.[2] Because of the product in (2.1.1), any scaling (adjustment of the transmit power) or normalization can be split between $a[k]$ and $g_{\rm T}(t)$. We select the data symbols $a[k]$ to have the unit volts (i.e., dimensionless after normalization) and, hence, $g_{\rm T}(t)$ to be dimensionless. But this requires the transfer function $G_{\rm T}(f)$ to have the unit of time, i.e., seconds. In order to handle only dimensionless transfer functions, we rewrite the pulse shape as

$$g_{\rm T}(t) = T \cdot h_{\rm T}(t) , \tag{2.1.3}$$

[2]The power of a signal would be watts (W) if a one-ohm resistor is considered, and the power spectral density of normalized signals has the unit Hz^{-1} = s.

where $h_\mathrm{T}(t)$ is the impulse response of the transmit filter. Thus, in this book, the continuous-time PAM transmit signal is given by ($*$ denotes convolution)

$$s(t) = T \cdot \sum_k a[k] \cdot h_\mathrm{T}(t-kT) = \left(\sum_k Ta[k]\delta(t-kT) \right) * h_\mathrm{T}(t) \,. \qquad (2.1.4)$$

In summary, the transmitter thus consists of a mapper from binary data to real or complex-valued symbols $a[k]$. These symbols, multiplied by T, are then assigned to the weights of Dirac impulses, which is the transition from the discrete-time sequence to a continuous-time signal. This pulse train is finally filtered by the transmit filter $H_\mathrm{T}(f) \triangleq \mathcal{F}\{h_\mathrm{T}(t)\}$ in order to obtain the desired transmit signal $s(t)$.

The factor "T" can also be explained from a different point of view: sampling, i.e., the transition from a continuous-time signal to a discrete-time sequence, corresponds to periodic continuation of the initial spectrum divided by T. Thus, it is reasonable that the inverse operation comes along with a multiplication of the signals by T.

The signal $s(t)$ is then transmitted over a linear, dispersive channel, characterized by its transfer function $H'_\mathrm{C}(f)$ or, equivalently, by its impulse response $h'_\mathrm{C}(t) \triangleq \mathcal{F}^{-1}\{H'_\mathrm{C}(f)\}$. In addition to the linear distortion, the channel introduces noise, which is assumed to be stationary, Gaussian, additive—effective at the channel output—and independent of the transmitted signal. The average PSD of the noise $n'_0(t)$ is denoted by $\Phi_{n'_0 n'_0}(f) \triangleq \mathcal{F}_\tau\{\mathrm{E}\{n'_0(t+\tau)\cdot n'^{*}_0(t)\}\}$. Thus, the signal

$$r'(t) = s(t) * h'_\mathrm{C}(t) + n'_0(t) \qquad (2.1.5)$$

is present at the receiver input.

Assuming the power spectral density $\Phi_{n'_0 n'_0}(f)$ of the noise to be strictly positive within the transmission band $\mathcal{B} = \{f \mid H_\mathrm{T}(f) \neq 0\}$, a modified transmission model can be set up for analysis. Due to white thermal noise, which is ever present, this assumption is always justified in practice and imposes no restriction. Without loss of information, a *(continuous-time) noise whitening filter* can be placed at the first stage of the receiver. This filter with transfer function

$$H_\mathrm{W}(f) \triangleq \sqrt{\frac{N_0}{\Phi_{n'_0 n'_0}(f)}} \cdot \mathrm{e}^{\mathrm{j}b(f)} \,, \qquad f \in \mathcal{B} \qquad (2.1.6)$$

converts the channel noise with PSD $\Phi_{n'_0 n'_0}(f)$ into *white noise*, i.e., its PSD is constant over the frequency with value $\Phi_{n_0 n_0}(f) \triangleq \Phi_{n'_0 n'_0}(f) \cdot |H_\mathrm{W}(f)|^2 = N_0$. The corresponding autocorrelation function is a Dirac pulse. Therefore, N_0 is an arbitrary constant with the dimension of a PSD. Since for the effect of whitening only $|H_\mathrm{W}(f)|^2$ is relevant, the phase $b(f)$ of the filter can be adjusted conveniently.

In this book, all derivations are done for complex-valued signals in the equivalent baseband. Here, for white complex-valued noise corresponding to a real-valued physical process the PSD of real and imaginary part are both equal to $N_0/2$, hence in total N_0 [Tre71, Hub92b, Hub93b]. When treating baseband signaling only, the real part is present. In this case, without further notice, the constant N_0 always has to be replaced by $N_0/2$.

For the subsequent analysis, it is reasonable to combine the whitening filter $H_{\mathrm{W}}(f)$ with the channel filter $H'_{\mathrm{C}}(f)$, which results in a new channel transfer function $H_{\mathrm{C}} \triangleq H'_{\mathrm{C}}(f) \cdot H_{\mathrm{W}}(f)$ and an *additive white Gaussian noise (AWGN)* process $n_0(t)$ with PSD $\Phi_{n_0n_0}(f) = N_0$. This procedure reflects the well-known fact that the effects of intersymbol interference and colored noise are interchangeable [Gal68, Bla87]. In practice, of course, the whitening filter is realized as part of the receive filter. Figure 2.2 shows the equivalence when applying a noise whitening filter.

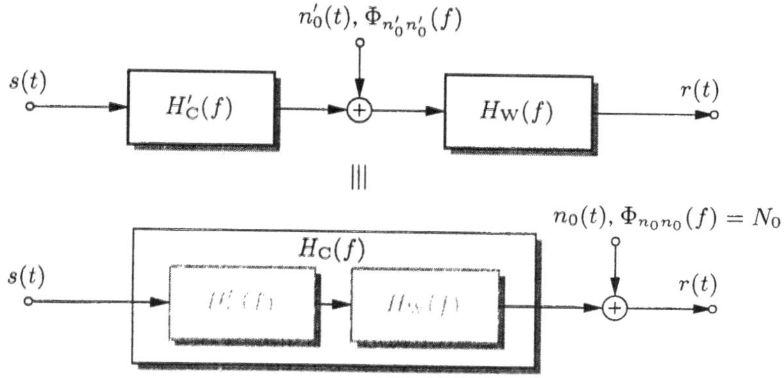

Fig. 2.2 Channel and noise whitening filter.

This (preprocessed) receive signal $r(t) \triangleq s(t) * h_{\mathrm{C}}(t) + n_0(t)$ is first passed through a receive filter $H_{\mathrm{R}}(f)$ and then sampled with frequency $1/T$, i.e., the symbol rate, resulting in the discrete-time receive sequence $\langle y[k] \rangle$. Here, as we do not regard synchronization algorithms, we always assume a correct (optimum) sampling phase. Since only so-called T-spaced sampling is considered, any fractional-spaced processing, e.g., for correction of a sampling phase offset, is equivalently incorporated into the continuous-time receive filter. The final transmission model is depicted in Figure 2.3.

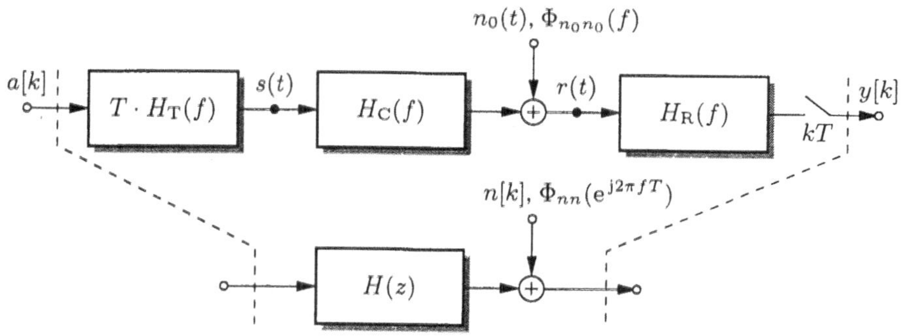

Fig. 2.3 Continuous-time transmission model and discrete-time representation.

Because both the transmitted data $a[k]$ and the sampled receive filter output $y[k]$ are T-spaced, in summary, a discrete-time model can be set up. The discrete-time transfer function $H(z)$ of the signal, evaluated on the unit circle, is given by

$$\begin{aligned} H(e^{j2\pi fT}) &= \frac{1}{T}\sum_{\mu} T H_{\mathrm{T}}(f-\tfrac{\mu}{T})H_{\mathrm{C}}(f-\tfrac{\mu}{T})H_{\mathrm{R}}(f-\tfrac{\mu}{T}) \\ &= \sum_{\mu} H_{\mathrm{T}}(f-\tfrac{\mu}{T})H_{\mathrm{C}}(f-\tfrac{\mu}{T})H_{\mathrm{R}}(f-\tfrac{\mu}{T})\,, \end{aligned} \tag{2.1.7a}$$

and the PSD of the discrete-time noise sequence $\langle n[k]\rangle$ reads

$$\Phi_{nn}(e^{j2\pi fT}) = \frac{1}{T}\sum_{\mu} \Phi_{n_0n_0}(f-\tfrac{\mu}{T})\left|H_{\mathrm{R}}(f-\tfrac{\mu}{T})\right|^2 . \tag{2.1.7b}$$

The discrete-time model is given in Figure 2.3, too.

After having set up the transmission scenario, in the following sections we have to discuss how to choose the receive filter and to recover data from the filtered signal $\langle y[k]\rangle$. We start from basic principles and proceed to more elaborate and better performing receiver concepts. Section 2.4 summarizes the resultant discrete-time models.

2.2 LINEAR EQUALIZATION

The most evident approach to equalization is to look for a linear receive filter, at the output of which information carried by one symbol can be recovered independently of previous or succeeding symbols by a simple threshold device. Since the end-to-end transfer function is $TH_{\mathrm{T}}(f)H_{\mathrm{C}}(f)$, system theory suggests *total linear equalization* via

$$H_{\mathrm{R}}(f) = \frac{1}{H_{\mathrm{T}}(f)H_{\mathrm{C}}(f)}\,, \tag{2.2.1}$$

which equalizes the transmission system to have a Dirac impulse response.

But this strategy is neither power efficient, nor is it required. First, if the transmit signal, i.e., $H_{\mathrm{T}}(f)$, is band-limited, all spectral components outside this band should be rejected by the receive filter. As only noise is present in these frequency ranges, a dramatic reduction of the noise bandwidth is achieved, and in fact only this limits the noise power to a finite value. Second, if, for example, the channel gain has deep notches or regions with high attenuation, the receive filter will highly amplify these frequency ranges. Another example is channels with low-pass characteristics, e.g., copper wires, where the receive filter is high-pass. But such receive filters lead to noise enhancement, which becomes worse as the channel gain tends to zero. For channels with spectral zeros, i.e., $H_{\mathrm{C}}(f) = 0$ exists for some f within the signaling band $\mathcal{B}$, total linear equalization is impossible, as the receive filter is not stable and noise enhancement tends to infinity.

2.2.1 Zero-Forcing Linear Equalization

The above problems can be overcome, if we remember that we transmit discrete-time data and sample the signal after the receive filter. Hence, only the considered sampling instances kT have to be ISI free. This requires the end-to-end impulse response $g_o(t)$ (overall impulse), including pulse-shaping filter, channel, and receive filter, to have equidistant zeros spaced by T. Assuming proper scaling of the receive filter, we demand

$$g_o(t) \triangleq \mathcal{F}^{-1}\{G_o(f)\} = \begin{cases} 1\,, & t=0 \\ 0\,, & t=kT, k \in \mathbb{Z} \\ \text{arbitrary}\,, & \text{else} \end{cases}\,, \tag{2.2.2}$$

where $G_o(f) \triangleq TH_T(f)H_C(f)H_R(f)$ has been used.

For such impulse responses, the Nyquist's criterion gives us the following constraint on the end-to-end transfer function $G_o(f)$ of the cascade transmit filter, channel, and receive filter (for a proof, see, e.g., [Pro01, Bla90]):

Theorem 2.1: *Nyquist's Criterion*

For the impulse response $g_o(t)$ to satisfy

$$g_o(t=kT) = \begin{cases} 1\,, & k=0 \\ 0\,, & \text{else} \end{cases} \tag{2.2.3}$$

it is necessary and sufficient that for its Fourier transform $G_o(f) = \mathcal{F}\{g_o(t)\}$ the following holds:

$$\frac{1}{T}\sum_{\mu=-\infty}^{\infty} G_o(f-\tfrac{\mu}{T}) = 1\,. \tag{2.2.4}$$

If the overall transfer function $G_o(f)$ satisfies Nyquist's criterion, the discrete-time impulse response $h[k] \triangleq \mathcal{Z}^{-1}\{H(z)\} = h(kT)$ is ISI free. Moreover, according to (2.1.7a), $H(e^{j2\pi fT}) = \frac{1}{T}\sum_\mu G_o(f-\frac{\mu}{T})$ holds, and thus the respective spectrum $H(z=e^{j2\pi fT})$ is flat.

It is noteworthy that a pulse $p(t)$, whose autocorrelation $\int p(\tau+t)p^*(\tau)\,d\tau$ satisfies Nyquist's criterion is called an *orthogonal pulse* or a *square-root Nyquist pulse* [Hub92b, And99, FU98].

Optimization Assuming the transmit filter $H_T(f)$ to be fixed and the channel $H_C(f)$ to be known, the task is to select a receive filter $H_R(f)$ such that the cascade of these three systems is Nyquist. Because Nyquist pulses are not uniquely determined—there are infinitely many—the remaining degree of freedom can be used for optimizing system performance.

Obviously, an *optimal* receiver would minimize the *bit error rate*. But, except for binary transmission, this criterion usually leads to mathematical problems, which no longer can be handled analytically. Moreover, the solution depends on the specific mapping. Thus, a common approach is to regard the *signal-to-noise ratio (SNR)* as an appropriate measure instead. As the noise has a Gaussian probability density function (pdf), the SNR is directly related to the *symbol error rate* via the complementary Gaussian integral function (Gaussian probability of error function)

$$\mathrm{Q}(x) = \frac{1}{\sqrt{2\pi}} \int_{x}^{\infty} \mathrm{e}^{-t^2/2}\,\mathrm{d}t\,. \tag{2.2.5}$$

Since the discrete-time signal transfer function is fixed, equivalently to maximizing the SNR, we minimize the variance of the discrete-time noise sequence. Thereby, in order to get a compact representation, we restrict the derivation to white noise $n_0(t)$ (cf. Figure 2.3) with PSD $\Phi_{n_0n_0}(f) = N_0$. As explained above, this is always possible without loss of generality. In summary, the optimization problem for $H_\mathrm{R}(f)$ can be stated as follows:

- Minimize the noise variance:

$$\sigma_n^2 = T \int_{-\frac{1}{2T}}^{\frac{1}{2T}} \Phi_{nn}(\mathrm{e}^{\mathrm{j}2\pi fT})\,\mathrm{d}f = T \int_{-\frac{1}{2T}}^{\frac{1}{2T}} \frac{1}{T} \sum_{\mu} N_0 \left| H_\mathrm{R}(f - \tfrac{\mu}{T}) \right|^2 \mathrm{d}f\,, \tag{2.2.6}$$

- subject to the additional constraint of an overall Nyquist characteristic:

$$\sum_{\mu} H_\mathrm{T}(f - \tfrac{\mu}{T}) H_\mathrm{C}(f - \tfrac{\mu}{T}) H_\mathrm{R}(f - \tfrac{\mu}{T}) \stackrel{!}{=} 1\,, \quad \forall f \in (-\tfrac{1}{2T}, \tfrac{1}{2T}]\,. \tag{2.2.7}$$

This problem can be solved analytically using *calculus of variations* and the *method of Lagrange multipliers* (e.g., [Hay96, Appendix C]). As the additional constraint (2.2.7) is not in integral form, it can be fulfilled independently for each frequency out of $(-\frac{1}{2T}, \frac{1}{2T}]$, the so-called *set of Nyquist frequencies* or *Nyquist interval*. Defining the real function $\lambda(\mathrm{e}^{\mathrm{j}2\pi fT})$ of Lagrange multipliers, i.e., each frequency bin has its own multiplier, we can set up a Lagrange function depending on $f \in (-\frac{1}{2T}, \frac{1}{2T}]$:

$$\begin{aligned} L(f) &= \sum_{\mu} \left| H_\mathrm{R}(f - \tfrac{\mu}{T}) \right|^2 - \lambda(\mathrm{e}^{\mathrm{j}2\pi fT}) \sum_{\mu} H_\mathrm{T}(f - \tfrac{\mu}{T}) H_\mathrm{C}(f - \tfrac{\mu}{T}) H_\mathrm{R}(f - \tfrac{\mu}{T}) \\ &= \sum_{\mu} \left(\left| H_\mathrm{R}(f - \tfrac{\mu}{T}) \right|^2 - \lambda(\mathrm{e}^{\mathrm{j}2\pi fT}) H_\mathrm{T}(f - \tfrac{\mu}{T}) H_\mathrm{C}(f - \tfrac{\mu}{T}) H_\mathrm{R}(f - \tfrac{\mu}{T}) \right). \end{aligned} \tag{2.2.8}$$

The optimal receive filter $H_\mathrm{R}(f)$ is then a *stationary point* of the Lagrange function (2.2.8). To determine this point, we add a (small) deviation $\varepsilon \cdot V(f)$ to the optimal solution $H_\mathrm{R}(f)$. Note that $\varepsilon \cdot V(f)$ is complex-valued, since all transfer functions are also complex quantities. In the optimum, the partial derivative of $L(f)$ with respect to $\varepsilon \in \mathbb{C}$ has to be zero:

$$\left.\frac{\partial L(f)}{\partial \varepsilon}\right|_{\varepsilon=0} \stackrel{!}{=} 0\,, \qquad \forall f \in (-\tfrac{1}{2T}, \tfrac{1}{2T}]\,.$$

Using the *Wirtinger Calculus* (see Appendix A for details) for derivation with respect to a complex variable, we have[3]

$$\begin{aligned}
\left.\frac{\partial L(f)}{\partial \varepsilon}\right|_{\varepsilon=0} &= \frac{\partial}{\partial \varepsilon}\Bigg(\sum_{\mu}\Big[\left|H_{\mathrm{R}}(f-\tfrac{\mu}{T})+\varepsilon V(f-\tfrac{\mu}{T})\right|^2 - \lambda(\mathrm{e}^{\mathrm{j}2\pi fT})H_{\mathrm{T}}(f-\tfrac{\mu}{T}) \\
&\qquad \cdot H_{\mathrm{C}}(f-\tfrac{\mu}{T})\cdot\left(H_{\mathrm{R}}(f-\tfrac{\mu}{T})+\varepsilon V(f-\tfrac{\mu}{T})\right)\Big]\Bigg)\Bigg|_{\varepsilon=0} \\
&= \Bigg(\sum_{\mu}\Big[2\left(H_{\mathrm{R}}^*(f-\tfrac{\mu}{T})+\varepsilon^* V^*(f-\tfrac{\mu}{T})\right)V(f-\tfrac{\mu}{T}) \\
&\qquad - \lambda(\mathrm{e}^{\mathrm{j}2\pi fT})H_{\mathrm{T}}(f-\tfrac{\mu}{T})H_{\mathrm{C}}(f-\tfrac{\mu}{T})V(f-\tfrac{\mu}{T})\Big]\Bigg)\Bigg|_{\varepsilon=0} \\
&= \sum_{\mu}\Big[2H_{\mathrm{R}}^*(f-\tfrac{\mu}{T}) - \lambda(\mathrm{e}^{\mathrm{j}2\pi fT})H_{\mathrm{T}}(f-\tfrac{\mu}{T})H_{\mathrm{C}}(f-\tfrac{\mu}{T})\Big]V(f-\tfrac{\mu}{T}) \\
&\stackrel{!}{=} 0\,.
\end{aligned} \tag{2.2.9}$$

For any deviation $V(f)$, (2.2.9) can be met if

$$2H_{\mathrm{R}}^*(f-\tfrac{\mu}{T}) - \lambda(\mathrm{e}^{\mathrm{j}2\pi fT})H_{\mathrm{T}}(f-\tfrac{\mu}{T})H_{\mathrm{C}}(f-\tfrac{\mu}{T}) = 0\,,$$

or equivalently, since $\lambda(\mathrm{e}^{\mathrm{j}2\pi fT})$ is real,

$$H_{\mathrm{R}}(f-\tfrac{\mu}{T}) = \frac{\lambda(\mathrm{e}^{\mathrm{j}2\pi fT})}{2}H_{\mathrm{T}}^*(f-\tfrac{\mu}{T})H_{\mathrm{C}}^*(f-\tfrac{\mu}{T})\,. \tag{2.2.10}$$

In the last step, the Lagrange multiplier is determined via the additional constraint (2.2.7), which now reads $\forall f \in (-\frac{1}{2T}, \frac{1}{2T}]$

$$\sum_{\mu} H_{\mathrm{T}}(f-\tfrac{\mu}{T})H_{\mathrm{C}}(f-\tfrac{\mu}{T})\cdot\frac{\lambda(\mathrm{e}^{\mathrm{j}2\pi fT})}{2}H_{\mathrm{T}}^*(f-\tfrac{\mu}{T})\cdot H_{\mathrm{C}}^*(f-\tfrac{\mu}{T}) = 1\,.$$

Since the Lagrange multiplier function $\lambda(\mathrm{e}^{\mathrm{j}2\pi fT})$ is periodic with $1/T$ it is not affected by summation over the shifted replica of the spectra, and we obtain

$$\frac{\lambda(\mathrm{e}^{\mathrm{j}2\pi fT})}{2} = \frac{1}{\sum_{\mu}\left|H_{\mathrm{T}}(f-\tfrac{\mu}{T})H_{\mathrm{C}}(f-\tfrac{\mu}{T})\right|^2}\,. \tag{2.2.11}$$

[3] z^* denotes the complex conjugate of $z = x + \mathrm{j}y$: $z^* = (x+\mathrm{j}y)^* = x - \mathrm{j}y$.

Finally, inserting (2.2.11) into (2.2.10) and substituting $f - \frac{\mu}{T}$ by f yields the optimum receive filter

$$H_{\mathrm{R}}^{(\mathrm{ZF\text{-}LE})}(f) = \frac{H_{\mathrm{T}}^{*}(f)H_{\mathrm{C}}^{*}(f)}{\sum\limits_{\mu}\left|H_{\mathrm{T}}(f-\frac{\mu}{T})H_{\mathrm{C}}(f-\frac{\mu}{T})\right|^{2}} \,. \tag{2.2.12}$$

Because this filter is optimum in the sense of noise variance under the constraint of an overall Nyquist impulse response, this receive filter is called *optimum Nyquist filter (ONF)*. As the intersymbol interference is forced to be zero, this strategy is also called *optimum zero-forcing linear equalization (ZF-LE)*. Here we prefer the latter term.

The following theorem summarizes the result:

Theorem 2.2: *Optimum Zero-Forcing Linear Equalization (ZF-LE)*

Let the transmit filter $H_{\mathrm{T}}(f)$, a channel with transfer function $H_{\mathrm{C}}(f)$, and additive white noise be given. The optimal linear receive filter which results in intersymbol-interference-free samples (zero-forcing linear equalization, ZF-LE) and minimal noise variance, called the *optimum Nyquist filter*, is given by

$$H_{\mathrm{R}}^{(\mathrm{ZF\text{-}LE})}(f) = \frac{H_{\mathrm{T}}^{*}(f)H_{\mathrm{C}}^{*}(f)}{\sum\limits_{\mu}\left|H_{\mathrm{T}}(f-\frac{\mu}{T})H_{\mathrm{C}}(f-\frac{\mu}{T})\right|^{2}} \,. \tag{2.2.13}$$

Often, the additive channel noise is nonwhite, but has PSD $\Phi_{n_0'n_0'}(f)$. To apply the above result, we first imagine a noise whitening filter $H_{\mathrm{W}}(f)$ and cascade it with the channel $H_{\mathrm{C}}'(f)$. The optimal Nyquist filter is then designed from the channel transfer function $H_{\mathrm{C}}(f) = H_{\mathrm{C}}'(f) \cdot H_{\mathrm{W}}(f) = H_{\mathrm{C}}'(f) \cdot \sqrt{N_0/\Phi_{n_0'n_0'}(f)}\; \mathrm{e}^{\mathrm{j}b(f)}$. In the last step, we combine the continuous-time noise whitening filter and the optimum Nyquist filter into the final receive filter. This yields

$$\begin{aligned} H_{\mathrm{R}}(f) &= \sqrt{\frac{N_0}{\Phi_{n_0'n_0'}(f)}}\,\mathrm{e}^{\mathrm{j}b(f)} \\ &\quad \cdot \frac{H_{\mathrm{T}}^{*}(f)H_{\mathrm{C}}'^{*}(f)\sqrt{N_0/\Phi_{n_0'n_0'}(f)}\mathrm{e}^{-\mathrm{j}b(f)}}{\sum\limits_{\mu}\left|H_{\mathrm{T}}(f-\frac{\mu}{T})H_{\mathrm{C}}'(f-\frac{\mu}{T})\sqrt{N_0/\Phi_{n_0'n_0'}(f-\frac{\mu}{T})}\mathrm{e}^{-\mathrm{j}b(f-\frac{\mu}{T})}\right|^{2}} \\ &= \frac{H_{\mathrm{T}}^{*}(f)H_{\mathrm{C}}'^{*}(f)/\Phi_{n_0'n_0'}(f)}{\sum\limits_{\mu}\left|H_{\mathrm{T}}(f-\frac{\mu}{T})H_{\mathrm{C}}'(f-\frac{\mu}{T})\right|^{2}/\Phi_{n_0'n_0'}(f-\frac{\mu}{T})} \,. \end{aligned} \tag{2.2.14}$$

Subsequently, for the derivations we always assume a transmission model with white noise. A possible coloring of the noise is equivalently accounted for—via a noise whitening filter, see above—in the channel transfer function $H_{\mathrm{C}}(f)$. After combining this continuous-time noise whitening filter and the receiver filter for white noise, the actual receiver filter results.

Discussion A close look at (2.2.12) (or (2.2.14) for colored noise) reveals that the optimal Nyquist filter conceptionally consists of two parts. First, the *matched filter*

$$H_{\mathrm{T}}^{*}(f)H_{\mathrm{C}}^{*}(f)$$

for the cascade of transmit filter and channel is present. It is well known that in PAM transmission over the additive white Gaussian noise AWGN channel, the matched filter is optimum for signal detection. In the next subsection, we will show that for linear distorting channels a filter matched to the cascade $H_{\mathrm{T}}(f)H_{\mathrm{C}}(f)$ should always be the first stage as well. Furthermore, this allows T-spaced sampling without loss of information on the transmitted data (see Section 2.5), although the sampling theorem [Pap77] usually is not satisfied here.

The second part

$$\frac{1}{\sum_{\mu}\left|H_{\mathrm{T}}(f-\frac{\mu}{T})H_{\mathrm{C}}(f-\frac{\mu}{T})\right|^2}$$

is periodic in f with period $1/T$ and, thus, a discrete-time filter. If sampling is done right after the matched filter, the data symbols are transmitted through the cascade $TH_{\mathrm{T}}(f)H_{\mathrm{C}}(f)\cdot H_{\mathrm{T}}^{*}(f)H_{\mathrm{C}}^{*}(f)$, and hence, after sampling, the transfer function $\sum_{\mu}\left|H_{\mathrm{T}}(f-\frac{\mu}{T})H_{\mathrm{C}}(f-\frac{\mu}{T})\right|^2$ is effective. Thus, the discrete-time part of the optimal Nyquist filter ideally cancels the intersymbol interference.

Last, it should be noted that optimum linear ZF equalization only exists if the periodic continuation of $|H_{\mathrm{T}}(f)H_{\mathrm{C}}(f)|^2$ is strictly positive. Thus, it is irrelevant which period μ contributes. The only requirement is that for each $f\in(-\frac{1}{2T},\frac{1}{2T}]$ at least at one frequency position $f-\frac{\mu}{T}$ out of the *set of Nyquist-equivalent frequencies* $\mathcal{F}(f)=\{f-\frac{\mu}{T}\mid\mu\in\mathbb{Z}\}$ [Eri73] transmission is possible, i.e., $\forall f\in(-\frac{1}{2T},\frac{1}{2T}]\ \exists\mu\in\mathbb{Z}$ such that $H_{\mathrm{T}}(f-\frac{\mu}{T})H_{\mathrm{C}}(f-\frac{\mu}{T})\neq 0$. In other words, at least one *full set of Nyquist frequencies* (a set of measure $\frac{1}{T}$) is required. However, transmission can also be done in disjoint frequency bands. The folded spectrum can only be zero if $H_{\mathrm{T}}(f)H_{\mathrm{C}}(f)$ has periodic (period $1/T$) zeros. For example, this is true when the time-domain pulses are rectangular ($H_{\mathrm{T}}(f)\sim\sin(\pi fT)/(\pi fT)$) with duration T and the channel has a zero at DC, e.g., due to transformer coupling.

Example 2.1: *Optimum Zero-Forcing Linear Equalization*

This example is aimed to visualize the various spectra and time-domain signals when applying zero-forcing linear equalization. The parameters for symbol spacing T, the transmit filter $H_{\mathrm{T}}(f)$, and the channel $H_{\mathrm{C}}(f)$ are given in Appendix B and reflect a typical digital subscriber lines scenario. Here, the simplified down-stream scenario with white Gaussian noise is regarded. The cable length is $\ell=3$ km.

First, at the top of Figure 2.4 the normalized squared magnitude of the cascade $H(f)\triangleq H_{\mathrm{T}}(f)H_{\mathrm{C}}(f)$ of transmit filter $H_{\mathrm{T}}(f)$ and channel filter $H_{\mathrm{C}}(f)$ is plotted. Applying the matched filter at the receiver, this overall transfer function is visible. Due to the T-spaced sampling, the spectrum is repeated periodically, resulting in the shape plotted in the middle. Since the discrete-time part has to equalize this transfer function, this function also serves as the denominator of the optimal ZF linear equalizer. This discrete-time part is plotted on the bottom of the figure.

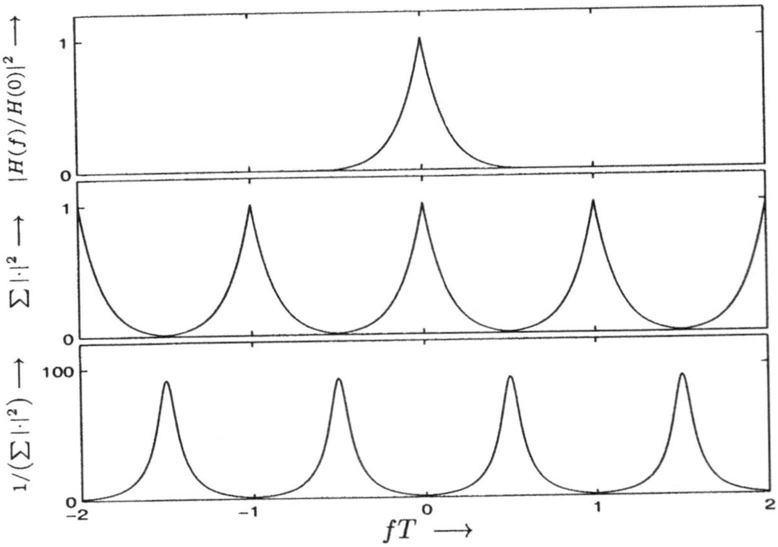

Fig. 2.4 Top: squared magnitude of the cascade $H_{\mathrm{T}}(f)H_{\mathrm{C}}(f)$. Bottom: Periodic continuation.

The magnitude of the optimal ZF linear equalizer (equation (2.2.12)) is plotted in Figure 2.5. It is noticeable that due to the low-pass characteristics of the channel, the receive filter has essentially high-pass characteristics. Because channel attenuation increases with frequency, it is preferable to suppress signal components approximately above (for negative frequencies, of course, below) Nyquist frequency $\pm\frac{1}{2T}$. In the region of $\pm\frac{1}{2T}$ the receive filter has to highly amplify the receive signal.

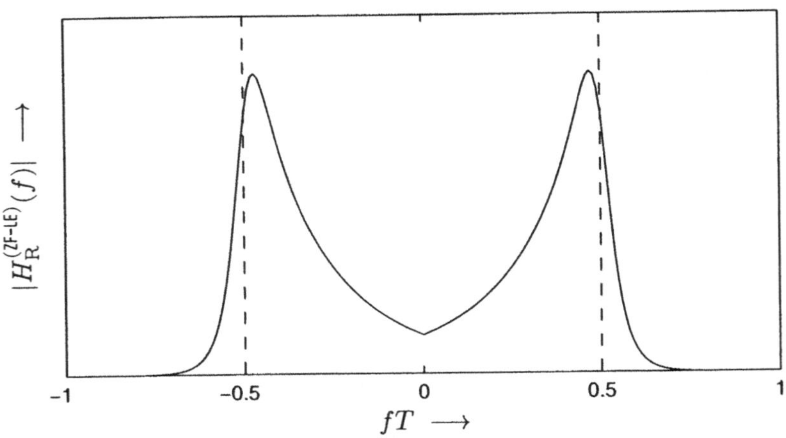

Fig. 2.5 Magnitude of the optimal ZF linear equalizer.

Figure 2.6 shows the magnitude of the end-to-end cascade with normalized transfer function $G_{\mathrm{o}}(f)/T = H_{\mathrm{T}}(f)H_{\mathrm{C}}(f)H_{\mathrm{R}}(f)$. As demanded for symbol-by-symbol detection of the data, the cascade exhibits the Nyquist characteristic, i.e., it has symmetric slopes (symmetric with respect to the marked points $(\pm\frac{1}{2T}, \frac{1}{2})$), which guarantee that the periodic sum results in the constant 1.

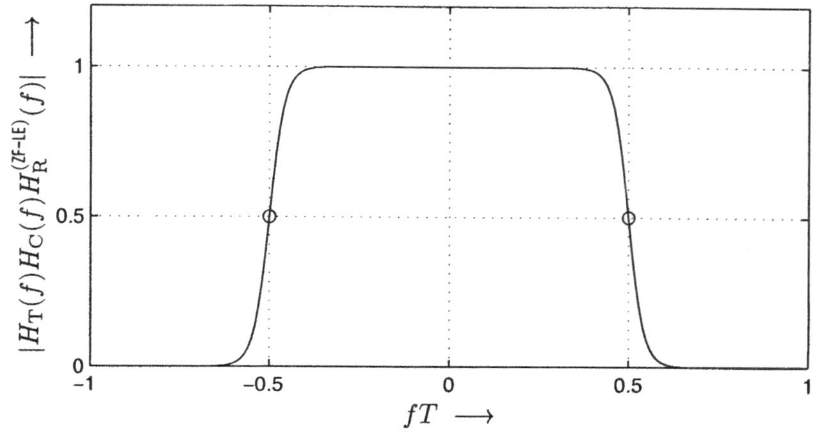

Fig. 2.6 Magnitude of the end-to-end cascade $G_{\rm o}(f)/T = H_{\rm T}(f)H_{\rm C}(f)H_{\rm R}^{(\rm ZF\text{-}LE)}(f)$.

In Figure 2.7 the respective time-domain signal—visible at the output of the receive filter—is plotted. Here, the Nyquist characteristic is visible, too: the impulse response has zeros uniformly spaced by T; marked by circles.

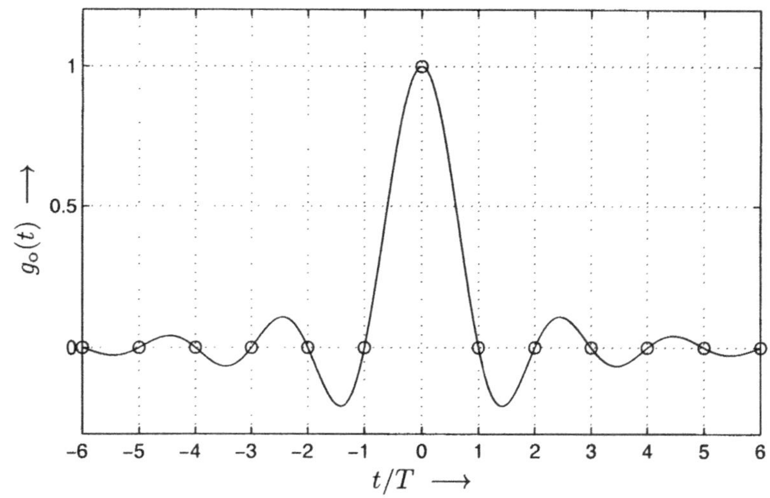

Fig. 2.7 Time-domain pulse $g_{\rm o}(t)$ at the output of the receive filter. Circles: optimal sampling instants.

Finally, the squared magnitude $|H_{\rm R}^{(\rm ZF\text{-}LE)}(f)|^2$ of the receive filter is sketched on the top of Figure 2.8. Since we have assumed white channel noise, this shape is identical to the PSD of the continuous-time noise at the output of the receive filter. After T-spaced sampling, the periodic noise PSD $\Phi_{nn}(e^{j2\pi fT})$ given in Figure 2.8 results. It is noticeable that the noise is extremely colored—the spectral components are highly concentrated around the Nyquist frequency.

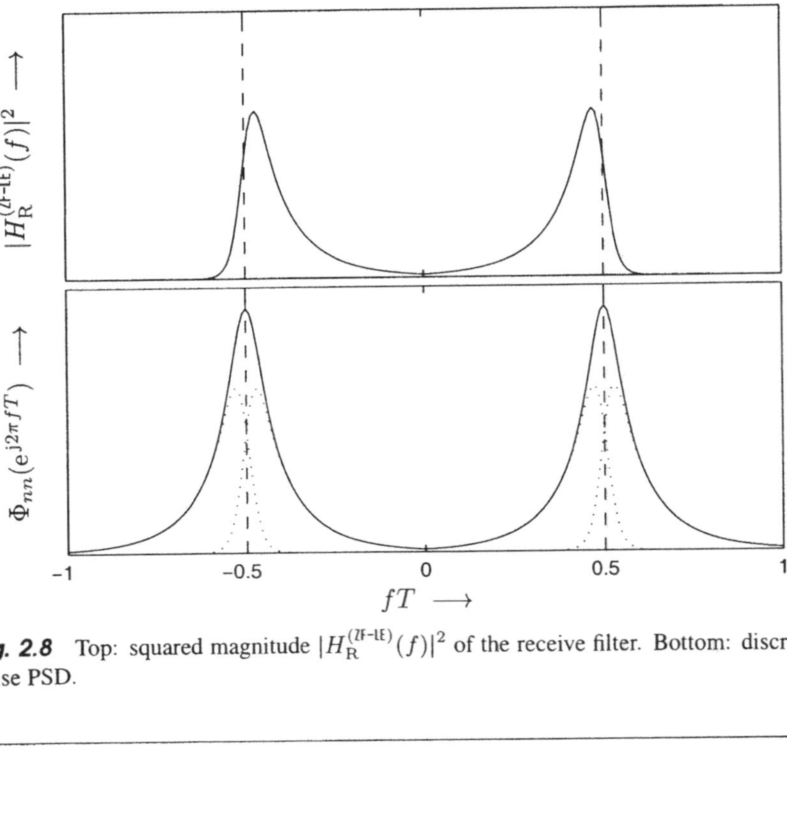

Fig. 2.8 Top: squared magnitude $|H_R^{(ZF-LE)}(f)|^2$ of the receive filter. Bottom: discrete-time noise PSD.

Signal-to-Noise Ratio After having optimized the receive filter, we now calculate the achievable performance. First, we derive the SNR at the decision point and then give the loss compared to transmission over the ISI-free AWGN channel.

Since the discrete-time end-to-end transfer function is 1, signal power is equal to σ_a^2, and we only have to regard noise power σ_n^2. Due to the receive filter $H_R(f)$, the noise sequence $\langle n[k] \rangle$ is not white, but colored. Regarding Figure 2.3 and equation (2.2.12), the noise PSD reads

$$
\begin{aligned}
\Phi_{nn}^{(ZF-LE)}(e^{j2\pi fT}) &= \frac{N_0}{T} \sum_{\mu} \left| H_R^{(ZF-LE)}(f - \tfrac{\mu}{T}) \right|^2 \\
&= \frac{N_0}{T} \sum_{\mu} \frac{\left| H_T(f - \frac{\mu}{T}) H_C(f - \frac{\mu}{T}) \right|^2}{\left| \sum_{\nu} \left| H_T(f - \frac{\mu}{T} - \frac{\nu}{T}) H_C(f - \frac{\mu}{T} - \frac{\nu}{T}) \right|^2 \right|^2},
\end{aligned}
$$

and, since the denominator is periodic in f (period $1/T$) and hence independent of μ, we have

$$
\begin{aligned}
&= \frac{N_0}{T} \frac{\sum_\mu \left|H_\mathrm{T}(f-\frac{\mu}{T})H_\mathrm{C}(f-\frac{\mu}{T})\right|^2}{\left|\sum_\mu \left|H_\mathrm{T}(f-\frac{\mu}{T})H_\mathrm{C}(f-\frac{\mu}{T})\right|^2\right|^2} \\
&= \frac{N_0}{T} \frac{1}{\sum_\mu \left|H_\mathrm{T}(f-\frac{\mu}{T})H_\mathrm{C}(f-\frac{\mu}{T})\right|^2} \,. \qquad (2.2.15)
\end{aligned}
$$

Thus, the noise variance is

$$
\begin{aligned}
\sigma_n^2 &= T \int_{-\frac{1}{2T}}^{\frac{1}{2T}} \Phi_{nn}^{(\mathrm{ZF\text{-}LE})}(\mathrm{e}^{\mathrm{j}2\pi fT})\,\mathrm{d}f \\
&= N_0 \int_{-\frac{1}{2T}}^{\frac{1}{2T}} \frac{1}{\sum_\mu \left|H_\mathrm{T}(f-\frac{\mu}{T})H_\mathrm{C}(f-\frac{\mu}{T})\right|^2}\,\mathrm{d}f \,. \qquad (2.2.16)
\end{aligned}
$$

Hence, the signal-to-noise ratio when applying ZF linear equalization reads:

$$
\begin{aligned}
\mathrm{SNR}^{(\mathrm{ZF\text{-}LE})} &= \frac{\sigma_a^2}{\sigma_n^2} \\
&= \frac{\sigma_a^2}{N_0 \int_{-\frac{1}{2T}}^{\frac{1}{2T}} \left(\sum_\mu \left|H_\mathrm{T}(f-\frac{\mu}{T})H_\mathrm{C}(f-\frac{\mu}{T})\right|^2\right)^{-1} \mathrm{d}f} \,. \qquad (2.2.17)
\end{aligned}
$$

It is common (e.g., [Hub92b]) to introduce the *spectral signal-to-noise ratio* (or *channel SNR function* [FU98])

$$
\mathrm{SNR}(f) \triangleq \frac{T\sigma_a^2|H_\mathrm{T}(f)H_\mathrm{C}(f)|^2}{N_0} \qquad (2.2.18)
$$

at the receiver input and its *folded version*:

$$
\widetilde{\mathrm{SNR}}(\mathrm{e}^{\mathrm{j}2\pi fT}) \triangleq \sum_\mu \mathrm{SNR}(f-\tfrac{\mu}{T}) \,. \qquad (2.2.19)
$$

By this, the SNR can be expressed by

$$
\mathrm{SNR}^{(\mathrm{ZF\text{-}LE})} = \frac{1}{T\int_{-\frac{1}{2T}}^{\frac{1}{2T}} \frac{1}{\widetilde{\mathrm{SNR}}(\mathrm{e}^{\mathrm{j}2\pi fT})}\,\mathrm{d}f} \,, \qquad (2.2.20)
$$

which is the *harmonic mean* [BS98] over the *folded spectral SNR*.

Assuming a signal constellation $\mathcal{A}$ with zero mean, variance σ_a^2, and spacing of the points equal to 2, and considering that for complex signals in each dimension the noise power $\sigma_n^2/2$ is active, the symbol error rate can be well approximated from (2.2.17) as follows [Pro01]:

$$\mathrm{SER}^{(\text{ZF-LE})} \approx \text{const} \cdot \mathrm{Q}\left(\sqrt{2\frac{\mathrm{SNR}^{(\text{ZF-LE})}}{\sigma_a^2}}\right) , \tag{2.2.21}$$

where again $\mathrm{Q}(x) = \frac{1}{\sqrt{2\pi}} \int_x^{\infty} \mathrm{e}^{-t^2/2}\, \mathrm{d}t$ has been used, and the constant depends on the actual constellation.

Let us now compare the achievable performance with that of transmission over an *ISI-free additive white Gaussian noise channel*, where $H_\mathrm{C}(f) = 1$ and $\Phi_{n_0 n_0}(f) = N_0$. There are several ways of doing this comparison. Here, we mainly concentrate on a fixed *receive power*. More precisely, assuming also a given signal constellation, the *receive energy per information bit* is assumed to be equal for all competing schemes. In the literature, other benchmarks can be found as well. In some situations it is preferable to compare performance on the basis of a given *transmit power* (see, e.g., [LSW68]). This also takes the attenuation of the channel into account and does not only look at the loss due to the introduced intersymbol interference.

When transmitting over the AWGN channel, we assume a transmit filter $H_\mathrm{T}(f)$ with *square-root Nyquist characteristic*, i.e., $|H_\mathrm{T}(f)|^2$ corresponds to a Nyquist pulse. The additive noise is white with (two-sided) PSD N_0 in the equivalent complex baseband. Using the optimal matched-filter receiver $H_\mathrm{R}(f) = \frac{T}{E_\mathrm{T}} \cdot H_\mathrm{T}^*(f)$ [Pro01], where

$$E_\mathrm{T} = \int_{-\infty}^{\infty} |Th_\mathrm{T}(t)|^2 \,\mathrm{d}t = \int_{-\infty}^{\infty} |TH_\mathrm{T}(f)|^2 \,\mathrm{d}f \tag{2.2.22}$$

is the energy of the transmit pulse $Th_\mathrm{T}(t)$, the discrete-time AWGN model results:

$$y[k] = a[k] + n[k] \; . \tag{2.2.23}$$

Considering (2.1.7b) and that $H_\mathrm{T}(f)$ is a square-root Nyquist pulse, the noise sequence $\langle n[k] \rangle$ is white with variance $\sigma_n^2 = N_0/E_\mathrm{T}$.

For the dispersionless channel, the energy E_b per information bit calculates to

$$E_\mathrm{b} = \frac{\sigma_a^2 E_\mathrm{T}}{R_\mathrm{m}} = \frac{\sigma_a^2}{R_\mathrm{m}} \cdot \int_{-\infty}^{\infty} |TH_\mathrm{T}(f)|^2 \,\mathrm{d}f \; , \tag{2.2.24}$$

where R_m is the rate of the modulation (number of information bits per transmitted symbol). With it, the symbol error rate is given by

$$\begin{aligned} \mathrm{SER}^{(\text{AWGN})} &\approx \text{const} \cdot \mathrm{Q}\left(\sqrt{2\frac{E_\mathrm{T}}{N_0}}\right) \\ &= \text{const} \cdot \mathrm{Q}\left(\sqrt{\frac{2R_\mathrm{m}}{\sigma_a^2}\frac{E_\mathrm{b}}{N_0}}\right) . \end{aligned} \tag{2.2.25}$$

Now, taking into account that on the ISI channel the receive energy per information bit is given by

$$E_b = \frac{\sigma_a^2}{R_m} \cdot \int_{-\infty}^{\infty} |TH_T(f)H_C(f)|^2 \, df \tag{2.2.26a}$$

$$= \frac{TN_0}{R_m} \cdot \int_{-\infty}^{\infty} \mathrm{SNR}(f) \, df = \frac{TN_0}{R_m} \cdot \int_{-\frac{1}{2T}}^{\frac{1}{2T}} \widetilde{\mathrm{SNR}}(e^{j2\pi fT}) \, df \,, \tag{2.2.26b}$$

rewriting the argument of the Q-function in (2.2.21) leads to

$$2\frac{\mathrm{SNR}^{(\mathrm{ZF\text{-}LE})}}{\sigma_a^2} = \frac{2}{\sigma_a^2 \cdot T \int_{-\frac{1}{2T}}^{\frac{1}{2T}} \left(\widetilde{\mathrm{SNR}}(e^{j2\pi fT})\right)^{-1} df}$$

$$= \frac{1}{T \int_{-\frac{1}{2T}}^{\frac{1}{2T}} \widetilde{\mathrm{SNR}}(e^{j2\pi fT}) \, df \cdot T \int_{-\frac{1}{2T}}^{\frac{1}{2T}} \left(\widetilde{\mathrm{SNR}}(e^{j2\pi fT})\right)^{-1} df} \cdot \frac{2R_m}{\sigma_a^2} \frac{E_b}{N_0} \,. \tag{2.2.27}$$

Keeping (2.2.21) in mind, a comparison of (2.2.27) with the argument of (2.2.25) reveals the loss for transmission over an ISI channel with zero-forcing linear equalization compared to a dispersionless channel (matched-filter bound). The factor virtually lowering the signal-to-noise ratio is called the *SNR attenuation factor* [LSW68]. The results are stated in the following theorem (cf. also [LSW68, eq. (5.55)]):

Theorem 2.3: *Signal-to-Noise Ratio of ZF Linear Equalization*

When using zero-forcing linear equalization at the receiver, the signal-to-noise ratio is given by the *harmonic mean* over the folded spectral SNR

$$\mathrm{SNR}^{(\mathrm{ZF\text{-}LE})} = \frac{1}{T \int_{-\frac{1}{2T}}^{\frac{1}{2T}} \frac{1}{\widetilde{\mathrm{SNR}}(e^{j2\pi fT})} \, df} \,, \tag{2.2.28}$$

and the degradation (based on equal receive power) compared to transmission over a dispersionless channel reads

$$\vartheta^2_{(\mathrm{ZF\text{-}LE})} = \left(T \int_{-\frac{1}{2T}}^{\frac{1}{2T}} \widetilde{\mathrm{SNR}}(e^{j2\pi fT}) \, df \cdot T \int_{-\frac{1}{2T}}^{\frac{1}{2T}} \left(\widetilde{\mathrm{SNR}}(e^{j2\pi fT})\right)^{-1} df \right)^{-1} . \tag{2.2.29}$$

Here, $\widetilde{\mathrm{SNR}}(e^{j2\pi fT})$ is the folded spectral SNR at the receiver input.

Note that the SNR attenuation factor ϑ^2 is always less than 1; equality only holds if and only if $\widetilde{\mathrm{SNR}}(\mathrm{e}^{\mathrm{j}2\pi fT}) = \text{const}$. This fact follows from the relation between arithmetic mean (first integral in (2.2.29)) and harmonic mean (second integral in (2.2.29)) [BB91].

If performance should be compared based on equal *transmit power*, in the above formula only E_b (equation (2.2.26)) has to be replaced by equation (2.2.24), which is the transmit energy per information bit. Hence, the loss is then given by

$$\vartheta'^2 = \vartheta^2 \cdot \frac{\int_{-\infty}^{\infty} |TH_\mathrm{T}(f)H_\mathrm{C}(f)|^2 \, \mathrm{d}f}{\int_{-\infty}^{\infty} |TH_\mathrm{T}(f)|^2 \, \mathrm{d}f} \,. \tag{2.2.30}$$

In situations where noise power is proportional to transmit power, e.g., for pure self-NEXT environments (see Appendix B), this loss based on equal transmit power makes no sense.

Example 2.2: *Loss of Optimum Zero-Forcing Linear Equalization*

Continuing the above example, the loss of ZF linear equalization is plotted in Figure 2.9 for the DSL down-stream example (white noise) over the cable length. For details on the transmission model, see Appendix B. The solid line gives the loss measured on equal receive power, whereas the dashed-dotted line corresponds to equal transmit power.

Since the transmit filter is not square-root Nyquist, even for cable length $\ell = 0$ km a (small) loss occurs. Because ϑ' includes the average line attenuation, which increases dramatically over the field length, both quantities diverge. For the present example, the loss due to introduced ISI and ZF linear equalization ranges from 4 dB up to 10 dB for cable lengths between 2 and 4 km.

For comparison, Figure 2.10 sketches the loss of the up-stream scenario. Here, a self-NEXT dominated environment is present and an assessment based on equal transmit power fails. Compared to the loss of the down-stream example, due to the colored noise—which increases the variations of the attenuation within the transmission band—a much larger loss results. The continuous-time noise whitening filter introduces additional ISI, and so even for $\ell = 0$ km a huge loss occurs. (Please disregard for the moment that the NEXT model is no longer valid for a very short cable length.)

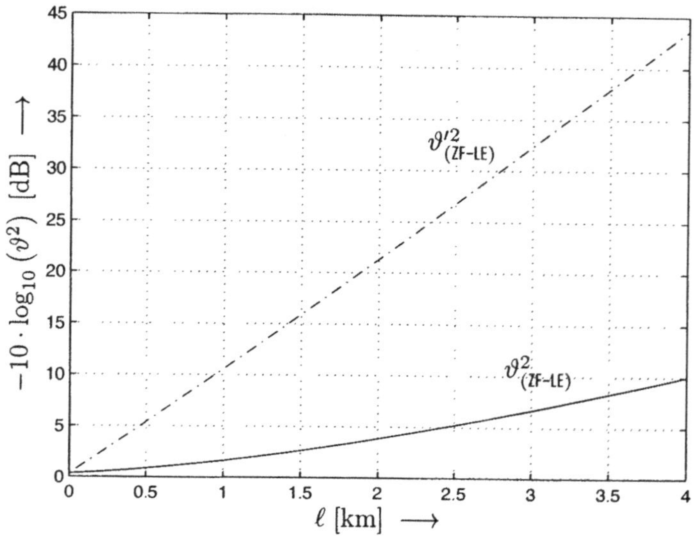

Fig. 2.9 Loss of optimum ZF linear equalization for the DSL down-stream example. Solid: $\vartheta^2_{(\text{ZF-LE})}$; dashed-dotted: $\vartheta'^2_{(\text{ZF-LE})}$.

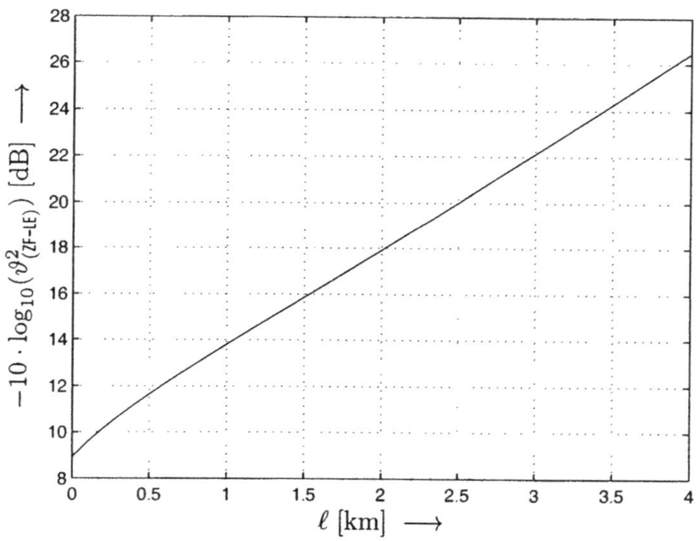

Fig. 2.10 Loss $\vartheta^2_{(\text{ZF-LE})}$ of optimum ZF linear equalization for the DSL up-stream example.

2.2.2 A General Property of the Receive Filter

The derivation of the receive filter for optimal linear zero-forcing results in a combination of a matched filter followed by T-spaced sampling, and a discrete-time filter. We will now show that this is a general principle, and that the cascade of both filters is always optimum. A simple proof for this was given by T. Ericson [Eri71], and here we will follow his presentation.

First, let $H_{\mathrm{R}}^{(\mathrm{c})}(f)$ be a given (continuous-time) receive filter for the cascade $H_{\mathrm{T}}(f)H_{\mathrm{C}}(f)$ of PAM transmit filter and channel. Then, the discrete-time end-to-end transfer function (2.1.7a) is

$$H(\mathrm{e}^{\mathrm{j}2\pi fT})\Big|_{H_{\mathrm{R}}^{(\mathrm{c})}(f)} = \sum_{\mu} H_{\mathrm{T}}(f-\tfrac{\mu}{T})H_{\mathrm{C}}(f-\tfrac{\mu}{T})H_{\mathrm{R}}^{(\mathrm{c})}(f-\tfrac{\mu}{T}) \quad (2.2.31\mathrm{a})$$

and the noise PSD equation (2.1.7b) is given as

$$\Phi_{nn}(\mathrm{e}^{\mathrm{j}2\pi fT})\Big|_{H_{\mathrm{R}}^{(\mathrm{c})}(f)} = \frac{N_0}{T}\cdot\sum_{\mu}\left|H_{\mathrm{R}}^{(\mathrm{c})}(f-\tfrac{\mu}{T})\right|^2 . \quad (2.2.31\mathrm{b})$$

Now, we replace the above receive filter $H_{\mathrm{R}}^{(\mathrm{c})}(f)$ by a matched filter cascaded with a discrete-time filter, i.e., the receive filter now has the form

$$H_{\mathrm{R}}^{(\mathrm{d})}(f) = H_{\mathrm{T}}^{*}(f)H_{\mathrm{C}}^{*}(f)\cdot F(\mathrm{e}^{\mathrm{j}2\pi fT}) , \quad (2.2.32)$$

where $F(\mathrm{e}^{\mathrm{j}2\pi fT})$ is the discrete-time, i.e., frequency-periodic filter. In this case, the discrete-time end-to-end transfer function and the noise PSD reads

$$\begin{aligned} H(\mathrm{e}^{\mathrm{j}2\pi fT})\Big|_{H_{\mathrm{R}}^{(\mathrm{d})}(f)} &= \sum_{\mu} H_{\mathrm{T}}(f-\tfrac{\mu}{T})H_{\mathrm{C}}(f-\tfrac{\mu}{T}) \\ &\qquad\cdot H_{\mathrm{T}}^{*}(f-\tfrac{\mu}{T})H_{\mathrm{C}}^{*}(f-\tfrac{\mu}{T})\cdot F(\mathrm{e}^{\mathrm{j}2\pi(f-\frac{\mu}{T})T}) \\ &= F(\mathrm{e}^{\mathrm{j}2\pi fT})\cdot\sum_{\mu}\left|H_{\mathrm{T}}(f-\tfrac{\mu}{T})H_{\mathrm{C}}(f-\tfrac{\mu}{T})\right|^2 , \end{aligned} \quad (2.2.33\mathrm{a})$$

and

$$\begin{aligned} \Phi_{nn}(\mathrm{e}^{\mathrm{j}2\pi fT})\Big|_{H_{\mathrm{R}}^{(\mathrm{d})}(f)} &= \frac{N_0}{T}\sum_{\mu}\left|H_{\mathrm{T}}^{*}(f-\tfrac{\mu}{T})H_{\mathrm{C}}^{*}(f-\tfrac{\mu}{T})\cdot F(\mathrm{e}^{\mathrm{j}2\pi(f-\frac{\mu}{T})T})\right|^2 \\ &= \frac{N_0}{T}\cdot\left|F(\mathrm{e}^{\mathrm{j}2\pi fT})\right|^2\cdot\sum_{\mu}\left|H_{\mathrm{T}}(f-\tfrac{\mu}{T})\cdot H_{\mathrm{C}}(f-\tfrac{\mu}{T})\right|^2 . \end{aligned} \quad (2.2.33\mathrm{b})$$

If, for all frequencies, the periodic continuation of $|H_{\mathrm{T}}(f)H_{\mathrm{C}}(f)|^2$ is nonzero, we can choose the discrete-time part according to

$$F(\mathrm{e}^{\mathrm{j}2\pi fT}) = \frac{\sum_{\mu} H_{\mathrm{T}}(f-\frac{\mu}{T})H_{\mathrm{C}}(f-\frac{\mu}{T})H_{\mathrm{R}}^{(\mathrm{c})}(f-\frac{\mu}{T})}{\sum_{\mu}\left|H_{\mathrm{T}}(f-\frac{\mu}{T})H_{\mathrm{C}}(f-\frac{\mu}{T})\right|^2} . \quad (2.2.34)$$

Inserting (2.2.34) in (2.2.33a) reveals that for this choice in both situations the same end-to-end transfer function results, i.e., $H(\mathrm{e}^{\mathrm{j}2\pi fT})\Big|_{H_{\mathrm{R}}^{(\mathrm{c})}(f)} = H(\mathrm{e}^{\mathrm{j}2\pi fT})\Big|_{H_{\mathrm{R}}^{(\mathrm{d})}(f)}$.

However, considering the noise PSD and using the Cauchy-Schwarz inequality $\left|\sum_\mu a_\mu b_\mu\right|^2 \le \sum_\mu |a_\mu|^2 \cdot \sum_\mu |b_\mu|^2$ (e.g., [BB91]), we obtain

$$\begin{aligned}
\Phi_{nn}(\mathrm{e}^{\mathrm{j}2\pi fT})\Big|_{H_{\mathrm{R}}^{(\mathrm{d})}(f)} &= \frac{N_0}{T} \cdot \frac{\left|\sum_\mu H_{\mathrm{T}}(f-\frac{\mu}{T})H_{\mathrm{C}}(f-\frac{\mu}{T})H_{\mathrm{R}}^{(\mathrm{c})}(f-\frac{\mu}{T})\right|^2}{\sum_\mu \left|H_{\mathrm{T}}(f-\frac{\mu}{T})H_{\mathrm{C}}(f-\frac{\mu}{T})\right|^2} \\
&\le \frac{N_0}{T} \cdot \frac{\sum_\mu \left|H_{\mathrm{T}}(f-\frac{\mu}{T})H_{\mathrm{C}}(f-\frac{\mu}{T})\right|^2 \cdot \sum_\mu \left|H_{\mathrm{R}}^{(\mathrm{c})}(f-\frac{\mu}{T})\right|^2}{\sum_\mu \left|H_{\mathrm{T}}(f-\frac{\mu}{T})H_{\mathrm{C}}(f-\frac{\mu}{T})\right|^2} \\
&= \frac{N_0}{T} \cdot \sum_\mu \left|H_{\mathrm{R}}^{(\mathrm{c})}(f-\frac{\mu}{T})\right|^2 \\
&= \Phi_{nn}(\mathrm{e}^{\mathrm{j}2\pi fT})\Big|_{H_{\mathrm{R}}^{(\mathrm{c})}(f)} .
\end{aligned} \qquad (2.2.35)$$

Now, let $H_{\mathrm{R}}^{(\mathrm{c})}(f)$ be a given receive filter, which we assume to be optimum with respect to some desired criterion of goodness. It is reasonable to consider only criteria, where—assuming two transmission systems that have the same signal transfer function—the one with the lower noise power is judged to be better. But, in each case, replacing the receive filter by $H_{\mathrm{R}}^{(\mathrm{d})}(f)$ according to (2.2.32) and (2.2.34), without affecting the desired signal, the noise power at the output of the receive filter could be reduced by

$$T \int_{-\frac{1}{2T}}^{\frac{1}{2T}} \left(\Phi_{nn}(\mathrm{e}^{\mathrm{j}2\pi fT})\Big|_{H_{\mathrm{R}}^{(\mathrm{c})}(f)} - \Phi_{nn}(\mathrm{e}^{\mathrm{j}2\pi fT})\Big|_{H_{\mathrm{R}}^{(\mathrm{d})}(f)} \right) df \ge 0 . \qquad (2.2.36)$$

But then, $H_{\mathrm{R}}^{(\mathrm{d})}(f)$ has to be judged better than $H_{\mathrm{R}}^{(\mathrm{c})}(f)$. As this contradicts our assumption, the possible noise reduction, i.e., the integral in (2.2.36), has to be zero. Taking (2.2.35) into consideration, and since power spectral densities are nonnegative, this is only possible if $\Phi_{nn}(\mathrm{e}^{\mathrm{j}2\pi fT})\Big|_{H_{\mathrm{R}}^{(\mathrm{d})}(f)} = \Phi_{nn}(\mathrm{e}^{\mathrm{j}2\pi fT})\Big|_{H_{\mathrm{R}}^{(\mathrm{c})}(f)}$. From the Cauchy-Schwarz inequality we know that equality in (2.2.35) holds if and only if

$$H^{(\mathrm{c})}(f-\tfrac{\mu}{T}) = P(\mathrm{e}^{\mathrm{j}2\pi fT}) \cdot H_{\mathrm{T}}^*(f-\tfrac{\mu}{T})H_{\mathrm{C}}^*(f-\tfrac{\mu}{T}) . \qquad (2.2.37)$$

For each value of f a complex-valued factor, independent of μ, is admitted. Thus $P(\mathrm{e}^{\mathrm{j}2\pi fT})$ is a periodic function in f, and the receive filter can be decomposed into a matched filter and a discrete-time filter. For optimum performance the receive filter has to be of the form (2.2.32), and $P(\mathrm{e}^{\mathrm{j}2\pi fT})$ can be identified as $F(\mathrm{e}^{\mathrm{j}2\pi fT})$. Since there is no other possibility to obtain the same signal transfer function, the optimum receive filter is unique.

The main result of this section is summarized in the following statement [Eri71].

Theorem 2.4: Decomposition of the Optimal Receive Filter

When transmitting over a linear, distorting channel, for any reasonable criterion of goodness, the optimal receive filter always consists of the cascade of a matched filter, T-spaced sampling, and discrete-time postfiltering.

It is noteworthy that the cascade $H_{\mathrm{T}}^*(f)H_{\mathrm{C}}^*(f)\cdot P(\mathrm{e}^{\mathrm{j}2\pi fT})$ can either be implemented as one analog front-end filter followed by T-spaced sampling, or as the continuous-time matched filter $H_{\mathrm{T}}^*(f)H_{\mathrm{C}}^*(f)$, followed by sampling and a succeeding digital (discrete-time) filter $P(\mathrm{e}^{\mathrm{j}2\pi fT})$.

Regarding the matched-filter front-end and T-spaced sampling, we arrive at an end-to-end discrete-time transfer function

$$H^{(\mathrm{MF})}(\mathrm{e}^{\mathrm{j}2\pi fT}) = \sum_{\mu} \left| H_{\mathrm{T}}(f - \tfrac{\mu}{T}) H_{\mathrm{C}}(f - \tfrac{\mu}{T}) \right|^2 , \qquad (2.2.38a)$$

and the PSD of the discrete-time noise sequence reads

$$\Phi_{nn}^{(\mathrm{MF})}(\mathrm{e}^{\mathrm{j}2\pi fT}) = \frac{N_0}{T} \sum_{\mu} \left| H_{\mathrm{T}}(f - \tfrac{\mu}{T}) H_{\mathrm{C}}(f - \tfrac{\mu}{T}) \right|^2 . \qquad (2.2.38b)$$

It is noteworthy that, for the matched-filter front-end, the signal transfer function and the noise PSD are proportional

$$\Phi_{nn}^{(\mathrm{MF})}(\mathrm{e}^{\mathrm{j}2\pi fT}) = \frac{N_0}{T} H^{(\mathrm{MF})}(\mathrm{e}^{\mathrm{j}2\pi fT}) , \qquad (2.2.38c)$$

and both quantities are real (not necessarily even) functions. Figure 2.11 sketches the discrete-time equivalent to PAM transmission with matched-filter front-end at the receiver.

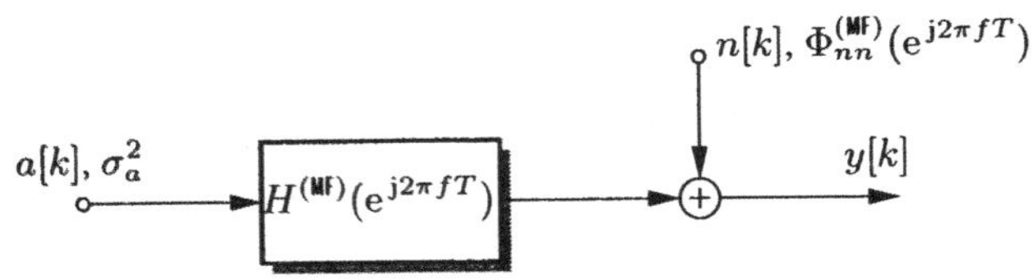

Fig. 2.11 Equivalent discrete-time model for PAM transmission with matched-filter front-end at the receiver.

Simply ignoring the ISI at the output of the matched filter, i.e., regarding a single transmitted pulse, the signal-to-noise ratio for the matched-filter receiver can be given as

$$\mathrm{SNR}^{(\mathrm{MF})} = \frac{\sigma_a^2 \left| h^{(\mathrm{MF})}[0] \right|^2}{\sigma_n^2}$$

$$= \frac{\sigma_a^2 \left| T \int_{-\frac{1}{2T}}^{\frac{1}{2T}} H^{(\mathrm{MF})}(\mathrm{e}^{\mathrm{j}2\pi fT})\, \mathrm{d}f \right|^2}{T \int_{-\frac{1}{2T}}^{\frac{1}{2T}} \Phi^{(\mathrm{MF})}(\mathrm{e}^{\mathrm{j}2\pi fT})\, \mathrm{d}f} ,$$

regarding (2.2.38a) and (2.2.38b), we have

$$= \frac{T\sigma_a^2}{N_0} T \int_{-\frac{1}{2T}}^{\frac{1}{2T}} \sum_{\mu} \left| H_{\mathrm{T}}(f - \tfrac{\mu}{T}) H_{\mathrm{C}}(f - \tfrac{\mu}{T}) \right|^2 \mathrm{d}f ,$$

and using (2.2.18) and (2.2.19), we arrive at

$$= T \int_{-\frac{1}{2T}}^{\frac{1}{2T}} \widetilde{\mathrm{SNR}}(\mathrm{e}^{\mathrm{j}2\pi fT})\, \mathrm{d}f . \qquad (2.2.39)$$

This quantity is often called *matched-filter bound* [LSW68], because the transmission of a single pulse provides the maximum achievable SNR. Intersymbol interference due to sequential pulses can only decrease the performance. In summary, we have

Theorem 2.5: *Signal-to-Noise Ratio for Matched-Filter Receiver*

When applying the matched filter at the receiver, the signal-to-noise ratio (matched-filter bound) is given by the *arithmetic mean* over the folded spectral signal-to-noise ratio at the receiver input

$$\mathrm{SNR}^{(\mathrm{MF})} = T \int_{-\frac{1}{2T}}^{\frac{1}{2T}} \widetilde{\mathrm{SNR}}(\mathrm{e}^{\mathrm{j}2\pi fT})\, \mathrm{d}f . \qquad (2.2.40)$$

Because of Theorem 2.4, the front-end matched filter is subsequently fixed, i.e.,

$$H_{\mathrm{R}}(f) = H_{\mathrm{T}}^{*}(f) H_{\mathrm{C}}^{*}(f) \cdot F(\mathrm{e}^{\mathrm{j}2\pi fT}) . \qquad (2.2.41)$$

Thus, only the discrete-time part remains for optimization. In the sequel we show that dropping the zero-forcing criterion can improve system performance. But, in order to get a neat exposition, in the next subsection, first an important general principle is reviewed.

2.2.3 MMSE Filtering and the Orthogonality Principle

Consider the filtering problem depicted in Figure 2.12. The input sequence $\langle a[k] \rangle$ is passed through a noisy system and the output signal $\langle y[k] \rangle$ is observed. Based on this observation, a filter $W(z)$ should produce estimates $r[k]$ of the initial samples $a[k]$ [Hay96].

This is a classical setting of PAM transmission, where the receiver has to recover data from a distorted signal corrupted by noise. For brevity of notation, for the moment, we do not regard a possible delay of the estimated data signal $\langle r[k] \rangle$ with respect to the reference signal $\langle a[k] \rangle$. The system under study is assumed to equivalently compensate for such a delay.

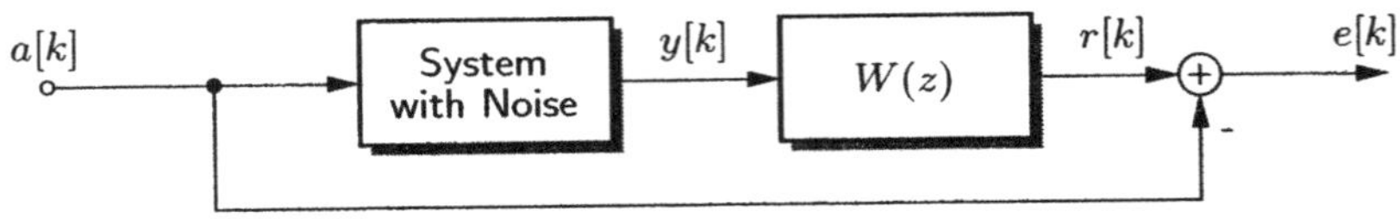

Fig. 2.12 Block diagram of the filtering problem.

Defining the *estimation error* by

$$e[k] = r[k] - a[k] \,, \tag{2.2.42}$$

our aim is now to optimize the filter $W(z)$, so that the *mean-squared value* of this error signal is minimized

$$\mathsf{E}\left\{|e[k]|^2\right\} \to \min \,, \tag{2.2.43}$$

i.e., the estimation is optimum is the sense of *minimum mean-squared error (MMSE)*. We later derive general predications for this setting. First, we restrict ourselves to a causal *finite impulse response (FIR)* filter $W(z) = \sum_{k=0}^{q} w[k] z^{-k}$, whose order is denoted by q.

Derivation of the Optimal Solution In (adaptive) filter theory, it is more convenient and usual to consider the complex conjugate of the tap weights $w[k]$. Then, using the vectors

$$\boldsymbol{w} = \begin{bmatrix} w^*[0] \\ w^*[1] \\ \vdots \\ w^*[q] \end{bmatrix} \,, \qquad \boldsymbol{y}[k] = \begin{bmatrix} y[k] \\ y[k-1] \\ \vdots \\ y[k-q] \end{bmatrix} \,, \tag{2.2.44}$$

the filter output (estimate) is given by the scalar product ($\cdot^{\mathsf{H}}$: Hermitian transpose, i.e., complex conjugation and transposition)

$$r[k] = \sum_{\kappa=0}^{q} w[\kappa] \cdot y[k-\kappa] = \boldsymbol{w}^{\mathsf{H}} \boldsymbol{y}[k] \,, \tag{2.2.45}$$

and the error reads

$$e[k] = \boldsymbol{w}^{\mathsf{H}}\boldsymbol{y}[k] - a[k] \,. \tag{2.2.46}$$

The mean-squared error, which is a function of the conjugate tap-weight vector, $\boldsymbol{w}$, is thus given as

$$\begin{aligned}
J(\boldsymbol{w}) &\triangleq \mathsf{E}\left\{|e[k]|^2\right\} \\
&= \mathsf{E}\left\{e[k]\cdot e^*[k]\right\} \\
&= \mathsf{E}\left\{\left(\boldsymbol{w}^{\mathsf{H}}\boldsymbol{y}[k] - a[k]\right)\cdot\left(\boldsymbol{y}^{\mathsf{H}}[k]\boldsymbol{w} - a^*[k]\right)\right\} \\
&= \mathsf{E}\left\{|a[k]|^2\right\} - \boldsymbol{w}^{\mathsf{H}}\cdot\mathsf{E}\left\{\boldsymbol{y}[k]a^*[k]\right\} \\
&\qquad\qquad - \mathsf{E}\left\{a[k]\boldsymbol{y}^{\mathsf{H}}[k]\right\}\cdot\boldsymbol{w} + \boldsymbol{w}^{\mathsf{H}}\cdot\mathsf{E}\left\{\boldsymbol{y}[k]\boldsymbol{y}^{\mathsf{H}}[k]\right\}\cdot\boldsymbol{w} \\
&= \sigma_a^2 - \boldsymbol{\phi}_{ya}^{\mathsf{H}}\boldsymbol{w} - \boldsymbol{w}^{\mathsf{H}}\boldsymbol{\phi}_{ya} + \boldsymbol{w}^{\mathsf{H}}\boldsymbol{\Phi}_{yy}\boldsymbol{w} \,,
\end{aligned} \tag{2.2.47}$$

where the additional definitions

$$\sigma_a^2 = \mathsf{E}\left\{|a[k]|^2\right\} \qquad \boldsymbol{\phi}_{ya} = \mathsf{E}\left\{\boldsymbol{y}[k]a^*[k]\right\} \qquad \boldsymbol{\Phi}_{yy} = \mathsf{E}\left\{\boldsymbol{y}[k]\boldsymbol{y}^{\mathsf{H}}[k]\right\} \tag{2.2.48}$$

have been used.

Now, the optimum filter vector $\boldsymbol{w}_{\mathrm{opt}}$ is a stationary point of the cost function $J(\boldsymbol{w})$. Applying Wirtinger Calculus[4] (see Appendix A), we arrive at (it turns out to be more convenient to use the derivative with respect to $\boldsymbol{w}^*$)

$$\frac{\partial}{\partial \boldsymbol{w}^*}J(\boldsymbol{w}) = 0 - 0 - \boldsymbol{\phi}_{ya} + \boldsymbol{\Phi}_{yy}\boldsymbol{w} \overset{!}{=} \boldsymbol{0} \,, \tag{2.2.49}$$

which leads to the *Wiener-Hopf equations* or the *normal equations* [Hay96]

$$\boldsymbol{\Phi}_{yy}\boldsymbol{w} = \boldsymbol{\phi}_{ya} \,. \tag{2.2.50}$$

Hence, the solution of (2.2.50) provides, in the MMSE sense, the optimal filter vector, namely the *Wiener filter* [Hay96]

$$\boldsymbol{w}_{\mathrm{opt}} = \boldsymbol{\Phi}_{yy}^{-1}\boldsymbol{\phi}_{ya} \,. \tag{2.2.51}$$

Finally, using (2.2.47) and (2.2.51), the corresponding minimum mean-squared error is given as

$$\begin{aligned}
J_{\min} \triangleq J(\boldsymbol{w}_{\mathrm{opt}}) &= \sigma_a^2 - \boldsymbol{\phi}_{ya}^{\mathsf{H}}\boldsymbol{w}_{\mathrm{opt}} - \boldsymbol{w}_{\mathrm{opt}}^{\mathsf{H}}\boldsymbol{\phi}_{ya} + \boldsymbol{w}_{\mathrm{opt}}^{\mathsf{H}}\boldsymbol{\Phi}_{yy}\boldsymbol{w}_{\mathrm{opt}} \\
&= \sigma_a^2 - \boldsymbol{\phi}_{ya}^{\mathsf{H}}\boldsymbol{\Phi}_{yy}^{-1}\boldsymbol{\phi}_{ya} - \boldsymbol{w}_{\mathrm{opt}}^{\mathsf{H}}\boldsymbol{\phi}_{ya} + \boldsymbol{w}_{\mathrm{opt}}^{\mathsf{H}}\boldsymbol{\phi}_{ya} \\
&= \sigma_a^2 - \boldsymbol{\phi}_{ya}^{\mathsf{H}}\boldsymbol{\Phi}_{yy}^{-1}\boldsymbol{\phi}_{ya} \,.
\end{aligned} \tag{2.2.52}$$

[4] Alternatively, the optimal solution can be obtained by inspection, if (2.2.47) is rewritten in a quadratic form.

Discussion and Interpretation In order to get an interpretation of the optimal solution, we rewrite the normal equations (2.2.50) using the definitions (2.2.48). In the optimum we have

$$\mathsf{E}\left\{\boldsymbol{y}[k]\boldsymbol{y}[k]^{\mathsf{H}}\right\}\boldsymbol{w}_{\text{opt}} = \mathsf{E}\left\{\boldsymbol{y}[k]a^*[k]\right\} , \tag{2.2.53}$$

or, moving all terms to the left-hand side,

$$\mathsf{E}\left\{\boldsymbol{y}[k]\big(\underbrace{\boldsymbol{y}[k]^{\mathsf{H}}\boldsymbol{w}_{\text{opt}} - a^*[k]}_{e^*_{\text{opt}}[k]}\big)\right\} = \mathsf{E}\left\{\boldsymbol{y}[k]e^*_{\text{opt}}[k]\right\} = \mathbf{0} , \tag{2.2.54}$$

which is equivalent to

$$\mathsf{E}\left\{y[k-\kappa]e^*_{\text{opt}}[k]\right\} = 0 , \quad \kappa = 0, 1, \ldots, q . \tag{2.2.55}$$

In words, (2.2.54) and (2.2.55) reveal that for attaining the minimum value of the mean-squared error, it is necessary and sufficient that the estimation error $e_{\text{opt}}[k]$ is *uncorrelated* to the observed signal within the time span of the estimation filter $W(z)$.

An alternative expression is obtained by multiplying (2.2.54) from the left with the constant vector $\boldsymbol{w}^{\mathsf{H}}_{\text{opt}}$, which results in

$$\boldsymbol{w}^{\mathsf{H}}_{\text{opt}}\,\mathsf{E}\left\{\boldsymbol{y}[k]e^*_{\text{opt}}[k]\right\} = 0 , \tag{2.2.56}$$

or, since $r_{\text{opt}}[k] = \boldsymbol{w}^{\mathsf{H}}_{\text{opt}}\boldsymbol{y}[k]$ (cf. (2.2.45)), we arrive at

$$\mathsf{E}\left\{r_{\text{opt}}[k]e^*_{\text{opt}}[k]\right\} = 0 . \tag{2.2.57}$$

Equation (2.2.57) states that in the optimum, the estimate $r[k]$—which is a linear combination of $y[k-\kappa]$, $\kappa = 0, 1, \ldots, q$—also has to be uncorrelated with the error signal $e[k]$.

Finally, if we drop the restriction to a *causal* FIR filter and assume a (two-sided) *infinite impulse response (IIR)* filter $W(z)$, the *cross-correlation* $\phi_{ye}[\kappa]$ between the observed signal $\langle y[k]\rangle$ and error signal $\langle e[k]\rangle$ has to vanish identically:

$$\phi_{ye}[\kappa] \triangleq \mathsf{E}\left\{y[k+\kappa]e^*[k]\right\} = 0 , \quad \forall\kappa \in \mathbb{Z} . \tag{2.2.58}$$

Note that because of the symmetry property $\phi_{ey}[\kappa] = \phi^*_{ye}[-\kappa]$, the cross-correlation $\phi_{ey}[\kappa]$ is zero, too, and the cross-PSD $\Phi_{ey}(z) = \mathcal{Z}\left\{\phi_{ey}[\kappa]\right\} = \sum_\kappa \phi_{ey}[\kappa]z^{-\kappa}$ also vanishes.

The main result of this section, known as the *Orthogonality Principle*, is stated in the following theorem. A geometric interpretation thereof is given on page 42.

> **Theorem 2.6: *Orthogonality Principle***
>
> When estimating a desired signal $a[k]$ from an observation $y[k]$ via a linear filter, the mean-square value of the estimation error $e[k]$ is only minimal, if the estimate $r[k]$, given as the filter output, and the error signal $e[k] = r[k] - a[k]$ are uncorrelated, i.e., *orthogonal* to each other. In the optimum, the observation $y[k]$ is also orthogonal to the error signal $e[k]$.

2.2.4 MMSE Linear Equalization

We are now in a position to derive the linear receive filter, which minimizes the mean-squared error. It is shown that by tolerating some residual intersymbol interference, the signal-to-noise ratio can be improved over ZF linear equalization.

Because of Theorem 2.4 we use the matched-filter front-end, followed by T-spaced sampling and a discrete-time filter $F(z)$ •—○ $\langle f[k]\rangle$. The end-to-end discrete-time transfer function $H^{(\mathrm{MF})}(\mathrm{e}^{\mathrm{j}2\pi fT})$ and the PSD $\Phi_{nn}^{(\mathrm{MF})}(\mathrm{e}^{\mathrm{j}2\pi fT})$ of the discrete-time noise sequence are given in (2.2.38a) and (2.2.38b), respectively.

Now, for optimization, only the discrete-time part $F(z)$ remains, which is assumed to be IIR and to have a two-sided impulse response. Resorting to the problem formulation of Figure 2.12 and identifying $F(z)$ by $W(z)$, the aim of the optimization is to minimize the mean-squared value of the error sequence

$$e[k] = r[k] - a[k] = y[k] * f[k] - a[k] \,. \tag{2.2.59}$$

Having the Orthogonality Principle in mind, the linear filter $F(z)$ is optimal, if the cross-correlation sequence $\phi_{ey}[\kappa]$ vanishes. Multiplying (2.2.59) by $y^*[k-\kappa]$, $\kappa \in \mathbb{Z}$, and taking the expected value yields

$$\begin{aligned}
\mathrm{E}\{e[k]y^*[k-\kappa]\} &= \mathrm{E}\{(y[k] * f[k]) \cdot y^*[k-\kappa]\} - \mathrm{E}\{a[k]y^*[k-\kappa]\} \\
&= \mathrm{E}\{\textstyle\sum_{k'} y[k-k']f[k']y^*[k-\kappa]\} - \mathrm{E}\{a[k]y^*[k-\kappa]\} \\
&= \textstyle\sum_{k'} \mathrm{E}\{y[k-k']y^*[k-\kappa]\}\, f[k'] - \mathrm{E}\{a[k]y^*[k-\kappa]\} \\
&= \textstyle\sum_{k'} \phi_{yy}[k'-\kappa] f[k'] - \phi_{ay}[\kappa] \,,
\end{aligned}$$

respectively

$$\phi_{ey}[\kappa] \;=\; \phi_{yy}[\kappa] * f[\kappa] - \phi_{ay}[\kappa] \stackrel{!}{=} 0 \,. \tag{2.2.60}$$

Taking the discrete-time Fourier transform, the constraint on $F(z)$ is transformed into

$$\Phi_{ey}(\mathrm{e}^{\mathrm{j}2\pi fT}) = \Phi_{yy}(\mathrm{e}^{\mathrm{j}2\pi fT}) \cdot F(\mathrm{e}^{\mathrm{j}2\pi fT}) - \Phi_{ay}(\mathrm{e}^{\mathrm{j}2\pi fT}) \stackrel{!}{=} 0 \,, \tag{2.2.61}$$

where, from basic system theory (e.g. [Pap91]) and considering that $\langle a[k]\rangle$ is a white sequence with variance σ_a^2 (cf. (2.1.2)), the PSDs are given by

$$\Phi_{yy}(\mathrm{e}^{\mathrm{j}2\pi fT}) \;=\; \sigma_a^2 \left(H^{(\mathrm{MF})}(\mathrm{e}^{\mathrm{j}2\pi fT})\right)^2 + \frac{N_0}{T} H^{(\mathrm{MF})}(\mathrm{e}^{\mathrm{j}2\pi fT}) \tag{2.2.62a}$$

$$\Phi_{ay}(\mathrm{e}^{\mathrm{j}2\pi fT}) \;=\; \sigma_a^2 H^{(\mathrm{MF})}(\mathrm{e}^{\mathrm{j}2\pi fT}) \,. \tag{2.2.62b}$$

Here, we have made use of the fact that $H^{(\mathrm{MF})}(\mathrm{e}^{\mathrm{j}2\pi fT})$ is real-valued (cf. (2.2.38)).

Using (2.2.62a), (2.2.62b), and solving (2.2.61) for the desired filter $F(z)$, we have

$$F(\mathrm{e}^{\mathrm{j}2\pi fT}) \;=\; \frac{\Phi_{ay}(\mathrm{e}^{\mathrm{j}2\pi fT})}{\Phi_{yy}(\mathrm{e}^{\mathrm{j}2\pi fT})}$$

$$= \frac{\sigma_a^2 H^{(\mathrm{MF})}(\mathrm{e}^{\mathrm{j}2\pi fT})}{\sigma_a^2 \left(H^{(\mathrm{MF})}(\mathrm{e}^{\mathrm{j}2\pi fT})\right)^2 + \frac{N_0}{T} H^{(\mathrm{MF})}(\mathrm{e}^{\mathrm{j}2\pi fT})}$$

$$= \frac{1}{H^{(\mathrm{MF})}(\mathrm{e}^{\mathrm{j}2\pi fT}) + \frac{N_0}{T\sigma_a^2}} . \tag{2.2.63}$$

Finally, combining $F(z)$ with the front-end matched filter results in the total receive filter for MMSE linear equalization:

Theorem 2.7: *MMSE Linear Equalization*

Let the transmit filter $H_\mathrm{T}(f)$, a channel with transfer function $H_\mathrm{C}(f)$, and additive noise with PSD N_0 be given. The linear receive filter, which is optimum in the MMSE sense, is given by

$$H_\mathrm{R}^{(\mathrm{MMSE\text{-}LE})}(f) = \frac{H_\mathrm{T}^*(f) H_\mathrm{C}^*(f)}{\sum\limits_\mu \left| H_\mathrm{T}(f - \frac{\mu}{T}) H_\mathrm{C}(f - \frac{\mu}{T}) \right|^2 + \frac{N_0}{T\sigma_a^2}} , \tag{2.2.64}$$

where σ_a^2 is the variance of the transmit data sequence.

Discussion A comparison of (2.2.64) and (2.2.12) indicates an additional term $\frac{N_0}{T\sigma_a^2}$ in the denominator of the transfer function of the MMSE linear equalizer. This term ensures that the denominator is strictly positive. Thus, in contrast to ZF linear equalization, even if the periodic sum has spectral zeros, the MMSE linear equalizer always exists and a stable implementation is possible.

Additionally, it is interesting to observe the asymptotic behavior of the filter. As the signal-to-noise ratio tends to infinity, the term $\frac{N_0}{T\sigma_a^2}$ vanishes, and the filters for both criteria, zero-forcing and minimum mean-squared error, become identical. Consequently, for $N_0 = 0$, the MMSE linear equalizer also completely eliminates the intersymbol interference. Thus, in the high SNR region, the receive filter primarily concentrates on the signal distortion. Conversely, for very low SNRs, the sum in the denominator can be neglected and $H_\mathrm{R}(f) \approx \frac{T\sigma_a^2}{N_0} H_\mathrm{T}^*(f) H_\mathrm{C}^*(f)$, i.e., the matched-filter receiver results. If the SNR tends to zero, the transfer function of the receive filter vanishes, too. Therefore, we can state that in the low SNR region, the receive filter basically concentrates on the noise and tries to maximize the instantaneous SNR without looking at the ISI.

Signal-to-Noise Ratio In order to calculate the achievable SNR, we have to consider the effects of the noise and the additional distortion due to the residual ISI. Using the end-to-end impulse response $h[k] \circ\!\!-\!\!\bullet\, H^{(\mathrm{MMSE\text{-}LE})}(e^{j2\pi fT})$ with

$$\begin{aligned} H^{(\mathrm{MMSE\text{-}LE})}(e^{j2\pi fT}) &= \sum_\mu H_T(f-\tfrac{\mu}{T})H_C(f-\tfrac{\mu}{T})H_R(f-\tfrac{\mu}{T}) \\ &= \frac{\sum_\mu \left|H_T(f-\tfrac{\mu}{T})H_C(f-\tfrac{\mu}{T})\right|^2}{\sum_\mu \left|H_T(f-\tfrac{\mu}{T})H_C(f-\tfrac{\mu}{T})\right|^2 + \frac{N_0}{T\sigma_a^2}} \\ &\overset{(2.2.18)}{\underset{(2.2.19)}{=}} \frac{\widetilde{\mathrm{SNR}}(e^{j2\pi fT})}{\widetilde{\mathrm{SNR}}(e^{j2\pi fT})+1}\,, \end{aligned} \tag{2.2.65}$$

the error signal is given by[5]

$$e[k] = (a[k]*h[k]+n[k]) - a[k] = a[k]*(h[k]-\delta[k]) + n[k]\,. \tag{2.2.66}$$

For conciseness, we define the abbreviation $\Sigma(f) \triangleq \widetilde{\mathrm{SNR}}(e^{j2\pi fT})$ and remark that this quantity is real. Because transmitted signal and noise are statistically independent of each other, the variance of the error signal calculates to

$$\begin{aligned} \sigma_e^2 &= \sigma_a^2 T \int_{-\frac{1}{2T}}^{\frac{1}{2T}} \left|H^{(\mathrm{MMSE\text{-}LE})}(e^{j2\pi fT}) - 1\right|^2 \mathrm{d}f + T\int_{-\frac{1}{2T}}^{\frac{1}{2T}} \frac{1}{T}\sum_\mu N_0 \left|H_R(f-\tfrac{\mu}{T})\right|^2 \mathrm{d}f \\ &= \sigma_a^2 T \int_{-\frac{1}{2T}}^{\frac{1}{2T}} \left(\frac{\Sigma(f)}{\Sigma(f)+1} - 1\right)^2 \mathrm{d}f + \sigma_a^2 T \int_{-\frac{1}{2T}}^{\frac{1}{2T}} \frac{\Sigma(f)}{(\Sigma(f)+1)^2}\,\mathrm{d}f \\ &= \sigma_a^2 T \int_{-\frac{1}{2T}}^{\frac{1}{2T}} \frac{1+\Sigma(f)}{(\Sigma(f)+1)^2}\,\mathrm{d}f \\ &= \sigma_a^2 T \int_{-\frac{1}{2T}}^{\frac{1}{2T}} \frac{1}{\Sigma(f)+1}\,\mathrm{d}f\,, \end{aligned} \tag{2.2.67}$$

and hence, the SNR for MMSE linear equalization reads

$$\mathrm{SNR}^{(\mathrm{MMSE\text{-}LE})} = \frac{\sigma_a^2}{\sigma_e^2} = \frac{1}{T\int_{-\frac{1}{2T}}^{\frac{1}{2T}} \frac{1}{\widetilde{\mathrm{SNR}}(e^{j2\pi fT})+1}\,\mathrm{d}f}\,. \tag{2.2.68}$$

[5] $\delta[k] = \begin{cases} 1, & k=0 \\ 0, & \text{else} \end{cases}$ is the discrete-time unit pulse.

The derived linear MMSE equalizer is optimum with respect to the signal-to-error ratio σ_a^2/σ_e^2. However, for the coefficient $h[0]$, on which the decision is based,

$$h[0] = T \int_{-\frac{1}{2T}}^{\frac{1}{2T}} H^{(\text{MMSE-LE})}(e^{j2\pi fT})\, df \leq 1 \tag{2.2.69}$$

holds. This follows from (2.2.65), where we see that the real-valued signal transfer function is bounded by $0 \leq H^{(\text{MMSE-LE})}(e^{j2\pi fT}) \leq 1$. Equality, and thus $h[0] = 1$, is achieved only in the limit for high SNR.

Hence, part of the desired signal is falsely apportioned to the error sequence. When using a standard slicer—where the decision levels are optimized for the original signal constellation—the decision rule is *biased*. In turn, for nonbinary signaling, performance is not optimum but can be improved by removing this bias, i.e., treating $(1 - h[0])a[k]$ as a signal rather than as part of the intersymbol interference. This is done by simply scaling the signal by $1/h[0]$ prior to the decision.[6] Then, the error signal is given by

$$e'[k] = \left(a[k] * \frac{h[k]}{h[0]} + \frac{n[k]}{h[0]}\right) - a[k] = a[k] * \left(\frac{h[k]}{h[0]} - \delta[k]\right) + \frac{n[k]}{h[0]}, \tag{2.2.70}$$

and for its variance we have

$$\begin{aligned}
\sigma_{e'}^2 &= \sigma_a^2 T \int_{-\frac{1}{2T}}^{\frac{1}{2T}} \left(\frac{1}{h[0]}\frac{\Sigma(f)}{\Sigma(f)+1} - 1\right)^2 df + \frac{\sigma_a^2 T}{h^2[0]} \int_{-\frac{1}{2T}}^{\frac{1}{2T}} \frac{\Sigma(f)}{(\Sigma(f)+1)^2}\, df \\
&= \sigma_a^2 \left(\frac{1}{h^2[0]} T \int_{-\frac{1}{2T}}^{\frac{1}{2T}} \frac{\Sigma^2(f)}{(\Sigma(f)+1)^2}\, df - \frac{2}{h[0]} \underbrace{T \int_{-\frac{1}{2T}}^{\frac{1}{2T}} \frac{\Sigma(f)}{\Sigma(f)+1}\, df}_{=h[0]} + 1\right) \\
&\quad + \frac{\sigma_a^2 T}{h^2[0]} \int_{-\frac{1}{2T}}^{\frac{1}{2T}} \frac{\Sigma(f)}{(\Sigma(f)+1)^2}\, df \\
&= \frac{1}{h^2[0]} \sigma_a^2 \underbrace{T \int_{-\frac{1}{2T}}^{\frac{1}{2T}} \frac{\Sigma^2(f)+\Sigma(f)}{(\Sigma(f)+1)^2}\, df}_{=h[0]} - \sigma_a^2 \\
&= \frac{\sigma_a^2}{T \int_{-\frac{1}{2T}}^{\frac{1}{2T}} \frac{\Sigma(f)}{\Sigma(f)+1}\, df} - \sigma_a^2 \,.
\end{aligned} \tag{2.2.71}$$

[6]Alternatively, the decision levels can be scaled by $h[0]$. Here we prefer to use a fixed slicer and to scale the signals.

The SNR for *unbiased MMSE linear equalization* then takes on the bulky form

$$\mathrm{SNR}^{(\mathsf{MMSE\text{-}LE,u})} = \frac{\sigma_a^2}{\sigma_{e'}^2} = \left(\left(T \int_{-\frac{1}{2T}}^{\frac{1}{2T}} \frac{\widetilde{\mathrm{SNR}}(\mathrm{e}^{\mathrm{j}2\pi fT})}{\widetilde{\mathrm{SNR}}(\mathrm{e}^{\mathrm{j}2\pi fT}) + 1} \, \mathrm{d}f \right)^{-1} - 1 \right)^{-1} . \quad (2.2.72)$$

With subsequent manipulations, we can identify the relation between the signal-to-noise ratios $\mathrm{SNR}^{(\mathsf{MMSE\text{-}LE})}$ and $\mathrm{SNR}^{(\mathsf{MMSE\text{-}LE,u})}$ to be ($\int(\cdot)$ abbreviates $T\int_{-\frac{1}{2T}}^{\frac{1}{2T}} (\cdot)\, \mathrm{d}f$)

$$\begin{aligned}
\mathrm{SNR}^{(\mathsf{MMSE\text{-}LE,u})} &= \frac{1}{\frac{1}{\int \frac{\Sigma(f)}{\Sigma(f)+1}} - 1} \\
&= \frac{\int \frac{\Sigma(f)}{\Sigma(f)+1}}{\left(1 - \int \frac{\Sigma(f)}{\Sigma(f)+1}\right)} \frac{\int \frac{1}{\Sigma(f)+1}}{\int \frac{1}{\Sigma(f)+1}} ,
\end{aligned}$$

and since $ab = (1-a)(1-b)$, if $a+b=1$, and $\int \frac{\Sigma(f)}{\Sigma(f)+1} + \int \frac{1}{\Sigma(f)+1} = \int \frac{\Sigma(f)+1}{\Sigma(f)+1} = T\int_{-\frac{1}{2T}}^{\frac{1}{2T}} 1 \, \mathrm{d}f = 1$,

$$\begin{aligned}
&= \frac{\left(1 - \int \frac{\Sigma(f)}{\Sigma(f)+1}\right)}{\left(1 - \int \frac{\Sigma(f)}{\Sigma(f)+1}\right)} \frac{\left(1 - \int \frac{1}{\Sigma(f)+1}\right)}{\int \frac{1}{\Sigma(f)+1}} \\
&= \frac{1}{\int \frac{1}{\Sigma(f)+1}} - 1 \\
&= \mathrm{SNR}^{(\mathsf{MMSE\text{-}LE})} - 1 . \qquad (2.2.73)
\end{aligned}$$

Thus, removing the bias by appropriate scaling results in a signal-to-noise ratio which is less by 1 compared to the SNR of the biased MMSE receiver. However, for communications, $\mathrm{SNR}^{(\mathsf{MMSE\text{-}LE,u})}$ is the relevant quantity. Moreover, with respect to error probability, the unbiased decision rule is optimum. Before showing that this is a general principle, we write down the SNR of MMSE linear equalization and its loss compared to transmission over the AWGN channel. Following the steps in Section 2.2.1 and using (2.2.68) and (2.2.73), we arrive at the following theorem.

Theorem 2.8: *Signal-to-Noise Ratio of Unbiased MMSE Linear Equalization*

When using unbiased minimum mean-squared error linear equalization at the receiver, the signal-to-noise ratio is given by

$$\mathrm{SNR}^{(\mathrm{MMSE\text{-}LE,u})} = \frac{1}{T\int_{-\frac{1}{2T}}^{\frac{1}{2T}} \frac{1}{\widetilde{\mathrm{SNR}}(\mathrm{e}^{\mathrm{j}2\pi fT})+1}\,\mathrm{d}f} - 1\,, \tag{2.2.74}$$

and the degradation (based on equal receive power) compared to transmission over an ideal channel reads

$$\vartheta^2_{(\mathrm{MMSE\text{-}LE,u})} = \left(T\int_{-\frac{1}{2T}}^{\frac{1}{2T}} \widetilde{\mathrm{SNR}}(\mathrm{e}^{\mathrm{j}2\pi fT})\,\mathrm{d}f\right)^{-1} \cdot \left(\left(T\int_{-\frac{1}{2T}}^{\frac{1}{2T}} \left(\widetilde{\mathrm{SNR}}(\mathrm{e}^{\mathrm{j}2\pi fT})+1\right)^{-1} \mathrm{d}f\right)^{-1} - 1\right). \tag{2.2.75}$$

Again, $\widetilde{\mathrm{SNR}}(\mathrm{e}^{\mathrm{j}2\pi fT})$ is the folded spectral signal-to-noise ratio at the receiver input.

Finally, we note that a derivation of the MMSE linear equalizer from scratch—without a prior decomposition of the receive filter into matched filter and discrete-time filter—can already be found in [Smi65]. Moreover, by applying the tools of Section 2.2.3 it is straightforward to obtain finite-length results (see also [Pro01]).

General Result on Unbiased Receivers In this section, we illustrate that the SNR relationship given above is very general and related to the MMSE criterion known from estimation theory. The exposition is similar to that in [CDEF95].

Assume a discrete-time, ISI-free additive noise channel which outputs

$$y[k] = a[k] + n[k]\,. \tag{2.2.76}$$

The data-carrying symbol $a[k]$ (zero mean, variance σ_a^2) is drawn from a finite signal set $\mathcal{A}$, and $n[k]$ (variance σ_n^2) is the additive noise term, independent of the transmitted symbols.

The receive sample $y[k]$ could be directly fed to a slicer, which produces estimates of $a[k]$. The signal-to-noise ratio for this *unbiased* decision rule is

$$\mathrm{SNR_u} = \frac{\sigma_a^2}{\sigma_n^2}\,. \tag{2.2.77}$$

It is noteworthy that for the additive noise channel without intersymbol interference, the unbiased MMSE solution is identical to the zero-forcing solution.

We now include scaling of the received signal by a real-valued gain factor g prior to the threshold device. This case is depicted in Figure 2.13.

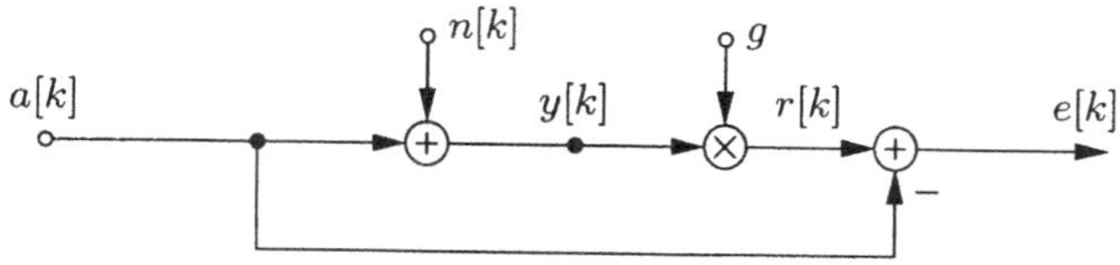

Fig. 2.13 Illustration of the SNR optimization problem.

The error signal is then given by

$$e[k] = g \cdot (a[k] + n[k]) - a[k] = (g-1)\, a[k] + g\, n[k] \,, \tag{2.2.78}$$

and the signal-to-error power ratio, which is dependent on g, reads

$$\mathrm{SNR}(g) = \frac{\sigma_a^2}{\sigma_e^2} = \frac{\sigma_a^2}{(g-1)^2\sigma_a^2 + g^2\sigma_n^2} \,. \tag{2.2.79}$$

The MMSE optimization problem is to find the g which minimizes the error variance or, respectively, maximizes the SNR. Differentiation of the denominator of the SNR with respect to g yields

$$\frac{\mathrm{d}}{\mathrm{d}g}\left((g-1)^2\sigma_a^2 + g^2\sigma_n^2\right) = 2(g-1)\sigma_a^2 + 2g\sigma_n^2 \stackrel{!}{=} 0 \,, \tag{2.2.80}$$

with the solution

$$g_{\mathrm{opt}} = \frac{\sigma_a^2}{\sigma_a^2 + \sigma_n^2} \,. \tag{2.2.81}$$

The proof is straightforward that for this g the SNR is

$$\begin{aligned} \mathrm{SNR_b} &\triangleq \mathrm{SNR}(g_{\mathrm{opt}}) \\ &= \frac{\sigma_a^2}{\sigma_n^2} + 1 \\ &= \mathrm{SNR_u} + 1 \,. \end{aligned} \tag{2.2.82}$$

Hence, an "optimal" scaling of the signal virtually increases the SNR by one. The receiver is optimum in the sense of estimation theory. But with respect to error rate, i.e., from a communications point of view, it is not optimum. Since the data signal is attenuated by $g < 1$, the slicer no longer fits and the decision rule is biased. (Only for bipolar binary transmission can any scaling of the received signal be tolerated.) Thus, given a receiver designed on principles from estimation theory, performance can be improved by scaling the signal prior to the decision device, and consequently compensating for the bias.

This observation is summarized in the following theorem.

Theorem 2.9: *Biased versus Unbiased MMSE Receiver*

Designing an MMSE equalizer based on the Orthogonality Principle will lead to a bias, i.e., the coefficient on which decision is based is smaller than one. Removing this bias by appropriate scaling improves symbol error rate. The signal-to-noise ratio of the unbiased MMSE reciever—the only one relevant in digital communications—is smaller by one compared to that of the biased MMSE receiver. .

The apparent discrepancy can be solved by recalling that error rate is only a monotone function of the signal-to-noise ratio if the pdf of the noise term is always of the same type. For example, for Gaussian noise, SNR and error rate are related by the Q-function. For the unbiased detection rule, the error $e[k]$ is identical to the additive noise $n[k]$, and thus has this particular pdf. However, in the biased receiver the pdf of the error $e[k]$ is a scaled and shifted version of the pdf of $n[k]$. In particular, the mean value is dependent on the actual data symbol $a[k]$. Because of this, the SNR of the biased and unbiased receiver cannot be compared directly. Moreover, for $\sigma_n^2 \to \infty$ the optimal scale factor g_{opt} goes to zero. This leads to the strange result $\text{SNR}_\text{b} = 1$, even though no data signal (or noise signal) is present at the decision point.

Figure 2.14 visualizes the relationship of the signal by interpreting the signals as vectors in a two-dimensional signal space. Here, the length of the vectors corresponds to their respective variances. First, the transmit signal $\langle a[k] \rangle$ is independent (and thus

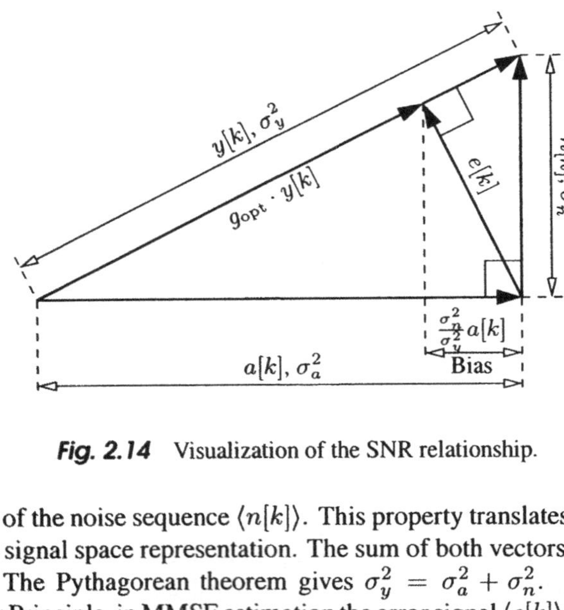

Fig. 2.14 Visualization of the SNR relationship.

uncorrelated) of the noise sequence $\langle n[k] \rangle$. This property translates to perpendicular vectors in the signal space representation. The sum of both vectors gives the receive signal $y[k]$. The Pythagorean theorem gives $\sigma_y^2 = \sigma_a^2 + \sigma_n^2$. By virtue of the Orthogonality Principle, in MMSE estimation the error signal $\langle e[k] \rangle$ is uncorrelated to the observation $\langle y[k]\rangle$. Furthermore, since $e[k] = g_{\text{opt}}y[k] - a[k]$, these three signals also constitute a right-angled triangle in signal space. Moreover, with $g_{\text{opt}} = \sigma_a^2/\sigma_y^2$ or $1 - g_{\text{opt}} = \sigma_n^2/\sigma_y^2$, respectively, taking the intercept theorems and the relations of similar triangles into consideration gives the bias as the projection of the intersection of $y[k]$ and $e[k]$ onto the $a[k]$ axis. From basic geometry, we have

$$\frac{\sigma_a^2}{\sigma_e^2} = \frac{\sigma_y^2}{\sigma_n^2}, \tag{2.2.83}$$

and the SNR relation is simply

$$\text{SNR}_\text{u} = \frac{\sigma_a^2}{\sigma_n^2} = \frac{\sigma_y^2 - \sigma_n^2}{\sigma_n^2} = \frac{\sigma_y^2}{\sigma_n^2} - 1 \overset{(2.2.83)}{=} \frac{\sigma_a^2}{\sigma_e^2} - 1 = \text{SNR}_\text{b} - 1\,. \tag{2.2.84}$$

2.2.5 Joint Transmitter and Receiver Optimization

So far the transmit pulse shape $h_\text{T}(t)$ was assumed to be fixed and optimization was restricted to the receiver side. We now address the problem of joint optimization of transmitter and receiver, cf. [Smi65, BT67, LSW68, ST85]. Therefore, for brevity, we concentrate on the zero-forcing solution. As shown above, at least for high signal-to-noise ratios the global optimum is very nearly achieved.

The following derivation is done in two steps: First, a problem dual to Section 2.2.1 is considered, i.e., the optimization of the transmitter given the receive filter. Then, both results are combined to get the final solution.

Transmitter Optimization Analogous to Section 2.2.1, we now fix the receive filter $H_\text{R}(f)$ and choose the transmit filter, such that the end-to-end cascade is Nyquist. The remaining degree of freedom is used to minimize transmit power. Thus, the optimization problem for $H_\text{T}(f)$ can be stated as follows:

- Minimize the average transmit power:

$$S = \sigma_a^2 \cdot T \int_{-\infty}^{\infty} |H_\text{T}(f)|^2 \,\text{d}f = \sigma_a^2 \cdot T \int_{-\frac{1}{2T}}^{\frac{1}{2T}} \sum_\mu \left|H_\text{T}(f - \tfrac{\mu}{T})\right|^2 \,\text{d}f\,, \tag{2.2.85}$$

- subject to the additional constraint of an end-to-end Nyquist characteristic:

$$\sum_\mu H_\text{T}(f - \tfrac{\mu}{T}) H_\text{C}(f - \tfrac{\mu}{T}) H_\text{R}(f - \tfrac{\mu}{T}) \overset{!}{=} 1\,, \quad \forall f \in (-\tfrac{1}{2T}, \tfrac{1}{2T}]\,. \tag{2.2.86}$$

Like above, this problem can be solved by the Lagrange multipliers method. With the real function $\lambda(\text{e}^{\text{j}2\pi fT})$ of Lagrange multipliers, and defining the real-valued constant $\zeta \overset{\Delta}{=} \sigma_a^2 T$, the Lagrange function reads:

$$L(f) = \sum_\mu \left(\zeta \left|H_\text{T}(f - \tfrac{\mu}{T})\right|^2 - \lambda(\text{e}^{\text{j}2\pi fT}) H_\text{T}(f - \tfrac{\mu}{T}) H_\text{C}(f - \tfrac{\mu}{T}) H_\text{R}(f - \tfrac{\mu}{T})\right)\,, \tag{2.2.87}$$

leading to the optimum

$$H_{\rm T}(f-\tfrac{\mu}{T}) = \frac{\lambda(e^{j2\pi fT})}{2\zeta} H_{\rm C}^*(f-\tfrac{\mu}{T}) \cdot H_{\rm R}^*(f-\tfrac{\mu}{T}) \,. \tag{2.2.88}$$

Joint Optimization Remember, the best receive filter for a given transmit filter is of the form (cf. (2.2.10))

$$H_{\rm R}(f-\tfrac{\mu}{T}) = \frac{\lambda(e^{j2\pi fT})}{2} H_{\rm T}^*(f-\tfrac{\mu}{T}) \cdot H_{\rm C}^*(f-\tfrac{\mu}{T}) \,. \tag{2.2.89}$$

For joint transmitter and receiver optimization, conditions (2.2.88) and (2.2.89) have to be met simultaneously. This requires either

(a) $H_{\rm T}(f-\tfrac{\mu}{T}) = H_{\rm R}(f-\tfrac{\mu}{T}) = 0$

or

(b) $H_{\rm T}(f-\tfrac{\mu}{T}) = \frac{\lambda(e^{j2\pi fT})}{2\zeta} H_{\rm C}^*(f-\tfrac{\mu}{T}) \cdot \frac{\lambda(e^{j2\pi fT})}{2} \cdot H_{\rm T}(f-\tfrac{\mu}{T}) \cdot H_{\rm C}(f-\tfrac{\mu}{T}) \,.$

Solution (b) leads to (here, $H_{\rm T}(f-\tfrac{\mu}{T}) \neq 0$)

$$1 = \frac{\lambda^2(e^{j2\pi fT})}{4\zeta} \cdot \left|H_{\rm C}(f-\tfrac{\mu}{T})\right|^2 \tag{2.2.90}$$

and thus

$$\lambda(e^{j2\pi fT}) = \frac{\pm 2\sqrt{\zeta}}{\left|H_{\rm C}(f-\tfrac{\mu}{T})\right|} \,. \tag{2.2.91}$$

Because the Lagrange multiplier is periodic in frequency, for each frequency f, (2.2.91) can only be satisfied for one value of μ (one frequency out of the set of Nyquist-equivalent frequencies)—except for the special case of a periodic channel transfer function. For all other μ the trivial solution $H_{\rm T}(f-\tfrac{\mu}{T}) = H_{\rm R}(f-\tfrac{\mu}{T}) = 0$ must be used. Thus, for each f, the transmit and receive filters are only nonzero for one particular value of μ. For this point we have from (2.2.86)

$$H_{\rm T}(f-\tfrac{\mu}{T}) \cdot H_{\rm C}(f-\tfrac{\mu}{T}) \cdot H_{\rm R}(f-\tfrac{\mu}{T}) = 1 \,, \tag{2.2.92}$$

and combining (2.2.92), (2.2.91), (2.2.88), and (2.2.89) results in (eliminating $H_{\rm T}(f)$)

$$\frac{\pm 2\sqrt{\zeta}}{\left|H_{\rm C}(f-\tfrac{\mu}{T})\right| \cdot 2\zeta} \cdot H_{\rm C}^*(f-\tfrac{\mu}{T}) \cdot H_{\rm R}^*(f-\tfrac{\mu}{T}) \cdot H_{\rm C}(f-\tfrac{\mu}{T}) \cdot H_{\rm R}(f-\tfrac{\mu}{T}) = 1$$

$$\pm\frac{1}{\sqrt{\zeta}} \left|H_{\rm C}(f-\tfrac{\mu}{T})\right| \cdot \left|H_{\rm R}(f-\tfrac{\mu}{T})\right|^2 = 1 \,,$$

thus, using $\lambda \triangleq \sqrt[4]{\zeta}$,

$$\left|H_{\rm R}(f-\tfrac{\mu}{T})\right| = \frac{\lambda}{\sqrt{\left|H_{\rm C}(f-\tfrac{\mu}{T})\right|}} \,, \tag{2.2.93}$$

and (eliminating $H_R(f)$)

$$H_T(f-\tfrac{\mu}{T})\cdot H_C(f-\tfrac{\mu}{T})\cdot \frac{\pm 2\sqrt{\zeta}}{\left|H_C(f-\tfrac{\mu}{T})\right|\cdot 2} H_T^*(f-\tfrac{\mu}{T})\cdot H_C^*(f-\tfrac{\mu}{T}) = 1$$

$$\pm\sqrt{\zeta}\cdot\left|H_C(f-\tfrac{\mu}{T})\right|\cdot\left|H_T(f-\tfrac{\mu}{T})\right|^2 = 1\,,$$

thus

$$\left|H_T(f-\tfrac{\mu}{T})\right| = \frac{1}{\lambda\sqrt{\left|H_C(f-\tfrac{\mu}{T})\right|}}\,. \tag{2.2.94}$$

The constant λ allows for an adjustment of the transmit power.

A close look on (2.2.93) and (2.2.94) reveals that the task of linear equalization of $H_C(f)$ is split equally among transmitter and receiver. Both transmit and receive filter magnitudes are proportional to the square-root of $\frac{1}{|H_C(f)|}$.

Finally, (2.2.92) gives a constraint on the phases of the filters

$$\arg\left\{H_T(f-\tfrac{\mu}{T})\right\} + \arg\left\{H_C(f-\tfrac{\mu}{T})\right\} + \arg\left\{H_R(f-\tfrac{\mu}{T})\right\} = 0\,. \tag{2.2.95}$$

Hence, the phase of, e.g., the transmit filter $H_T(f)$ can be chosen arbitrarily, as long as it is compensated for by the phase of the receive filter $H_R(f)$.

The last open point is the question: Which frequency out of the set of Nyquist-equivalent frequencies should be used for transmission? It is intuitively clear that it is optimum to use that period μ, for which the amplitude of the channel transfer function $|H_C(f-\frac{\mu}{T})|$ is maximum. Let

$$\mathcal{F} = \left\{f \;\middle|\; |H_C(f)| \geq \left|H_C(f-\tfrac{\mu}{T})\right|\,,\; f\in\mathbb{R},\, \mu\in\mathbb{Z}\right\} \tag{2.2.96}$$

be the set of frequencies f for which the channel gain is maximum over all periodically shifted frequencies $f-\frac{\mu}{T}$, $\mu\in\mathbb{Z}$. Note that for each $f\in\mathcal{F}$, $f-\frac{\mu}{T}\notin\mathcal{F}$, $\mu\in\mathbb{Z}\setminus\{0\}$—for each f there exists one and only one $\mu\in\mathbb{Z}$ with $f-\frac{\mu}{T}\in\mathcal{F}$—the measure of the set $\mathcal{F}$ is $\frac{1}{T}$, i.e., $\int_{\mathcal{F}}\mathrm{d}f = \frac{1}{T}$. Hence, as it is indispensable, a full set of Nyquist frequencies or a Nyquist interval is present for transmission. Each frequency component of the data is transmitted only once, using the "best" point out of all periods. The set $\mathcal{F}$ is sometimes called a *generalized Nyquist interval* [Eri73].

It is noteworthy that the optimum transmit spectrum is limited to a band of width $1/T$. It can be shown [Eri73] that for a broad class of optimization criteria, the transmit and receive filters are always strictly band-limited.

The above derivations are valid for a fixed symbol duration T. In order to achieve the best performance, this parameter has to be optimized, too. For a fixed data rate this gives an optimal exchange between signal bandwidth and number of signaling levels, and hence the required SNR. We discuss this topic in more detail in the section on decision-feedback equalization.

In summary, the optimal transmit and receive filters for zero-forcing linear equalization are given in the following theorem.

Theorem 2.10: *Optimal Transmit and Receive Filter for ZF Linear Equalization*

Let the channel with transfer function $H_C(f)$ and additive white noise be given. The optimal design of linear transmit and receive filters which results in intersymbol-interference free samples and minimal noise variance is given by

$$|H_T(f)| = \begin{cases} \frac{1}{\lambda\sqrt{|H_C(f)|}}, & f \in \mathcal{F} \\ 0, & \text{else} \end{cases}, \tag{2.2.97}$$

$$|H_R(f)| = \begin{cases} \frac{\lambda}{\sqrt{|H_C(f)|}}, & f \in \mathcal{F} \\ 0, & \text{else} \end{cases}, \tag{2.2.98}$$

and

$$\arg\{H_T(f)\} + \arg\{H_C(f)\} + \arg\{H_R(f)\} = 0 .$$

The constant λ is chosen so that the desired transmit power is guaranteed and the support of the filters is defined by

$$\mathcal{F} = \left\{ f \;\middle|\; |H_C(f)| \geq |H_C(f - \tfrac{\mu}{T})| \,,\; f \in \mathbb{R},\, \mu \in \mathbb{Z} \right\} . \tag{2.2.99}$$

Because of zero-forcing linear equalization, the equivalent discrete-time channel model has signal transfer function 1, and the PSD of the discrete-time noise sequence reads

$$\begin{aligned} \Phi_{nn}^{(\mathrm{LE,opt})}(e^{j2\pi fT}) &= \frac{N_0}{T} \sum_{\mu} \begin{cases} \left|\lambda/\sqrt{|H_C(f - \frac{\mu}{T})|}\right|^2, & f - \frac{\mu}{T} \in \mathcal{F} \\ 0, & \text{else} \end{cases} \\ &= \frac{N_0}{T} \lambda^2 \frac{1}{|H_C(\psi(f))|}, \end{aligned} \tag{2.2.100}$$

where we have used the function

$$\psi : f \mapsto \psi(f) = f - \tfrac{\mu}{T} \quad \text{with } \mu \text{ such that} \quad \psi(f) \in \mathcal{F} , \tag{2.2.101}$$

which maps every frequency point to its Nyquist-equivalent frequency out of the set $\mathcal{F}$. As the transmitter and receiver are optimized, the discrete-time equivalent model depends only on the original channel transfer function $H_C(f)$.

It should be noted that the joint optimization using the MMSE criterion can be found in [Smi65] and [BT67]. The results are similar, and, for high signal-to-noise ratios, tend to that based on the ZF criterion.

Example 2.3: *Optimal Transmit Filter*

An example for transmit filter optimization is given in Figure 2.15. At the top, the magnitude of the channel filter $H_{\mathrm{C}}(f)$ is displayed. For this channel, the shaded region depicts the support of the optimum transmit and receive filters (cf. (2.2.99)). Having this, at the bottom of the figure, the magnitude of the optimal transmit and receive filters—except for the scaling, they are identical—is shown. In this example, it is best to use (one-sided) three noncontiguous frequency bands for transmission. The band around $f = 0$ is omitted; instead, the band starting at $f = 1/T$ is more appropriate. Similarly, instead of the band right below $f = 1/(2T)$, the period at $f = 3/(2T)$ is used.

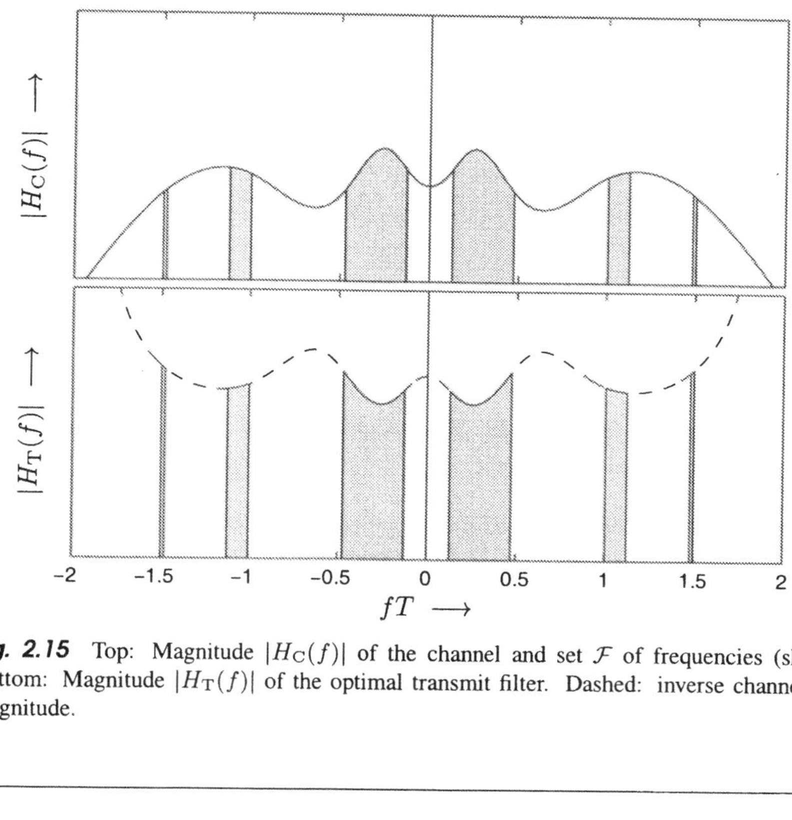

Fig. 2.15 Top: Magnitude $|H_{\mathrm{C}}(f)|$ of the channel and set $\mathcal{F}$ of frequencies (shaded). Bottom: Magnitude $|H_{\mathrm{T}}(f)|$ of the optimal transmit filter. Dashed: inverse channel filter magnitude.

Signal-to-Noise Ratio Because the above result is only a special case of ZF linear equalization, using (2.2.97) the SNR of (2.2.17) calculates to

$$\mathrm{SNR}^{(\mathrm{LE,opt})} = \frac{\sigma_a^2}{N_0 \int_{-\frac{1}{2T}}^{\frac{1}{2T}} \left(\sum_{\mu} \left| \begin{cases} \frac{1}{\lambda\sqrt{H_{\mathrm{C}}(f-\frac{\mu}{T})}}, & f-\frac{\mu}{T}\in\mathcal{F} \\ 0, & \text{else} \end{cases} H_{\mathrm{C}}(f-\tfrac{\mu}{T}) \right|^2 \right)^{-1} df}$$

$$= \frac{\sigma_a^2}{N_0 \int\limits_{-\frac{1}{2T}}^{\frac{1}{2T}} \begin{Bmatrix} \frac{\lambda^2}{\sum_\mu |H_C(f-\frac{\mu}{T})|}, & f-\frac{\mu}{T}\in\mathcal{F} \\ 0, & \text{else} \end{Bmatrix} \mathrm{d}f} . \quad (2.2.102)$$

Because $H_T(f - \frac{\mu}{T})$ is only nonzero for one specific value of μ, the sum can be dropped and instead be integrated over the set $\mathcal{F}$ of frequencies. Hence, we have

$$\mathrm{SNR}^{\mathrm{LE,opt}} = \frac{\sigma_a^2}{N_0 \int\limits_{\mathcal{F}} \frac{\lambda^2}{|H_C(f)|} \, \mathrm{d}f} . \quad (2.2.103)$$

Finally, for equal receive power, the loss compared to transmission over an ISI free channel is given by (cf. (2.2.29))

$$\begin{aligned} \vartheta^2_{(\mathrm{LE,opt})} &= \left(\int\limits_{f\in\mathcal{F}} \left| T \frac{H_C(f)}{\lambda\sqrt{|H_C(f)|}} \right|^2 \mathrm{d}f \cdot \int\limits_{f\in\mathcal{F}} \left(\left| \frac{H_C(f)}{\lambda\sqrt{|H_C(f)|}} \right|^2 \right)^{-1} \mathrm{d}f \right)^{-1} \\ &= \left(\frac{T^2}{\lambda^2} \int\limits_{f\in\mathcal{F}} |H_C(f)| \, \mathrm{d}f \cdot \int\limits_{f\in\mathcal{F}} \frac{1}{\frac{1}{\lambda^2}|H_C(f)|} \, \mathrm{d}f \right)^{-1} \\ &= \left(T \int\limits_{f\in\mathcal{F}} |H_C(f)| \, \mathrm{d}f \cdot T \int\limits_{f\in\mathcal{F}} \frac{1}{|H_C(f)|} \, \mathrm{d}f \right)^{-1} . \end{aligned} \quad (2.2.104)$$

This result is summarized in the following theorem (cf. [LSW68, eq. (5.80), p. 121] for the comparison based on equal transmit power):

Theorem 2.11: *Loss of Optimized Transmission with ZF Linear Equalization*

When using zero-forcing linear equalization at the receiver jointly with the optimal transmit spectrum, for equal receive power the degradation compared to transmission over an ideal channel is given by

$$\vartheta^2_{(\mathrm{LE,opt})} = \left(T \int\limits_{f\in\mathcal{F}} |H_C(f)| \, \mathrm{d}f \cdot T \int\limits_{f\in\mathcal{F}} |H_C(f)|^{-1} \, \mathrm{d}f \right)^{-1} . \quad (2.2.105)$$

Here, the additive white Gaussian noise channel has transfer function $H_C(f)$, and the support $\mathcal{F}$ of the transmit filter is defined by (2.2.99).

2.3 NOISE PREDICTION AND DECISION-FEEDBACK EQUALIZATION

In the last section, strategies for linear equalization of the distorted PAM signal have been discussed. The advantage of these procedures is that intersymbol interference is eliminated completely (or at least to a great extent in the case of MMSE equalization), and thus a simple symbol-by-symbol threshold decision will recover data.

We now show, starting from the drawbacks of linear equalization, how to improve system efficiency by nonlinear equalization. The gain of noise prediction and decision-feedback equalization, respectively, is given, and the performance is compared to bounds from information theory.

2.3.1 Noise Prediction

Consider zero-forcing linear equalization. Then, the end-to-end discrete-time model is given by an additive Gaussian noise channel, where

$$y[k] = a[k] + n[k] \,. \tag{2.3.1}$$

However, due to the receive filter $H_{\mathrm{R}}(f)$, the noise sequence $\langle n[k]\rangle$ is not white, but colored. It is only when the cascade $H_{\mathrm{T}}(f)H_{\mathrm{C}}(f)$ has square-root Nyquist characteristics, i.e., $|H_{\mathrm{T}}(f)H_{\mathrm{C}}(f)|^2$ corresponds to a Nyquist pulse, that the noise will be white. From (2.2.15) and taking (2.2.19) into consideration, the noise PSD reads

$$\begin{aligned}\Phi_{nn}^{(\mathsf{ZF\text{-}LE})}(\mathrm{e}^{\mathrm{j}2\pi fT}) &= \frac{N_0}{T}\,\frac{1}{\sum\limits_{\mu}\left|H_{\mathrm{T}}(f-\frac{\mu}{T})H_{\mathrm{C}}(f-\frac{\mu}{T})\right|^2} \\ &= \frac{\sigma_a^2}{\widetilde{\mathrm{SNR}}(\mathrm{e}^{\mathrm{j}2\pi fT})}\,,\end{aligned} \tag{2.3.2}$$

with the corresponding autocorrelation sequence

$$\phi_{nn}^{(\mathsf{ZF\text{-}LE})}[\kappa] \;\circ\!\!-\!\!\bullet\; \Phi_{nn}^{(\mathsf{ZF\text{-}LE})}(\mathrm{e}^{\mathrm{j}2\pi fT}) \,. \tag{2.3.3}$$

Since the PSD is not constant, the autocorrelation sequence has nonzero terms for $\kappa \neq 0$. Thus, subsequent samples are correlated, i.e., they are statistically dependent. This means that if past samples are known, we can calculate a *prediction* of the following noise sample. If this prediction is subtracted from the received signal, only the *prediction error* remains as noise at the decision point, cf. [BP79, GH85, HM85].

Figure 2.16 sketches the noise prediction structure. First, the threshold device produces estimates $\hat{a}[k]$ of the data symbols $a[k]$. Subtracting these estimated symbols from the receive signal $y[k]$ (cf. (2.3.1)) gives estimates $\hat{n}[k]$ of the noise samples $n[k]$. As long as the decisions are correct, the noise estimates coincide with the actual noise samples. Then, using the p past values $\hat{n}[k-\kappa]$, $\kappa = 1, 2, \ldots, p$, via the *linear prediction filter*

$$P(z) = \sum_{k=1}^{p} p[k] z^{-k} \,, \tag{2.3.4}$$

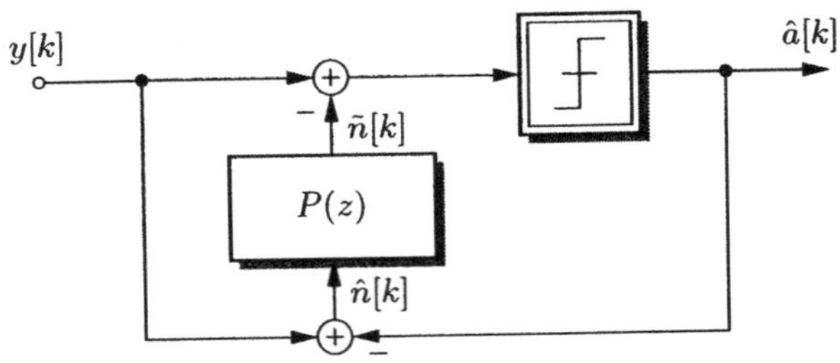

Fig. 2.16 Noise prediction structure.

the prediction

$$\tilde{n}[k] \triangleq \sum_{\kappa=1}^{p} \rho[\kappa] \cdot \hat{n}[k-\kappa] \tag{2.3.5}$$

of the current noise sample $n[k]$ is calculated. Finally, the prediction is subtracted from the receive signal. The coefficient $\rho[0]$ of the FIR filter $P(z)$ has to be zero, as for calculating a prediction the current noise estimate is indeed not yet available.

Now, the residual noise sequence $\langle n'[k] \rangle$ with

$$n'[k] \triangleq n[k] - \tilde{n}[k] \tag{2.3.6}$$

is present at the slicer input. Hence, the aim of system optimization is to build a predictor which minimizes the variance of this residual noise sequence. In order to get an analytic result, we have to assume that the data estimates $\hat{a}[k]$ are correct, i.e., $\hat{a}[k] = a[k]$, which in turn gives correct noise estimates $\hat{n}[k] = n[k]$. Using the vectors

$$\boldsymbol{\rho} = \begin{bmatrix} \rho^*[1] \\ \rho^*[2] \\ \vdots \\ \rho^*[p] \end{bmatrix}, \qquad \boldsymbol{n}[k] = \begin{bmatrix} n[k-1] \\ n[k-2] \\ \vdots \\ n[k-p] \end{bmatrix}, \tag{2.3.7}$$

where, for convenience, again the complex conjugates $\rho^*[k]$, $k = 1, 2, \ldots, p$, of the tap weights are comprised (cf. Section 2.2.3), the residual noise reads

$$n'[k] = n[k] - \boldsymbol{\rho}^{\mathsf{H}} \boldsymbol{n}[k] \,. \tag{2.3.8}$$

Since we want to minimize the variance of the residual noise sequence, the cost function, depending on the tap-weight vector $\boldsymbol{\rho}$, is given by

$$\begin{aligned} J(\boldsymbol{\rho}) &\triangleq \mathsf{E}\left\{ |n'[k]|^2 \right\} \\ &= \mathsf{E}\left\{ \left(n[k] - \boldsymbol{\rho}^{\mathsf{H}} \boldsymbol{n}[k]\right) \cdot \left(n^*[k] - \boldsymbol{n}^{\mathsf{H}}[k] \boldsymbol{\rho}\right) \right\} \\ &= \mathsf{E}\left\{ |n[k]|^2 \right\} - \boldsymbol{\rho}^{\mathsf{H}} \cdot \mathsf{E}\left\{ \boldsymbol{n}[k] n^*[k] \right\} \\ &\qquad - \mathsf{E}\left\{ n[k] \boldsymbol{n}^{\mathsf{H}}[k] \right\} \cdot \boldsymbol{\rho} + \boldsymbol{\rho}^{\mathsf{H}} \cdot \mathsf{E}\left\{ \boldsymbol{n}[k] \boldsymbol{n}^{\mathsf{H}}[k] \right\} \cdot \boldsymbol{\rho} \\ &= \sigma_n^2 - \boldsymbol{\rho}^{\mathsf{H}} \boldsymbol{\phi}_{nn} - \boldsymbol{\phi}_{nn}^{\mathsf{H}} \boldsymbol{\rho} + \boldsymbol{\rho}^{\mathsf{H}} \boldsymbol{\Phi}_{nn} \boldsymbol{\rho} \,, \end{aligned} \tag{2.3.9}$$

where the definitions[7]

$$\begin{aligned}
\sigma_n^2 &= \mathrm{E}\left\{|n[k]|^2\right\} && && (2.3.10a)\\
\boldsymbol{\phi}_{nn} &= \mathrm{E}\left\{\boldsymbol{n}[k]n^*[k]\right\} &=& \left[\phi_{nn}^{(\mathrm{ZF\text{-}LE})}[-i]\right]_{i=1,\ldots,p} && (2.3.10b)\\
\boldsymbol{\Phi}_{nn} &= \mathrm{E}\left\{\boldsymbol{n}[k]\boldsymbol{n}^{\mathsf{H}}[k]\right\} &=& \left[\phi_{nn}^{(\mathrm{ZF\text{-}LE})}[j-i]\right]_{\substack{i=0,\ldots,p-1\\ j=0,\ldots,p-1}} && (2.3.10c)
\end{aligned}$$

have been used.

Using the Wirtinger Calculus, the vector $\boldsymbol{\rho}$, which minimizes $J(\boldsymbol{\rho})$ has to satisfy

$$\frac{\partial}{\partial \boldsymbol{\rho}^*} J(\boldsymbol{\rho}) = 0 - \boldsymbol{\phi}_{nn} - 0 + \boldsymbol{\Phi}_{nn}\boldsymbol{\rho} \overset{!}{=} \mathbf{0}\,, \tag{2.3.11}$$

or equivalently

$$\boldsymbol{\Phi}_{nn}\boldsymbol{\rho} = \boldsymbol{\phi}_{nn}\,. \tag{2.3.12}$$

This set of equations is called the *Yule-Walker equations* and its solution reads

$$\boldsymbol{\rho}_{\mathrm{opt}} = \boldsymbol{\Phi}_{nn}^{-1}\boldsymbol{\phi}_{nn}\,. \tag{2.3.13}$$

Defining for the moment the abbreviation $\phi[\kappa] \triangleq \phi_{nn}^{(\mathrm{ZF\text{-}LE})}[\kappa]$, the Yule-Walker equations (after complex conjugation and taking the symmetry property of autocorrelation sequences $\phi[-\kappa] = \phi^*[\kappa]$ into consideration) read in detail

$$\begin{bmatrix} \phi[0] & \phi[-1] & \cdots & \phi[-p+1] \\ \phi[1] & \ddots & \ddots & \vdots \\ \vdots & \ddots & \ddots & \phi[-1] \\ \phi[p-1] & \cdots & \phi[1] & \phi[0] \end{bmatrix} \cdot \begin{bmatrix} \rho[1] \\ \rho[2] \\ \vdots \\ \rho[p] \end{bmatrix} = \begin{bmatrix} \phi[1] \\ \phi[2] \\ \vdots \\ \phi[p] \end{bmatrix} . \tag{2.3.14}$$

Please note the similarity and difference between the Wiener-Hopf (2.2.50) and the Yule-Walker equations (2.3.12). For the Wiener-Hopf equations, the right-hand side is the cross-correlation vector, whereas in the Yule-Walker equations, the autocorrelation vector is present.

In addition, the Toeplitz structure of the correlation matrix $\boldsymbol{\Phi}_{nn}$—the terms on the diagonals of the matrix are identical—allows us to solve this set of equations efficiently by the *Levinson-Durbin algorithm*. Details are given, e.g., in [Pro01] and [Hay96].

Finally, using (2.3.13) and (2.3.9), the variance of the residual noise sequence for an optimally adjusted linear predictor is given by

$$\begin{aligned}
\sigma_{n'}^2 = J_{\min} \triangleq J(\boldsymbol{\rho}_{\mathrm{opt}}) &= \sigma_n^2 - \boldsymbol{\phi}_{nn}^{\mathsf{H}}\boldsymbol{\rho}_{\mathrm{opt}} && (2.3.15a)\\
&= \sigma_n^2 - \boldsymbol{\rho}_{\mathrm{opt}}^{\mathsf{H}}\boldsymbol{\Phi}_{nn}^{\mathsf{H}}\boldsymbol{\rho}_{\mathrm{opt}}\,. && (2.3.15b)
\end{aligned}$$

[7] $[x_{ij}]_{\substack{i=i_l,\ldots,i_u\\ j=j_l,\ldots,j_u}}$ denotes a matrix with elements x_{ij}, whose row index i ranges from i_l to i_u, and whose column index j ranges from j_l to j_u. Only one index is given for vectors.

The reduction of the effective noise power, i.e., the performance improvement, is usually quantified by the *prediction gain*

$$G_{\mathrm{p}} \triangleq \frac{\sigma_n^2}{\sigma_{n'}^2} . \tag{2.3.16}$$

Since the correlation matrix is (a) Hermitian ($\boldsymbol{\Phi}_{nn}^{\mathsf{H}} = \boldsymbol{\Phi}_{nn}$) and (b) positive definite (only in rare situations is it nonnegative definite) [Hay96], the Hermitian form $\boldsymbol{\rho}^{\mathsf{H}}\boldsymbol{\Phi}_{nn}^{\mathsf{H}}\boldsymbol{\rho}$ is strictly positive. Hence, from (2.3.15b) we get $\sigma_{n'}^2 \leq \sigma_n^2$. In other words, prediction always provides a gain.

We finally note that, starting with zero-forcing linear equalization, the performance of noise prediction can be calculated by combining (2.2.29) and (2.3.16).

Decision-Feedback Equalization The *noise prediction (NP) structure* of Figure 2.16 can be implemented in the given way. But the separate realization of linear equalizer and noise prediction is somewhat disadvantageous [Hub87]. This can be overcome by the following idea: Since $P(z)$ is a linear filter, subtraction and filtering can be interchanged, i.e., the signals $\langle y[k]\rangle$ and $\langle \hat{a}[k]\rangle$, respectively, are filtered separately. The results are then subtracted from or added to $y[k]$. Thus, the noise estimates $\hat{n}[k]$ no longer appear explicitly, and, for the moment, the prediction filter $P(z)$ has to be implemented twice, see Figure 2.17.

Finally, defining the *prediction-error filter*

$$H^{(\mathrm{NP})}(z) \triangleq 1 - P(z) , \tag{2.3.17}$$

i.e., $h[0] = 1$ and $h[k] = -p[k]$, $k = 1, 2, \ldots, p$, and zero else, the structure shown on the bottom of Figure 2.17 results. Then, the optimal linear ZF equalizer and the prediction-error filter can be combined into a single receive filter. Because of the ISI-free equalization, it is obvious that the discrete-time end-to-end transfer function for the data symbols $a[k]$ is now given by $H^{(\mathrm{NP})}(\mathrm{e}^{\mathrm{j}2\pi fT})$—intersymbol interference is introduced. In order to enable a threshold decision, this ISI produced by the prediction error filter has to be compensated. Let the decision produce the estimate $\hat{a}[k]$ of the current data symbol. Then, the "tail" of the response corresponding to this sample also is known. Subtracting these postcursors rids the next sample from ISI. This successive elimination of ISI is done by filtering the decision $\hat{a}[k]$ with $P(z) = -(H^{(\mathrm{NP})}(z) - 1)$ and feeding it back to the input of the slicer. That is why this strategy is called *(zero-forcing) decision-feedback equalization.* The idea of using previous decisions to improve system performance has been known for a very long time. Price [Pri72] lists references dating back to the year 1919.

It is of note that only the desired signal is affected by the feedback part, as, assuming correct decisions, the slicer completely eliminates the noise. Hence, the PSD of the noise at the input of the decision device for noise prediction calculates to

$$\begin{aligned}\Phi_{nn}^{(\mathrm{NP})}(\mathrm{e}^{\mathrm{j}2\pi fT}) &= \Phi_{nn}^{(\mathrm{ZF\text{-}LE})}(\mathrm{e}^{\mathrm{j}2\pi fT}) \cdot \left|H^{(\mathrm{NP})}(\mathrm{e}^{\mathrm{j}2\pi fT})\right|^2 \\ &= \frac{N_0}{T} \frac{\left|\sum_k h[k]\mathrm{e}^{-\mathrm{j}2\pi fT}\right|^2}{\sum_\mu \left|H_{\mathrm{T}}(f-\frac{\mu}{T})H_{\mathrm{C}}(f-\frac{\mu}{T})\right|^2} .\end{aligned} \tag{2.3.18}$$

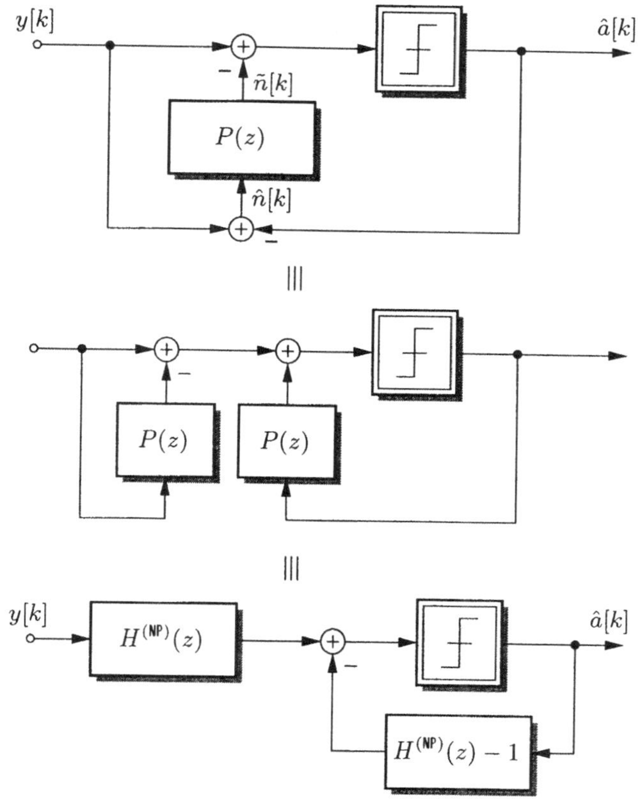

Fig. 2.17 Modifications of the noise prediction structure.

Properties of the Prediction-Error Filter We now discuss an interesting property of the prediction-error filter $H^{(\mathsf{NP})}(z)$, or equivalently, of the discrete-time end-to-end transfer function for the data sequence $\langle a[k] \rangle$. A fuller elaboration can be found in [Hay96].

Theorem 2.12: *Minimum-Phase Property of the Prediction-Error Filter*

The prediction-error filter $H^{(\mathsf{NP})}(z)$ is minimum-phase, i.e., all zeros lie inside the unit circle.

The proof follows the idea of Pakula and Kay [PK83]: Let $z_{0,i}$, $i = 1, 2, \ldots, p$, be the p zeros of $H^{(\mathsf{NP})}(z)$. Since $H^{(\mathsf{NP})}(z)$ is monic (i.e., $h[0] = 1$), we have

$$H^{(\mathsf{NP})}(z) = \prod_{i=1}^{p} \left(1 - z_{0,i} z^{-1}\right) . \tag{2.3.19}$$

Now, let us assume that $H^{(\mathrm{NP})}(z)$ possesses (at least) one zero $z_{0,j}$ outside the unit circle, i.e., $|z_{0,j}| > 1$ for some j, and thus is nonminimum-phase. Thus we can write:

$$\begin{aligned} H^{(\mathrm{NP})}(z) &= \left(1 - z_{0,j} \cdot z^{-1}\right) \cdot \prod_{\substack{i=1 \\ i \neq j}}^{p} \left(1 - z_{0,i} \cdot z^{-1}\right) \\ &\triangleq \left(1 - z_{0,j} \cdot z^{-1}\right) \cdot H'(z) \,. \end{aligned} \tag{2.3.20}$$

The variance of the residual noise sequence $\langle n^{(\mathrm{NP})}[k] \rangle = \langle n^{(\mathrm{ZF\text{-}LE})}[k] * h[k] \rangle$ then reads

$$\begin{aligned} \sigma^2_{n,(\mathrm{NP})} &= T \int_{-\frac{1}{2T}}^{\frac{1}{2T}} \left| H^{(\mathrm{NP})}(\mathrm{e}^{\mathrm{j}2\pi fT}) \right|^2 \cdot \Phi_{nn}^{(\mathrm{ZF\text{-}LE})}(\mathrm{e}^{\mathrm{j}2\pi fT}) \,\mathrm{d}f \qquad (2.3.21) \\ &= T \int_{-\frac{1}{2T}}^{\frac{1}{2T}} \left| 1 - z_{0,j} \mathrm{e}^{-\mathrm{j}2\pi fT} \right|^2 \cdot \left| H'(\mathrm{e}^{\mathrm{j}2\pi fT}) \right|^2 \cdot \Phi_{nn}^{(\mathrm{ZF\text{-}LE})}(\mathrm{e}^{\mathrm{j}2\pi fT}) \,\mathrm{d}f \,. \end{aligned}$$

Considering the first term of the integrand, we have

$$\begin{aligned} \left|1 - z_{0,j} \cdot \mathrm{e}^{-\mathrm{j}2\pi fT}\right|^2 &= |z_{0,j}|^2 \cdot \left| \frac{1}{z_{0,j}} - \mathrm{e}^{-\mathrm{j}2\pi fT} \right|^2 \\ &= |z_{0,j}|^2 \cdot \left| \mathrm{e}^{\mathrm{j}2\pi fT} - \frac{1}{z^*_{0,j}} \right|^2 , \qquad \text{since } |x|^2 = |-x^*|^2 \\ &= \underbrace{|z_{0,j}|^2}_{>1} \cdot \left| 1 - \frac{1}{z^*_{0,j}} \mathrm{e}^{-\mathrm{j}2\pi fT} \right|^2 , \qquad \text{since } |\mathrm{e}^{-\mathrm{j}2\pi fT}|^2 = 1 \\ &> \left| 1 - \frac{1}{z^*_{0,j}} \mathrm{e}^{-\mathrm{j}2\pi fT} \right|^2 \qquad \text{for } |z_{0,j}| > 1 \,. \end{aligned} \tag{2.3.22}$$

Thus, replacing $z_{0,j}$ (with $|z_{0,j}| > 1$) by its complex-conjugate reciprocal $\frac{1}{z^*_{0,j}}$ (mirroring the zero at the unit circle), the residual noise power is decreased by the factor $|z_{0,j}|^2 > 1$.

If $H^{(\mathrm{NP})}(z)$ would be nonminimum-phase, then by replacing the zeros outside the unit circle by their conjugate reciprocal the residual noise power could be further decreased. But this contradicts the fact that the prediction filter is adjusted for minimum variance of the error signal, and the Yule-Walker equations result in the optimal solution. Hence, $H^{(\mathrm{NP})}(z)$ is minimum-phase. *q.e.d.*

A second property of the prediction-error filter is as follows: Prediction uses the correlations between adjacent samples of the noise process at the input. Based on these dependencies, estimates are generated and, in the prediction-error filter, subtracted from the input signal. In this way, the correlations of the residual error signal are reduced. Increasing the order of the filter, prediction gets more and more complete. In the limit, if all dependencies are exploited, an uncorrelated, i.e., white, error sequence results. Thus, the prediction-error filter is a *whitening filter* for the input process. We summarize:

Theorem 2.13: Noise Whitening Property of the Prediction-Error Filter

With increasing order of the predictor, the residual noise sequence at the output of the prediction-error filter tends to a white sequence.

Asymptotic Prediction Gain If the order of the prediction filter is sufficiently high, the residual noise sequence $\langle n^{(\mathsf{NP})}[k]\rangle$ will be white. Thus, from (2.3.18) we have

$$\Phi_{nn}^{(\mathsf{ZF\text{-}LE})}(e^{j2\pi fT}) \cdot \left|H^{(\mathsf{NP})}(e^{j2\pi fT})\right|^2 = \sigma^2_{n,(\mathsf{NP})} = \text{const} . \tag{2.3.23}$$

Taking the logarithm of (2.3.23) and integrating over one period yields[8]

$$T\int_{-\frac{1}{2T}}^{\frac{1}{2T}} \log\left(\Phi_{nn}^{(\mathsf{ZF\text{-}LE})}(e^{j2\pi fT})\right)\, df + T\int_{-\frac{1}{2T}}^{\frac{1}{2T}} \log\left(\left|H^{(\mathsf{NP})}(e^{j2\pi fT})\right|^2\right)\, df$$
$$= \log\left(\sigma^2_{n,(\mathsf{NP})}\right) . \tag{2.3.24}$$

Because of (2.3.19), the (pole-)zero representation of $H^{(\mathsf{NP})}(z)\cdot H^{(\mathsf{NP})*}(1/z^*)$, the analytic continuation of $\left|H^{(\mathsf{NP})}(e^{j2\pi fT})\right|^2$, reads

$$H^{(\mathsf{NP})}(z)\cdot H^{(\mathsf{NP})*}(1/z^*) = 1\cdot\prod_{i=1}^{p}\left(1-z_{0,i}z^{-1}\right)\left(1-z^*_{0,i}z\right) . \tag{2.3.25}$$

In [OS75] it is proved that the *cepstrum* at time instant zero of such transfer functions, which is exactly the second integral in (2.3.24), is $\log 1 = 0$. This leads to the important result:

$$T\int_{-\frac{1}{2T}}^{\frac{1}{2T}} \log\left(\Phi_{nn}^{(\mathsf{ZF\text{-}LE})}(e^{j2\pi fT})\right)\, df = \log\left(\sigma^2_{n,(\mathsf{NP})}\right) . \tag{2.3.26}$$

Solving this equation for $\sigma^2_{n,(\mathsf{NP})}$, and then regarding (2.3.2), we arrive at

[8]Unless otherwise stated, $\log(\cdot)$ denotes the natural logarithm.

Theorem 2.14: *Minimum Noise Variance After Noise Prediction*

Using noise prediction with infinite order, the variance of the residual noise sequence is given by

$$\sigma^2_{n,(\mathrm{NP})} = \frac{N_0}{T} \cdot \exp\left\{ T \int_{-\frac{1}{2T}}^{\frac{1}{2T}} \log\left(\frac{1}{\sum_{\mu} \left| H_{\mathrm{T}}(f-\frac{\mu}{T}) H_{\mathrm{C}}(f-\frac{\mu}{T})\right|^2} \right) \mathrm{d}f \right\} . \tag{2.3.27}$$

Knowing that $\sigma_n^2 = T \int_{-\frac{1}{2T}}^{\frac{1}{2T}} \Phi_{nn}^{(\mathrm{ZF\text{-}LE})}(\mathrm{e}^{\mathrm{j}2\pi fT})\,\mathrm{d}f$ holds, the *ultimate prediction gain* is given by

$$G_{\mathrm{p},\infty} = \frac{T \int_{-\frac{1}{2T}}^{\frac{1}{2T}} \Phi_{nn}^{(\mathrm{ZF\text{-}LE})}(\mathrm{e}^{\mathrm{j}2\pi fT})\,\mathrm{d}f}{\exp\left\{ T \int_{-\frac{1}{2T}}^{\frac{1}{2T}} \log\left(\Phi_{nn}^{(\mathrm{ZF\text{-}LE})}(\mathrm{e}^{\mathrm{j}2\pi fT})\right) \mathrm{d}f \right\}} . \tag{2.3.28}$$

Approximating the integrals by sums, the numerator and the denominator read

$$\frac{1}{N} \sum_{i=0}^{N-1} \Phi_{nn}^{(\mathrm{ZF\text{-}LE})}(\mathrm{e}^{\mathrm{j}2\pi i/N})$$

and

$$\exp\left\{ \frac{1}{N} \sum_{i=0}^{N-1} \log\left(\Phi_{nn}^{(\mathrm{ZF\text{-}LE})}(\mathrm{e}^{\mathrm{j}2\pi i/N})\right)\right\} = \sqrt[N]{\prod_{i=0}^{N-1} \Phi_{nn}^{(\mathrm{ZF\text{-}LE})}(\mathrm{e}^{\mathrm{j}2\pi i/N})} ,$$

respectively. We see that the numerator has the form of an *arithmetic mean*, while the denominator has the form of a *geometric mean*. The integrals in (2.3.28) are simply the respective means for continuous functions. Hence, the ultimate prediction gain is the quotient of the *arithmetic mean* and *geometric mean* of the noise PSD after zero-forcing linear equalization. Because the geometric mean of nonnegative functions is always smaller than the arithmetic mean [BB91], the prediction gain (in dB) is always positive. Moreover, if the periodic sum $\sum_{\mu} \left| H_{\mathrm{T}}(f-\frac{\mu}{T}) H_{\mathrm{C}}(f-\frac{\mu}{T})\right|^2$ has spectral zeros and ZF linear equalization does not exist, the numerator in (2.3.28) is unbounded, and thus the gain tends to infinity.

Example 2.4: *Noise Prediction and Prediction Gain*

For the simplified DSL up-stream transmission scheme (self-NEXT-dominated environment), Figure 2.18 shows the prediction gain G_{p} over the order p of the prediction filter (cf. also [GH85, Hub92b]). The cable length is chosen to be 3.0 km. Already for a small order

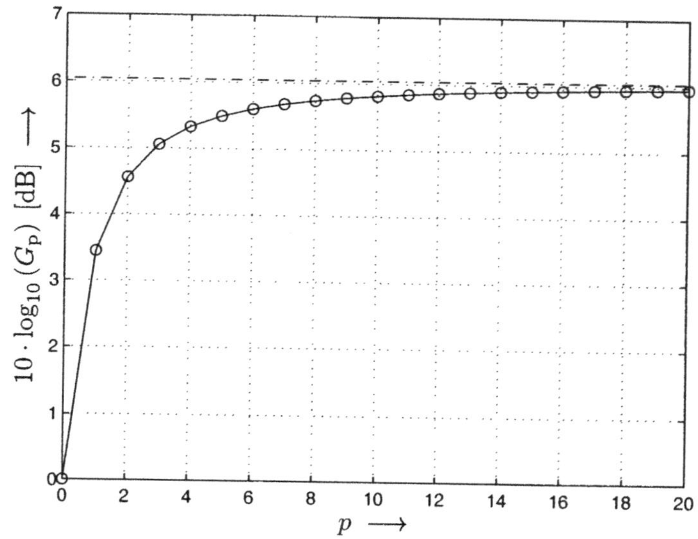

Fig. 2.18 Prediction gain G_p versus order p of the prediction filter. Dashed: Asymptotic prediction gain $G_{\mathrm{p},\infty}$.

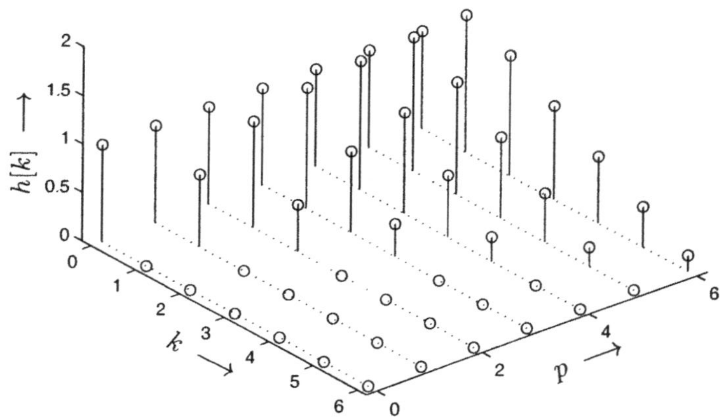

Fig. 2.19 Impulse responses of the FIR prediction-error filter (noise whitening filter) $H^{(\mathrm{NP})}(z)$ over the order p.

of p, significant gains can be achieved. For orders above approximately 10, no essential improvement is visible, and the gain converges to the asymptotic prediction gain (indicated by the dashed line), which in this example equals 6.04 dB.

The respective impulse responses of the prediction-error filter $H^{(\mathrm{NP})}(z)$ are depicted in Figure 2.19. Here, p ranges from 0 (identical to linear ZF equalization) through 6.

For comparision, in Figure 2.20, the ultimate prediction gain is calculated for different cable lengths. As with increasing length, the variations of the spectral line attenuation within the transmission band grow, prediction can provide more and more gain.

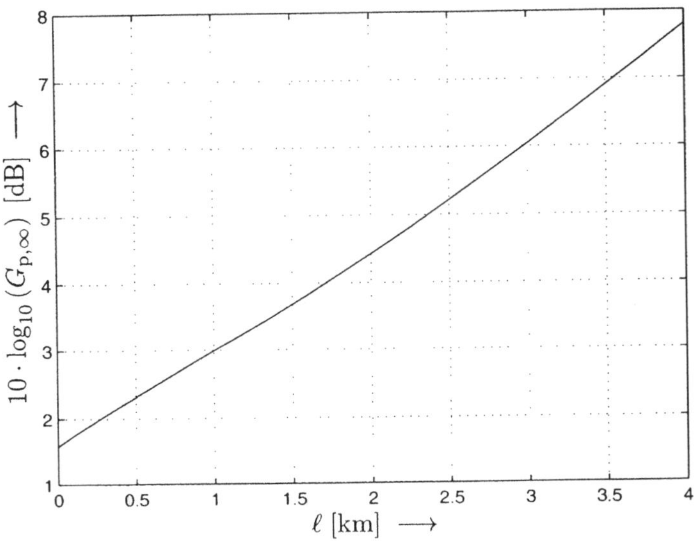

Fig. 2.20 Asymptotic prediction gain $G_{\mathrm{p},\infty}$ over the cable length.

Noise-Predictive Decision-Feedback Equalization With respect to implementation, noise prediction structure and decision-feedback equalization are two extreme points. The noise prediction strategy requires a ZF linear equalization front-end and uses a single discrete-time filter for prediction. The DFE structure implements two separate filters, a feedforward and a feedback filter. As we will see later (Section 2.3.3), these filters may be different, and each receiver front-end—as long as the matched filter is present—is applicable. This relaxes the requirements on the analog receiver components. In DFE, the feedforward part has to fulfill two major tasks: first, it has to whiten the noise, and second, it has to guarantee a minimum-phase end-to-end impulse (both properties have been proved above).

In Figure 2.21 a combination of both extremes is depicted, called *noise-predictive decision-feedback equalization* [GH89]. Here, three different filters are used, each filter having its special task. Now, the feedforward filter $F(z)$ only has to produce a minimum-phase impulse. The tail of this impulse is then canceled via the feedback

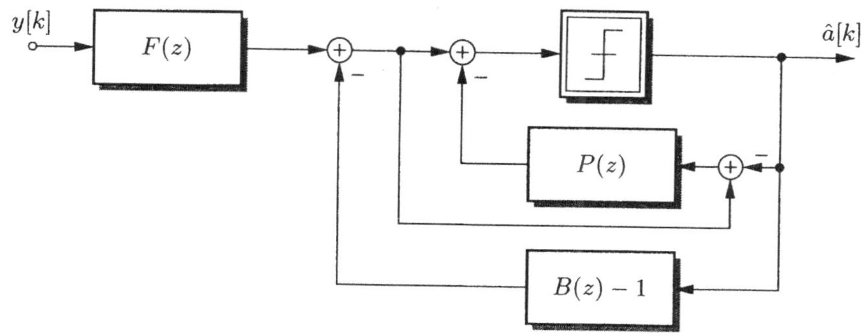

Fig. 2.21 Structure of noise-predictive decision-feedback equalization.

filter $B(z) - 1$. Noise prediction, i.e., whitening of the residual noise sequence, is done by the subsidiary prediction filter $P(z)$.

Assuming sufficient orders of the filters, it is easy to prove that this 3-filter structure is equivalent to the 2-filter DFE structure using a feedforward filter $F(z)\,(1 - P(z))$ and a feedback filter $B(z)\,(1 - P(z)) - 1$. The advantage of noise-predictive decision-feedback equalization is that in an adaptive implementation different algorithms for the adjustment of the coefficients can be used. According to [GH89], $F(z)$ and $B(z)$ are preferably updated using the ZF algorithm, whereas for $P(z)$ the LMS algorithm is more appropriate. This separate filter adaptation results in a larger region where convergence is possible. Further realization aspects can be found in detail in [GH89].

2.3.2 Zero-Forcing Decision-Feedback Equalization

In the last subsection, zero-forcing decision-feedback equalization has been derived from linear equalization followed by the noise prediction structure. Now, we omit this detour and directly calculate the optimal filters. Now, only infinite-length results are regarded.

With Theorem 2.4 in mind, we choose the matched-filter front-end as the starting point. After T-spaced sampling the discrete-time transfer function $H^{(\mathsf{MF})}(\mathrm{e}^{\mathrm{j}2\pi fT})$ and noise power spectral density $\Phi_{nn}^{(\mathsf{MF})}(\mathrm{e}^{\mathrm{j}2\pi fT})$ given in (2.2.38a) and (2.2.38b), respectively, hold. Remember that both functions are proportional to each other: $\Phi_{nn}^{(\mathsf{MF})}(\mathrm{e}^{\mathrm{j}2\pi fT}) = \frac{N_0}{T} H^{(\mathsf{MF})}(\mathrm{e}^{\mathrm{j}2\pi fT})$.

Because the noise is only affected by the feedforward filter, and white noise at the decision point is desired, the discrete-time part $F(z)$ of the receive filter (cf. (2.2.41)) has to serve as a noise whitening filter, cf. Theorem 2.13. Thus, at its output,

$$\Phi_{nn}^{(\mathsf{MF})}(\mathrm{e}^{\mathrm{j}2\pi fT}) \cdot \left|F(\mathrm{e}^{\mathrm{j}2\pi fT})\right|^2 \stackrel{!}{=} \sigma_{n,(\mathsf{ZF\text{-}DFE})}^2 = \mathrm{const} \tag{2.3.29}$$

should hold. Note that because the continuous-time channel noise is assumed to be white, the total receive filter has to have square-root Nyquist characteristics in order to obtain a white discrete-time noise sequence after sampling.

To solve the above problem, we write the scaled PSD—to be precise, the analytic continuation of the function $\frac{\sigma_a^2 T}{N_0}\Phi_{nn}^{(\mathsf{MF})}(\mathrm{e}^{\mathrm{j}2\pi fT})$ evaluated on the unit circle, to the whole plane of complex numbers—in the following way[9]

$$\Phi_{hh}(z) \triangleq \sigma_a^2 H^{(\mathsf{MF})}(z) \triangleq \sigma_g^2 \cdot G(z) \cdot G^*(z^{-*}) \,. \tag{2.3.30}$$

Here, $G(z)$ is induced to be causal, monic, i.e., $G(z) = 1 + \sum_{k\geq 1} g[k]z^{-k}$, and minimum-phase. Such filters (causal, monic, and minimum-phase) are sometimes called *canonical* [CDEF95]. Then, $G^*(z^{-*})$ is anticausal, monic, and maximum-phase, i.e., anticanonical. Because $G(z)$ is monic, a scaling factor σ_g^2 is required to meet (2.3.30).

It can be shown (e.g. [Pap91]) that the factorization according to (2.3.30) is possible if the PSD $\Phi_{hh}(\mathrm{e}^{\mathrm{j}2\pi fT})$ satisfies the *Paley-Wiener condition*

$$T \int_{-\frac{1}{2T}}^{\frac{1}{2T}} \left|\log\left(\Phi_{hh}(\mathrm{e}^{\mathrm{j}2\pi fT})\right)\right| \, \mathrm{d}f < \infty \,. \tag{2.3.31}$$

The interpretation of equation (2.3.30) is as follows: Given a white noise sequence with variance σ_g^2—the so-called *innovations sequence*—a random sequence with PSD $\Phi_{hh}(\mathrm{e}^{\mathrm{j}2\pi fT})$ is obtained by filtering this sequence with $G(z)$, the so-called *innovations filter* [Pap91], see Figure 2.22.

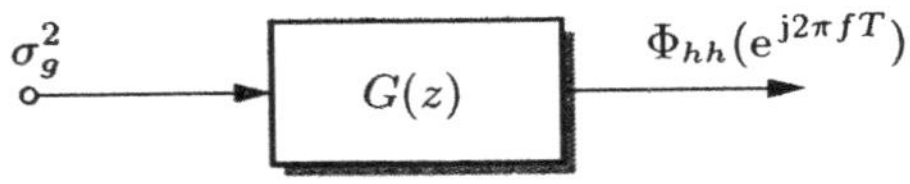

Fig. 2.22 Generation of a noise process with the given PSD from white noise by filtering.

Now, if we can express the PSD in the above form, by choosing

$$F(z) = \frac{\sigma_a^2}{\sigma_g^2}\frac{1}{G^*(z^{-*})} \,, \tag{2.3.32}$$

we obtain a causal and minimum-phase end-to-end transfer function for the data symbols. Furthermore, the scaling of $F(z)$ is chosen to obtain a monic impulse response. We have

$$H^{(\mathsf{ZF\text{-}DFE})}(z) = H^{(\mathsf{MF})}(z) \cdot \frac{\sigma_a^2}{\sigma_g^2}\frac{1}{G^*(z^{-*})} = G(z) \,. \tag{2.3.33}$$

The noise PSD, also denoted by the superscript (ZF-DFE), reads

$$\Phi_{nn}^{(\mathsf{ZF\text{-}DFE})}(z) = \Phi_{nn}^{(\mathsf{MF})}(z) \cdot |F(z)|^2$$

[9] We write z^{-*} instead of the correct, but more intricate expression $(z^*)^{-1}$.

$$
\begin{aligned}
&= \frac{N_0}{T}\frac{\sigma_g^2}{\sigma_a^2} G(z) G^*(z^{-*}) \cdot \left(\frac{\sigma_a^2}{\sigma_g^2}\right)^2 \frac{1}{G^*(z^{-*}) G(z)} \\
&= \frac{\sigma_a^2}{\sigma_g^2}\frac{N_0}{T} \, .
\end{aligned}
\tag{2.3.34}
$$

Thus, the feedforward part forces the end-to-end transfer function to $H^{(\text{ZF-DFE})}(z)$ and produces white noise with variance $\sigma^2_{n,(\text{DFE})} = \frac{\sigma_a^2}{\sigma_g^2}\frac{N_0}{T}$.

As seen in Figure 2.17, after the decision, the feedback part cancels the intersymbol interference introduced by the feedforward filter. It is obvious that this is only possible if the feedforward impulse response is causal. Moreover, the normalization for a monic impulse response guarantees the use of a slicer without further scaling. The last question to be answered is why the end-to-end impulse response has to be minimum-phase. A mathematical explanation was given in the last section. It has been shown that, for nonminimum-phase responses, the noise power can be reduced by replacing the zeros outside the unit circle by their reciprocal conjugates. Only if all zeros lie inside the unit circle the remaining noise power is minimum.

A more illustrative reason is as follows: It is well-known that among all causal and stable impulse responses $h[k]$, which have the same squared magnitude spectrum $|H(e^{j2\pi fT})|^2$, the one which is minimum-phase, $h_{\text{mp}}[k]$, has most of its energy concentrated in the first few samples [Pap77]:

$$
\sum_{k=0}^{\kappa} |h_{\text{mp}}[k]|^2 \ge \sum_{k=0}^{\kappa} |h[k]|^2 \, , \quad \forall \kappa \ge 0 \, . \tag{2.3.35}
$$

In particular ($\kappa = 0$), the first tap, $h_{\text{mp}}[0]$, is maximum for the minimum-phase impulse. But this is the reference tap, on which the decision is based—the rest of the impulse is canceled and does not contribute to signal detection. Thus, for the minimum-phase transfer function, the highest signal-to-noise ratio is achieved.

Spectral Factorization The calculation of the prediction-error filter, and thus of the transfer function $H^{(\text{ZF-DFE})}(z)$ via the Yule-Walker equations was given in Section 2.3.1. This procedure is impractical for infinite-length results; instead we have to solve the factorization problem (2.3.30). A method for doing this *discrete-time spectral factorization* is now developed, following [Hub92b, FU98]; see also [OSS68, And73].

The task can be paraphrased as follows: Given the function $\Phi_{hh}(z)$, find the variance σ_g^2 and a transfer function $G(z)$ which fulfill

$$
\Phi_{hh}(z) = \sigma_g^2 \cdot G(z) \cdot G^*(z^{-*}) \, . \tag{2.3.36}
$$

By taking the logarithm of (2.3.36), we obtain:

$$
\log\left(\Phi_{hh}(z)\right) = \log\left(\sigma_g^2\right) + \log\left(G(z)\right) + \log\left(G^*(z^{-*})\right) \, . \tag{2.3.37}
$$

Since $G(z)$ is forced to be stable, monic, and minimum-phase, its pole-zero representation reads

$$G(z) = \frac{\prod_i (1 - a_i z^{-1})}{\prod_j (1 - b_j z^{-1})} , \quad \text{with} \quad |a_i|, |b_j| < 1 . \tag{2.3.38}$$

Using $\log(1 - xz^{-1}) = -\sum_{k=1}^{\infty} \frac{x^k}{k} z^{-k}$, $|x| < 1$ [BS98], we can write

$$\begin{aligned} \hat{G}(z) &\triangleq \log(G(z)) = \sum\nolimits_i \log(1 - a_i z^{-1}) - \sum\nolimits_j \log(1 - b_j z^{-1}) \\ &= -\sum\nolimits_{k=1}^{\infty} \left(\sum\nolimits_i \frac{a_i^k}{k} - \sum\nolimits_j \frac{b_j^k}{k} \right) z^{-k} . \end{aligned} \tag{2.3.39}$$

As in (2.3.39), only terms z^{-k}, $k > 0$, appear, and the corresponding time-domain signal,

$$\hat{g}[k] \circ\!\!-\!\!\bullet\, \hat{G}(z) \triangleq \log\left(G(z)\right) , \tag{2.3.40}$$

called the *cepstrum*, is strictly causal [OSS68].

Similarly, the left-hand side of (2.3.37) can be expressed as

$$\log\left(\Phi_{hh}(z)\right) = \sum_k \hat{\varphi}[k] z^{-k} . \tag{2.3.41}$$

Since this series converges for $z = \mathrm{e}^{\mathrm{j}2\pi fT}$, the coefficients $\hat{\varphi}_i$ are given by

$$\hat{\varphi}[k] = \mathcal{Z}^{-1}\left\{\log\left(\Phi_{hh}(z)\right)\right\} = T \int_{-\frac{1}{2T}}^{\frac{1}{2T}} \log\left(\Phi_{hh}(\mathrm{e}^{\mathrm{j}2\pi fT})\right) \mathrm{e}^{+\mathrm{j}2\pi fTk} \, \mathrm{d}f . \tag{2.3.42}$$

The requested factorization (2.3.36) can now be obtained by grouping the terms with negative and positive indices, respectively, of the above series. In the series expansion of $\log(G(z))$ only terms with negative exponents appear. Thus the terms with positive index k in (2.3.41) correspond to $\log(G(z))$. Moreover, we can identify $\hat{g}[k] = \hat{\varphi}[k]$, $k > 0$. Conversely, the terms with negative index belong to $G^*(z^{-*})$. Finally, the coefficient with index zero gives $\log\left(\sigma_g^2\right)$. In summary

$$\log\left(G(z)\right) = \sum\nolimits_{k>0} \hat{\varphi}[k] z^{-k} \tag{2.3.43a}$$

$$\log\left(\sigma_g^2\right) = \hat{\varphi}[0] \tag{2.3.43b}$$

$$\log\left(G^*(z^{-*})\right) = \sum\nolimits_{k<0} \hat{\varphi}[k] z^{-k} . \tag{2.3.43c}$$

In particular, we obtain

$$\sigma_g^2 = \exp\left\{ T \int_{-\frac{1}{2T}}^{\frac{1}{2T}} \log\left(\Phi_{hh}(\mathrm{e}^{\mathrm{j}2\pi fT})\right) \, \mathrm{d}f \right\} , \tag{2.3.44}$$

which together with (2.3.30), (2.3.34), and (2.2.38a) confirms the result of (2.3.27).

In order to obtain an explicit expression for the coefficients $g[k]$ of the requested function $G(z)$, we take the derivative of $\hat{G}(z)$ with respect to z, which yields

$$\frac{\mathrm{d}}{\mathrm{d}z}\hat{G}(z) = \frac{\mathrm{d}}{\mathrm{d}z}\log\left(G(z)\right) = \frac{1}{G(z)} \cdot \frac{\mathrm{d}}{\mathrm{d}z}G(z)\,, \tag{2.3.45}$$

or, equivalently, after multiplication by $-z \cdot G(z)$,

$$-z \cdot \frac{\mathrm{d}}{\mathrm{d}z}G(z) = -z \cdot \frac{\mathrm{d}}{\mathrm{d}z}\hat{G}(z) \cdot G(z)\,. \tag{2.3.46}$$

Using the correspondence $k \cdot x[k] \circ\!\!-\!\!\bullet -z \cdot \frac{\mathrm{d}}{\mathrm{d}z}X(z)$, with $x[k] \circ\!\!-\!\!\bullet X(z)$, and considering that $G(z)$ and $\hat{G}(z)$ correspond to causal sequences, (2.3.46) reads in the time domain

$$k \cdot g[k] = \sum_{i=0}^{k-1}(k-i) \cdot \hat{g}[k-i] \cdot g[i]\,.$$

Solving for $g[k]$ and again using $\hat{g}[k] = \hat{\varphi}[k]$, $k > 0$, leads to a recursive expression for the desired coefficients

$$g[0] = 1 \qquad \text{and} \qquad g[k] = \sum_{i=0}^{k-1}\frac{k-i}{k}\,g[i]\hat{\varphi}[k-i]\,. \tag{2.3.47}$$

By construction, $\langle g[k] \rangle$ is minimum-phase.

Whitened Matched Filter The cascade of matched filter $H_\mathrm{T}^*(f)H_\mathrm{C}^*(f)$ and discrete-time noise whitening filter gives the overall receive filter. If sampling is moved to the output of this filter, it can be implemented as a single analog front-end filter. Because this filter (a) is the optimum receiver front-end for ZF-DFE, (b) *whitens* the discrete-time noise sequence, and (c) is *matched* to the transfer function for the data it is called *zero-forcing whitened matched filter (ZF-WMF)* [For72, And73].

Recall that if the channel noise is colored, the first stage of the receiver consists of a continuous-time noise whitening filter. Thus, in general, the WMF has the structure of Figure 2.23.

Using (2.3.32), the transfer function of the ZF-WMF is given by

$$H_\mathrm{R}^{(\mathrm{WMF})}(f) = H_\mathrm{T}^*(f)H_\mathrm{C}^*(f) \cdot \frac{\sigma_a^2}{\sigma_g^2}\frac{1}{G^*(\mathrm{e}^{\mathrm{j}2\pi fT})}\,, \tag{2.3.48}$$

and thus, with (2.2.38a) and (2.3.30), the power transfer function reads

$$\begin{aligned}\left|H_\mathrm{R}^{(\mathrm{WMF})}(f)\right|^2 &= |H_\mathrm{T}^*(f)H_\mathrm{C}^*(f)|^2 \cdot \left|\frac{\sigma_a^2}{\sigma_g^2}\right|^2 \frac{1}{G(\mathrm{e}^{\mathrm{j}2\pi fT})G^*(\mathrm{e}^{\mathrm{j}2\pi fT})} \\ &= \frac{\sigma_a^2}{\sigma_g^2}\frac{|H_\mathrm{T}(f)H_\mathrm{C}(f)|^2}{\sum_\mu \left|H_\mathrm{T}(f-\frac{\mu}{T})H_\mathrm{C}(f-\frac{\mu}{T})\right|^2}\,. \end{aligned} \tag{2.3.49}$$

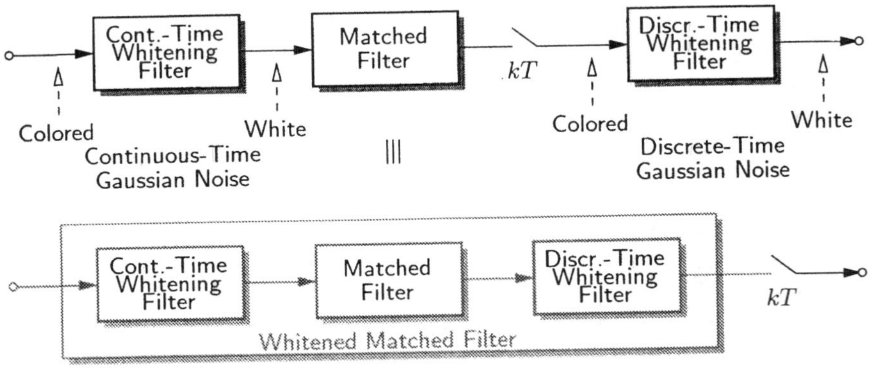

Fig. 2.23 Decomposition of the whitened matched filter.

Note that the power transfer function of the zero-foring whitened matched filter is proportional to the end-to-end transfer function when applying optimal zero-forcing linear equalization, i.e., $|H_{\rm R}^{(\rm WMF)}(f)|^2 = \text{const} \cdot H_{\rm T}(f)H_{\rm C}(f)H_{\rm R}^{(\rm ZF\text{-}LE)}(f)$; cf. Section 2.2. As already anticipated, the ZF-WMF has square-root Nyquist characteristics; its power transfer function thus is Nyquist (cf. page 15), i.e.,

$$\sum_{\mu} \left| H_{\rm R}^{(\rm WMF)}(f - \tfrac{\mu}{T}) \right|^2 = \frac{\sigma_a^2}{\sigma_g^2} = \text{const} . \tag{2.3.50}$$

We recapitulate this important result in the following theorem.

Theorem 2.15: *Zero-Forcing Whitened Matched Filter*

The power transfer function of the *zero-forcing whitened matched filter (ZF-WMF)*, which has Nyquist characteristics, is given by

$$\left| H_{\rm R}^{(\rm WMF)}(f) \right|^2 = \frac{\sigma_a^2}{\sigma_g^2} \frac{|H_{\rm T}(f)H_{\rm C}(f)|^2}{\sum_\mu |H_{\rm T}(f-\frac{\mu}{T})H_{\rm C}(f-\frac{\mu}{T})|^2} . \tag{2.3.51}$$

Regarding the channel noise, we have the following situation: At the input of the WMF, *continuous-time additive white Gaussian noise* is present. The noise power spectral density is constant with value N_0 and, hence, the noise power is infinite. (As already noted, we regard signals in the equivalent low-pass domain. For real-valued signals, the two-sided power spectral density has value $N_0/2$.) The WMF limits the (equivalent) noise bandwidth to the Nyquist range of width $1/T$.

Because of the specific scaling—the overall gain of the data symbols is fixed at one—the noise PSD is additionally scaled by the factor σ_a^2/σ_g^2. Thus, the power of the *discrete-time additive white Gaussian noise* at the filter output is finite and reads $(N_0/T) \cdot (\sigma_a^2/\sigma_g^2)$. Note that in the literature (e.g., [For72, Hub92b]) other normalizations of the whitened matched filter are also common.

Example 2.5: *Zero-Forcing Whitened Matched Filter*

This example continues Example 2.1, from page 19. Again, the DSL down-stream scenario (white channel noise) is considered, cf. Appendix B.

First, Figure 2.24 sketches the impulse responses of the discrete-time noise whitening filter (calculated from (2.3.14) and (2.3.17)) for different orders p. In addition, the result of spectral factorization is shown. As can be seen, for small orders p, the impulse responses already closely resemble the asymptotic one, which is obtained from spectral factorization.

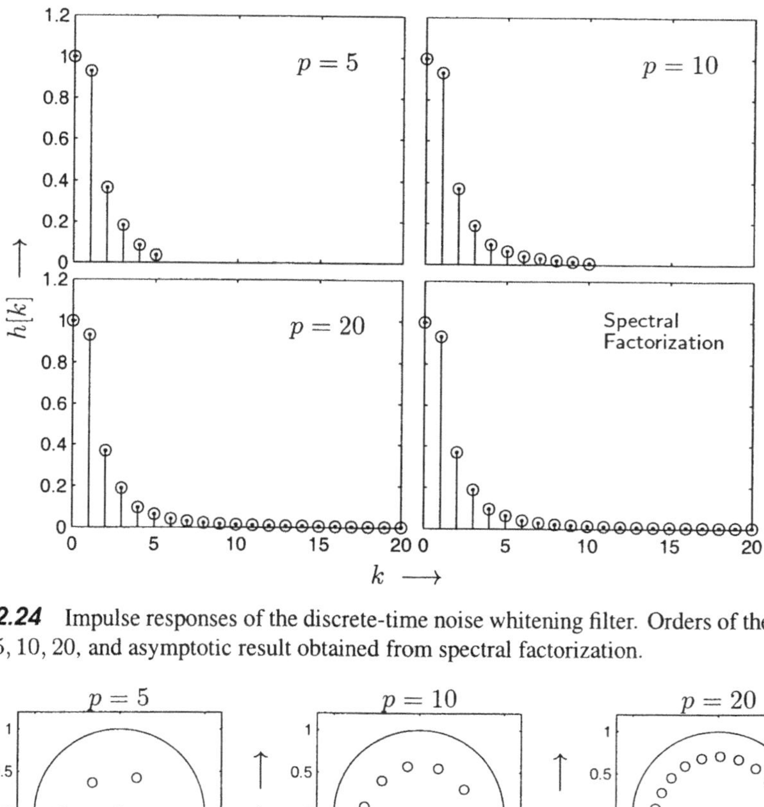

Fig. 2.24 Impulse responses of the discrete-time noise whitening filter. Orders of the filters: $p = 5, 10, 20$, and asymptotic result obtained from spectral factorization.

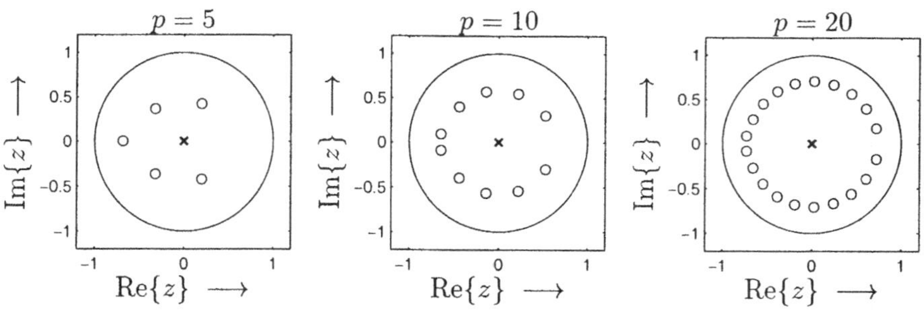

Fig. 2.25 Pole-zero diagrams of the discrete-time noise whitening filter. Orders of the filters: $p = 5, 10, 20$.

The respective pole-zero diagrams are shown in Figure 2.25. As stated in Theorem 2.12, the prediction error-filters are minimum-phase—all zeros lie inside the unit circle. Because spectral factorization results in an infinitely long impulse response modeled by a nonrecursive filter, the corresponding pole-zero diagram is omitted.

Figure 2.26 shows the impulse response of the whitened matched receive filter $h_{\mathrm{R}}^{(\mathrm{WMF})}(t) \circ\!\!-\!\!\bullet\, H_{\mathrm{R}}^{(\mathrm{WMF})}(f)$ for the present situation. Since its transfer function (2.3.48) is the complex conjugate of a product of causal and minimum-phase filters, the impulse response is strictly anticausal and maximum-phase. Note that the squared magnitude of this receive filter has Nyquist characteristics and is proportional to the transfer function of the overall cascade when applying optimal ZF linear equalization. The respective plot (Figure 2.6) was given in Example 2.1.

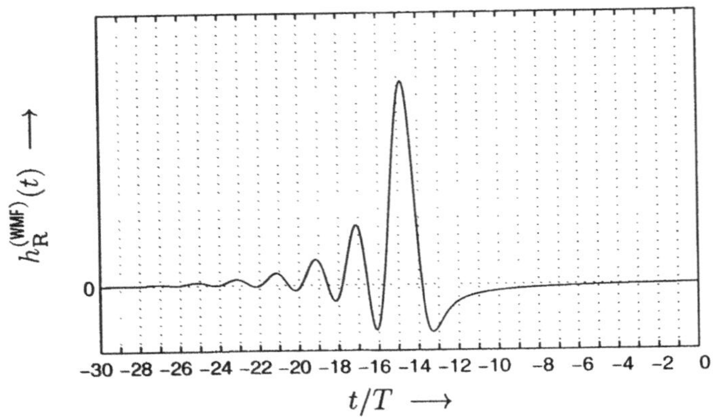

Fig. 2.26 Impulse response $h_{\mathrm{R}}^{(\mathrm{WMF})}(t)$ of the whitened matched receive filter.

The corresponding overall impulse response $h^{(\mathrm{WMF})}(t)$ (including transmit filter, channel, and receive filter) when using the ZF-WMF is depicted in Figure 2.27. Notice that the reference tap equals one, and after T-spaced sampling (circles), the strictly causal discrete-time impulse given in Figure 2.24 results.

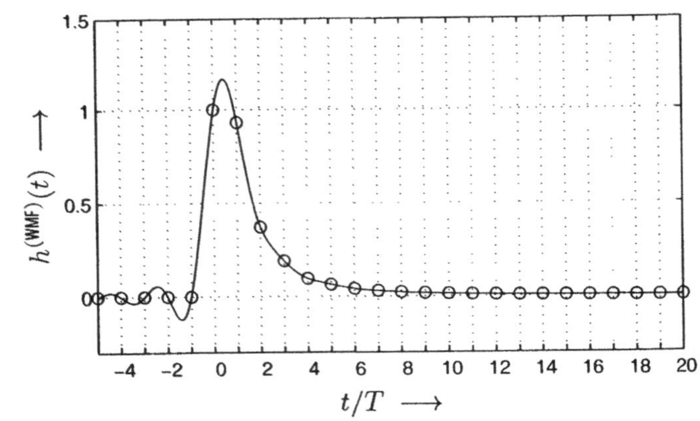

Fig. 2.27 Overall impulse response when using the whitened matched filter front-end.

Signal-to-Noise Ratio The feedback part of the zero-forcing DFE ideally cancels the "tail" of the discrete-time end-to-end impulse response. Thus, considering the data symbols $a[k]$, and assuming absence of decision errors, an ISI-free AWGN channel results. As shown above, the scaling of the data symbols equals one, and the noise variance is $\sigma_n^2 = \frac{N_0}{T}\frac{\sigma_a^2}{\sigma_g^2}$. Using (2.3.44), (2.3.30), and (2.2.38a), the signal-to-noise ratio is thus given as

$$\begin{aligned} \mathrm{SNR}^{(\mathrm{ZF\text{-}DFE})} &= \frac{\sigma_a^2}{\sigma_n^2} \\ &= \frac{T\sigma_a^2}{N_0} \cdot \exp\left\{ T \int_{-\frac{1}{2T}}^{\frac{1}{2T}} \log\left(\sum_{\mu} \left| H_{\mathrm{T}}(f - \tfrac{\mu}{T}) H_{\mathrm{C}}(f - \tfrac{\mu}{T}) \right|^2 \right) \mathrm{d}f \right\} . \end{aligned} \tag{2.3.52}$$

Using (2.2.19), the SNR (2.3.52) can also be expressed by

$$\mathrm{SNR}^{(\mathrm{ZF\text{-}DFE})} = \exp\left\{ T \int_{-\frac{1}{2T}}^{\frac{1}{2T}} \log\left(\widetilde{\mathrm{SNR}}(\mathrm{e}^{\mathrm{j}2\pi fT}) \right) \mathrm{d}f \right\} , \tag{2.3.53}$$

which is the *geometric mean* over the folded spectral signal-to-noise ratio.

It is noteworthy that this result assumes correct decisions. The effect of decision errors, which then propagate through the feedback filter and affect the subsequent samples, is ignored. Only for some special cases, usually binary signaling and very short impulse responses, error propagation and degradation of performance can be treated analytically. Therefore, complex Markov models are employed. For details see, e.g., [And77] or [AB93] and the references therein.

From (2.3.52) the loss of ZF-DFE compared to an ISI-free AWGN channel can be given. The derivation is again analogous to that in Section 2.2.1, and gives

$$\begin{aligned} 2\frac{\mathrm{SNR}^{(\mathrm{ZF\text{-}DFE})}}{\sigma_a^2} &= \frac{2}{\sigma_a^2} \cdot \exp\left\{ T \int_{-\frac{1}{2T}}^{\frac{1}{2T}} \log\left(\widetilde{\mathrm{SNR}}(\mathrm{e}^{\mathrm{j}2\pi fT}) \right) \mathrm{d}f \right\} \\ &= \frac{\exp\left\{ T \int_{-\frac{1}{2T}}^{\frac{1}{2T}} \log\left(\widetilde{\mathrm{SNR}}(\mathrm{e}^{\mathrm{j}2\pi fT}) \right) \mathrm{d}f \right\}}{T \int_{-\frac{1}{2T}}^{\frac{1}{2T}} \widetilde{\mathrm{SNR}}(\mathrm{e}^{\mathrm{j}2\pi fT}) \, \mathrm{d}f} \cdot \frac{2R_{\mathrm{m}}}{\sigma_a^2} \frac{E_{\mathrm{b}}}{N_0} . \end{aligned} \tag{2.3.54}$$

A comparison of (2.3.54) with the argument of (2.2.25) reveals the loss for transmission over an ISI channel with ZF-DFE compared to a dispersionless channel (matched-filter bound).

Theorem 2.16: *Signal-to-Noise Ratio of ZF Decision-Feedback Equalization*

When using zero-forcing decision-feedback equalization at the receiver, the signal-to-noise ratio is given by the *geometric mean* over the folded spectral SNR

$$\mathrm{SNR}^{(\mathrm{ZF\text{-}DFE})} = \exp\left\{ T \int_{-\frac{1}{2T}}^{\frac{1}{2T}} \log\left(\widetilde{\mathrm{SNR}}(\mathrm{e}^{\mathrm{j}2\pi fT})\right) \mathrm{d}f \right\}, \qquad (2.3.55)$$

and the degradation (based on equal receive power) compared to transmission over an ideal channel reads

$$\vartheta^2_{(\mathrm{ZF\text{-}DFE})} = \left(T \int_{-\frac{1}{2T}}^{\frac{1}{2T}} \widetilde{\mathrm{SNR}}(\mathrm{e}^{\mathrm{j}2\pi fT})\, \mathrm{d}f \right)^{-1} \cdot \exp\left\{ T \int_{-\frac{1}{2T}}^{\frac{1}{2T}} \log\left(\widetilde{\mathrm{SNR}}(\mathrm{e}^{\mathrm{j}2\pi fT})\right) \mathrm{d}f \right\}. \qquad (2.3.56)$$

Again, $\widetilde{\mathrm{SNR}}(\mathrm{e}^{\mathrm{j}2\pi fT})$ is the folded spectral signal-to-noise ratio at the receiver input. Ideal, i.e., error-free, decisions are assumed.

In order to conclude this paragraph on SNRs, following [KH99], we give a rule of thumb of how an easy-to-calculate estimate—actually a lower bound—of the SNR can be obtained. For that we assume the symbol spacing T to be optimized, and that within the transmission band $\mathcal{B}$ of width $B = \frac{1}{T}$, the folded spectral SNR is well approximated by the spectral SNR. Then, from (2.3.55) and (2.2.18) the signal-to-noise ratio expressed in decibels reads

$$\mathrm{SNR}^{(\mathrm{DFE})}_{\mathrm{dB}} \triangleq 10 \cdot \log_{10}\left(\mathrm{SNR}^{(\mathrm{DFE})}\right) \geq 10 \cdot \frac{1}{B} \int_{\mathcal{B}} \log_{10}\left(\frac{\bar{\Phi}_{ss}(f) \cdot |H_{\mathrm{C}}(f)|^2}{\Phi_{nn}(f)} \right) \mathrm{d}f\,. \qquad (2.3.57)$$

Here, nonwhite channel noise with PSD $\Phi_{nn}(f)$ is assumed, and the average transmit PSD $\bar{\Phi}_{ss}(f) = \sigma_a^2 T |H_{\mathrm{T}}(f)|^2$ is used. Separating the terms under the logarithm, the SNR of ZF-DFE can be approximated by

$$\mathrm{SNR}^{(\mathrm{DFE})}_{\mathrm{dB}} \geq \mathrm{ET} - \mathrm{EL} - \mathrm{EN}\,, \qquad (2.3.58)$$

where

$$\mathrm{ET} \triangleq \frac{1}{B} \int_{\mathcal{B}} 10 \cdot \log_{10}\left(\frac{\bar{\Phi}_{ss}(f)}{1\ \mathrm{W/Hz}} \right) \mathrm{d}f \qquad (2.3.59a)$$

is the equivalent transmit spectral density,

$$\mathrm{EL} \triangleq -\frac{1}{B}\int_{\mathcal{B}} 20 \cdot \log_{10}\left(|H_{\mathrm{C}}(f)|\right)\, \mathrm{d}f \qquad (2.3.59b)$$

is the equivalent insertion loss of the cable,[10] and

$$\mathrm{EN} \triangleq \frac{1}{B}\int_{\mathcal{B}} 10 \cdot \log_{10}\left(\frac{\Phi_{nn}(f)}{1\ \mathrm{W/Hz}}\right) \mathrm{d}f \qquad (2.3.59c)$$

is the equivalent noise spectral density of the channel.

Note that the equivalent quantities are calculated by averaging the spectra after they have been converted to decibels. They can be interpreted as PSDs of white signals and constant-gain systems, which have the same effect on the SNR as the term under consideration.

If an analytical expression for the spectral SNR is available, the above integrals can be solved. In [Hub93a] this is done for a crosstalk dominated environment (cf., e.g., the DSL up-stream example). Here, the spectral SNR can be written as

$$\mathrm{SNR}(f) = \frac{10^{-\ell\,\alpha(f)/10}}{|H_{\mathrm{X}}(f)|^2} \,. \qquad (2.3.60)$$

The cable is characterized by its length ℓ and the spectral attenuation per km, which, without loss of generality, can be modeled with sufficient accuracy by (cf. also Appendix B).

$$\alpha(f) = \sum_{i=0}^{n-1} a_i \left(\frac{f}{1\,\mathrm{MHz}}\right)^{b_i}\,, \qquad a_i, b_i \in \mathbb{R}\,. \qquad (2.3.61)$$

The crosstalk transfer function is well approximated by (e.g., [Wel89], [ANSI98, Annex B])

$$|H_{\mathrm{X}}(f)|^2 = K_{\mathrm{X}} \cdot \left(\frac{|f|}{1\,\mathrm{MHz}}\right)^{X}\,, \qquad (2.3.62)$$

where K_{X} and X are constants. Then the SNR (in dB) of ideal ZF-DFE is bounded by

$$\begin{aligned}\mathrm{SNR}_{\mathrm{dB}}^{(\mathrm{DFE})} \;\geq\; & -\ell \cdot A\left(\frac{1}{2T\cdot 1\,\mathrm{MHz}}\right) - 10\cdot\log_{10}(K_{\mathrm{X}}) \\ & - X\cdot 10\cdot \log_{10}\left(\frac{1}{2T\cdot 1\,\mathrm{MHz}}\right) + X\cdot 10\cdot\log_{10}(\mathrm{e})\,, \qquad (2.3.63)\end{aligned}$$

if $A(x)$, the normalized integral function of the line attenuation per km, is defined as

$$A(x) = \frac{1}{x\cdot 1\,\mathrm{MHz}} \int_0^{x\cdot 1\,\mathrm{MHz}} \alpha(f')\, \mathrm{d}f' = \sum_{i=0}^{n-1} a_i \frac{x^{b_i}}{1+b_i}\,. \qquad (2.3.64)$$

[10]The insertion loss (in dB) of a cable is defined as $-20\cdot\log_{10}\left(|H_{\mathrm{C}}(f)|\right)$.

Example 2.6: SNR and Loss of ZF Decision-Feedback Equalization

Continuing the above Example, Figure 2.28 plots the loss when transmitting over an ISI channel and using ZF-DFE. The comparison is again based on equal receive power, and the DSL up-stream example is used.

For reference, the loss when using ZF linear equalization is shown as well (dashed line). At each point, the reduction of the loss equals the asymptotic prediction gain $G_{\mathrm{p},\infty}$.

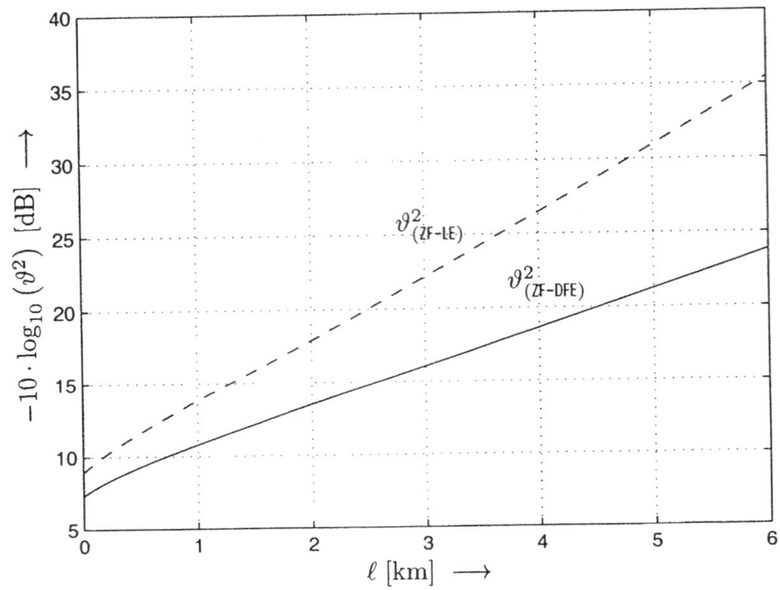

Fig. 2.28 Solid: Loss $\vartheta^2_{(\text{ZF-DFE})}$ (in dB) of ZF-DFE over the cable length for the DSL up-stream example. Dashed: Loss $\vartheta^2_{(\text{ZF-LE})}$ when using ZF linear equalization.

In order to illustrate the approximative SNR calculation, the transmit PSD $\bar{\Phi}_{ss}(f)$, the channel transfer magnitude $|H_{\mathrm{C}}(f)|$, and the noise PSD $\Phi_{nn}(f)$ are plotted in Figure 2.29 for the DSL up-stream. The horizontal range shown corresponds to the Nyquist interval (one-sided), since $\frac{1}{2T} = 386.66$ kHz. The ordinate is scaled in "dBm/Hz," an often used but unfortunately not physically interpretable measure. It is related to power (in dB relative to 1 W) by X dBm/Hz = X dB $-10 \cdot \log_{10}(1/(T \cdot 1\ \text{kHz}))$; in the present case ($\frac{1}{T} = 770.66$ kHz), X dBm/Hz = X dBm -28.8687 dB.

The equivalent quantities work out to

$$\begin{aligned} \text{ET} &= -15.8934\ \text{dB} \\ \text{EL} &= 28.5586\ \text{dB} \\ \text{EN} &= -69.2931\ \text{dB} \end{aligned}$$

This results in an SNR equal to

$$\text{SNR}_{\text{dB}}^{(\text{DFE})} \geq -15.8934 - 28.5586 - (-69.2931) = 24.8411\ \text{dB}\ ,$$

which is very close to the exact solution of 25.3631 dB.

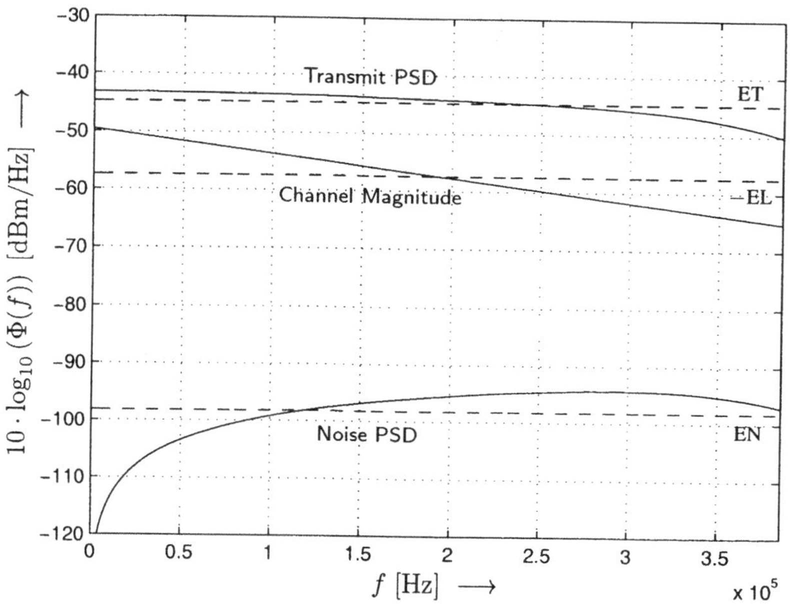

Fig. 2.29 Solid: Transmit PSD $\bar{\Phi}_{ss}(f)$, channel transfer magnitude $|H_C(f)|$, and noise PSD $\Phi_{nn}(f)$ for the DSL up-stream example. Dashed: Equivalent quantities (all in dBm/Hz).

For the calulation according to [Hub93a], we note that in this example

$$A(x) = 6.9 + 13.4 \cdot \frac{x^{0.99}}{1.99}\,, \qquad \text{and} \qquad |H_X(f)|^2 = 0.8356\ 10^{-4} \cdot \left(\frac{|f|}{1\,\text{MHz}}\right)^{1.5}$$

are valid. With $\frac{1}{2T \cdot 1\,\text{MHz}} = 0.3853$, we obtain

$$\begin{aligned} \text{SNR}_{\text{dB}}^{(\text{DFE})} &\geq -3.0 \cdot A(0.3853) - 10 \cdot \log_{10}(0.8356\ 10^{-4}) \\ &\quad - 15 \cdot \log_{10}(0.3866) + 15 \cdot 0.4343 \\ &= -28.5587 + 40.7800 + 6.2125 + 6.5144 \\ &= 24.9482 \text{ dB}\,, \end{aligned}$$

which differs from the above results only slightly.

Optimal Transmit Filter We now optimize the transmit filter such that the signal-to-noise ratio (2.3.52) is maximized. This is done in two steps.

First, the periodic sum under the integral is considered. Let $f \in (-\frac{1}{2T}, \frac{1}{2T}]$ be a given frequency point. To contribute to the integrand at this frequency point, any of the replica $f - \frac{\mu}{T}$ can be used. Thus, we divide the total "power"[11] S in this point such that $S = \sum_\mu S_\mu$, and the contribution to the integral is $\sum_\mu S_\mu |H_{\rm C}(f - \frac{\mu}{T})|^2$. Assume, $|H_{\rm C}(f - \frac{\mu_{\max}}{T})|$ is maximum over all $|H_{\rm C}(f - \frac{\mu}{T})|$, then we have $\sum_\mu S_\mu |H_{\rm C}(f - \frac{\mu}{T})|^2 < \sum_\mu S_\mu |H_{\rm C}(f - \frac{\mu_{\max}}{T})|^2 = S \cdot |H_{\rm C}(f - \frac{\mu_{\max}}{T})|^2$. Hence, placing the power in the replica, for which the magnitude of the channel transfer function is maximum, is superior to all other power distributions.

Thus, as for optimal linear equalization (Section 2.2.5), for each f, choose that μ, for which $\left|H_{\rm C}(f - \frac{\mu}{T})\right|$ is maximum. Hence,

$$\mathcal{F} = \left\{ f \;\middle|\; |H_{\rm C}(f)| \geq \left|H_{\rm C}(f - \tfrac{\mu}{T})\right| , \; \mu \in \mathbb{Z} \right\} \tag{2.3.65}$$

again is the support of the transmit filter $H_{\rm T}(f)$ (cf. (2.2.96)).

In the second step, taking (2.3.52) into account, we now arrive at the following optimization problem:

- Minimize

$$\exp\left\{ T \int_{f \in \mathcal{F}} \log\left(|H_{\rm T}(f) H_{\rm C}(f)|^2 \right) \, \mathrm{d}f \right\} , \tag{2.3.66}$$

- subject to the additional constraint

$$S = \sigma_a^2 T \int_{f \in \mathcal{F}} |H_{\rm T}(f)|^2 \, \mathrm{d}f = \text{const} . \tag{2.3.67}$$

Since $\exp\{x\}$ is a strictly monotonic increasing function, instead of regarding $\exp\{x\}$, we maximize its argument x. Defining λ as the Lagrange multiplier, we can set up the following real-valued Lagrange function

$$L = \int_{f \in \mathcal{F}} \log\left(|H_{\rm T}(f)|^2 \, |H_{\rm C}(f)|^2 \right) \, \mathrm{d}f - \lambda \cdot \sigma_a^2 T \int_{f \in \mathcal{F}} |H_{\rm T}(f)|^2 \, \mathrm{d}f \, . \tag{2.3.68}$$

In order to determine the optimum transmit spectrum $|H_{\rm T}(f)|^2$, we add a real-valued deviation $\varepsilon V(f)$ to the optimal solution and take the partial derivative of L with respect to ε. This gives

$$\left.\frac{\partial L}{\partial \varepsilon}\right|_{\varepsilon=0} = \frac{\partial}{\partial \varepsilon} \left(\int_{f \in \mathcal{F}} \log\left(\left(|H_{\rm T}(f)|^2 + \varepsilon V(f) \right) |H_{\rm C}(f)|^2 \right) \mathrm{d}f \right.$$

[11] Strictly speaking, to obtain powers, we have to regard finite frequency intervals and not only frequency points.

$$
\begin{aligned}
&\left. - \lambda \cdot \sigma_a^2 T \int_{f \in \mathcal{F}} \left(|H_{\mathrm{T}}(f)|^2 + \varepsilon V(f) \right) \mathrm{d}f \right) \Bigg|_{\varepsilon=0} \\
&= \left(\int_{f \in \mathcal{F}} \frac{V(f) \cdot |H_{\mathrm{C}}(f)|^2}{\left(|H_{\mathrm{T}}(f)|^2 + \varepsilon V(f) \right) \cdot |H_{\mathrm{C}}(f)|^2} \right. \\
&\left. \quad - \lambda \sigma_a^2 T \int_{f \in \mathcal{F}} V(f) \, \mathrm{d}f \right) \Bigg|_{\varepsilon=0} \\
&= \int_{f \in \mathcal{F}} \left(\frac{1}{|H_{\mathrm{T}}(f)|^2} - \lambda \sigma_a^2 T \right) V(f) \, \mathrm{d}f \stackrel{!}{=} 0 \, .
\end{aligned}
\tag{2.3.69}
$$

Since this equation has to hold for all functions $V(f)$, the optimal transmit spectrum has to meet

$$
\frac{1}{|H_{\mathrm{T}}(f)|^2} = \lambda \sigma_a^2 T \, . \tag{2.3.70}
$$

Finally considering the additional constraint of a given total transmit power, the multiplier λ can be eliminated and the optimal transmit filter is given by

$$
|H_{\mathrm{T}}(f)|^2 = \frac{S}{\sigma_a^2} \, , \qquad \forall f \in \mathcal{F} \, , \tag{2.3.71}
$$

i.e., the transmit PSD is flat within the support $\mathcal{F}$ and takes on the value $S \cdot T$.

Theorem 2.17: *Optimal Transmit Filter for ZF-DFE*

The optimal linear transmit filter for zero-forcing decision-feedback equalization at the receiver is given by

$$
|H_{\mathrm{T}}(f)|^2 = \mathrm{const} \, , \qquad \forall f \in \mathcal{F} \, . \tag{2.3.72}
$$

The constant is chosen so that a desired transmit power is guaranteed, and the support $\mathcal{F}$ of the filter is as defined in (2.3.65).

Self-NEXT Environment The situation for optimization of the transmit spectrum changes completely when a pure self-NEXT environment is considered. Here, the folded signal-to-noise ratio $\widetilde{\mathrm{SNR}}(\mathrm{e}^{\mathrm{j}2\pi fT})$ is independent of the transmit filter $H_{\mathrm{T}}(f)$, and reads

$$
\widetilde{\mathrm{SNR}}(\mathrm{e}^{\mathrm{j}2\pi fT}) = \sum_{\mu} \frac{|H_{\mathrm{C}}(f - \frac{\mu}{T})|^2}{|H_{\mathrm{X}}(f - \frac{\mu}{T})|^2} \, , \tag{2.3.73}
$$

where $H_{\mathrm{X}}(f)$ is the NEXT transfer function. Thus, since noise is proportional to the signal, no optimization with respect to the *shape* of the transmit spectrum is possible. But, as only those periods μ for which the transmit filter is nonzero contribute to the above sum, the *support* of $H_{\mathrm{T}}(f)$ is still a relevant parameter. Since the SNR is a monotonic function of the folded SNR, this quantity should be as large as possible. As all terms in (2.3.73) are positive, the sum should comprise as much as possible periods μ. Hence, for maximum SNR the support of the transmit spectrum should be as broad as possible. In a pure self-NEXT environment transmit pulses with arbitrary spectral shape but infinite bandwidth are optimal [Hub93a].

Example 2.7: *Loss of Optimized ZF Decision-Feedback Equalization*

The gain of optimizing the transmit spectrum for ZF-DFE is visualized in Figure 2.30. Here the DSL down-stream example (white channel noise, cf. Appendix B) is used. Because the cable is strictly low-pass, the support $\mathcal{F}$ of the transmit filter is simply the first Nyquist set of frequencies: $\mathcal{F} = [-\frac{1}{2T}, \frac{1}{2T}]$. As one can see, over the whole cable range, an optimized spectrum provides only small gains of approximately 1 to 1.5 dB. Because the optimum transmit filter has square-root Nyquist characteristics, the loss tends to zero as the cable length approaches zero.

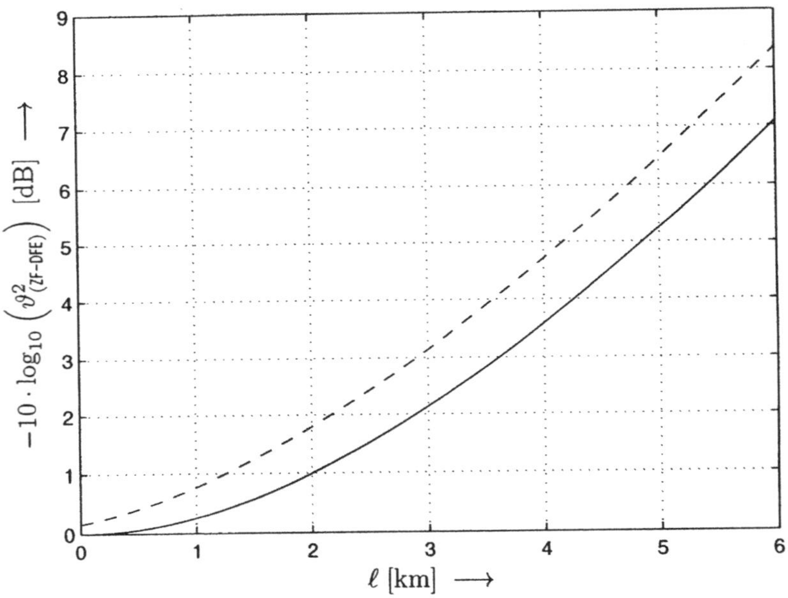

Fig. 2.30 Loss $\vartheta^2_{(\mathrm{ZF\text{-}DFE})}$ (in dB) over the cable length for the DSL down-stream example. Solid: optimized ZF-DFE. Dashed: ZF-DFE with fixed transmit filter according to Appendix B, (B.2.2).

Optimal Symbol Frequency To fully optimize PAM transmission, in the last step, the symbol duration T has to be selected. If, as usual, a fixed data rate (in bits per second) is desired, this optimization leads to an optimal exchange of signal bandwidth versus number of points $M = |\mathcal{A}|$ of the PAM signal constellation $\mathcal{A}$.

On the one hand, considering the equations for the SNR of DFE and the optimum transmit spectrum, it is obvious that these quantities then are functions of the symbol spacing T. In particular, the bandwidth of the transmit signal is governed by the symbol frequency $1/T$. On the other hand, from (2.2.21), we see that the relevant parameter for the error rate (assuming the minimum spacing of the signal points to be 2) is SNR/σ_a^2, i.e., the SNR normalized by the variance of the signal constellation. This variance, in turn, is a function of the number M of signal points. Thus, for minimizing the error rate we have to maximize

$$\zeta \triangleq \mathrm{SNR}^{(\mathrm{DFE})}(T)/\sigma_a^2(M) \,. \tag{2.3.74}$$

As is commonly known, increasing the number of signal points also increases the required SNR (e.g., [Pro01]).

Assuming a fixed data rate $1/T_\mathrm{b}$, where T_b is the (average) duration of one bit, the symbol spacing T and number of signal points are related by

$$T = T_\mathrm{b} \cdot \log_2(M) \,, \tag{2.3.75}$$

and the only remaining parameter is the cardinality M of the signal constellation, or, equivalently, the rate $R_\mathrm{m} = \log_2(M)$ of the modulation. Usually there is an optimal exchange between occupied bandwidth $1/T$ and size M of the signal alphabet. Starting with binary signaling, the required SNR is minimum, but the bandwidth is maximum. Especially in DSL applications, where the channel is typically low-pass, it is advantageous to increase M, i.e., to reduce the bandwidth, and hence avoid regions of high attenuation. If the gain in SNR is larger than the increase in required SNR, a net gain can be utilized. At some point, the gain due to bandwidth reduction becomes less compared to the increase in required SNR; beyond this point, going to even larger constellations is counterproductive.

Example 2.8: *Optimization of the Symbol Frequency for DFE*

The optimization of the symbol frequency via the number of signal levels is shown in Figure 2.31 for the DSL up-stream example. Here, we consider only integer modulation rates R_m which allow for simple mapping of binary data to signal points. For cable lengths ranging from 1 km to 3 km, the SNR divided by the variance of the signal constellation is plotted. Additionally, a normalization to the case of binary signaling ($R_\mathrm{m} = \log_2(M) = 1$) and cable length of 1 km is performed.

For short cables, quaternary transmission is preferable. As the cable length ℓ increases, going to higher rates, i.e., $M = 8$, is rewarding, since the fluctuations of the attenuation within the transmission band increase. Additionally, increasing ℓ decreases the SNR, since the insertion loss of the cable increases linearly with the length. In summary, for the present situation, rates R_m between 2 and 3 bits per symbol (4-ary to 8-ary ASK) are optimum.

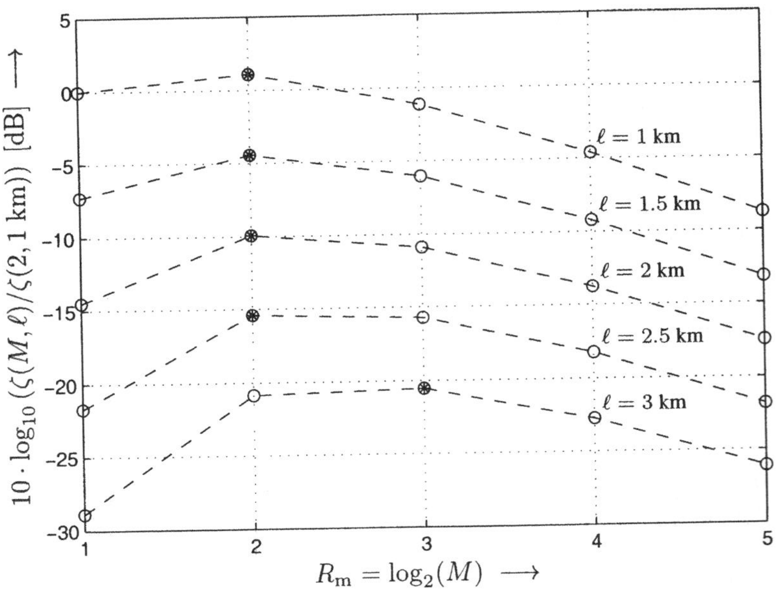

Fig. 2.31 Normalized SNR according to (2.3.74) over the rate of the modulation. The optimum (integer) rate is marked.

Decision-Feedback Equalization and Channel Capacity Inserting the optimal transmit filter (2.3.71) in (2.3.52) yields the maximum signal-to-noise ratio

$$\mathrm{SNR}^{(\text{ZF-DFE})} = \exp\left\{ T \int_{f \in \mathcal{F}} \log\left(\frac{S \cdot T}{N_0} |H_\mathrm{C}(f)|^2 \right) \mathrm{d}f \right\} . \tag{2.3.76}$$

Changing the basis of both the logarithm and the exponential function from e to 2, and taking the binary logarithm of (2.3.76) leads to

$$\log_2\left(\mathrm{SNR}^{(\text{ZF-DFE})}\right) = T \int_{f \in \mathcal{F}} \log_2\left(\frac{S \cdot T}{N_0} |H_\mathrm{C}(f)|^2 \right) \mathrm{d}f . \tag{2.3.77}$$

This equation has a very interesting interpretation.

Left-hand side:

The ZF-DFE produces an overall AWGN channel with signal-to-noise ratio $\mathrm{SNR}^{(\text{ZF-DFE})}$. From the basic information theory we know that the channel capacity (in bits per channel use) of the AWGN channel with i.i.d. Gaussian distributed input equals $\log_2(1 + \mathrm{SNR})$, which, in turn, for high SNR is well approximated by $\log_2(\mathrm{SNR})$. Thus, the left-hand side is simply an approximation of the capacity of the discrete-time channel created by zero-forcing decision-feedback equalization.

Right-hand side:

The transmission is done over the underlying continuous-time channel with spectral signal-to-noise ratio $\frac{S \cdot T}{N_0} |H_C(f)|^2$ within the support $\mathcal{F}$. Hence, its channel capacity (in bits per second) reads [Gal68, CT91]

$$\int_{f \in \mathcal{F}} \log_2 \left(1 + \frac{S \cdot T}{N_0} |H_C(f)|^2\right) \, df \, .$$

Since the channel is used once per T seconds, multiplying by T results in the capacity measured in bits per use. For high SNRs the "1" under the logarithm can again be dropped, and the water-filling solution for the transmit spectrum will tend to a brick-wall shape.

In summary, equation (2.3.77) states that the capacity usable by DFE is approximately equal to the capacity of the underlying channel, i.e.,

$$C_{\text{ZF-DFE}} \approx C_{\text{underlying channel}} \; ; \tag{2.3.78}$$

equality is achieved asymptotically for high SNRs. Thus, in the sense of information theory, decision-feedback equalization is optimum. With this technique and Gaussian distributed symbols, the entire capacity for given transmit power of the underlying channel can be used. This result was first derived by Price [Pri72]; at this point, it remains open how to come close to this fundamental limit in practice. A more detailed treatment on this topic—especially on the loss associated with ZF-DFE compared to the actual capacity—can be found in [BLM93, BLM96].

Because of its optimality and low complexity, the decision-feedback equalizer can be seen as a *canonical structure* for equalization. However, the above result assumes correct decisions. Unfortunately, decision-feedback equalization suffers from error propagation, which degrades performance, especially at low SNRs (high error rates). Moreover, channel coding—which is indispensable for approaching capacity—cannot be applied in a straightforward manner, since DFE requires zero-delay decisions. But in channel coding decisions are based on the observation of whole blocks/sequences of symbols. We return to this point in Chapter 3.

2.3.3 Finite-Length MMSE Decision-Feedback Equalization

We have seen that for linear equalization optimizing the filters with respect to the MMSE criterion leads to a gain over the ZF solution. Consequently, decision-feedback equalization now is designed for minimizing the mean-squared error, called *minimum mean-squared error decision-feedback equalization (MMSE-DFE)*. MMSE-DFE was first considered by Monsen [Mon71]. Thereby, he mainly concentrated on infinite-length filters.

In ZF-DFE, when starting from optimal ZF linear equalization, except for the leading coefficient, feedforward and feedback filter are identical. To get an additional degree of freedom, this restriction is dropped. Moreover, the starting point is now the matched-filter front-end. The T-spaced discrete-time model is thus given by (2.2.38a) and (2.2.38b). As for ZF-DFE, first we derive results for finite-length filters, then we address the asymptotic case.

It is convenient to consider definition (2.3.30) and express all quantities using

$$\Phi_{hh}(z) \triangleq \mathcal{Z}\{\varphi_{hh}[k]\} = \sum_{k=-\infty}^{+\infty} \varphi_{hh}[k]z^{-k} \,. \tag{2.3.79}$$

In particular, the signal transfer function is given by $H^{(\mathrm{MF})}(z) = \Phi_{hh}(z)/\sigma_a^2$, and the noise PSD reads $\Phi_{nn}^{(\mathrm{MF})}(z) = \frac{N_0}{T\sigma_a^2}\Phi_{hh}(z)$. From Section 2.3.2 we know that the noise after the matched filter can be modeled as being generated from white noise with variance $\sigma_n^2 = \frac{\sigma_g^2}{\sigma_a^2}\frac{N_0}{T}$, filtered with $G(z)$. Variance σ_g^2 and filter $G(z)$ are obtained, e.g., from spectral factorization, cf. (2.3.30). In Figure 2.32 the relevant signals and quantities are collocated.

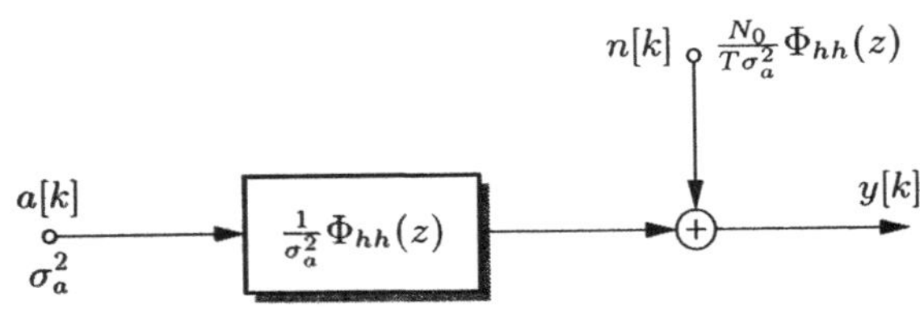

Fig. 2.32 Transmission model for the matched-filter front-end.

Optimization Figure 2.33 sketches the receiver structure. The DFE part of the receiver consists of a feedforward filter $F(z)$ and a feedback filter $B(z) - 1$. For finite-length results, we assume the feedforward filter to be causal and FIR of order q_f, i.e., $F(z) = \sum_{k=0}^{q_f} f[k]z^{-k}$. The feedback filter $B(z) - 1$ is causal and has a monic FIR polynomial $B(z) = 1 + \sum_{k=1}^{q_b} b[k]z^{-k}$ of order q_b. As an additional degree of freedom for minimizing the MSE, a delay k_0 for producing the estimates $\hat{a}[k]$ of the transmitted amplitude coefficients $a[k]$ is admitted. This delay could equivalently be modeled as a noncausal, two-sided feedforward filter $F(z)$. The derivation in this section follows the development of [Kam94]. Similar approaches can be found, e.g., in [AC95] and [GH96].

Using the above definitions, the error signal at the slicer input can be written as

$$e[k] = \sum_{\kappa=0}^{q_f} f[\kappa]y[k-\kappa] - \sum_{\kappa=1}^{q_b} b[\kappa]a[k-k_0-\kappa] - a[k-k_0] \,. \tag{2.3.80}$$

Once again, we have to assume that the q_b previous decisions are correct, i.e., $\hat{a}[\kappa] = a[\kappa]$, $\kappa = k - k_0 - q_b, \ldots, k - k_0$. The aim of the optimization is to determine filters $F(z)$ and $B(z)$ such that the variance of the error signal at the slicer input is minimum

$$\mathsf{E}\{|e[k]|^2\} \to \min \,. \tag{2.3.81}$$

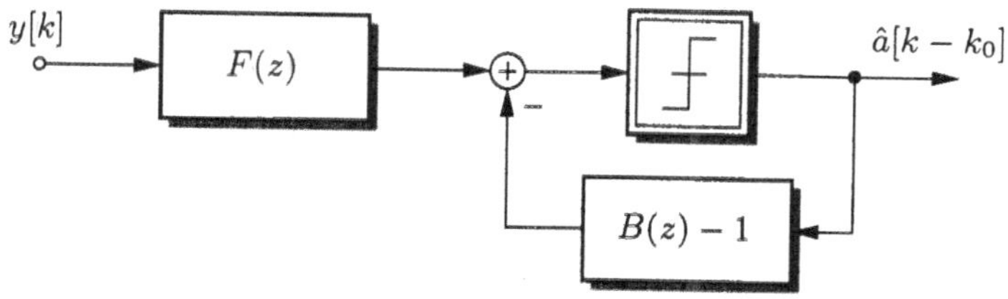

Fig. 2.33 Receiver structure for MMSE-DFE.

To get a compact representation, we resort to vector notation, using the definitions

$$\boldsymbol{f} = \begin{bmatrix} f^*[0] \\ f^*[1] \\ \vdots \\ f^*[q_f] \end{bmatrix}, \qquad \boldsymbol{b} = \begin{bmatrix} b^*[1] \\ b^*[2] \\ \vdots \\ b^*[q_b] \end{bmatrix},$$

$$\boldsymbol{y}[k] = \begin{bmatrix} y[k] \\ y[k-1] \\ \vdots \\ y[k-q_f] \end{bmatrix}, \qquad \boldsymbol{a}[k] = \begin{bmatrix} a[k-k_0-1] \\ a[k-k_0-2] \\ \vdots \\ a[k-k_0-q_b] \end{bmatrix}. \tag{2.3.82}$$

By this, the error signal (2.3.80) can be expressed as

$$e[k] = \boldsymbol{f}^{\mathsf{H}}\boldsymbol{y}[k] - \boldsymbol{b}^{\mathsf{H}}\boldsymbol{a}[k] - a[k-k_0], \tag{2.3.83}$$

and straightforward manipulations give the *mean-squared error (MSE)* as

$$\begin{aligned} \mathsf{E}\left\{|e[k]|^2\right\} &= \mathsf{E}\left\{|\boldsymbol{f}^{\mathsf{H}}\boldsymbol{y}[k] - \boldsymbol{b}^{\mathsf{H}}\boldsymbol{a}[k] - a[k-k_0]|^2\right\} \\ &= \boldsymbol{f}^{\mathsf{H}}\boldsymbol{\Phi}_{yy}\boldsymbol{f} - \boldsymbol{f}^{\mathsf{H}}\boldsymbol{\Phi}_{ya}\boldsymbol{b} - \boldsymbol{f}^{\mathsf{H}}\boldsymbol{\phi}_{ya} - \boldsymbol{b}^{\mathsf{H}}\boldsymbol{\Phi}_{ya}^{\mathsf{H}}\boldsymbol{f} + \boldsymbol{b}^{\mathsf{H}}\boldsymbol{\Phi}_{aa}\boldsymbol{b} + \boldsymbol{b}^{\mathsf{H}}\boldsymbol{\phi}_{aa} \\ &\quad - \boldsymbol{\phi}_{ya}^{\mathsf{H}}\boldsymbol{f} + \boldsymbol{\phi}_{aa}^{\mathsf{H}}\boldsymbol{b} + \sigma_a^2 . \end{aligned} \tag{2.3.84}$$

Here the correlation matrices and vectors calculate to ($\boldsymbol{I}$: identity matrix)

$$\begin{aligned} \boldsymbol{\Phi}_{yy} &= \mathsf{E}\left\{\boldsymbol{y}[k]\boldsymbol{y}^{\mathsf{H}}[k]\right\} \\ &= \left[\tfrac{1}{\sigma_a^2}(\varphi_{hh}[\nu] * \varphi_{hh}[\nu])\big|_{\nu=k-\kappa} + \tfrac{N_0}{T\sigma_a^2}\varphi_{hh}[k-\kappa]\right]_{\substack{\kappa=0,\ldots,q_f \\ k=0,\ldots,q_f}} \end{aligned} \tag{2.3.85a}$$

$$\begin{aligned} \boldsymbol{\Phi}_{aa} &= \mathsf{E}\left\{\boldsymbol{a}[k]\boldsymbol{a}^{\mathsf{H}}[k]\right\} \\ &= \sigma_a^2\boldsymbol{I}_{q_b \times q_b} \end{aligned} \tag{2.3.85b}$$

$$\begin{aligned} \boldsymbol{\Phi}_{ya} &= \mathsf{E}\left\{\boldsymbol{y}[k]\boldsymbol{a}^{\mathsf{H}}[k]\right\} \\ &= \left[\varphi_{hh}[k_0+k-\kappa]\right]_{\substack{\kappa=0,\ldots,q_f \\ k=1,\ldots,q_b}} \end{aligned} \tag{2.3.85c}$$

$$\boldsymbol{\phi}_{ya} = \mathsf{E}\left\{\boldsymbol{y}[k]a^*[k-k_0]\right\}$$

$$= [\varphi_{hh}[k_0 - \kappa]]_{\kappa=0,\ldots,q_f} \tag{2.3.85d}$$

$$\phi_{aa} = \mathsf{E}\{a[k]a^*[k-k_0]\}$$
$$= \mathbf{0}_{q_b\times 1}\,, \quad k_0 > 0\,. \tag{2.3.85e}$$

Note, these quantities are only valid for the matched-filter front-end and an i.i.d. data sequence $\langle a[k]\rangle$.

At the optimum, the gradients of $\mathsf{E}\{|e[k]|^2\}$ with respect to the vectors $\boldsymbol{f}$ and $\boldsymbol{b}$ of the filter coefficients have to be zero. Using Wirtinger Calculus, we have

$$\frac{\partial}{\partial \boldsymbol{f}^*}\mathsf{E}\{|e[k]|^2\} = \boldsymbol{\Phi}_{yy}\boldsymbol{f} - \boldsymbol{\Phi}_{ya}\boldsymbol{b} - \boldsymbol{\phi}_{ya} = \mathbf{0} \tag{2.3.86a}$$

$$\frac{\partial}{\partial \boldsymbol{b}^*}\mathsf{E}\{|e[k]|^2\} = -\boldsymbol{\Phi}_{ya}^{\mathsf{H}}\boldsymbol{f} + \boldsymbol{\Phi}_{aa}\boldsymbol{b} + \boldsymbol{\phi}_{aa} = \mathbf{0} \tag{2.3.86b}$$

or compactly

$$\begin{bmatrix} \boldsymbol{\Phi}_{yy} & \boldsymbol{\Phi}_{ya} \\ \boldsymbol{\Phi}_{ya}^{\mathsf{H}} & \boldsymbol{\Phi}_{aa} \end{bmatrix} \cdot \begin{bmatrix} \boldsymbol{f} \\ -\boldsymbol{b} \end{bmatrix} = \begin{bmatrix} \boldsymbol{\phi}_{ya} \\ \mathbf{0}_{q_b\times 1} \end{bmatrix}. \tag{2.3.87}$$

To solve this set of equations, we multiply (2.3.86b) by $\boldsymbol{\Phi}_{ya}\boldsymbol{\Phi}_{aa}^{-1} = \frac{1}{\sigma_a^2}\boldsymbol{\Phi}_{ya}$ from the left, and add up equations (2.3.86). This results in

$$\boldsymbol{f}_{\text{opt}} = \left(\boldsymbol{\Phi}_{yy} - \frac{1}{\sigma_a^2}\boldsymbol{\Phi}_{ya}\boldsymbol{\Phi}_{ya}^{\mathsf{H}}\right)^{-1}\boldsymbol{\phi}_{ya}\,, \tag{2.3.88a}$$

and

$$\boldsymbol{b}_{\text{opt}} = \frac{1}{\sigma_a^2}\boldsymbol{\Phi}_{ya}^{\mathsf{H}}\boldsymbol{f}_{\text{opt}}\,. \tag{2.3.88b}$$

At this point, one important observation can be made. Because of the above definitions, the signal impulse response for the matched-filter front-end is $h^{(\mathsf{MF})}[k] = \mathcal{Z}^{-1}\{H^{(\mathsf{MF})}(z)\} = \frac{1}{\sigma_a^2}\varphi_{hh}[k]$. The end-to-end impulse response from the input of the pulse-shaping filter to the slicer input is then given by $h^{(\mathsf{MMSE\text{-}DFE})}[k] \triangleq h^{(\mathsf{MF})}[k] * f[k]$. Regarding the complex conjugate of (2.3.88b) and inserting the definitions, we have

$$\begin{bmatrix} b_{\text{opt}}[1] \\ b_{\text{opt}}[2] \\ \vdots \\ b_{\text{opt}}[q_b] \end{bmatrix} = \frac{1}{\sigma_a^2}\cdot \begin{bmatrix} \varphi_{hh}[k_0+1] & \varphi_{hh}[k_0] & \cdots & \varphi_{hh}[k_0+1-q_f] \\ \varphi_{hh}[k_0+2] & \ddots & \ddots & \vdots \\ \vdots & \ddots & & \vdots \\ \varphi_{hh}[k_0+q_b] & \cdots & \cdots & \varphi_{hh}[k_0+q_b-q_f] \end{bmatrix} \cdot \begin{bmatrix} f_{\text{opt}}[0] \\ f_{\text{opt}}[1] \\ \vdots \\ f_{\text{opt}}[q_f] \end{bmatrix}. \tag{2.3.89}$$

The right-hand side is exactly the convolution of $h^{(\mathsf{MF})}[k]$ and $f[k]$, written in matrix form. Thus, within its support, the impulse response $b[k]$ of the feedback filter equals a segment of the end-to-end impulse response $h^{(\mathsf{MMSE\text{-}DFE})}[k]$ seen by the data sequence. Starting with the sample for delay k_0+1, the intersymbol interference contributed by the following q_b samples is eliminated completely. Only the precursors $h^{(\text{MMSE-DFE})}[k]$ for $k = -\infty, \ldots, k_0 - 1$ and the postcursors for $k = k_0 + q_b + 1, \ldots, \infty$ remain.

Inserting the optimal solution (2.3.88) into (2.3.84), the minimum mean-squared error is given by

$$\mathsf{E}\{|e[k]|^2\} = \sigma_a^2 - \boldsymbol{\phi}_{ya}^{\mathsf{H}} \boldsymbol{f}_{\text{opt}} = \sigma_a^2 \left(1 - \boldsymbol{\phi}_{ya}^{\mathsf{H}} \left(\sigma_a^2 \boldsymbol{\Phi}_{yy} - \boldsymbol{\Phi}_{ya} \boldsymbol{\Phi}_{ya}^{\mathsf{H}}\right)^{-1} \boldsymbol{\phi}_{ya}\right), \tag{2.3.90}$$

and the signal-to-noise ratio of MMSE-DFE calculates to

$$\text{SNR}^{(\text{MMSE-DFE})} = \frac{1}{1 - \boldsymbol{\phi}_{ya}^{\mathsf{H}} \left(\sigma_a^2 \boldsymbol{\Phi}_{yy} - \boldsymbol{\Phi}_{ya} \boldsymbol{\Phi}_{ya}^{\mathsf{H}}\right)^{-1} \boldsymbol{\phi}_{ya}}. \tag{2.3.91}$$

Unbiased Receiver In Section 2.2.4 we have seen that generally the MMSE solution is biased. By removing this bias, the SNR is decreased by one, but performance is increased. This fact, of course, also holds for MMSE-DFE.

To see this, we first implicitly define a $(q_f + 1) \times (q_f + 1)$ matrix $\boldsymbol{\Psi}^{-1}$ by[12]

$$\boldsymbol{\Psi}^{-1} + \boldsymbol{\phi}_{ya}^{\mathsf{H}} \boldsymbol{\phi}_{ya} \triangleq \sigma_a^2 \boldsymbol{\Phi}_{yy} - \boldsymbol{\Phi}_{ya} \boldsymbol{\Phi}_{ya}^{\mathsf{H}}, \tag{2.3.92}$$

and use the *Sherman-Morrison formula* [PTVF92] (here in its special form known as *Woodbury's identity*) to rewrite $\left(\boldsymbol{\Psi}^{-1} + \boldsymbol{\phi}_{ya}^{\mathsf{H}} \boldsymbol{\phi}_{ya}\right)^{-1}$ as a function of $\boldsymbol{\Psi}$ and $\boldsymbol{\phi}_{ya}$:

$$\begin{aligned} \left(\boldsymbol{\Psi}^{-1} + \boldsymbol{\phi}_{ya}^{\mathsf{H}} \boldsymbol{\phi}_{ya}\right)^{-1} &= \boldsymbol{\Psi} - \boldsymbol{\Psi} \boldsymbol{\phi}_{ya} \left(1 + \boldsymbol{\phi}_{ya}^{\mathsf{H}} \boldsymbol{\Psi} \boldsymbol{\phi}_{ya}\right)^{-1} \boldsymbol{\phi}_{ya}^{\mathsf{H}} \boldsymbol{\Psi} \\ &= \boldsymbol{\Psi} - \frac{1}{1 + \boldsymbol{\phi}_{ya}^{\mathsf{H}} \boldsymbol{\Psi} \boldsymbol{\phi}_{ya}} \boldsymbol{\Psi} \boldsymbol{\phi}_{ya} \boldsymbol{\phi}_{ya}^{\mathsf{H}} \boldsymbol{\Psi}. \end{aligned} \tag{2.3.93}$$

With this, the optimum vector $\boldsymbol{f}_{\text{opt}}$ (2.3.88a) can be expressed by

$$\boldsymbol{f}_{\text{opt}} = \sigma_a^2 \boldsymbol{\Psi} \boldsymbol{\phi}_{ya} \left(1 - \frac{\boldsymbol{\phi}_{ya}^{\mathsf{H}} \boldsymbol{\Psi} \boldsymbol{\phi}_{ya}}{1 + \boldsymbol{\phi}_{ya}^{\mathsf{H}} \boldsymbol{\Psi} \boldsymbol{\phi}_{ya}}\right), \tag{2.3.94}$$

and the coefficient $h^{(\text{MMSE-DFE})}[k_0]$ in the decision point is

$$\begin{aligned} h^{(\text{MMSE-DFE})}[k_0] &= \frac{1}{\sigma_a^2} \boldsymbol{f}_{\text{opt}}^{\mathsf{H}} \boldsymbol{\phi}_{ya} \\ &= \left(1 - \frac{\boldsymbol{\phi}_{ya}^{\mathsf{H}} \boldsymbol{\Psi} \boldsymbol{\phi}_{ya}}{1 + \boldsymbol{\phi}_{ya}^{\mathsf{H}} \boldsymbol{\Psi} \boldsymbol{\phi}_{ya}}\right) \boldsymbol{\phi}_{ya}^{\mathsf{H}} \boldsymbol{\Psi} \boldsymbol{\phi}_{ya} \\ &= \frac{\boldsymbol{\phi}_{ya}^{\mathsf{H}} \boldsymbol{\Psi} \boldsymbol{\phi}_{ya}}{1 + \boldsymbol{\phi}_{ya}^{\mathsf{H}} \boldsymbol{\Psi} \boldsymbol{\phi}_{ya}}. \end{aligned} \tag{2.3.95}$$

[12]Note that the matrix $\boldsymbol{\Psi}^{-1}$ coincides with the matrix $\boldsymbol{V}$ in [Ger96].

Since the matrix $\boldsymbol{\Psi}$ has the form of a correlation matrix and thus is positive definite, the quadratic form is real-valued and positive. Hence, $h^{(\text{MMSE-DFE})}[k_0] < 1$ holds and (2.3.95) gives the attenuation of the reference tap. For removing this bias the signal at the slicer input has to be scaled by

$$\frac{1 + \boldsymbol{\phi}_{ya}^{\mathsf{H}} \boldsymbol{\Psi} \boldsymbol{\phi}_{ya}}{\boldsymbol{\phi}_{ya}^{\mathsf{H}} \boldsymbol{\Psi} \boldsymbol{\phi}_{ya}} > 1 .$$

This increases performance and results in an SNR [CDEF95] (cf. also Section 2.2.4)

$$\text{SNR}^{(\text{MMSE-DFE,u})} = \text{SNR}^{(\text{MMSE-DFE})} - 1 = \frac{\boldsymbol{\phi}_{ya}^{\mathsf{H}} \left(\sigma_a^2 \boldsymbol{\Phi}_{yy} - \boldsymbol{\Phi}_{ya} \boldsymbol{\Phi}_{ya}^{\mathsf{H}} \right)^{-1} \boldsymbol{\phi}_{ya}}{1 - \boldsymbol{\phi}_{ya}^{\mathsf{H}} \left(\sigma_a^2 \boldsymbol{\Phi}_{yy} - \boldsymbol{\Phi}_{ya} \boldsymbol{\Phi}_{ya}^{\mathsf{H}} \right)^{-1} \boldsymbol{\phi}_{ya}} . \tag{2.3.96}$$

Completing this section, we note that the unbiased MMSE-DFE solution is identical to the so-called *maximum-SNR DFE (Max-SNR-DFE)*. Therefore, the optimization is done for a direct maximization of the SNR, i.e., taking both the coefficient in the decision point and the variance of the residual distortion (noise plus unprocessed ISI) into account. Details on this equivalence are given in [Ger96].

Example 2.9: *Finite-Length MMSE Decision-Feedback Equalization*

In order to visualize the various effects of MMSE-DFE, this example calculates the finite-length MMSE decision-feedback equalizer. Once again, we consider the simplified DSL down-stream (white noise) scenario.

First, Figure 2.34 shows the normalized autocorrelation sequence $\varphi_{hh}[\kappa]$, which is proportional to the end-to-end impulse response of the data sequence and proportional to the noise autocorrelation sequence, when applying the matched-filter front-end.

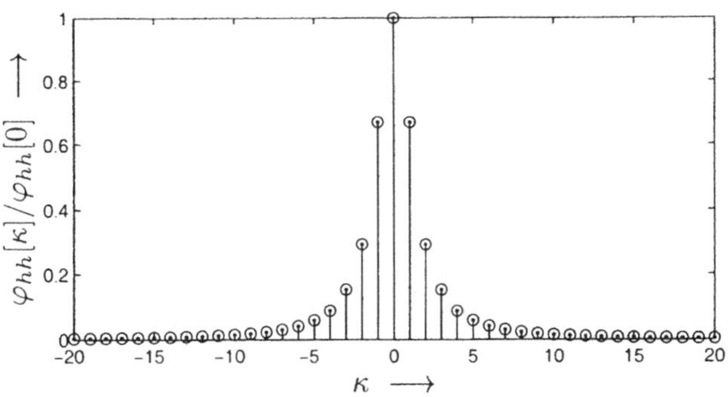

Fig. 2.34 Normalized autocorrelation sequence $\varphi_{hh}[\kappa]$.

For this situation, the MMSE-DFE solution is calculated, assuming a feedforward filter $F(z)$ of order $q_f = 20$, and a feedback filter $B(z) - 1$ of order $q_b = 10$. The decision delay is fixed to be $k_0 = 10$, and a SNR of $10 \cdot \log_{10}\left(\frac{T\sigma_a^2}{N_0}\right) = 10$ dB was chosen. Figure 2.35 plots the resulting impulse responses of the feedforward and feedback filters.

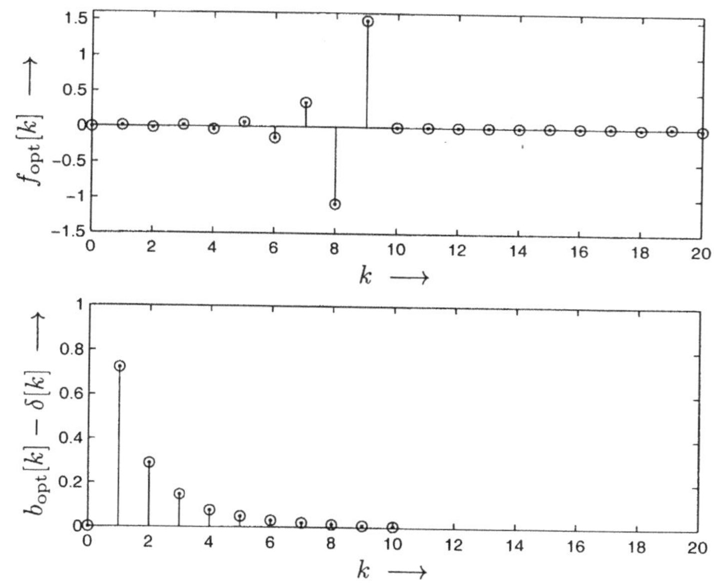

Fig. 2.35 Impulse responses of feedforward filter $F(z)$, $q_f = 20$, and feedback filter $B(z) - 1$, $q_b = 10$, for MMSE-DFE. Decision delay: $k_0 = 10$; Signal-to-noise ratio: $10 \cdot \log_{10}\left(\frac{T\sigma_a^2}{N_0}\right) = 10$ dB.

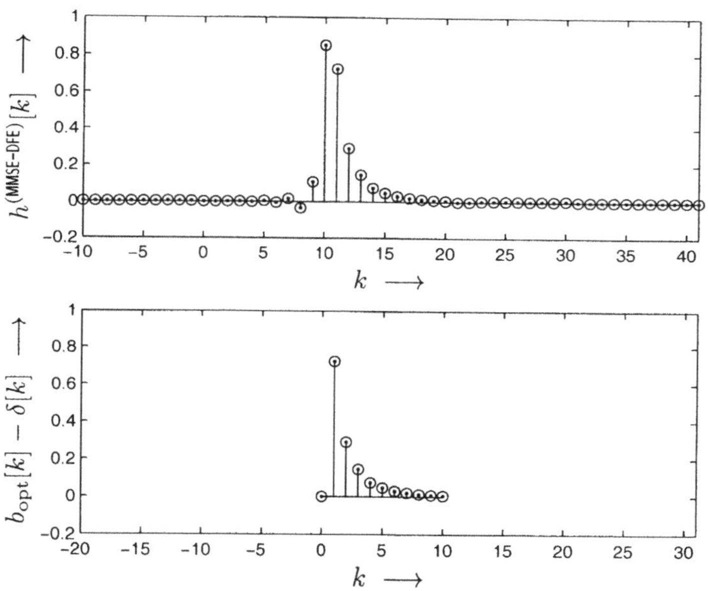

Fig. 2.36 Top: end-to-end impulse response experienced by the data, Bottom: impulse response of the feedback filter.

Applying these filters, the end-to-end impulse response experienced by the data sequence (prior to subtraction of the feedback part) is sketched in Figure 2.36 (top). On the bottom of this figure, the impulse response of the feedback filter is repeated, aligned such that it has the correct timing with respect to the end-to-end impulse response and decision delay $k_0 = 10$. A close look reveals that within its time span, the feedback filter completely eliminates ISI, and mainly unprocessed precursors remain. Furthermore, the bias is clearly visible. The reference tap on which the decision is based is $h^{(\text{MMSE-DFE})}[k_0 = 10] = 0.85$, and thus smaller than one.

In order to examine this effect in more detail, the end-to-end impulse responses and the respective feedback filters for different SNRs are compiled in Figure 2.37. Here, the parameters $q_f = 40$, $q_b = 20$, and $k_0 = 20$ are selected. With increasing SNR, the precursors vanish and—since the feedforward filter has sufficient order—the impulse response approaches the ZF solution (cf. Figure 2.24). Moreover, the bias tends to zero, i.e., the reference tap goes up to one.

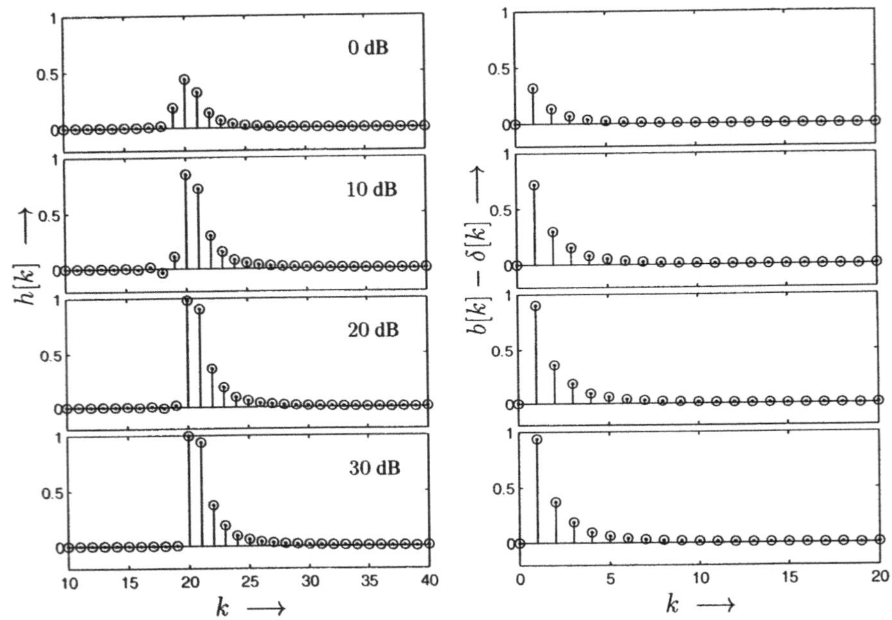

Fig. 2.37 Left: end-to-end impulse response $h[k] = h^{(\text{MMSE-DFE})}[k]$ experienced by the data, Right: impulse response $b[k] - \delta[k]$ of the feedback filter. Top to bottom: Signal-to-noise ratio $10 \cdot \log_{10}\left(\frac{T\sigma_a^2}{N_0}\right) = 0, 10, 20, 30$ dB.

Finally, the dependency on the decision delay is assessed in Figure 2.38. For k_0 ranging from -5 through 20, the achievable signal-to-noise ratio $\text{SNR}^{(\text{MMSE-DFE,u})}$ for unbiased MMSE-DFE is calculated. The filters have order $q_f = 10$ and $q_b = 5$, and a ratio $10 \cdot \log_{10}\left(\frac{T\sigma_a^2}{N_0}\right) = 20$ dB is used. Between $k_0 = 4$ and $k_0 = 11$, the SNR changes only slightly. But if the decision delay is too small or too large, the time span of the feedforward filter is exceeded and the SNR degrades dramatically. In addition, the signal-to-noise ratio $\text{SNR}^{(\text{MMSE-DFE})}$ for biased MMSE-DFE is shown. Note that due to the SNR definition (2.3.91), here $\text{SNR} \geq 1$ always holds, and thus the SNR in dB is always positive and saturates at 0 dB.

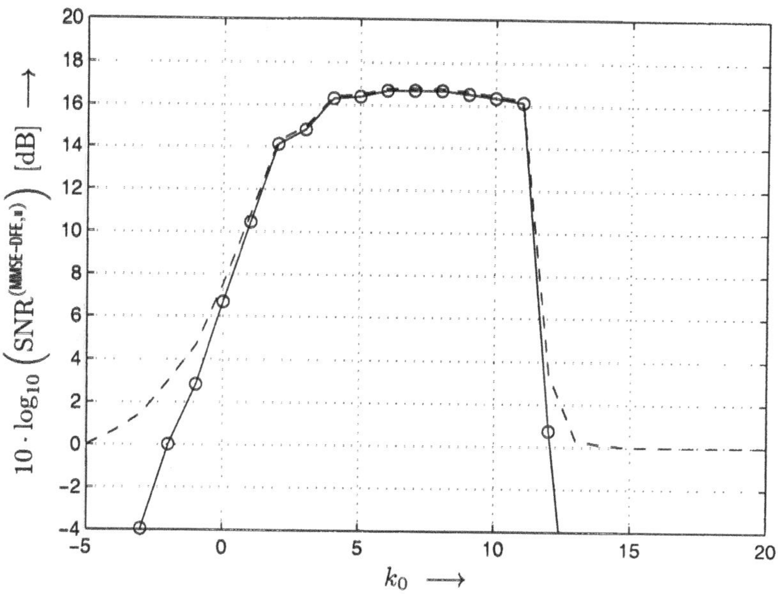

Fig. 2.38 Signal-to-noise ratio for MMSE-DFE over the decision delay k_0. Solid: unbiased MMSE-DFE; Dashed: biased MMSE-DFE.

2.3.4 Infinite-Length MMSE Decision-Feedback Equalization

After having derived the finite-length MMSE-DFE, we now turn to the asymptotic case and regard infinite-length filters. Here, the feedforward filter $F(z)$ is assumed to be two-sided. Since we now admit noncausal infinite impulse responses, without loss of optimality, the decision delay k_0 can be set to zero. The feedback filter $B(z) - 1$ is also IIR, but, of course, has to be strictly causal. The exposition follows [CDEF95].

Optimization From Figures 2.32 and 2.33, the error sequence, expressed by its z-transform, reads, if correct decisions are assumed

$$E(z) = Y(z)F(z) - A(z)(B(z) - 1) - A(z) = Y(z)F(z) - A(z)B(z) \,. \tag{2.3.97}$$

For optimization, we first imagine that the feedback part $B(z)$ is given. Then the feedforward filter $F(z)$ has to be chosen such that the mean-squared error is minimized. For solving this problem, we can resort to the Orthogonality Principle (Section 2.2.3). Recalling the problem statement of Figure 2.12, only the reference signal has to be changed to $a[k] * b[k]$. Thus, $F(z)$ is the optimum linear predictor for the sequence $A(z)B(z)$ based on the observation $Y(z)$.

By virtue of the Orthogonality Principle the error sequence $E(z)$ has to be orthogonal to the observation $Y(z)$, i.e., the cross-correlation has to be zero

$$\Phi_{ey}(z) = \Phi_{yy}(z)F(z) - \Phi_{ay}(z)B(z) \stackrel{!}{=} 0\,. \tag{2.3.98}$$

Here, the obvious definitions

$$\Phi_{ey}(z) \bullet\!\!-\!\!\circ \mathrm{E}\{e[k+\kappa]y^*[k]\} \qquad \Phi_{ay}(z) \bullet\!\!-\!\!\circ \mathrm{E}\{a[k+\kappa]y^*[k]\} \tag{2.3.99}$$

have been used. Because of the matched-filter front-end (equations (2.2.38a) and (2.2.38b)) and an i.i.d. data sequence $\langle a[k]\rangle$, the cross PSDs calculate to

$$\Phi_{ay}(z) = \Phi_{hh}(z) \tag{2.3.100a}$$

and

$$\begin{aligned}\Phi_{yy}(z) &= \sigma_a^2\left(H^{(\mathrm{MF})}(z)\right)^2 + \Phi_{nn}^{(\mathrm{MF})}(z)\\ &= \frac{1}{\sigma_a^2}\Phi_{hh}(z)\underbrace{\left(\Phi_{hh}(z) + \frac{N_0}{T}\right)}\\ &\triangleq \frac{1}{\sigma_a^2}\Phi_{hh}(z) \qquad \Phi_{ff}(z)\,.\end{aligned} \tag{2.3.100b}$$

Thus, solving (2.3.98) for $F(z)$, we have

$$F(z) = B(z)\Phi_{ay}(z)/\Phi_{yy}(z) = B(z)\sigma_a^2/\Phi_{ff}(z)\,, \tag{2.3.101}$$

and the error sequence is given by:

$$E(z) = B(z)\left[\sigma_a^2\frac{Y(z)}{\Phi_{ff}(z)} - A(z)\right] \triangleq B(z)E'(z)\,. \tag{2.3.102}$$

For the PSD of the newly defined error sequence $e'[k] \circ\!\!-\!\!\bullet E'(z)$, we obtain

$$\begin{aligned}\Phi_{e'e'}(z) &= \sigma_a^4\frac{\Phi_{yy}(z)}{(\Phi_{ff}(z))^2} - 2\sigma_a^2\frac{\Phi_{ay}(z)}{\Phi_{ff}(z)} + \sigma_a^2\\ &= \sigma_a^2\frac{\Phi_{hh}(z)\Phi_{ff}(z)}{(\Phi_{ff}(z))^2} - 2\sigma_a^2\frac{\Phi_{hh}(z)}{\Phi_{ff}(z)} + \sigma_a^2\\ &= \sigma_a^2\frac{-\Phi_{hh}(z) + \Phi_{ff}(z)}{\Phi_{ff}(z)}\\ &= \sigma_a^2\frac{N_0}{T}\frac{1}{\Phi_{ff}(z)}\,.\end{aligned} \tag{2.3.103}$$

From prediction theory we know that in the optimum, $\langle e[k]\rangle$ is a white sequence (cf. Theorem 2.13). Hence, regarding (2.3.102), $B(z)$ has to be the whitening filter for $E'(z)$. For this, similar to (2.3.30), a spectral factorization can be defined as

$$\Phi_{ff}(z) \triangleq \sigma_a^2 H^{(\mathrm{MF})}(z) + \frac{N_0}{T} = \sigma_g^2\cdot G(z)\cdot G^*(z^{-*})\,, \tag{2.3.104}$$

where $G(z)$ is again forced to be causal, monic, and minimum-phase, i.e., canonical. Since the feedback filter should also exhibit these properties, we choose

$$B(z) = G(z) , \tag{2.3.105a}$$

which, regarding (2.3.101) and (2.3.104), corresponds to the feedforward filter

$$\begin{aligned} F(z) &= G(z)\sigma_a^2/(\sigma_g^2 G(z)G^*(z^{-*})) \\ &= \frac{\sigma_a^2}{\sigma_g^2}\frac{1}{G^*(z^{-*})} . \end{aligned} \tag{2.3.105b}$$

Using these filters, the PSD of the error sequence $\langle e[k]\rangle$ calculates to

$$\Phi_{nn}^{(\text{MMSE-DFE})}(z) = \frac{\sigma_a^2}{\sigma_g^2}\frac{N_0}{T} , \tag{2.3.106}$$

which also gives the minimum mean-squared error. The variance σ_g^2 is obtained from (2.3.44) as

$$\sigma_g^2 = \exp\left\{T\int_{-\frac{1}{2T}}^{\frac{1}{2T}} \log\left(\Phi_{ff}(e^{j2\pi fT})\right)\, df\right\} . \tag{2.3.107}$$

Finally, with the definition of $\Phi_{ff}(z)$ and (2.2.38a), the signal-to-noise ratio for MMSE-DFE reads

$$\begin{aligned} \text{SNR}^{(\text{MMSE-DFE})} &= \frac{\sigma_a^2}{\mathrm{E}\{|e[k]|^2\}} \\ &= \frac{\sigma_g^2}{N_0/T} \\ &= \exp\left\{T\int_{-\frac{1}{2T}}^{\frac{1}{2T}} \log\left(\frac{T\sigma_a^2}{N_0}\sum_\mu \left|H_\text{T}(f-\tfrac{\mu}{T})H_\text{C}(f-\tfrac{\mu}{T})\right|^2+1\right) df\right\} \\ &= \exp\left\{T\int_{-\frac{1}{2T}}^{\frac{1}{2T}} \log\left(\widetilde{\text{SNR}}(e^{j2\pi fT})+1\right) df\right\} . \end{aligned} \tag{2.3.108}$$

It is noteworthy that the results for MMSE-DFE are almost identical to that of ZF-DFE. The spectral factorization is only amended by the additional term N_0/T, which in turn leads to the $+1$ in the argument of the logarithm. Hence, once again, for high SNRs, $\frac{\sigma_a^2}{N_0/T}$ the results for the MMSE criterion tend to the ZF solution.

Mean-Square Whitened Matched Filter Like for ZF-DFE, the cascade of the matched filter $H_\mathrm{T}^*(f)H_\mathrm{C}^*(f)$ and discrete-time feedforward filter $F(z)$ establishes the total receive filter. If sampling is moved to the output of this filter, it can be implemented as a single analog front-end filter. Because this filter with transfer function

$$H^{(\text{MS-WMF})}(f) = H^{(\text{MF})}(f) \cdot \frac{\sigma_a^2}{\sigma_g^2} \frac{1}{G^*(\mathrm{e}^{\mathrm{j}2\pi fT})} \tag{2.3.109}$$

is optimized according to the MSE criterion, we call this filter a *mean-square whitened matched filter (MS-WMF)* [CDEF95].

It is straightforward to show that the power transfer function of the MS-WMF is not Nyquist, but has the following form:

Theorem 2.18: *Mean-Square Whitened Matched Filter*

The power transfer function of the *mean-square whitened matched filter (MS-WMF)* is given by

$$\left|H_\mathrm{R}^{(\text{MS-WMF})}(f)\right|^2 = \frac{\sigma_a^2}{\sigma_g^2} \frac{|H_\mathrm{T}(f)H_\mathrm{C}(f)|^2}{\sum_\mu \left|H_\mathrm{T}(f-\frac{\mu}{T})H_\mathrm{C}(f-\frac{\mu}{T})\right|^2 + \frac{N_0}{T\sigma_a^2}}\,. \tag{2.3.110}$$

Since the left-hand side of the factorization problem (2.3.104) is strictly positive for $N_0 > 0$, unlike the ZF solution, $G(z)$ is always well defined. Hence, the MS-WMF is guaranteed to exist. Note that the ZF-WMF does not exist if the folded squared magnitude $\sum_\mu \left|H_\mathrm{T}(f-\frac{\mu}{T})H_\mathrm{C}(f-\frac{\mu}{T})\right|^2$ is zero within intervals of nonzero measure; but it exists only when isolated spectral nulls are present.

Unbiased MMSE-DFE Over all, using the MS-WMF, the filtered receive sequence calculates to

$$\begin{aligned}
R(z) &= \frac{\sigma_a^2}{\sigma_g^2}\frac{1}{G^*(z^{-*})}Y(z) \\
&= \frac{\sigma_a^2}{\sigma_g^2}\frac{A(z)\Phi_{hh}(z)/\sigma_a^2 + N^{(\text{MF})}(z)}{G^*(z^{-*})} \\
&= \frac{1}{\sigma_g^2}\frac{A(z)\left(\Phi_{ff}(z) - \frac{N_0}{T}\right) + \sigma_a^2 N^{(\text{MF})}(z)}{G^*(z^{-*})} \\
&= \frac{1}{\sigma_g^2}\frac{A(z)\left(\sigma_g^2 G(z)G^*(z^{-*}) - \frac{N_0}{T}\right) + \sigma_a^2 N^{(\text{MF})}(z)}{G^*(z^{-*})} \\
&= A(z)G(z) + \underbrace{\frac{\sigma_a^2}{\sigma_g^2}\frac{N^{(\text{MF})}(z)}{G^*(z^{-*})} - \frac{N_0}{T\sigma_g^2}\frac{A(z)}{G^*(z^{-*})}}_{\triangleq E(z)} \\
&= A(z)G(z) + E(z)\,.
\end{aligned} \tag{2.3.111}$$

The MS-WMF output sequence may therefore be decomposed into three parts: First, the data sequence is filtered with the minimum-phase response $G(z)$, which gives the discrete-time channel model; second, an additive Gaussian noise sequence—generated by filtering the noise $\langle n^{(\mathrm{MF})}[k]\rangle$ after the matched filter—is present; and third, the residual, anticausal intersymbol interference, which is proportional to $A(z)/G^*(z^{-*})$. From the above we know that the error sequence $\langle e[k]\rangle$ is white with variance $\frac{\sigma_a^2}{\sigma_g^2}\frac{N_0}{T}$. Since, besides the additive noise, it contains residual intersymbol interference, its pdf usually is not (exactly) Gaussian and $\langle e[k]\rangle$ is statistically dependent on the data sequence $\langle a[k]\rangle$. As with high SNRs unprocessed ISI tends to zero, here the pdf of $e[k]$ approaches a Gaussian one.

Since $G^*(z^{-*})$ is monic and anticausal, its inverse $1/G^*(z^{-*})$ also has these properties. Thus, the biasing term $-\frac{N_0}{T\sigma_g^2}$ is present in the decision point. In order to get an unbiased receiver, we rewrite (2.3.111) as

$$
\begin{aligned}
R(z) &= A(z)G(z) - \frac{N_0}{T\sigma_g^2}\frac{A(z)}{G^*(z^{-*})} + \frac{\sigma_a^2}{\sigma_g^2}\frac{N^{(\mathrm{MF})}(z)}{G^*(z^{-*})} \\
&= A(z)G(z) - \frac{N_0}{T\sigma_g^2}A(z) \\
&\quad + \frac{N_0}{T\sigma_g^2}A(z)\left(1 - \frac{1}{G^*(z^{-*})}\right) + \frac{\sigma_a^2}{\sigma_g^2}\frac{N^{(\mathrm{MF})}(z)}{G^*(z^{-*})} \\
&\triangleq \frac{\sigma_g^2 - N_0/T}{\sigma_g^2}A(z)G'(z) \\
&\quad + \frac{N_0}{T\sigma_g^2}A(z)\left(1 - \frac{1}{G^*(z^{-*})}\right) + \frac{\sigma_a^2}{\sigma_g^2}\frac{N^{(\mathrm{MF})}(z)}{G^*(z^{-*})}\,,
\end{aligned}
\qquad (2.3.112)
$$

with

$$
G'(z) = \frac{\sigma_g^2 G(z) - N_0/T}{\sigma_g^2 - N_0/T}\,. \qquad (2.3.113)
$$

Note that like $G(z)$, $G'(z)$ is causal, minimum-phase, and monic, i.e., canonical. The feedback part of the DFE remains unchanged; it cancels the term $A(z)(G(z)-1)$. Thus, from the first part in (2.3.113), only the nondelayed term $1 - \frac{N_0/T}{\sigma_g^2}$ remains. To compensate for this bias, the signal at the slicer is scaled by

$$
\frac{\sigma_g^2}{\sigma_g^2 - N_0/T}\,, \qquad (2.3.114)
$$

see Figure 2.39.

By construction, the effective distortion sequence $e''[k] \triangleq e[k] + \frac{N_0}{T\sigma_g^2}a[k]$ after postcursor subtraction and prior to scaling is independent of $a[k]$, since $e''[k]$ contains only past and future sample of the data sequence $\langle a[k]\rangle$. Due to this independence and taking $e[k] = e''[k] - \frac{N_0}{T\sigma_g^2}a[k]$ into account, the variances of the sequences are

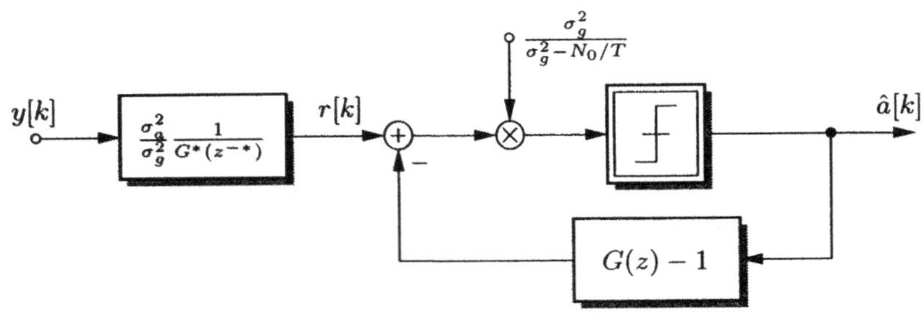

Fig. 2.39 Unbiased minimum mean-squared error decision-feedback equalizer.

related by $\sigma_e^2 = \sigma_{e''}^2 + (N_0/(T\sigma_g^2))^2 \sigma_a^2$, and the variance of $e''[k]$ calculates to

$$\sigma_{e''}^2 = \sigma_e^2 - \left(\frac{N_0}{T\sigma_g^2}\right)^2 \sigma_a^2 = \frac{\sigma_a^2}{\sigma_g^2}\frac{N_0}{T} - \left(\frac{N_0}{T\sigma_g^2}\right)^2 \sigma_a^2 = \frac{\sigma_a^2 N_0/T(\sigma_g^2 - N_0/T)}{\sigma_g^4} . \tag{2.3.115}$$

After scaling by $\frac{\sigma_g^2}{\sigma_g^2 - N_0/T}$, the MSE is thus

$$\frac{\sigma_a^2 N_0/T(\sigma_g^2 - N_0/T)}{\sigma_g^4} \cdot \left(\frac{\sigma_g^2}{\sigma_g^2 - N_0/T}\right)^2 = \frac{\sigma_a^2 N_0/T}{\sigma_g^2 - N_0/T} , \tag{2.3.116}$$

and the signal-to-noise ratio reads

$$\begin{aligned} \mathrm{SNR}^{(\text{MMSE-DFE,u})} &= \frac{\sigma_g^2 - N_0/T}{N_0/T} = \frac{\sigma_g^2}{N_0/T} - 1 \\ &= \mathrm{SNR}^{(\text{MMSE-DFE})} - 1 \\ &= \exp\left\{ T \int_{-\frac{1}{2T}}^{\frac{1}{2T}} \log\left(\widetilde{\mathrm{SNR}}(\mathrm{e}^{\mathrm{j}2\pi fT}) + 1\right) \mathrm{d}f \right\} - 1 . \end{aligned} \tag{2.3.117}$$

This result again supports the general statement in Section 2.2.4 concerning the relationship between unbiased and biased SNR.

For completeness, we state the SNR and the loss of unbiased MMSE-DFE compared to an AWGN channel:

Theorem 2.19: *Signal-to-Noise Ratio of MMSE-DFE*

When using unbiased minimum mean-squared error decision-feedback equalization at the receiver, the signal-to-noise ratio is given by

$$\mathrm{SNR}^{(\mathrm{MMSE\text{-}DFE},u)} = \exp\left\{T \int_{-\frac{1}{2T}}^{\frac{1}{2T}} \log\left(\widetilde{\mathrm{SNR}}(\mathrm{e}^{\mathrm{j}2\pi fT}) + 1\right) \, \mathrm{d}f\right\} - 1, \tag{2.3.118}$$

and the degradation (based on equal receive power) compared to transmission over an ideal channel reads

$$\vartheta^2_{(\mathrm{MMSE\text{-}DFE},u)} = \left(T \int_{-\frac{1}{2T}}^{\frac{1}{2T}} \widetilde{\mathrm{SNR}}(\mathrm{e}^{\mathrm{j}2\pi fT}) \, \mathrm{d}f\right)^{-1} \cdot \left(\exp\left\{T \int_{-\frac{1}{2T}}^{\frac{1}{2T}} \log\left(\widetilde{\mathrm{SNR}}(\mathrm{e}^{\mathrm{j}2\pi fT}) + 1\right) \, \mathrm{d}f\right\} - 1\right). \tag{2.3.119}$$

Again, $\widetilde{\mathrm{SNR}}(\mathrm{e}^{\mathrm{j}2\pi fT})$ is the folded spectral signal-to-noise ratio at the receiver input. Ideal, i.e., error-free, decisions are assumed.

Optimal Transmit Filter for MMSE-DFE After having derived the optimal receiver structure, we finally optimize the transmit spectrum for MMSE-DFE. This is again done in two steps. First, the periodic sum $\widetilde{\mathrm{SNR}}(\mathrm{e}^{\mathrm{j}2\pi fT})$ in (2.3.117) is cosidered. Using the same arguments as in Section 2.3.2, the power is best placed in that replica $f - \frac{\mu}{T}$, $\mu \in \mathbb{Z}$, for which the magnitude of the channel transfer function is maximum. Hence, the support of the transmit filter is again given by (cf. (2.2.96) and (2.3.65))

$$\mathcal{F} = \left\{ f \;\middle|\; |H_{\mathrm{C}}(f)| \geq |H_{\mathrm{C}}(f - \tfrac{\mu}{T})| \,,\ \mu \in \mathbb{Z} \right\} . \tag{2.3.120}$$

The remaining optimization problem is then paraphrased by

- Minimize

$$\exp\left\{ T \int_{f\in\mathcal{F}} \log\left(\frac{T\sigma_a^2}{N_0} |H_{\mathrm{T}}(f) H_{\mathrm{C}}(f)|^2 + 1 \right) \mathrm{d}f \right\} - 1 \,, \tag{2.3.121}$$

- subject to the additional constraint

$$S = \sigma_a^2 T \int_{f\in\mathcal{F}} |H_{\mathrm{T}}(f)|^2 \,\mathrm{d}f = \text{const} \,. \tag{2.3.122}$$

For optimization we can drop the "-1" in (2.3.121) and, since $\exp\{x\}$ is strictly monotonic increasing, only regard the argument x. With a Lagrange multiplier λ, the following real-valued Lagrange function can be set up

$$L = T \int_{f\in\mathcal{F}} \log\left(\frac{T\sigma_a^2}{N_0} |H_{\mathrm{T}}(f)|^2 \, |H_{\mathrm{C}}(f)|^2 + 1 \right) \mathrm{d}f - \lambda \cdot \sigma_a^2 T \int_{f\in\mathcal{F}} |H_{\mathrm{T}}(f)|^2 \,\mathrm{d}f \,. \tag{2.3.123}$$

For determination of the optimum transmit spectrum $|H_{\mathrm{T}}(f)|^2$, we add a real-valued deviation $\varepsilon V(f)$ to the optimal solution and take the partial derivative of L with respect to ε. This results in

$$\begin{aligned}
\left.\frac{\partial L}{\partial \varepsilon}\right|_{\varepsilon=0} &= \frac{\partial}{\partial \varepsilon}\Bigg(T \int_{f\in\mathcal{F}} \log\left(\frac{T\sigma_a^2}{N_0} \left(|H_{\mathrm{T}}(f)|^2 + \varepsilon V(f) \right) |H_{\mathrm{C}}(f)|^2 + 1 \right) \mathrm{d}f \\
&\qquad \left. - \lambda \cdot \sigma_a^2 T \int_{f\in\mathcal{F}} \left(|H_{\mathrm{T}}(f)|^2 + \varepsilon V(f) \right) \mathrm{d}f \Bigg) \right|_{\varepsilon=0} \\
&= \Bigg(T \int_{f\in\mathcal{F}} \frac{\frac{T\sigma_a^2}{N_0} V(f) \cdot |H_{\mathrm{C}}(f)|^2}{\frac{T\sigma_a^2}{N_0} \left(|H_{\mathrm{T}}(f)|^2 + \varepsilon V(f) \right) \cdot |H_{\mathrm{C}}(f)|^2 + 1} \\
&\qquad \left. - \lambda \sigma_a^2 T \int_{f\in\mathcal{F}} V(f) \,\mathrm{d}f \Bigg) \right|_{\varepsilon=0}
\end{aligned}$$

$$= T \int_{f\in\mathcal{F}} \left(\frac{|H_{\rm C}(f)|^2}{|H_{\rm T}(f)|^2\,|H_{\rm C}(f)|^2 + \frac{N_0}{T\sigma_a^2}} - \lambda\sigma_a^2 T \right) V(f)\,\mathrm{d}f \overset{!}{=} 0\,. \tag{2.3.124}$$

Since this equation has to be valid for all functions $V(f)$, the optimal transmit spectrum has to satisfy the following equation

$$\sigma_a^2 T\,|H_{\rm T}(f)|^2 + \frac{N_0}{|H_{\rm C}(f)|^2} = \text{const}\,, \qquad \forall f\in\mathcal{F}\,. \tag{2.3.125}$$

The constant is determined by regarding the additional constraint (2.3.122). Notice that equation (2.3.125) gives the optimal transmit PSD $\bar{\Phi}_{ss}(f) = \sigma_a^2 T\,|H_{\rm T}(f)|^2$ as the *water-pouring* or *water-filling solution* [Gal68], known from information theory. Transmit PSD and equivalent noise spectrum $\frac{N_0}{|H_{\rm C}(f)|^2}$ have to add up to a constant within the support $\mathcal{F}$.

It should finally be remarked that water pouring can result in a support of the transmit filter which is *not* a full-sized Nyquist interval. This is because the symbol spacing T has not yet been optimized. Performing this optimization will lead to a water-filling spectrum within a generalized Nyquist interval of width $1/T_{\rm opt}$.

Example 2.10: *Optimal Transmit Filter for MMSE-DFE*

We now turn back to the channel of Example 2.3. There, the optimal transmit filter for zero-forcing linear equalization was given. Here, we address the problem of finding the optimal transmit spectrum for MMSE-DFE.

First, in Figure 2.40 the magnitude of the channel is repeated and the support of the transmit filter (set $\mathcal{F}$ of frequencies), which is identical to that of optimal zero-forcing linear equalization, is given.

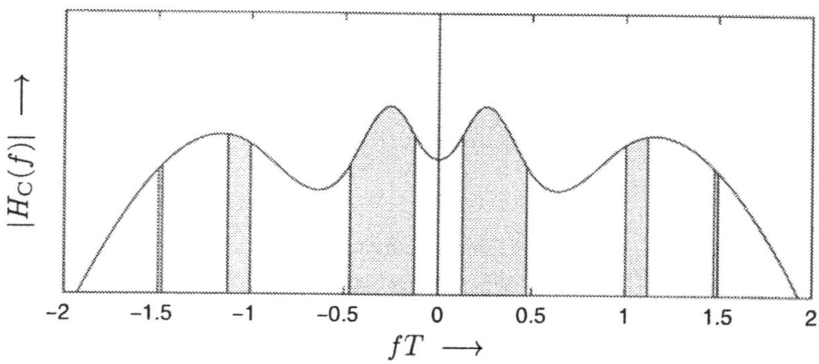

Fig. 2.40 Magnitude of $|H_{\rm C}(f)|$ of the channel and set $\mathcal{F}$ of frequencies (shaded).

The water-filling solution

$$|H_{\rm T}(f)|^2 + \frac{1}{|H_{\rm C}(f)|^2} = c\,, \qquad \forall f\in\mathcal{F}$$

within the support $\mathcal{F}$ is shown in Figure 2.41. In the left column, the squared reciprocal magnitude of the channel is plotted and the "water level" c is indicated. The difference between the constant c and the squared reciprocal magnitude gives the transmit power spectral density, which is shaded and repeated in the right column. From top to bottom, the transmit power is gradually increased. Please note the different scalings.

Noticeably, for low transmit power, a full Nyquist interval is not obtained. This is due to the nonoptimized symbol spacing T. As the total transmit power increases, the optimal PSD tends to become an on-off keyed shape—the transmit power is uniformly distributed within the support of the transmit filter.

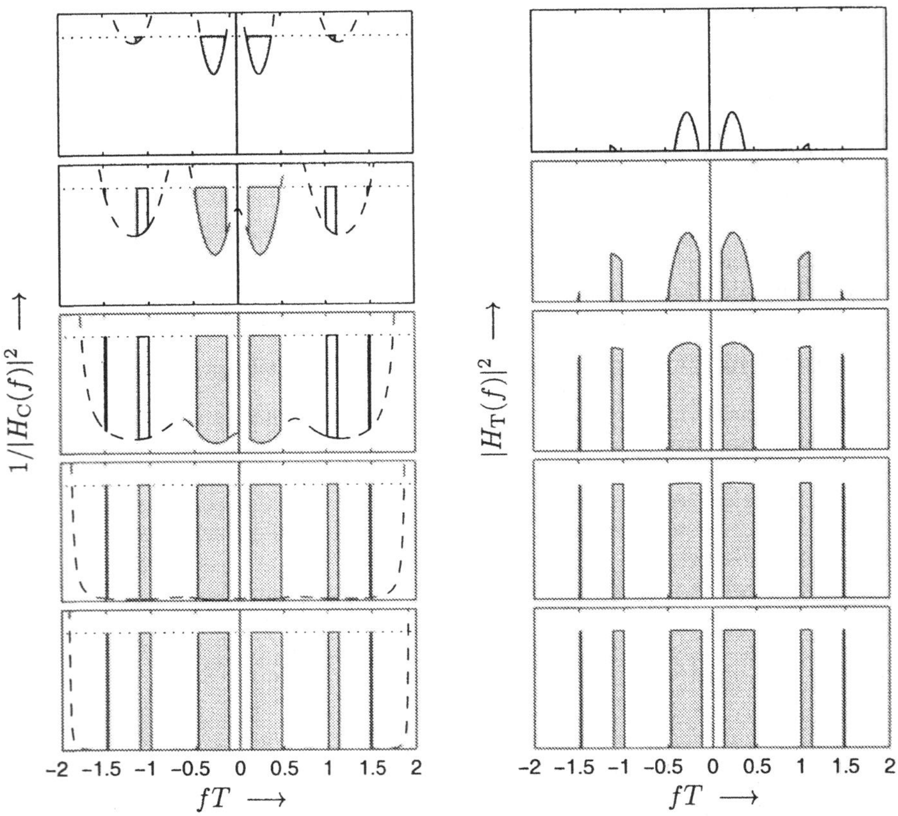

Fig. 2.41 Left column: squared reciprocal magnitude of the channel (dashed) and water-filling solution (shaded). Right column: optimal transmit power spectral density. Top to bottom: gradually increasing transmit power.

MMSE-DFE and Channel Capacity To conclude this section on MMSE-DFE, we generalize a result given above. From (2.3.117), the signal-to-noise ratio can now be written as

$$\mathrm{SNR}^{(\mathsf{MMSE\text{-}DFE,u})} = \exp\left\{ T \int\limits_{f\in\mathcal{F}} \log\left(\frac{T\sigma_a^2}{N_0} \left|H_{\mathrm{T}}(f)H_{\mathrm{C}}(f)\right|^2 + 1 \right) \,\mathrm{d}f \right\} - 1 \,, \tag{2.3.126}$$

or, after minor manipulations,

$$\log_2\left(1 + \mathrm{SNR}^{(\mathsf{MMSE\text{-}DFE,u})}\right) = T \int\limits_{f\in\mathcal{F}} \log_2\left(1 + \mathrm{SNR}(f)\right) \,\mathrm{d}f \,. \tag{2.3.127}$$

Finally, regard the fact that for an optimized symbol spacing, the transmit spectrum which maximizes the SNR is water-pouring and coincides with the water-pouring spectrum for maximizing mutual information, and thus achieves channel capacity [CDEF95]. Hence, the identity (2.3.127) states that the capacity[13] usable by unbiased MMSE-DFE and Gaussian transmit symbols is equal to the capacity of the underlying channel, i.e.,

$$C_{\text{MMSE-DFE,u}} = C_{\text{underlying channel}} \,, \tag{2.3.128}$$

which generalizes the asymptotic results of Section 2.3.2 concerning ZF-DFE to all SNRs. We conclude that MMSE decision-feedback equalization is optimum in the sense of information theory. A much more detailed discussion on this topic can be found in [CDEF95].

[13]To be precise, here the fact that for MMSE-DFE the additive error sequence (noise plus residual intersymbol interference) is not exactly Gaussian has been ignored.

2.4 SUMMARY OF EQUALIZATION STRATEGIES AND DISCRETE-TIME CHANNEL MODELS

This section summarizes the equalization strategies and the resultant equivalent discrete-time representations of PAM transmission over linear, distorting channels that have been presented. Moreover, some extensions of these channel descriptions are discussed. In the next chapter, these models are the starting point for the development and discussion of precoding schemes for channel equalization.

2.4.1 Summary of Equalization Strategies

Table 2.1 on pages 98 and 99 is a compilation of the equalization srategies developed in the last sections. Given the transmit filter $H_{\mathrm{T}}(f)$ and the underlying continuous-time channel with transfer function $H_{\mathrm{C}}(f)$ and additive white noise (two-sided PSD N_0 in the equivalent complex baseband domain—for baseband transmission N_0 has to be replaced by $N_0/2$, cf. Section 2.1), the respective receive filter $H_{\mathrm{R}}^{(\cdot)}(f)$ is given, and the end-to-end discrete-time signal transfer function

$$H^{(\cdot)}(e^{j2\pi fT}) = \sum_{\mu} H_{\mathrm{T}}(f-\tfrac{\mu}{T})H_{\mathrm{C}}(f-\tfrac{\mu}{T})H_{\mathrm{R}}^{(\cdot)}(f-\tfrac{\mu}{T}) \tag{2.4.1a}$$

and PSD of the additive discrete-time noise sequence

$$\Phi_{nn}^{(\cdot)}(e^{j2\pi fT}) = \frac{N_0}{T}\sum_{\mu}\left|H_{\mathrm{R}}^{(\cdot)}(f-\tfrac{\mu}{T})\right|^2 \tag{2.4.1b}$$

are stated. The superscript $(\cdot)$ denotes the respective equalization strategy. Additionally, the equation for calculation of the signal-to-noise ratio is given, and for decision-feedback equalization the required feedback filter $B(z)-1$ is noted.

The discrete-time channel models for the following receiver structures are of special interest:

(i) The matched-filter front-end, which is the starting point for various other equalization strategies and gives a reference (matched-filter bound),

(ii) Zero-forcing linear equalization as starting point for noise prediction, and

(iii) Noise prediction, which offers an optimal exchange of prediction gain versus the length of the impulse response of the feedforward filter in the equivalent ZF-DFE structure.

2.4.2 IIR Channel Models

Obviously, the channel model for noise prediction, which is calculated via the Yule-Walker equations is given by a *finite impulse response (FIR)* filter. But also for infinite-length DFE, the end-to-end representation including a whitened matched filter is obtained (from spectral factorization) in terms of the impulse response, i.e., a tapped-delay-line model. As for a stable impulse response, its amplitude has to fade away over time, an FIR model of sufficient order is always a good approximation (cf. Example 2.5 on page 65). But in some situations, it is more convenient to have a channel model, which we can think of as being realized by an *infinite impulse response (IIR)* filter. All-pole descriptions are of especial interest, because here equalization can be done using an FIR filter—we discuss this topic in the next chapter.

Now, methods for finding an appropriate IIR discrete-time channel model for PAM transmission over linear, dispersive channels are presented. As we have seen, in decision-feedback equalization, regardless of the actual receiver front-end, the discrete-time part always has to serve as a noise whitening filter. Hence, we now concentrate on this property.

All-Pole Model The problem of finding an all-pole (hence IIR) whitening filter is much more intricate than calculating an all-zero (FIR) filter, and it is equivalent to moving average spectral estimation. Here, we start from the FIR noise whitening filter (prediction-error filter) given in (2.3.17) and now denoted by $W(z)$. Using a slightly modified *Prony's method* (e.g., [MK93]), $W(z)$ is approximated by an all-pole filter $1/C(z)$, which is the desired all-pole whitening filter. This approach leads to an identical procedure to the approximated maximum-likelihood moving average estimation given by Durbin (see [Kay88]).

Let $W(z) = 1 + \sum_{k=1}^{q} w[k]z^{-k}$ be a nonrecursive, monic whitening filter of order q calculated from (2.3.14) and $C(z) = 1 + \sum_{k=1}^{p} c[k]z^{-k}$, the denominator of the desired system $1/C(z)$ of order p, for which

$$\frac{1}{C(z)} \approx W(z) \qquad \text{or} \qquad W(z) \cdot C(z) \approx 1 \tag{2.4.2}$$

should be valid. Hence, all coefficients, except the leading 1, of the product $W(z) \cdot C(z)$ have to vanish. Using the definitions of $W(z)$ and $C(z)$ and comparing the coefficients of the powers of z^{-1} of the product $W(z) \cdot C(z)$, the following over-determined system of equations for the $c[k]$'s, $k = 1, 2, \ldots, p$, results

$$\underbrace{\begin{bmatrix} w[1] \\ \vdots \\ w[q] \\ 0 \\ \vdots \\ 0 \end{bmatrix}}_{\boldsymbol{w}} + \underbrace{\begin{bmatrix} 1 & & 0 \\ w[1] & \ddots & \\ \vdots & \ddots & \\ w[q] & & 1 \\ & \ddots & w[1] \\ & & \vdots \\ 0 & & w[q] \end{bmatrix}}_{\boldsymbol{W}} \cdot \underbrace{\begin{bmatrix} c[1] \\ \vdots \\ c[p] \end{bmatrix}}_{\boldsymbol{c}} = \underbrace{\begin{bmatrix} f[1] \\ \vdots \\ f[q+p] \end{bmatrix}}_{\boldsymbol{f}} . \tag{2.4.3}$$

Here, $f[k]$ denotes the approximation error. Instead of minimizing the difference between $W(z)$ and $1/C(z)$, as usual, the squared error $\sum_{k=1}^{q+p} |f[k]|^2$ is minimized,

Table 2.1 Summary of equalization strategies and equivalent discrete-time channel models.

	Model	$H_{\mathrm{R}}^{(\cdot)}(f)$	$H^{(\cdot)}(\mathrm{e}^{\mathrm{j}2\pi fT})$
	AWGN	—	1
	Matched Filter (MF)	$H_{\mathrm{T}}^*(f)H_{\mathrm{C}}^*(f)$	$\sum_\mu \left\lvert H_{\mathrm{T}}(f-\frac{\mu}{T})H_{\mathrm{C}}(f-\frac{\mu}{T})\right\rvert^2$
Linear Equalization	ZF (ZF-LE)	$\frac{H_{\mathrm{T}}^*(f)H_{\mathrm{C}}^*(f)}{\sum_\mu \left\lvert H_{\mathrm{T}}(f-\frac{\mu}{T})H_{\mathrm{C}}(f-\frac{\mu}{T})\right\rvert^2}$	1
Linear Equalization	ZF, opt. (LE,opt)	$H_{\mathrm{R}}^{(\mathrm{ZF\text{-}LE})}(f)$, with $\lvert H_{\mathrm{T}}(f)\rvert = \frac{\mathrm{const}}{\sqrt{\lvert H_{\mathrm{C}}(f)\rvert}},\ f \in \mathcal{F}$	1
Linear Equalization	MMSE (MMSE-LE)	$\frac{H_{\mathrm{T}}^*(f)H_{\mathrm{C}}^*(f)}{\sum_\mu \left\lvert H_{\mathrm{T}}(f-\frac{\mu}{T})H_{\mathrm{C}}(f-\frac{\mu}{T})\right\rvert^2 + \frac{N_0}{T\sigma_a^2}}$	$\frac{\sum_\mu \left\lvert H_{\mathrm{T}}(f-\frac{\mu}{T})H_{\mathrm{C}}(f-\frac{\mu}{T})\right\rvert^2}{\sum_\mu \left\lvert H_{\mathrm{T}}(f-\frac{\mu}{T})H_{\mathrm{C}}(f-\frac{\mu}{T})\right\rvert^2 + \frac{N_0}{T\sigma_a^2}}$
Decision-Feedback Equalization	Noise Predict. (NP)	$H_{\mathrm{R}}^{(\mathrm{ZF\text{-}LE})}(f)$, followed by $H^{(\mathrm{NP})}(z)$	$1 - \sum_{k=1}^{p} \rho[k] z^{-k}$, $\rho[k]$ via Yule-Walker eqs. (2.3.14) Feedback Filter: $B(z) = H(z)$
Decision-Feedback Equalization	ZF (ZF-DFE)	$\left\lvert H_{\mathrm{R}}^{(\mathrm{WMF})}(f)\right\rvert^2 =$ $\frac{\sigma_a^2}{\sigma_g^2}\frac{\lvert H_{\mathrm{T}}(f)H_{\mathrm{C}}(f)\rvert^2}{\sum_\mu \left\lvert H_{\mathrm{T}}(f-\frac{\mu}{T})H_{\mathrm{C}}(f-\frac{\mu}{T})\right\rvert^2}$	$G(z) = 1 + \sum_{k>0} g[k]z^{-k}$, with $\sigma_g^2 G(z)G^*(z^{-*}) = \sigma_a^2 H^{(\mathrm{MF})}(z)$ Feedback Filter: $B(z) = G(z)$
Decision-Feedback Equalization	MMSE (MMSE-DFE)	$\left\lvert H_{\mathrm{R}}^{(\mathrm{MS\text{-}WMF})}(f)\right\rvert^2 =$ $\frac{\sigma_a^2}{\sigma_g^2}\frac{\lvert H_{\mathrm{T}}(f)H_{\mathrm{C}}(f)\rvert^2}{\sum_\mu \left\lvert H_{\mathrm{T}}(f-\frac{\mu}{T})H_{\mathrm{C}}(f-\frac{\mu}{T})\right\rvert^2 + \frac{N_0}{T\sigma_a^2}}$	$G(z) - \frac{N_0}{T\sigma_g^2}\frac{1}{G^*(z^{-*})}$, with $\sigma_g^2 G(z)G^*(z^{-*}) = \sigma_a^2 H^{(\mathrm{MF})}(z) + \frac{N_0}{T}$ Feedback Filter: $B(z) = G(z)$

$\Phi_{nn}^{(\cdot)}(e^{j2\pi fT})$	$\mathrm{SNR}^{(\cdot)}$
$\sigma_n^2 = \text{const}$	—
$\frac{N_0}{T} \sum_\mu \left\lvert H_T(f - \frac{\mu}{T}) H_C(f - \frac{\mu}{T}) \right\rvert^2$	$T \int_{-\frac{1}{2T}}^{\frac{1}{2T}} \widetilde{\mathrm{SNR}}(e^{j2\pi fT})\, df$ *Arithmetic* Mean Over $\widetilde{\mathrm{SNR}}$
$\frac{N_0}{T} \frac{1}{\sum_\mu \left\lvert H_T(f-\frac{\mu}{T}) H_C(f-\frac{\mu}{T}) \right\rvert^2}$	$\frac{1}{T \int_{-\frac{1}{2T}}^{\frac{1}{2T}} \left(\widetilde{\mathrm{SNR}}(e^{j2\pi fT})\right)^{-1} df}$ *Harmonic* Mean Over $\widetilde{\mathrm{SNR}}$
$\frac{N_0}{T} \lambda^2 \sum_\mu \frac{1}{\lvert H_C(f-\frac{\mu}{T}) \rvert}$ $= \frac{N_0}{T} \lambda^2 \frac{1}{\lvert H_C(\psi(f)) \rvert}$	$\frac{\sigma_a^2}{N_0 \int_{\mathcal{F}} \frac{\lambda^2}{\lvert H_C(f) \rvert} df}$
$\frac{N_0}{T} \frac{1}{\sum_\mu \left\lvert H_T(f-\frac{\mu}{T}) H_C(f-\frac{\mu}{T}) \right\rvert^2 + \frac{N_0}{T\sigma_a^2}}$	$\frac{1}{T \int_{-\frac{1}{2T}}^{\frac{1}{2T}} \left(\widetilde{\mathrm{SNR}}(e^{j2\pi fT}) + 1\right)^{-1} df}$
$\frac{N_0}{T} \frac{\left\lvert 1 - \sum_{k=1}^{p} b[k] e^{-j2\pi fTk} \right\rvert^2}{\sum_\mu \left\lvert H_T(f-\frac{\mu}{T}) H_C(f-\frac{\mu}{T}) \right\rvert^2}$	—
$\frac{N_0}{T} \frac{\sigma_a^2}{\sigma_g^2} = \text{const}$	$\exp\left\{ T \int_{-\frac{1}{2T}}^{\frac{1}{2T}} \log\left(\widetilde{\mathrm{SNR}}(e^{j2\pi fT})\right) df \right\}$ *Geometric* Mean Over $\widetilde{\mathrm{SNR}}$
$\frac{N_0}{T} \frac{\sigma_a^2}{\sigma_g^2} = \text{const}$	$\exp\left\{ T \int_{-\frac{1}{2T}}^{\frac{1}{2T}} \log\left(\widetilde{\mathrm{SNR}}(e^{j2\pi fT}) + 1\right) df \right\}$

i.e., we are looking for the least-squares solution. According to [BS98] the optimal coefficient vector $\boldsymbol{c}$ must satisfy the equation:

$$\boldsymbol{W}^{\mathsf{H}}\boldsymbol{W}\cdot\boldsymbol{c} = -\boldsymbol{W}^{\mathsf{H}}\cdot\boldsymbol{w}\,. \tag{2.4.4}$$

With the definition $w[k] = 0$, for $k < 0$ and $k > q$, and $w[0] = 1$, the products of matrices or vectors in (2.4.4) can be simplified to:

$$\begin{aligned}\boldsymbol{W}^{\mathsf{H}}\boldsymbol{W} &= \Big[w^*[m-l]\Big]_{\substack{l=1,\ldots,p\\ m=1,\ldots,q+p}}\cdot\Big[w[m-n]\Big]_{\substack{m=1,\ldots,q+p\\ n=1,\ldots,p}} \\ &= \Big[\sum_{m=0}^{q+p} w^*[m-l]\cdot w[m-n]\Big]_{\substack{l=1,\ldots,p\\ n=1,\ldots,p}} = \Big[\varphi_{ww}[l-n]\Big]_{\substack{l=1,\ldots,p\\ n=1,\ldots,p}}\,,\end{aligned} \tag{2.4.5a}$$

and

$$\begin{aligned}\boldsymbol{W}^{\mathsf{H}}\boldsymbol{w} &= \Big[w^*[m-l]\Big]_{\substack{l=1,\ldots,p\\ m=1,\ldots,q+p}}\cdot\Big[w[m]\Big]_{m=1,\ldots,q+p} \\ &= \Big[\sum_{m=0}^{q+p} w^*[m-l]\cdot w[m]\Big]_{l=1,\ldots,p} = \Big[\varphi_{ww}[l]\Big]_{l=1,\ldots,p}\,,\end{aligned} \tag{2.4.5b}$$

where $\varphi_{ww}[\kappa] = \sum_k w[k+\kappa]\cdot w^*[k]$, the system autocorrelation of $\langle w[k]\rangle$ has been used. From (2.4.3), the Yule-Walker equations (cf. (2.3.14)) result, now applied to a "noise sequence" with autocorrelation sequence $\varphi_{ww}[\kappa]$:

$$\sum_{l=1}^{p} c[l]\cdot\varphi_{ww}[k-l] = -\varphi_{ww}[k], \qquad k = 1,2,\ldots,p\,. \tag{2.4.6}$$

Because the Yule-Walker equations result in a strictly minimum-phase filter $C(z)$ (see Section 2.3.1 or [Tre76]), contrary to Prony's method in general, it is guaranteed that $H(z) = 1/C(z)$ can be implemented by a stable filter.

To summarize, an all-pole noise whitening filter can be calculated by the following procedure:

- The channel noise is assumed to be an autoregressive process. Applying the Yule-Walker equations, an FIR whitening filter $W(z)$ of sufficient order is determined.
- Then, $W(z)$ is regarded as an innovations filter for a new noise process (PSD proportional to $|W(e^{j2\pi fT})|^2$). For this process, once again using the Yule-Walker equations, the respective FIR whitening filter $C(z)$ is obtained.
- Finally, $H(z) = 1/C(z)$ is the all-pole noise whitening filter for the channel noise.

Thus, for computation of $H(z)$, the Yule-Walker equations are applied twice. In [Kay88] it is proved that this procedure is optimum in the sense of estimation theory.

Example 2.11: Comparison of FIR and IIR Noise Whitening Filters

Figure 2.42 sketches the impulse responses for FIR and IIR whitening filters, respectively, and for different orders p. Cable and noise parameters are again those of the DSL up-stream example. The calculation of $1/C(z)$ was based on a FIR whitening filter of order $q = 10$.

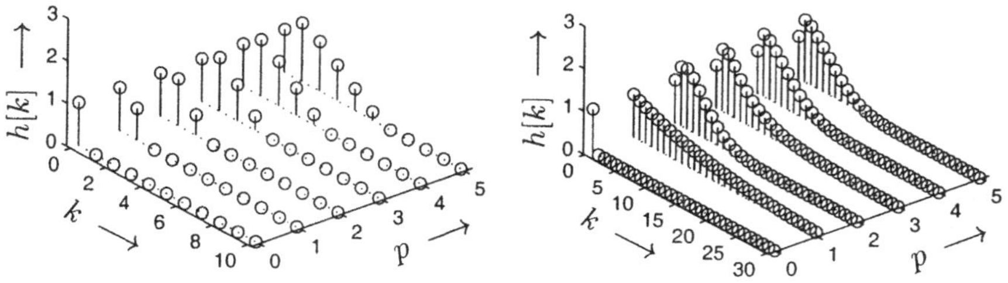

Fig. 2.42 Impulse responses for FIR (left) and IIR (right) whitening filters over the order p of the whitening filter.

Additionally, in Figure 2.43 the prediction gain G_{p} (in dB) is plotted over the order p of the whitening filter. For comparison, the curve using an FIR filter (cf. Figure 2.18) is repeated. Since the calculation of $1/C(z)$ was based on a FIR whitening filter of order $q = 10$, the gains are limited to this region. Because for small orders p the impulses already resemble the optimum one (using the whitened matched filter), the gains are superior compared to those of nonrecursive filters. Using a recursive noise whitening filter, orders 3 or 4 are sufficient.

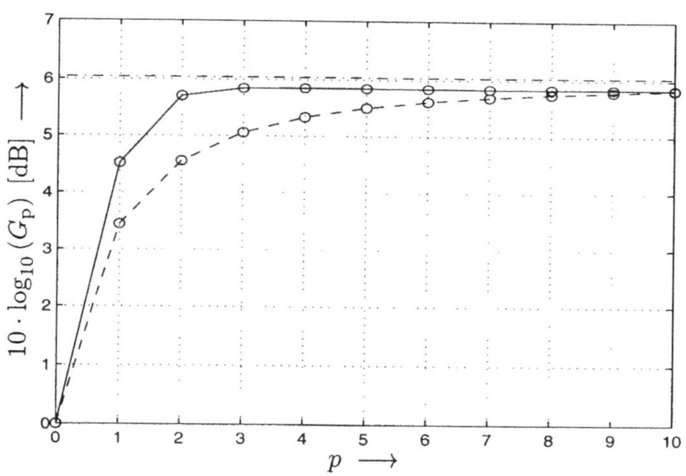

Fig. 2.43 Prediction gain over the order p of the whitening filter. Solid: IIR; dashed: FIR.

Pole-Zero Model In some situations, a discrete-time channel model with poles and zeros is preferred over an all-pole model. Using such an IIR filter allows to approximate notches or even spectral zeros much closer. Now, following the Appendix of [CEGH98], we sketch a possible way of finding a pole-zero noise whitening filter. The derivation and the solution are very similar to finite-length MMSE-DFE in Section 2.3.3.

Let $W(z) = 1 + \sum_k w[k]z^{-k}$ again be the given FIR noise whitening filter (prediction-error filter) which should be approximated by the IIR model with transfer function

$$H(z) = \frac{1 + \beta_1 z^{-1} + \cdots + \beta_\mu z^{-\mu}}{1 + \alpha_1 z^{-1} + \cdots + \alpha_\nu z^{-\nu}} \,. \tag{2.4.7}$$

For a causal impulse response, we require $\nu \le \mu$ [Sch94].

The initial noise sequence $\langle n[k] \rangle$ is filtered by $H(z)$ and results in the prediction-error sequence $\langle e[k] \rangle$, which is given by the following difference equation

$$e[k] = n[k] + \beta_1 n[k-1] + \cdots + \beta_\mu n[k-\mu] - \alpha_1 e[k-1] - \cdots - \alpha_\nu e[k-\nu] \,. \tag{2.4.8}$$

With the definitions of a pole-zero vector

$$\boldsymbol{\zeta} = \begin{bmatrix} -\beta_1^* & -\beta_2^* & \cdots & -\beta_\mu^* & \alpha_1^* & \alpha_2^* & \cdots & \alpha_\nu^* \end{bmatrix}^\mathsf{T} \tag{2.4.9a}$$

and a state vector

$$\boldsymbol{\theta}[k] = \begin{bmatrix} n[k-1] & \cdots & n[k-\mu] & e[k-1] & \cdots & e[k-\nu] \end{bmatrix}^\mathsf{T} , \tag{2.4.9b}$$

the error signal (2.4.8) can be expressed compactly as

$$e[k] = n[k] - \boldsymbol{\zeta}^\mathsf{H} \boldsymbol{\theta}[k] \,. \tag{2.4.10}$$

As we want to approximate a noise whitening filter, the aim of the system optimization can also be paraphrased by

$$\mathsf{E}\{|e[k]|^2\} \to \min \,. \tag{2.4.11}$$

The solution to this MMSE criterion reads (cf. Sections 2.2.3 and 2.3.3)

$$\boldsymbol{\Phi}\boldsymbol{\zeta} = \boldsymbol{\phi} \,, \tag{2.4.12}$$

where $\boldsymbol{\Phi}$ is a partitioned matrix

$$\boldsymbol{\Phi} \triangleq \mathsf{E}\left\{\boldsymbol{\theta}[k]\boldsymbol{\theta}^\mathsf{H}[k]\right\} \triangleq \begin{bmatrix} \boldsymbol{\Phi}_{nn} & \boldsymbol{\Phi}_{ne} \\ \boldsymbol{\Phi}_{ne}^\mathsf{H} & \boldsymbol{\Phi}_{ee} \end{bmatrix} , \tag{2.4.13a}$$

and $\boldsymbol{\phi}$ a partitioned column vector

$$\boldsymbol{\phi} \triangleq \mathsf{E}\left\{\boldsymbol{\theta} y^*[k]\right\} \triangleq \begin{bmatrix} \boldsymbol{\phi}_{nn} \\ \boldsymbol{\phi}_{en} \end{bmatrix} . \tag{2.4.13b}$$

For evaluation of $\boldsymbol{\Phi}$ and $\boldsymbol{\phi}$, a problem arises. Since we want to calculate the filter coefficients α_i and β_i, the autocorrelations $\varphi_{ee}[\kappa]$ and the cross-correlations $\varphi_{ne}[\kappa]$ are not known a priori. But as the filter $H(z)$ should approximate the FIR noise whitening filter $W(z)$, we can use this filter for estimating the respective correlations. The components of the matrix $\boldsymbol{\Phi}$ and the vector $\boldsymbol{\phi}$ can then be given as

$$\boldsymbol{\Phi}_{nn} = [\phi_{nn}[k-\kappa]]_{\substack{\kappa=1,\ldots,\mu \\ k=1,\ldots,\mu}} \tag{2.4.14a}$$

$$\boldsymbol{\Phi}_{ee} = [\phi_{ee}[k-\kappa]]_{\substack{\kappa=1,\ldots,\mu \\ k=1,\ldots,\mu}} \approx \sigma_e^2 \boldsymbol{I}_{\nu\times\nu} \tag{2.4.14b}$$

$$\boldsymbol{\Phi}_{ne} = [\phi_{ne}[k-\kappa]]_{\substack{\kappa=1,\ldots,\mu \\ k=1,\ldots,\mu}} \approx \left[\sum\nolimits_{i=0}^{p} w[i]\phi_{nn}[k-\kappa+i]\right]_{\substack{\kappa=1,\ldots,\mu \\ k=1,\ldots,\nu}} \tag{2.4.14c}$$

$$\boldsymbol{\phi}_{nn} = [\phi_{nn}[\kappa]]_{\kappa=1,\ldots,\mu} \tag{2.4.14d}$$

$$\boldsymbol{\phi}_{en} = [\phi_{en}[\kappa]]_{\kappa=1,\ldots,\nu} \approx \left[\sum\nolimits_{i=0}^{p} w[i]\phi_{nn}[\kappa+i]\right]_{\kappa=1,\ldots,\nu} . \tag{2.4.14e}$$

Finally, it should be remarked that in the literature various other strategies for approximating an FIR model by an IIR filter can be found, e.g., [ASC97].

2.4.3 Channels with Spectral Nulls

There are scenarios where the channel completely suppresses isolated frequency components. An example is transformer coupling of twisted pair lines which severely attenuates signal components at or close to DC. Moreover, high-pass filtering at the receiver may be desired if low-frequency noise (e.g., impulse noise) is present. In order to take such effects into account, we now extend our discrete-time channel model appropriately.

Continuous-Time Modeling First we consider the effect of spectral zeros in the cascade of transmit filter $H_{\mathrm{T}}(f)$ and continuous-time channel $H_{\mathrm{C}}(f)$. Since after matched filtering and sampling the periodic sum $\sum_\mu |H_{\mathrm{T}}(f-\frac{\mu}{T})H_{\mathrm{C}}(f-\frac{\mu}{T})|^2$ is effective, isolated zeros are of no interest. But if $H_{\mathrm{T}}(f)H_{\mathrm{C}}(f)$ has periodic zeros with period equal to $1/T$, the PSD

$$\Phi_{ff}(\mathrm{e}^{\mathrm{j}2\pi fT}) = \sigma_a^2 \sum_\mu \left|H_{\mathrm{T}}(f-\tfrac{\mu}{T})H_{\mathrm{C}}(f-\tfrac{\mu}{T})\right|^2 , \tag{2.4.15}$$

which has to be factored for calculating the whitened matched filter, has spectral zeros, each of even multiplicity. Then the Paley-Wiener condition is not satisfied, i.e., the integral (2.3.31) diverges, and a stable spectral factorization—hence the zero-forcing whitened matched filter—does not exist. Such cases occur, e.g., for DC coupling and rectangular time-domain transmit pulses, or pulses which are band-limited to an interval smaller than $2/T$.

If the MS-WMF (2.3.110), which can always be implemented in a stable way, is not wanted, the solution to the above problem is to treat the zeros separately [For72].

Therefore, we combine the periodic zeros into the (monic) polynomial

$$P(z) = \prod_i (1 - z_{0,i} z^{-1}) , \qquad |z_{0,i}| = 1 , \tag{2.4.16}$$

and dissect the continuous-time end-to-end transfer function in the following way

$$H_\mathrm{T}(f) H_\mathrm{C}(f) \triangleq P(\mathrm{e}^{\mathrm{j}2\pi fT}) \cdot H'(f) . \tag{2.4.17}$$

Factorization is then performed with respect to the transfer function $H'(f)$, i.e.,

$$\sigma_a^2 \sum_\mu \left| H'(f - \tfrac{\mu}{T}) \right|^2 = \sigma_g^2 G'(\mathrm{e}^{\mathrm{j}2\pi fT}) G'^*(\mathrm{e}^{\mathrm{j}2\pi fT}) . \tag{2.4.18}$$

The receiver now consists of the matched filter for $H'(f)$, followed by T-spaced sampling and the discrete-time noise whitening filter $\frac{\sigma_a^2}{\sigma_g^2} \frac{1}{G'^*(z^{-*})}$.

Finally, the end-to-end discrete-time channel model is given by

$$H(z) = P(z) G'(z) , \tag{2.4.19}$$

and additive white Gaussian noise. Thus, because of $P(z)$, the discrete-time equivalent model has spectral zeros. Note, the polynomial $P(z)$ is usually regarded as performing a so-called *partial-response encoding* [Hub92b].

Discrete-Time Modeling Beside the strategy of splitting off a partial-response encoder from the channel, we now address an approach to modeling spectral zeros in the discrete-time domain. This can be done by directly forcing $H(z)$ to have a spectral null. For the calculation of such models, we restrict ourselves to a spectral zero at DC, and our starting point is zero-forcing linear equalization.

As in Section 2.3.1, we are now interested in the optimum noise whitening filter, but subject to the additional constraint of a spectral zero. The solution can be determined in two steps [Hub92a].

First, we assume the discrete-time noise sequence $\langle n[k] \rangle$ at the output of the optimal Nyquist filter to be passed through a system with transfer function $1 - z^{-1}$. This filter generates the desired spectral zero but increases noise power, too. The optimum prediction-error filter $H_0(z)$ of order p fitted to the new noise sequence $\langle n'[k] \rangle$ is again calculated by applying the Yule-Walker equations (2.3.14), where the autocorrelation $\phi_{nn}^{(\mathrm{ZF\text{-}LE})}[\kappa]$ is replaced by

$$\phi_{nn}^{(\mathrm{ZF\text{-}LE})}[\kappa] * \Big(-\delta[\kappa+1] + 2\delta[\kappa] - \delta[\kappa-1] \Big) . \tag{2.4.20}$$

In the second step, $H_0(z)$ and the spectral zero are combined into the channel model (consequently of order $p+1$)

$$H(z) = \left(1 - z^{-1}\right) \cdot H_0(z). \tag{2.4.21}$$

For simplicity—in spite of the spectral null—for sufficient order p, the residual noise sequence can be assumed to be white, because the width of the spectral zero tends to zero as p increases.

The above approach is only applicable if an FIR whitening filter is desired. To extend this strategy to IIR filters, we would have to drop the restriction to all-pole models. Here, starting from $H_0(z)$ an all-pole approximation $C(z)$ is determined as explained in Section 2.4.2. Finally, $H(z) = (1 - z^{-1})/C(z)$ is the desired IIR whitening filter.

Note that the present procedure offers both a channel model with a fixed spectral zero and an optimal exchange of prediction gain versus length of the impulse response. We later use these discrete-time models frequently in the context of precoding schemes.

Example 2.12: *Discrete-Time Spectral Zero at DC*

Again, we continue the example of the simplified DSL up-stream scenario. In Figure 2.44 the prediction gain G_p is plotted over the order p of the prediction-error filter. The solid line corresponds to the conventional application of the Yule-Walker equations, whereas the dashed line indicates that a discrete-time spectral zero has been forced at DC. We can see that the additional constraint of the spectral null leads to some performance loss. For $p \to \infty$ the width of the spectral zero tends to zero and the ultimate prediction gain $G_{\mathrm{p},\infty}$ can be achieved.

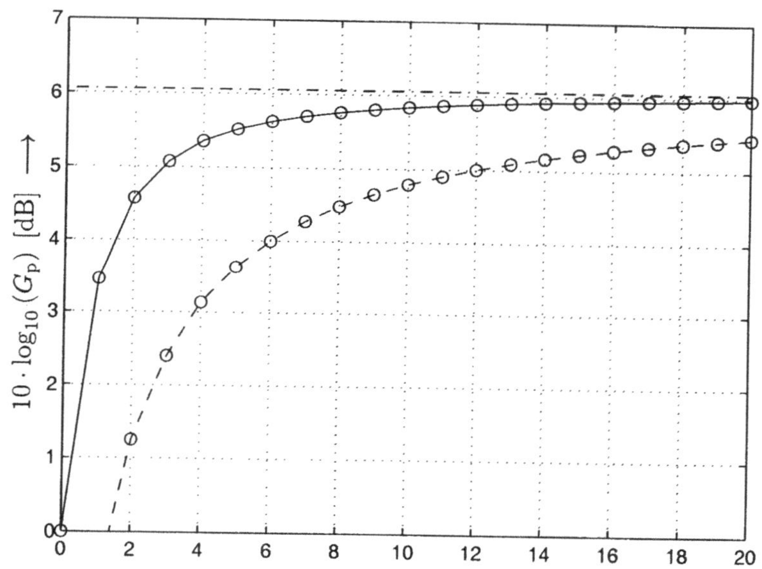

Fig. 2.44 Prediction gain over the order p of the whitening filter. Solid: conventional approach; dashed: discrete-time spectral zero forced at DC. Dashed-dotted: ultimate prediction gain $G_{\mathrm{p},\infty}$.

Figure 2.45 compares the impulse responses of the noise whitening filter for both cases. For achieving the DC zero, the sum over the taps has to vanish. Hence, for the DC-free impulses, the positive taps at the beginning have to be balanced by the tail of negative taps at the end. But as p increases, the first few taps of both cases become identical.

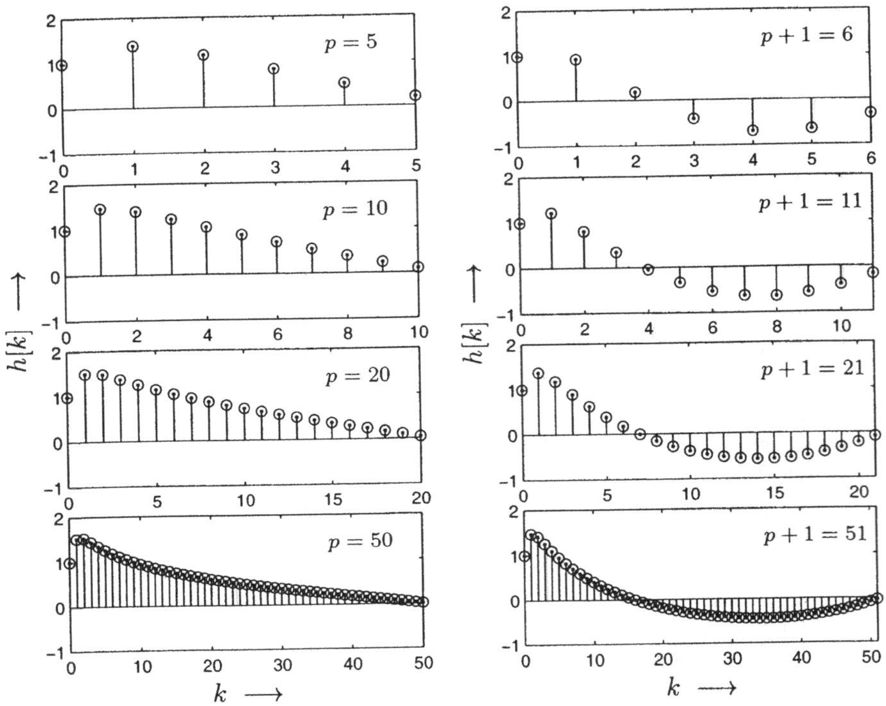

Fig. 2.45 Impulse responses of the noise whitening filter. Left column: conventional approach; right column: discrete-time spectral zero forced at DC. Top to bottom: prediction order $p = 5, 10, 20, 50$ ($+1$ for DC-free channels).

Finally, Figure 2.46, left column, shows the PSD of the noise after the whitening filter. The orders of the filters are $p = 5$, 10, 20, and 50. For reference, the PSD without prediction ($p = 0$), i.e., right after optimal Nyquist filtering, is plotted. The PSD is normalized to the variance $\sigma^2_{n,(\mathrm{NP})}$ (equation (2.3.27)), which is approached for infinite prediction order. In this case, a PSD constant to one (dotted line) results. It can be observed that increasing p results in an increasingly flat, i.e., white, PSD of the residual noise.

The right column of Figure 2.46 shows the noise PSDs for channels with a spectral zero at DC. The orders of the filters are again $p = 5$, 10, 20, and 50, and for reference, the PSD without prediction, but including the term $1 - z^{-1}$, is given. Again, the PSD is normalized to the variance $\sigma^2_{n'}$. As for increasing p, the width of the spectral zero tends to zero, the PSD converges to a constant, too.

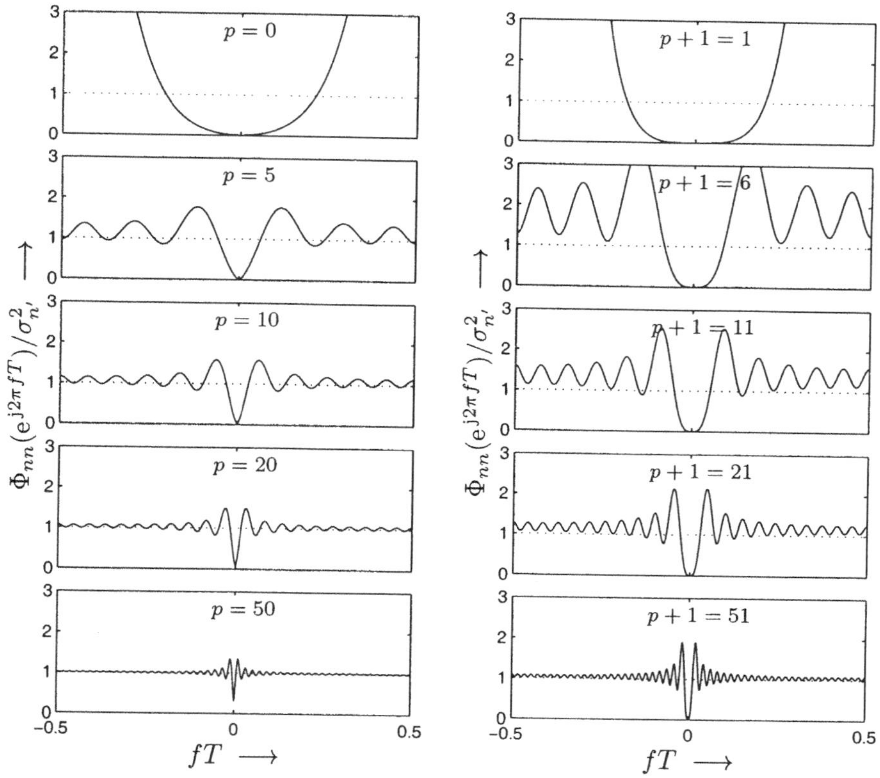

Fig. 2.46 Normalized PSD of the noise after the whitening filter. Dotted line: the asymptotic ($p \to \infty$) case ($\Phi_{nn}(e^{j2\pi fT})/\sigma_{n'}^2 = 1$). Left column: channels without spectral zero at DC; right column: channels with spectral zero at DC. Top to bottom: filter orders $p = 0$ (optimal Nyquist filtering, only) $5, 10, 20, 50$.

2.5 MAXIMUM-LIKELIHOOD SEQUENCE ESTIMATION

In the preceding discussion, we have improved system performance step-by-step by applying increasingly elaborate receiver concepts. All of the above structures have in common that a symbol-by-symbol decision is performed. In order to further improve performance, we now drop this strategy and regard whole sequences of symbols. In this book, only a brief look on *maximum-likelihood sequence estimation (MLSE)* is taken. A much more detailed treatment of this topic can be found, e.g., in [Pro01, Bla90, Hub92b] or in the original work of Forney [For72] and Ungerböck [Ung74].

2.5.1 Whitened-Matched-Filter Front-End

Consider digital transmission over a linear distorting channel and application of the (zero-forcing) whitened matched filter as receiver front-end. Then, as shown in Section 2.4, the end-to-end discrete-time channel model is given by its monic impulse response $h[k] \triangleq h^{(\mathrm{WMF})}[k]$ of some order p, and additive white Gaussian noise with variance σ_n^2.

This channel is driven by the input data sequence $\langle a[k]\rangle$, where the data symbols $a[k]$ are drawn from the signal set $\mathcal{A}$ of cardinality M. Thus, at the receiver the sequence $\langle y[k]\rangle$ with

$$y[k] = a[k] * h[k] + n[k] \,, \tag{2.5.1}$$

is observed, where $n[k]$ denotes the noise sample.

The task of *maximum-likelihood sequence estimation*[14] is to choose the transmit sequence $\langle a[k]\rangle$ as an estimate, for which the observed channel output $\langle y[k]\rangle$ is produced with the highest probability, i.e.,

$$\widehat{\langle a[k]\rangle} = \underset{\langle a[k]\rangle}{\operatorname{argmax}}\, \mathsf{f}_{\langle y[k]\rangle}\left(\langle y[k]\rangle \mid \langle a[k]\rangle\right) . \tag{2.5.2}$$

Here, $\mathsf{f}_{\cdot}(\cdot)$ denotes probability density functions. Since the noise samples at the output of the WMF are statistically independent, the logarithm of the likelihood function breaks up into the sum

$$\log \mathsf{f}_{\langle y[k]\rangle}\left(\langle y[k]\rangle \mid \langle a[k]\rangle\right) = \sum_k \log \mathsf{f}_{y[k]}\left(y[k] \mid \langle a[k]\rangle\right) . \tag{2.5.3}$$

Finally, if we consider the complex Gaussian density of the sample $y[k]$, and that the logarithm is strictly monotonic increasing, we arrive at

$$\widehat{\langle a[k]\rangle} = \underset{\langle a[k]\rangle}{\operatorname{argmax}} \sum_k \log \mathsf{f}_{y[k]}\left(y[k] \mid \langle a[k]\rangle\right)$$

[14]The denomination maximum-likelihood sequence *detection* would be more appropriate, because an MLSE algorithm indeed detects the sequence which has maximum probability [For96].

$$= \underset{\langle a[k]\rangle}{\operatorname{argmax}} \sum_k \log\left(\frac{1}{\pi\sigma_n^2}\cdot\exp\left\{-\frac{\left|y[k]-(a[\kappa]*h[\kappa])\big|_k\right|^2}{\sigma_n^2}\right\}\right),$$

and dropping all irrelevant terms, and considering $\operatorname{argmax}(-f(x)) = \operatorname{argmin}(f(x))$, yields

$$= \underset{\langle a[k]\rangle}{\operatorname{argmin}} \sum_k \left|y[k]-(a[\kappa]*h[\kappa])\big|_k\right|^2 . \tag{2.5.4}$$

In words, it turns out that MLSE searches for that transmit sequence $\langle a[k]\rangle$ for which the *squared Euclidean distance* between the corresponding noise-free channel output $\langle z[k]\rangle \triangleq \langle a[k]*h[k]\rangle$ and the observation $\langle y[k]\rangle$ is minimum. We will now show how to find this minimum with finite effort.

Because of the assumption of an FIR channel, the noiseless channel output $z[k]$ can be regarded as being generated by a finite-state machine [For73]. Since the tapped-delay line contains the p most recent input symbols $a[k-\kappa]$, $\kappa = 1, 2, \ldots, p$, and each sample can assume M different values, the machine has M^p states $S^{(i)}$, $i = 1, 2, \ldots, M^p$. Each state has M leaving branches, and M branches merge in each state. We define the state $S[k]$ at time k as the row vector of the p past symbols

$$S[k] \triangleq [a[k-1],\ a[k-2],\ \ldots,\ a[k-p]] . \tag{2.5.5}$$

Moreover, each input symbol $a[k]$ causes an output sample

$$\begin{aligned} z[k] &= a[k] + \sum_{\kappa=1}^{p} h[\kappa]a[k-\kappa] \\ &\triangleq \mathrm{f_o}(a[k], S[k]) \end{aligned} \tag{2.5.6a}$$

and a state transition to

$$\begin{aligned} S[k+1] &= [a[k],\ a[k-1],\ a[k-2],\ \ldots,\ a[k-p+1]] \\ &\triangleq \mathrm{f_s}(a[k], S[k]) . \end{aligned} \tag{2.5.6b}$$

The output function $\mathrm{f_o}(\cdot,\cdot)$ and the state transition function $\mathrm{f_s}(\cdot,\cdot)$ characterize the finite-state machine completely.

The *trellis diagram* of the finite-state machine graphically shows the evolution over time. At each time index k (horizontally arranged), all states are drawn vertically and all connections to preceding and succeeding states are shown. Figure 2.47 sketches an example for binary signaling and channel transfer function $H(z) = 1 + 0.5\,z^{-1} + 0.2\,z^{-2}$. The branches are labeled by the pair "$a[k] \mid z[k]$."

Now, let us turn back to the maximum-likelihood (ML) problem (2.5.4). Regarding the trellis diagram, each possible receive sequence $\langle z[k]\rangle$ is uniquely given by a *path* through the trellis. Hence, maximum-likelihood sequence estimation can be interpreted as finding the "best" path through the trellis, i.e., the path which has minimum squared Euclidean distance from the noisy observation $\langle y[k]\rangle$ [For73].

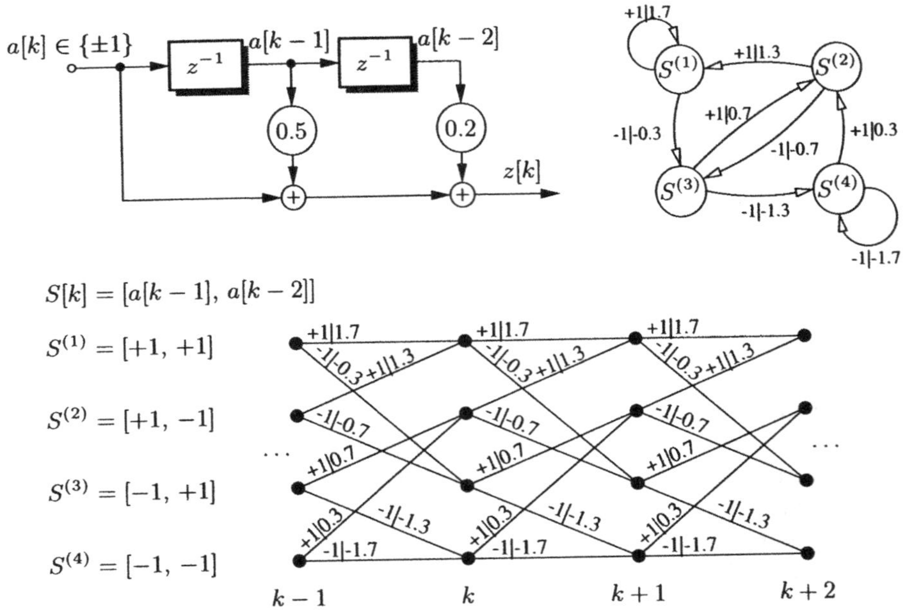

Fig. 2.47 Trellis diagram and state-transition diagram for binary signaling and channel $H(z) = 1 + 0.5\,z^{-1} + 0.2\,z^{-2}$.

Suppose we know that at time instant k the ML path passes through the state $S[k] = S^{(i)}$. Of all possible (partial) sequences $\langle \cdots a[k-2],\, a[k-1] \rangle$ which merge into this state, only the one with the least *path metric*

$$\Lambda[k] \triangleq \sum_{\kappa=-\infty}^{k} |y[\kappa] - z[\kappa]|^2$$

has to be considered. Obviously, all other paths being different from this *survivor path* can never become superior, and thus can be ignored. Hence, the survivor is the initial segment (up to the time index k) of the ML path. Unfortunately, we do not have a priori knowledge of the respective state. Since the ML path has to pass through exactly one of the M^p possible states, for each state $S^{(i)}$ we track its own survivor, i.e., the survivor conditioned to state $S^{(i)}$. Tracking of the survivors can be done iteratively. Given the survivors at time $k-1$, for the next time step, all M^p survivors are extended to their M successors. Then for each new state, the path with the lowest metric $\Lambda[k]$ is selected. When all survivors merge, i.e., have common history up to a certain time index, the ML sequence is detected up to the merging point [For72].

Hence, ML sequence estimation can be performed based on the samples of the output of the whitened matched filter using the *Viterbi algorithm* [Vit67, For73, Pro01]. The branch metric for transition from state $S[k] = S^{(i)}$ to state $S[k+1] =$

$S^{(j)} = \mathrm{f_s}(a[k], S^{(i)})$, induced by data symbol $a[k]$, is given by

$$\lambda^{(i \to j)}[k] = \left| y[k] - a[k] - \sum_{\kappa=1}^{p} h[\kappa] \cdot a[k - \kappa] \right|^2 , \tag{2.5.7}$$

with $S[k] = [a[k-1], a[k-2], \ldots, a[k-p]]$. As shown above, due to the whitened-matched-filter front-end (independence of subsequent noise samples), the path metric is additive.

To summarize, the following Viterbi algorithm performs maximum-likelihood sequence estimation based on the output samples of the whitened matched filter, as proposed by Forney in 1972 [For72].

Theorem 2.20: *Algorithm for Maximum-Likelihood Sequence Estimation*

Let $\langle y[k] \rangle$ be the observed sequence at the output of the whitened matched filter, and $H(z) = 1 + \sum_{k=1}^{p} h[k] z^{-k}$ the end-to-end transfer function for the transmitted data, drawn from an M-ary signal set. Furthermore, state $S^{(i)}$, $i = 1, 2, \ldots, M^p$, shall be defined as $S^{(i)} = [a^{(i)}[k-1], a^{(i)}[k-2], \ldots, a^{(i)}[k-p]]$, i.e., the register contents of the tapped-delay-line model.

The Viterbi algorithm for maximum-likelihood sequence estimation performs the following steps:

For each state $S[k+1] = S^{(j)}$, $j = 1, 2, \ldots, M^p$, at time index $k+1$:

- Calculate the metric for each of the M paths merging into state $S[k+1]$ at time $k+1$. For each branch, this is done by adding the branch metric

  $$\lambda^{(i \to j)}[k] = \left| y[k] - a[k] - \sum\nolimits_{\kappa=1}^{p} h[\kappa] \cdot a^{(i)}[k - \kappa] \right|^2$$

  to the accumulated state metric $\Lambda^{(i)}[k]$. The previous state $S[k] = S^{(i)}$ is implicitly defined by $S^{(j)} = \mathrm{f_s}(a[k], S^{(i)})$.
- Compare the M competing metrics $\Lambda^{(j)}[k+1]$ and
- Select the path with the lowest metric $\Lambda^{(j)}[k+1] = \Lambda^{(i)}[k] + \lambda^{(i \to j)}[k]$ as the survivor for state $S^{(j)}$ at time $k+1$.

After sufficient delay, when the survivors for all states have merged, the maximum-likelihood sequence is detected up to the merging point.

The performance of MLSE is determined by the fact of how well different paths can be distinguished. For moderate to high signal-to-noise ratios this is quantified by the *minimum squared Euclidean distance*

$$d^2_{\mathrm{E,min}} = \min_{\substack{\forall \langle z^{(m)}[k] \rangle, \langle z^{(n)}[k] \rangle \\ \langle z^{(m)}[k] \rangle \neq \langle z^{(n)}[k] \rangle}} \sum_k \left| z^{(m)}[k] - z^{(n)}[k] \right|^2 , \tag{2.5.8}$$

where minimization is carried out over all possible pairs of noiseless channel output sequences $\langle z[k] \rangle$. It is straightforward to prove that the squared Euclidean distance can also be interpreted as the energy of the difference between the respective continuous-time signals [For72, Hub92b, Pro01]. Since only the difference between signals is relevant in (2.5.8), the minimum distance can be searched over all possible *difference sequences*. Here, the samples can assume all possible differences between signal points $a[k] \in \mathcal{A}$. Beside the brute-forward search for the minimum squared Euclidean distance there exist sophisticated algorithms for its calculation. Examples are [Hub92b], a sequential algorithm working on the product trellis, [RWM93], a sequential search for "error events" (paths starting and ending in the zero state) in the difference trellis, and [Lar91], who has modified Dijkstra's algorithm [AHU74, Dij59] known from graph theory.

Finally, the minimum distance can be transformed into the loss (based on equal receive power) compared to the ISI-free AWGN channel and spacing of the signal points equal 2. Then we have

$$\vartheta^2_{(\mathsf{MLSE})} = \frac{4}{d^2_{\mathrm{E,min}} / \sum_{k \geq 0} |h[k]|^2} \; . \qquad (2.5.9)$$

As shown above, the complexity of MLSE is proportional to M^p, the number of states which have to be processed. In many practical situations, this number is very high, and hence MLSE becomes impractical. In digital subscriber line transmission, for example, the cardinality of the signal set is 8 or 16, and the length p of the impulse response after whitened matched filtering can be up to hundreds of samples. Although methods exist for reducing the number of states without sacrificing much performance—for *reduced-state sequence estimation (RSSE)*; see, e.g., [DH89, CE89, EQ88, Eyu88] or [Hub92b, SH96]—for our major application, we regard MLSE as much too complex, and thus consider only DFE and equivalent precoding methods (Chapter 3).

2.5.2 Alternative Derivation

To complete this section on maximum-likelihood sequence estimation, we sketch an alternative derivation, and show how to use the samples right after the *matched filter* instead of applying the whitened matched filter.

For the moment, we focus on finite-length sequences of N symbols. The question is how to detect such sequences in an optimum way. For $N \to \infty$ we arrive at MLSE for infinite-length sequences. The solution to the problem at hand is to combine all possible N-tuples of data symbols $a[k]$ and the overall impulse response $h(t) \triangleq h_{\mathrm{T}}(t) * h_{\mathrm{C}}(t)$ into the M^N different *signal elements*

$$\boldsymbol{a}^{(i)}(t) = \sum_{k=1}^{N} a^{(i)}[k] \cdot h(t - kT) \; , \quad i = 1, 2, \ldots, M^N \; , \qquad (2.5.10)$$

which have to be detected in white Gaussian noise. In order to emphasize, that the signal elements and the respective matched filters, which are continuous-time functions, are dependent on *vectors* of N symbols $a[k]$, they are denotated by boldface letters, i.e., as vectors.

From literature (e.g., [Pro01]), we know that this can be done optimally using a bank of matched filters. The impulse responses of these filters are given by

$$\boldsymbol{h}^{(i)}(t) = \boldsymbol{a}^{(i)^*}(-t) = \sum_{k=1}^{N} a^{(i)^*}[k] \cdot h^*(-t - kT) \,. \tag{2.5.11}$$

Using $E^{(i)} = \int_{-\infty}^{\infty} |\boldsymbol{a}^{(i)}(t)|^2 \, \mathrm{d}t$, the energy of signal element $\boldsymbol{a}^{(i)}(t)$, Figure 2.48 sketches the matched-filter demodulator.

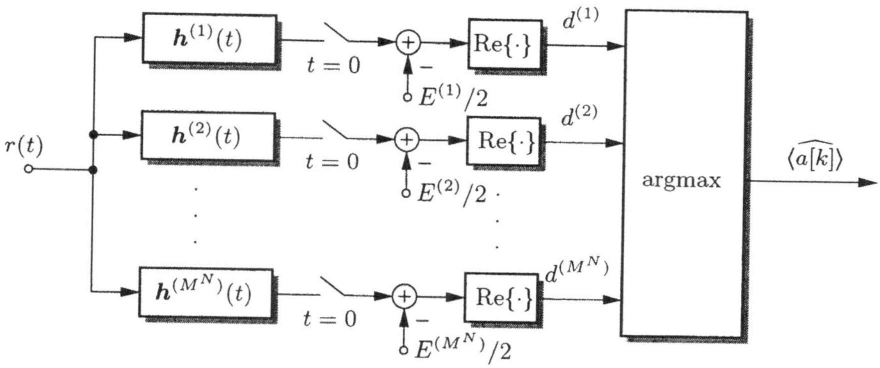

Fig. 2.48 Optimum receiver for the signal elements $\boldsymbol{a}^{(i)}(t)$.

The decision variable in branch i is then given as

$$\begin{aligned}
d^{(i)} &= \mathrm{Re}\left\{ \left(r(t) * \boldsymbol{h}^{(i)}(t) \right) \Big|_{t=0} - E^{(i)}/2 \right\} \\
&= \mathrm{Re}\left\{ \int_{-\infty}^{\infty} r(\tau) \cdot \sum_{k=1}^{N} a^{(i)^*}[k] \cdot h^*(\tau - kT) \, \mathrm{d}\tau \right. \\
&\qquad \left. - \frac{1}{2} \int_{-\infty}^{\infty} \left| \sum_{k=1}^{N} a^{(i)}[k] \cdot h(t - kT) \right|^2 \mathrm{d}t \right\} \\
&= \mathrm{Re}\left\{ \sum_{k=1}^{N} a^{(i)^*}[k] \int_{-\infty}^{\infty} r(\tau) \cdot h^*(\tau - kT) \, \mathrm{d}\tau \right. \\
&\qquad \left. - \frac{1}{2} \int_{-\infty}^{\infty} \sum_{k=1}^{N} \sum_{j=1}^{N} a^{(i)}[j] a^{(i)^*}[k] \cdot h(t - jT) h^*(t - kT) \, \mathrm{d}t \right\} \\
&= \mathrm{Re}\left\{ \sum_{k=1}^{N} a^{(i)^*}[k] \, y[k] \right.
\end{aligned}$$

$$
\left. -\frac{1}{2} \sum_{k=1}^{N} \sum_{j=1}^{N} a^{(i)}[j] a^{(i)*}[k] \cdot \int_{-\infty}^{\infty} h(t-jT) h^*(t-kT) \, \mathrm{d}t \right\}
$$

$$
= \operatorname{Re} \left\{ \sum_{k=1}^{N} a^{(i)*}[k] \, y[k] - \frac{1}{2} \sum_{k=1}^{N} \sum_{j=1}^{N} a^{(i)}[j] a^{(i)*}[k] \cdot h[k-j] \right\} , \tag{2.5.12}
$$

where $y[k]$ denotes the T-spaced samples after the matched filter (matched to the underlying "chips" $h(t)$)

$$
y[k] \triangleq \int_{-\infty}^{\infty} r(\tau) h^*(\tau - kT) \, \mathrm{d}\tau , \tag{2.5.13}
$$

and

$$
h[k] \triangleq h^{(\mathrm{MF})}[k] = \int_{-\infty}^{\infty} h(t+kT) h^*(t) \, \mathrm{d}t \tag{2.5.14}
$$

is the end-to-end discrete-time impulse response, including matched filter and T-spaced sampling.

Equation (2.5.12) has an important consequence: For MLSE, the samples after the matched filter are sufficient. It can be argued that the matched filter provides a lossless transition from continuous-time to discrete-time domain [Hub92b]. This fact again supports the optimality of the matched-filter front-end, which was derived in Section 2.2.2 based on the signal-to-noise ratio.

From (2.5.12) the maximum-likelihood decision rule can be written as

$$
[\hat{a}[1], \ldots, \hat{a}[N]] = \underset{[a[1], \ldots, a[N]] \in \mathcal{A}^N}{\operatorname{argmax}} \quad J([a[1], \ldots, a[N]]) , \tag{2.5.15}
$$

with (since the double sum in (2.5.12) is a Hermitian quadratic form, and hence real-valued)

$$
J([a[1], \ldots, a[N]]) = \sum_{k=1}^{N} 2 \operatorname{Re}\{a^*[k] y[k]\} - \sum_{k=1}^{N} \sum_{j=1}^{N} a^*[k] h[k-j] a[j] \, . \tag{2.5.16}
$$

In order to obtain a practical algorithm for MLSE, we first note that $J([a[1] \cdots a[n]])$ can be calculated recursively, namely

$$
\begin{aligned}
J([a[1] \cdots a[n]]) \;=\; & \sum_{k=1}^{n-1} 2 \operatorname{Re}\{a^*[k] y[k]\} - \sum_{k=1}^{n-1} \sum_{j=1}^{n-1} a^*[k] h[k-j] a[j] \\
& + 2 \operatorname{Re}\{a^*[n] y[n]\} - a^*[n] \sum_{j=1}^{n-1} h[n-j] a[j] \\
& - a[n] \sum_{k=1}^{n-1} h[k-n] a^*[k] - a^*[n] h[0] a[n]
\end{aligned}
$$

$$= J([a[1] \cdots a[n-1]]) \\ + \operatorname{Re}\left\{ a^*[n] \left(2y[n] - h[0]a[n] - 2\sum_{j=1}^{p} h[j]a[n-j] \right) \right\} . \tag{2.5.17}$$

In the last step we have used the fact that the support of the impulse response $h[k]$ is limited to the range[15] $k = -p, \ldots, p$. Hence, the update depends only on the past p samples $a[k]$.

Using (2.5.17), MLSE can again be performed using the Viterbi algorithm. This version of MLSE was proposed by Ungerböck in 1974 [Ung74]. The p past data samples again give the M^p possible states. To each state $S[k] = S^{(i)} = [a^{(i)}[k-1],\ a^{(i)}[k-2],\ \ldots,\ a^{(i)}[k-p]]$, an accumulated path metric $\Lambda^{(i)}[k] \triangleq J([\cdots a[k-1]])$ is associated. The metric increment for transition from state $S[k] = S^{(i)}$ to state $S[k+1] = S^{(j)} = \mathrm{f_s}(a[k], S^{(i)})$, induced by data symbol $a[k]$, now reads

$$\lambda^{(i\to j)}[k] = \operatorname{Re}\left\{ a^*[k] \left(2y[k] - h[0]a[k] - 2\sum_{j=1}^{p} h[j]a^{(i)}[k-j] \right) \right\} . \tag{2.5.18}$$

Please note, because of (2.5.15), the Viterbi algorithm now has to *maximize* the accumulated metric

$$\Lambda^{(j)}[k+1] = \Lambda^{(i)}[k] + \lambda^{(i\to j)}[k] . \tag{2.5.19}$$

With respect to performance both versions of MLSE, Forney's and Ungerböck's, are identical [Ung74]. The complexity is mainly determined by the number of states. Therefore, their complexity is of the same order. Since Euclidean distance requires a square operation, whereas Ungerböck's metric needs only to take the real part, there are slight implementation advantages for the second form. For further details, the reader is referred to [Ung74].

[15] From the spectral factorization problem (2.3.30) we see that if the WMF end-to-end impulse response is of order p, then the matched-filter impulse response spans the interval $k = -p, \ldots, p$.

REFERENCES

[AB93] S. A. Altekar and N. C. Beaulieu. Upper Bounds to the Error Probability of Decision Feedback Equalization. *IEEE Transactions on Information Theory*, IT-39, pp. 145–156, January 1993.

[AC95] N. Al-Dhahir and J. M. Cioffi. MMSE Decision-Feedback Equalizer: Finite-Length Results. *IEEE Transactions on Information Theory*, IT-41, pp. 961–975, July 1995.

[AHU74] A. V. Aho, J. E. Hopcroft, and J. D. Ullman. *The Design and Analysis of Computer Algorithms*. Addison-Wesley Publishing Company, Reading, MA, 1974.

[And73] I. N. Andersen. Sample-Whitened Matched Filters. *IEEE Transactions on Information Theory*, IT-19, pp. 653–660, 1973.

[And77] W. Andexser. *Untersuchung von Datenübertragungssystemen mit adaptiver Quantisierter Rückkopplung*. PhD thesis, Universität Stuttgart, Stuttgart, Germany, 1977. (In German.)

[And99] J. B. Anderson. *Digital Transmission Engineering*. IEEE Press, Piscataway, NJ, 1999.

[ANSI98] American National Standards Institute (ANSI). Network and Customer Installation Interfaces — Asymmetric Digital Subscriber Line (ADSL) Metallic Interface. Draft American National Standard for Telecommunications, December 1998.

[ASC97] N. Al-Dhahir, A. H. Sayed, and J. M. Cioffi. Stable Pole-Zero Modeling of Long FIR Filters with Application to the MMSE-DFE. *IEEE Transactions on Communications*, COM-45, pp. 508–513, May 1997.

[BB91] E. J. Borowski and J. M. Borwein. *The HarperCollins Dictionary of Mathematics*. HarperPerennial, New York, 1991.

[Bla87] R. E. Blahut. *Principles and Practice of Information Theory*. Addison-Wesley Publishing Company, Reading, MA, 1987.

[Bla90] R. E. Blahut. *Digital Transmission of Information*. Addison-Wesley Publishing Company, Reading, MA, 1990.

[BLM93] J. R. Barry, E. A. Lee, and D. G. Messerschmitt. Capacity Penalty Due to Ideal Zero-Forcing Decision-Feedback Equalization. In *Proceedings of the IEEE International Conference on Communications (ICC'93)*, pp. 422–427, Geneva, Switzerland, May 1993.

[BLM96] J. R. Barry, E. A. Lee, and D. G. Messerschmitt. Capacity Penalty Due to Ideal Zero-Forcing Decision-Feedback Equalization. *IEEE Transactions on Information Theory*, IT-42, pp. 1062–1071, July 1996.

[BP79] C. A. Belfiore and J. H. Park. Decision Feedback Equalization. *Proceedings of the IEEE*, 67, pp. 1143–1156, August 1979.

[BS98] I. N. Bronstein and K. A. Semendjajew. *Handbook of Mathematics*. Springer Verlag, Berlin, Heidelberg, Reprint of the third edition, 1998.

[BT67] T. Berger and D. W. Tufts. Optimum Pulse Amplitude Modulation — Part I: Transmitter-Receiver Design and Bounds from Information Theory, Part II: Inclusion of Timing Jitter. *IEEE Transactions on Information Theory*, IT-13, pp. 196–216, April 1967.

[CDEF95] J. M. Cioffi, G. P. Dudevoir, M. V. Eyuboğlu, and G. D. Forney. MMSE Decision-Feedback Equalizers and Coding—Part I: Equalization Results, Part II: Coding Results. *IEEE Transactions on Communications*, COM-43, pp. 2582–2604, October 1995.

[CE89] P. R. Chevillat and E. Eleftheriou. Decoding of Trellis-Encoded Signals in the Presence of Intersymbol Interference and Noise. *IEEE Transactions on Communications*, COM-37, pp. 669–676, July 1989.

[CEGH98] J. D. Coker, E. Eleftheriou, R. L. Galbraith, and W. Hirt. Noise-Predictive Maximum Likelihood (NPML) Detection. *IEEE Transactions on Magnetics*, MAG-34, pp. 110–117, January 1998.

[CT91] T. M. Cover and J. A. Thomas. *Elements of Information Theory*. John Wiley & Sons, Inc., New York, 1991.

[DH89] A. Duel-Hallen and C. Heegard. Delayed Decision-Feedback Sequence Estimation. *IEEE Transactions on Communications*, COM-37, pp. 428–436, May 1989.

[Dij59] E. W. Dijkstra. A Note on two Problems in Connexion with Graphs. *Numerische Mathematik*, 1, pp. 269–271, 1959.

[EQ88] M. V. Eyuboğlu and S. U. H. Qureshi. Reduced-State Sequence Estimation with Set Partitioning and Decision Feedback. *IEEE Transactions on Communications*, COM-36, pp. 13–20, January 1988.

[Eri71] T. Ericson. Structure of Optimum Receiving Filters in Data Transmission Systems. *IEEE Transactions on Information Theory*, IT-17, pp. 352–353, May 1971.

[Eri73] T. Ericson. Optimum PAM Filters Are Always Band Limited. *IEEE Transactions on Information Theory*, IT-19, pp. 570–573, July 1973.

[Eyu88] M. V. Eyuboğlu. Detection of Coded Modulation Signals on Linear, Severely Distorted Channels Using Decision-Feedback Noise Prediction with Interleaving. *IEEE Transactions on Communications*, COM-36, pp. 401–409, April 1988.

[For72] G. D. Forney. Maximum Likelihood Sequence Estimation of Digital Sequences in the Presence of Intersymbol Interference. *IEEE Transactions on Information Theory*, IT-18, pp. 363–378, May 1972.

[For73] G. D. Forney. The Viterbi Algorithm. *Proceedings of the IEEE*, 61, pp. 268–278, March 1973.

[For96] G. D. Forney. 1995 Shannon Lecture: Performance and Complexity. *IEEE Information Theory Newsletter*, 46, pp. 3, 4, 23–25, March 1996.

[Fra69] L. E. Franks. *Signal Theory*. Prentice-Hall, Inc., Englewood Cliffs, NJ, 1969.

[FU98] G. D. Forney and G. Ungerböck. Modulation and Coding for Linear Gaussian Channels. *IEEE Transactions on Information Theory*, IT-44, pp. 2384–2415, October 1998.

[Gal68] R. G. Gallager. *Information Theory and Reliable Communication*. John Wiley & Sons, Inc., New York, London, 1968.

[Ger96] W. Gerstacker. An Alternative Approach to Minimum Mean-Squared Error DFE with Finite Length Constraints. *Archiv für Elektronik und Übertragungstechnik (International Journal of Electronics and Communications)*, 50, pp. 27–31, January 1996.

[GH85] K. P. Graf and J. Huber. Fast optimale Empfänger für die Digitalsignalübertragung über symmetrische Leitungen. In *NTG–Fachtagung "Wege zum integrierten Kommunikationsnetz," (Fachbericht 88)*, pp. 66–73, Berlin, Germany, March 1985. (In German.)

[GH89] K. P. Graf and J. Huber. Design and Performance of an All-Digital Adaptive 2.048 Mbit/s Data Transmission System using Noise Prediction. In *Proceedings of the ISCAS'89*, pp. 1808–1812, Portland, OR, 1989.

[GH96] W. Gerstacker and J. Huber. Maximum SNR Decision-Feedback Equalization with FIR Filters: Filter Optimization and a Signal Processing Application. In *Proceedings of the IEEE International Conference on Communications (ICC'96)*, pp. 1188–1192, Dallas, TX, June 1996.

[Hay96] S. Haykin. *Adaptive Filter Theory*. Prentice-Hall, Inc., Englewood Cliffs, NJ, 3rd edition, 1996.

[HM85] M. L. Honig and D. G. Messerschmitt. *Adaptive Filters–Structures, Algorithms, and Applications*. Kluwer Academic Publishers, Boston, 3rd printing, 1985.

[Hub87] J. Huber. Detektoren und Optimalfilter für Digitalsignale mit Impulsinterferenzen. Teil I: Optimaldetektion und Whitened-Matched-Filter;

Teil II: Subotimale Empfänger. *Frequenz*, 41, pp. 161–167/189–196, June/August 1987. (In German.)

[Hub92a] J. Huber. Reichweitenabschätzung durch Kanalcodierung bei der digitalen Übertragung über symmetrische Leitungen. Internal Report, Lehrstuhl für Nachrichtentechnik, Universität Erlangen–Nürnberg, Erlangen, Germany, 1992.

[Hub92b] J. Huber. *Trelliscodierung*. Springer Verlag, Berlin, Heidelberg, 1992. (In German.)

[Hub93a] J. Huber. Distance-Gains by Multiple-Duplex Transmission, Coding and Shaping for HDSL. In *Proceedings of the IEEE International Conference on Communications (ICC'93)*, pp. 1820–1829, Geneva, May 1993.

[Hub93b] J. Huber. Signal- und Systemteoretische Grundlagen zur Vorlesung Nachrichtenübertragung. Skriptum, Lehrstuhl für Nachrichtentechnik II, Universität Erlangen–Nürnberg, Erlangen, Germany, 1993. (In German.)

[Kam94] K. D. Kammeyer. Time Truncation of Channel Impulse Responses by Linear Filtering: A Method to Reduce the Complexity of Viterbi Equalization. *Archiv für Elektronik und Übertragungstechnik (International Journal of Electronics and Communications)*, 48, pp. 237–243, May 1994.

[Kay88] S. M. Kay. *Modern Spectral Estimation: Theory and Application*. Prentice-Hall, Inc., Englewood Cliffs, NJ, 1988.

[KH99] M. Kimpe and J. Hausner. Equivalent Loss and Equivalent Noise. ETSI TM6, Edinburgh, UK, September 1999.

[Lar91] T. Larsson. A State-Space Partitioning Approach to Trellis Decoding. Chalmers University of Technology, Göteborg, Sweden, 1991.

[LSW68] R. W. Lucky, J. Salz, and E. J. Weldon, Jr. *Principles of Data Communications*. McGraw-Hill, New York, 1968.

[MK93] S. K. Mitra and J. F. Kaiser. *Handbook for Digital Signal Processing*. John Wiley & Sons, Inc., New York, 1993.

[Mon71] P. Monsen. Feedback Equalization for Fading Dispersive Channels. *IEEE Transactions on Information Theory*, IT-17, pp. 56–64, January 1971.

[OS75] A. V. Oppenheim and R. W. Schafer. *Digital Signal Processing*. Prentice-Hall, Inc., Englewood Cliffs, NJ, 1975.

[OSS68] A. V. Oppenheim, R. W. Schafer, and T. G. Stockham. Nonlinear Filtering of Multiplied and Convolved Signals. *Proceedings of the IEEE*, 56, pp. 1264–1291, August 1968.

[Pap77] A. Papoulis. *Signal Analysis*. McGraw-Hill, New York, 1977.

[Pap91] A. Papoulis. *Probability, Random Variables, and Stochastic Processes*. McGraw-Hill, New York, 3ed edition, 1991.

[PK83] L. Pakula and S. Kay. Simple Proofs of the Minimum Phase Property of the Prediction Error Filter. *IEEE Transactions on Acoustics, Speech, and Signal Processing*, ASSP-31, pp. 501, April 1983.

[Pri72] R. Price. Nonlinear Feedback Equalized PAM versus Capacity for Noisy Filter Channels. In *Proceedings of the IEEE International Conference on Communications (ICC'72)*, pp. 22.12–22.17, 1972.

[Pro01] J. G. Proakis. *Digital Communications*. McGraw-Hill, New York, 4th edition, 2001.

[PTVF92] W. H. Press, S. A. Teukolsky, W. T. Vetterling, and B. P. Flannery. *Numerical Recipes in C—The Art of Scientific Computing*. Cambridge University Press, Cambridge, 2nd edition, 1992.

[RWM93] S. A. Raghavan, J. K. Wolf, and L. B. Milstein. On the Performance Evaluation of ISI Channels. *IEEE Transactions on Information Theory*, IT-39, pp. 957–965, May 1993.

[Sch94] H. W. Schüßler. *Digitale Signalverarbeitung, Band I*. Springer Verlag, Berlin, Heidelberg, 4th edition, 1994. (In German.)

[SH96] B. Spinnler and J. Huber. Design of Hyper States for Reduced-State Sequence Estimation. *Archiv für Elektronik und Übertragungstechnik (International Journal of Electronics and Communications)*, 50, pp. 17–26, 1996.

[Smi65] J. W. Smith. The Joint Optimization of Transmitted Signal and Receiving Filter for Data Transmission Systems. *Bell System Technical Journal*, pp. 2363–2392, December 1965.

[ST85] G. Söder and K. Tröndle. *Digitale Übertragungssysteme*. Springer Verlag, Berlin, Heidelberg, 1985. (In German.)

[Tre71] H. L. van Trees. *Detection, Estimation, and Modulation Theory—Part III: Radar-Sonar Signal Processing and Gaussian Signals in Noise*. John Wiley & Sons, Inc., New York, 1971.

[Tre76] S. Tretter. *Introduction to Discrete-Time Signal Processing*. John Wiley & Sons, Inc., New York, 1976.

[Ung74] G. Ungerböck. Adaptive Maximum-Likelihood Receiver for Carrier-Modulated Data-Transmission Systems. *IEEE Transactions on Communications*, COM-22, pp. 624–636, May 1974.

[Vit67] A. J. Viterbi. Error Bounds for Convolutional Codes and an Asymptotically Optimum Decoding Algorithm. *IEEE Transactions on Information Theory*, IT-13, pp. 260–269, 1967.

[Wel89] H.-W. Wellhausen. Eigenschaften symmetrischer Kabel der Ortsnetze und generelle Übertragungsmöglichkeiten. *Der Fernmelde–Ingenieur*, 43. Jahrgang, pp. 1–51, October/November 1989. (In German.)

3

Precoding Schemes

Because of its optimality and its simple structure, *decision-feedback equalization* is considered as a *canonical receiver structure* [CDEF95]. In the last chapter we have seen that by using DFE the Shannon capacity of the underlying continuous-time channel can be utilized. Thus combining DFE and powerful channel coding schemes, developed for the AWGN channel, should lead to transmission close to channel capacity.

Unfortunately, DFE suffers from error propagation, and even worse, channel coding cannot be applied in a straightforward manner. For DFE zero-delay decisions are required, which is irreconcilable with the basic idea of channel coding, that is, to make decisions based only on the observation of whole sequences of symbols. From the literature (e.g., [Eyu88]) techniques are known which solve this problem by interleaving and eventually by iterative processing, but complexity is increased and huge signal delays are introduced. Here, we will not consider such methods.

In this chapter *precoding techniques* are discussed, i.e., channel equalization is done at the transmitter side. In contrast to *predistortion*, precoding is done in a nonlinear way. Employing precoding avoids the disadvantages of DFE: Coding techniques can be applied in the same way as for channels without ISI, and no error propagation occurs.[1]

Precoding is applicable if the channel transfer function $H(z)$ is known at the transmitter. For it, usually during start-up, of a two-way communication the channel

[1]This point is not strictly true; we will discuss this later on.

impulse response is estimated at the receiver. This can be done by adaptively adjusting a conventional DFE structure, e.g., by using known training sequences. Then, the feedback part of the DFE gives the desired impulse response. At the end of the start-up procedure, the (estimated) tap weights are then passed from the receiver to the transmitter.

If the channel is not (completely) known at the transmitter, one can still use precoding, now with a compromise setting. The residual intersymbol interference or a mismatch due to estimation errors is then removed by adaptive linear equalization at the receiver. In situations relevant in practice the loss due to mismatch is negligible; see [FGH95].

Basically, two precoding techniques are known: *Tomlinson-Harashima precoding (THP)*, which was proposed almost 30 years ago, and *flexible precoding (FLP)*, developed recently during the standards activities for the international telephone line modem standard ITU V.34. Subsequently, both schemes are explained and various topics, in particular the combination of precoding and channel coding, are discussed. The differences and dualities of THP and FLP are stated. Again, as in Chapter 2, the examples concentrate on fast digital transmission over twisted-pair lines, such as single-pair digital subscriber lines or asymmetric digital subscriber lines, but the general statements are valid for all kinds of transmission scenarios.

3.1 PRELIMINARIES

Consider the end-to-end discrete-time channel model when applying Forney's zero-forcing whitened matched filter as receiver front-end, or its approximation via noise prediction (Section 2.3.1). According to Section 2.3, the channel impulse response $\langle h[k] \rangle$ of some order p is *canonical*, i.e., causal, monic ($h[0] = 1$), and minimum-phase. The corresponding transfer function is denoted by $H(z) = 1 + \sum_{k=1}^{p} h[k]\, z^{-k}$. Although it is easier to think of FIR transfer functions, all subsequent derivations do hold for IIR (all-pole or pole-zero) channel models (Section 2.4.2), as well. The channel is driven by the i.i.d. zero-mean data symbols $a[k]$ with variance $\sigma_a^2 = \mathrm{E}\{|a[k]|^2\}$, and the additive zero-mean Gaussian noise sequence $\langle n[k] \rangle$ is white with variance $\sigma_n^2 = \mathrm{E}\{|n[k]|^2\}$. The discrete-time, possibly complex-valued channel model is summarized in Figure 3.1.

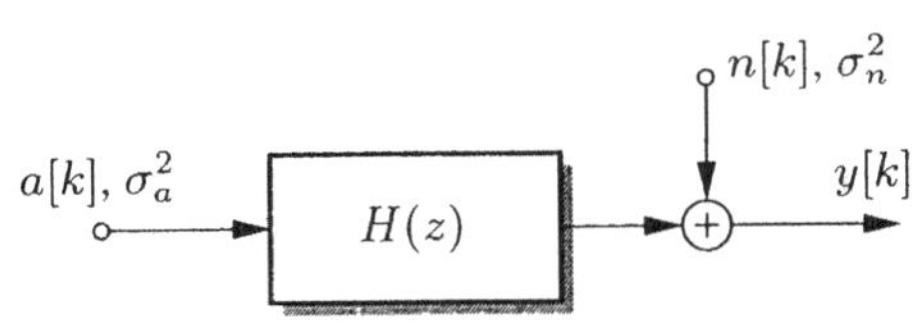

Fig. 3.1 Discrete-time channel model.

Recapitulating Chapter 2, an obvious approach to equalization is zero-forcing linear equalization (ZF-LE), depicted in the upper part of Figure 3.2. Here, due to the inverse channel filter $1/H(z)$, the end-to-end transfer function is 1, and a symbol-by-symbol threshold decision for the data symbols is possible. From basic system theory (e.g. [Pap77]) we know that the transfer function $1/H(z)$ can be realized by the feedback structure shown, with negative feedback $H(z) - 1$.

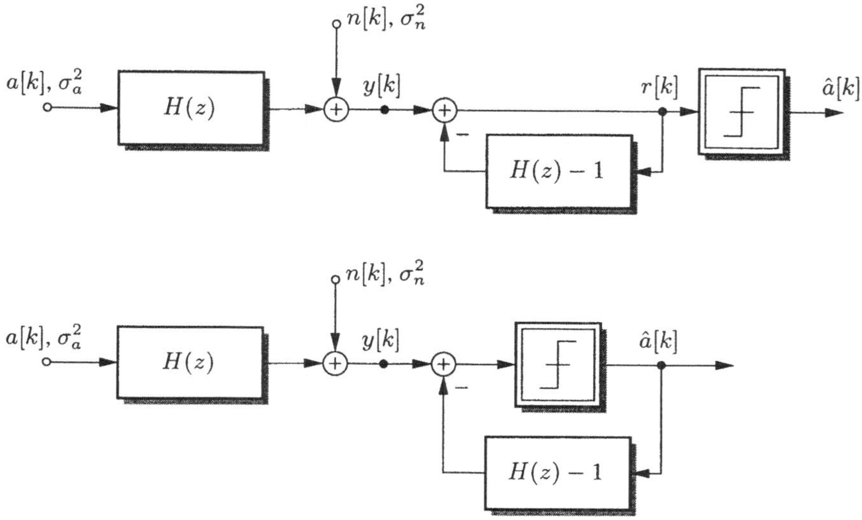

Fig. 3.2 Comparison of linear equalization and decision-feedback equalization.

The drawback of this strategy of decoupled equalization and decision is that noise power is enhanced by the factor

$$T \int_{-\frac{1}{2T}}^{\frac{1}{2T}} \frac{1}{|H(e^{j2\pi fT})|^2} \, df \,, \tag{3.1.1}$$

which, using (2.3.30), (2.3.34), (2.3.1), (2.2.38), and (2.3.28), can be identified with the ultimate prediction gain (elaborate!). This is obvious, since here we proceed in the opposite way as in Section 2.3.

Noise enhancement is avoided by decision-feedback equalization, cf. the bottom of Figure 3.2. Compared to ZF-LE, equalization and decision are combined into one unit—the slicer is incorporated into the feedback structure, implementing $1/H(z)$. Assuming correct decisions $\hat{a}[k] = a[k]$ ("ideal DFE" assumption), the additive noise is eliminated ("grounded") at the slicer. Without any noise, the feedback part cancels the postcursors of $\langle h[k] \rangle$, and at the input of the threshold device we have

$$a[k] * h[k] + n[k] - a[k] * (h[k] - \delta[k]) = a[k] + n[k] \,. \tag{3.1.2}$$

The signal-to-noise ratio is then simply (see Theorem 2.16 for the expression in terms of the parameters of the underlying continuous-time channel)

$$\mathrm{SNR}^{(\mathrm{ZF\text{-}DFE})} = \frac{\sigma_a^2}{\sigma_n^2} \,. \tag{3.1.3}$$

We now turn to (pre)equalization at the transmitter side, sketched in the upper part of Figure 3.3. Since we deal with linear systems, the inverse channel transfer

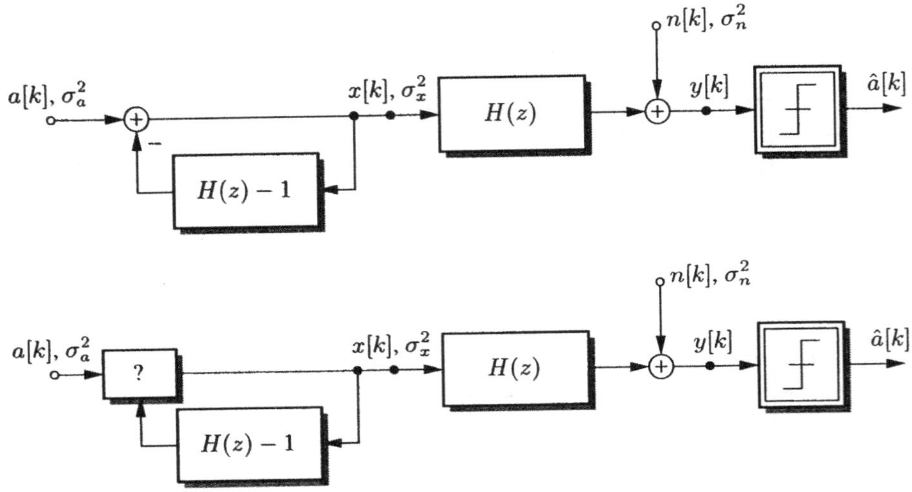

Fig. 3.3 Comparison of linear (pre)equalization at the transmitter side and precoding.

function can also be realized at the transmitter, leading to an overall ($a[k] \to y[k]$) ideal channel. One may argue that at the slicer the data symbols $a[k]$ are detectable in white Gaussian noise with variance σ_n^2, resulting in the same SNR as for DFE, but the transmit power is boosted due to the filter $1/H(z)$, again by the factor (3.1.1). A comparison based on *equal transmit power* reveals that linear preequalization is exactly as bad as linear equalization at the receiver.

The idea of *precoding* (bottom of Figure 3.3) is similar to the step from ZF-LE to DFE: replace the linear filter by a nonlinear feedback structure in order to avoid power enhancement. If $\sigma_x^2 = \sigma_a^2$ can be achieved, precoding will work as well as DFE. But since equalization is done at the transmitter, error propagation of DFE is avoided.

The following sections present two precoding strategies which meet the above requirements.

3.2 TOMLINSON-HARASHIMA PRECODING

A nonlinear technique employing modulo arithmetics usually called *Tomlinson-Harashima precoding (THP)* was introduced independently and almost simultaneously by Tomlinson [Tom71] and Harashima and Miyakawa [HM69, HM72]. First, we explain THP for uncoded baseband transmission, then we extend it to passband transmission, i.e., two-dimensional signal constellations. Later, we discuss various aspects of Tomlinson-Harashima precoding, in particular the combination with coded modulation.

3.2.1 Precoder

Tomlinson-Harashima precoding was originally proposed for use with an M-point one-dimensional PAM signal set $\mathcal{A} = \{\pm 1, \pm 3, \ldots, \pm(M-1)\}$, M even. For this constellation THP is almost identical to the inverse channel filter $1/H(z)$, displayed in Figure 3.3, except that an offset-free (symmetrical about the origin) *modulo-$2M$ adder* is used instead of the conventional adder [Tom71]. The operation of this modulo adder is as follows: If the result of the summation is greater than M, $2M$ is (repeatedly) subtracted until the result is less than M. If the result of the summation is less than $-M$, $2M$ is (repeatedly) added until the result is greater than or equal to $-M$. In other words, the conventional sum is reduced modulo $2M$ to the half-open interval[2] $[-M, M)$. The block diagram of the Tomlinson-Harashima precoder using a sawtooth nonlinearity for modulo reduction is sketched in Figure 3.4 (cf. [HM72]).

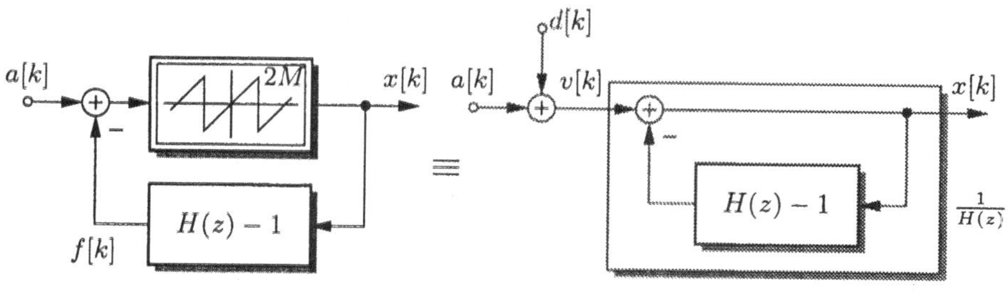

Fig. 3.4 Tomlinson-Harashima precoder and linearized description.

The effect of the modulo reduction can be characterized as follows [FE91, EF92]: A unique sequence $\langle d[k] \rangle$, $d[k] \in 2M\mathbb{Z}$—we call it *precoding sequence*—is added to the data sequence $\langle a[k] \rangle$ in order to create an *effective data sequence (EDS)* $\langle v[k] \rangle$, $v[k] = a[k] + d[k]$. This sequence is then filtered with the inverse of $H(z)$ (assume for the moment that $1/H(z)$ exists). The integers $d[k]$ are chosen so that the real-valued

[2]In principle, it is immaterial which of these two boundaries is included in the interval. But we will see that, using the present definition, a simple hardware implementation is straightforward.

channel symbols

$$x[k] = a[k] + d[k] - \sum_{\kappa=1}^{p} h[\kappa] \cdot x[k-\kappa] = v[k] - f[k] \tag{3.2.1}$$

fall into the interval $\mathcal{R} = [-M,\ M)$. Note that in practice the effective data sequence does not occur explicitly at the transmitter, and the values $d[k]$ are implicitly selected symbol-by-symbol by the memoryless modulo operation, which reduces $x[k]$ to $\mathcal{R}$. On the right, Figure 3.4 shows this *linearized description* of Tomlinson-Harashima precoding.

THP employs the principle of modulo precoding, a multiple-symbol representation based on congruent signal levels, which already can be found in [Len64, Kre66]. The congruent signal points are generated by extending the signal set $\mathcal{A}$ periodically to the set[3]

$$\mathcal{V} \triangleq \mathcal{A} + 2M\mathbb{Z} = \{a + d \mid a \in \mathcal{A},\ d \in 2M\mathbb{Z}\} . \tag{3.2.2}$$

Figure 3.5 shows the extended signal set $\mathcal{V}$ for $M = 4$.

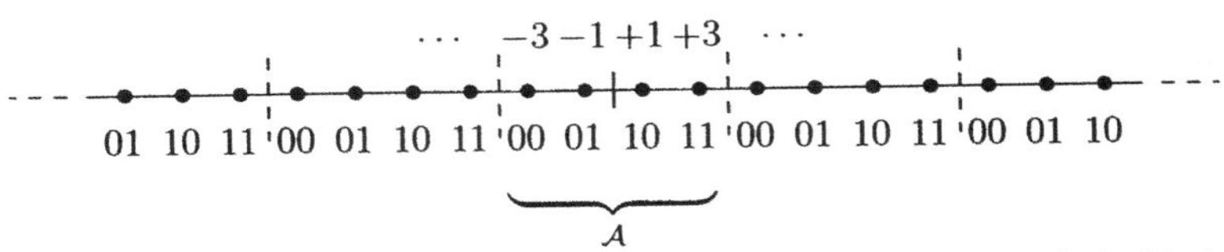

Fig. 3.5 Extended signal set $\mathcal{V}$. $M = 4$. Congruent signal levels are labeled with the same dibit.

From $\mathcal{V}$, the current effective data symbol is selected in THP, which is (a) congruent to the current $a[k]$, and (b) minimizes the magnitude of the corresponding channel symbol $x[k]$. Hence, THP is an extension of linear preequalization, now using the signal set $\mathcal{V}$ instead of $\mathcal{A}$. It is noteworthy that THP not only has to be matched to the channel, but also is closely tied to the actual signal constellation $\mathcal{A}$.

Considering the linearized description, we see that the precoder is matched to the discrete-time channel impulse response $\langle h[k] \rangle$ and that is presubtracts the intersymbol interference $f[k]$ due to postcursors. Hence, the receiver output sequence is given by

$$y[k] = v[k] + n[k] , \tag{3.2.3}$$

where $n[k]$ is again the white Gaussian noise sequence. Thus in the absence of noise, the effective data sequence $\langle v[k] \rangle$ appears at the input of the slicer. Reducing this signal to the interval $\mathcal{R} = [-M,\ M)$ by the same sawtooth device as was used in the precoder enables symbol-by-symbol threshold decision. Therefore, the same slicer as for DFE can be used. Modulo reduction and decision can be combined, too, i.e., producing decisions modulo $2M$.

[3]The denomination of sets is adopted from lattice theory: $a\mathbb{Z} + b \triangleq \{az + b \mid z \in \mathbb{Z}\}$, $a, b \in \mathbb{R}$, and $\mathcal{A} + \mathcal{B} \triangleq \{a + b \mid a \in \mathcal{A},\ b \in \mathcal{B}\}$; see also Appendix C.

Disregarding the modulo congruence, e.g., by reducing $y[k]$ to the region $\mathcal{R}$, THP transforms the ISI channel $H(z)$ to a memoryless one. However, due to the modulo operation, the additive noise sequence is signal dependent and no longer exactly Gaussian. But for moderate to high signal-to-noise ratios this effect can be neglected and the end-to-end behavior is well approximated by the AWGN model.

In any case, the modulo device of the precoder reduces the transmit signal to a well prescribed range. Regardless of the current channel impulse response, the precoder is stable in the bounded-input/bounded-output (BIBO) sense. Hence, like DFE, THP is also stable for channels with spectral zeros. In such cases, regardless of the unstable inverse channel filter $1/H(z)$, even the linear description of THP is still valid. Here, the PSD of the effective data sequence also exhibits these spectral zeros, and thus possible oscillations at the precoder are not excited.[4]

For implementation, the maximum possible magnitude of the channel output signal is of interest. All receiver input filters have to operate linearly within this dynamic range, i.e., arithmetics of the digital filters have to be carried out with high word length. Because the magnitude of the channel input is bounded by M, the channel output amplitude is restricted to

$$M \cdot \sum_{k=0}^{p} |h[k]| \,, \tag{3.2.4}$$

i.e., M times the absolute sum of the channel tap weights. Moreover, considering that the noiseless channel output $v[k]$ stems from the grid $2\mathbb{Z}+1$, rounding (3.2.4) to the next smaller odd integer gives the upper bound on the channel output magnitude [Hub92a]

$$|v[k]| \le V_{\max} \triangleq 2 \cdot \mathrm{INT}\left[\left(M \cdot \sum_{k=0}^{p} |h[k]| + 1\right) \Big/ 2\right] - 1 \,. \tag{3.2.5}$$

$\mathrm{INT}[x]$ denotes the integer part of x. Especially when $H(z)$ corresponds to a prediction-error filter which achieves high prediction gain, the dynamic range can be very large. In Chapter 5 we return to this problem, and show how to mitigate this effect.

3.2.2 Statistical Characteristics of the Transmit Signal

We will now address the statistical properties of the discrete-time transmit signal $\langle x[k] \rangle$. First, the probability density function (pdf) is given, after which the autocorrelation sequence is considered. Similar derivations can be found in [TF92].

Effect of a Nonlinear Modulo Device In order to calculate the pdf of the channel symbols $x[k]$, we first regard the effect of the modulo operation on its own.

[4] THP would even work with nonminimum-phase impulse responses, but here we are always restricted to minimum-phase ones, which result in the maximum signal-to-noise ratio.

This question is equivalent to that of the statistical properties of the quantization error of a uniform quantizer, which is known from the literature on digital signal processing. Here, we follow the exposition of [SS77].

Consider the sawtooth characteristics displayed in Figure 3.6. The output signal

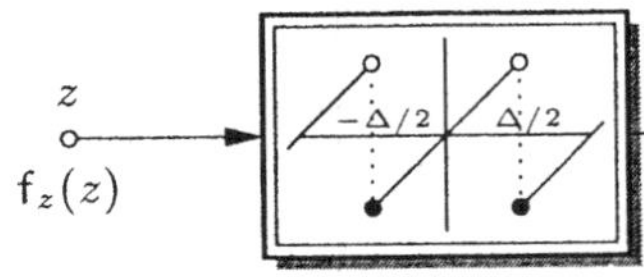

Fig. 3.6 Sawtooth characteristics of the modulo device.

x is given by reducing the input z to the interval $[-\Delta/2, \Delta/2)$.

The probability density $f_x(x)$ of x is given in terms of the pdf of z as

$$f_x(x) = \left(\sum_i f_z(x - i \cdot \Delta) \right) \cdot \mathrm{rect}(x/\Delta) \,, \tag{3.2.6}$$

where $\mathrm{rect}(x) = 1, -1/2 \leq x < 1/2$, zero elsewhere, is the rectangular pulse of unit width and unit height.

It is convenient to look at the periodically extended version of $f_x(x)$ (by simply ignoring the "rect" function in (3.2.6)) and express this periodic function as a Fourier series:

$$\sum_i f_z(x - i \cdot \Delta) \triangleq \sum_l \alpha_l \cdot e^{j\frac{2\pi}{\Delta} l x} \,, \tag{3.2.7}$$

with the Fourier coefficients

$$\begin{aligned} \alpha_l &= \frac{1}{\Delta} \int_{-\Delta/2}^{\Delta/2} \sum_i f_z(x - i \cdot \Delta) \cdot e^{-j\frac{2\pi}{\Delta} l x} \, dx \\ &= \frac{1}{\Delta} \int_{-\infty}^{\infty} f_z(x) \cdot e^{-j\frac{2\pi}{\Delta} l (x + i \cdot \Delta)} \, dx \\ &= \frac{1}{\Delta} \int_{-\infty}^{\infty} f_z(x) \cdot e^{-j\frac{2\pi}{\Delta} l x} \, dx \,. \end{aligned} \tag{3.2.8}$$

With the definition of the *characteristic function* ($\mathcal{F}^{-1}\{\cdot\}$: inverse Fourier transform)

$$C_z(\nu) \triangleq \mathrm{E}\{e^{j2\pi\nu z}\} = \int_{-\infty}^{\infty} f_z(z) \cdot e^{+j2\pi\nu z} \, dz = \mathcal{F}^{-1}\{f_z(z)\} \,, \tag{3.2.9}$$

$\alpha_l = \frac{1}{\Delta} \cdot C_z(-l/\Delta)$ holds. Taking $C_z(0) = \int_{-\infty}^{\infty} \mathrm{f}_z(z)\, \mathrm{d}z = 1$ into account, we arrive at

$$\mathrm{f}_x(x) = \begin{cases} \frac{1}{\Delta} + \frac{1}{\Delta} \sum\limits_{l, l \neq 0} C_z(l/\Delta) \cdot \mathrm{e}^{-\mathrm{j}\frac{2\pi}{\Delta} l x}, & -\Delta/2 \leq x < \Delta/2 \\ 0, & \text{else} \end{cases} . \tag{3.2.10}$$

Because of the unambiguousness of the Fourier transform, from (3.2.10) it follows that $\mathrm{f}_x(x)$ is uniformly distributed in $[-\Delta/2, \Delta/2)$ if and only if the characteristic function $C_z(\nu)$ of the input density has equidistant zeros, i.e.,

$$\mathrm{f}_x(x) = \begin{cases} \frac{1}{\Delta}, & -\Delta/2 \leq x < \Delta/2 \\ 0, & \text{else} \end{cases} , \tag{3.2.11a}$$

iff

$$C_z(l/\Delta) = \delta[l] = \begin{cases} 1, & l = 0 \\ 0, & l \in \mathbb{Z} \setminus \{0\} \end{cases} . \tag{3.2.11b}$$

Note that this result is *Nyquist's criterion* (cf. Theorem 2.1) applied to probability densities (instead of spectra) and characteristic functions (instead of signals over time).

Moreover, if (3.2.11) holds, the output symbols have zero mean, $\mathrm{E}\{x\} = 0$, and variance $\sigma_x^2 = \mathrm{E}\{x^2\} = \Delta^2/12$.

Probability Density of THP Transmit Signal We now apply the above result to THP. First, we note that, assuming an i.i.d. data sequence $\langle a[k] \rangle$, the current data symbol $a[k]$ and the postcursor sum $f[k] = \sum_{\kappa=1}^{p} h[\kappa] \cdot x[k-\kappa]$ (cf. Figure 3.4) are statistically independent. This follows from the fact that $f[k]$ only depends on past data symbols. Hence, the density of the symbol $z[k]$ at the input of the modulo device is given by the convolution of the respective densities of $a[k]$ and $-f[k]$ [Pap91]. This transforms to

$$C_z(\nu) = C_a(\nu) \cdot C_f^*(\nu) \tag{3.2.12}$$

for the characteristic functions.

The THP transmit signal is uniformly distributed over $[-M, M)$, if and only if $C_z(\frac{l}{2M}) = \delta[l]$. This is possible, e.g., if the data symbols are uniformly distributed over the same region. In this case we have

$$C_a(\nu) = \frac{\sin(2\pi M \nu)}{2\pi M \nu}, \qquad \text{hence} \qquad C_a(\tfrac{l}{2M}) = \frac{\sin(l\pi)}{l\pi} = \delta[l] . \tag{3.2.13}$$

For uniformly distributed data symbols the modulo device compensates exactly for the action of the convolution of the densities.

Now, let us turn to conventional ASK signal sets $\mathcal{A} = \{\pm 1, \pm 3, \ldots, \pm(M-1)\}$, M even. Here the characteristic function from equiprobable data symbol reads

$$C_a(\nu) = \int_{-\infty}^{\infty} \sum_{a \in \mathcal{A}} \tfrac{1}{M}\, \delta(\alpha - a) \cdot \mathrm{e}^{+\mathrm{j}2\pi\nu\alpha}\, \mathrm{d}\alpha = \frac{1}{M} \sum_{a \in \mathcal{A}} \mathrm{e}^{\mathrm{j}2\pi\nu a} . \tag{3.2.14}$$

With regard to only the discrete points $\nu = \frac{l}{2M}$, equation (3.2.14) evaluates to

$$
\begin{aligned}
C_a(\tfrac{l}{2M}) &= \frac{1}{M}\sum_{i=0}^{M-1} \mathrm{e}^{\mathrm{j}2\pi\frac{l}{2M}(2i-M+1)} \\
&= \frac{1}{M}\mathrm{e}^{-\mathrm{j}2\pi l\frac{M-1}{2M}} \cdot \sum_{i=0}^{M-1} \mathrm{e}^{\mathrm{j}\frac{2\pi}{M}li} ,
\end{aligned}
$$

and since the sum equals M, for $l \in M\mathbb{Z}$, and zero elsewhere [Sch94]

$$
\begin{aligned}
&= \frac{1}{M}\mathrm{e}^{-\mathrm{j}2\pi l\frac{M-1}{2M}} \cdot M \sum_{r=-\infty}^{\infty} \delta[l - rM] \\
&= \sum_{r=-\infty}^{\infty} (-1)^r \delta[l - rM] . \qquad (3.2.15)
\end{aligned}
$$

In summary the characteristic function reads

$$
C_a(\tfrac{l}{2M}) = \begin{cases} 1 , & l \in 2M\mathbb{Z} \\ -1 , & l \in 2M\mathbb{Z} + M \\ 0 , & \text{else} \end{cases} . \qquad (3.2.16)
$$

For $M \to \infty$ the density of the data symbols goes to a uniform distribution and the characteristic function approaches the discrete delta pulse. Figure 3.7 shows the correspondence for $M = 16$. Due to the even symmetry of $\mathrm{f}_a(a)$, the characteristic function is real-valued.

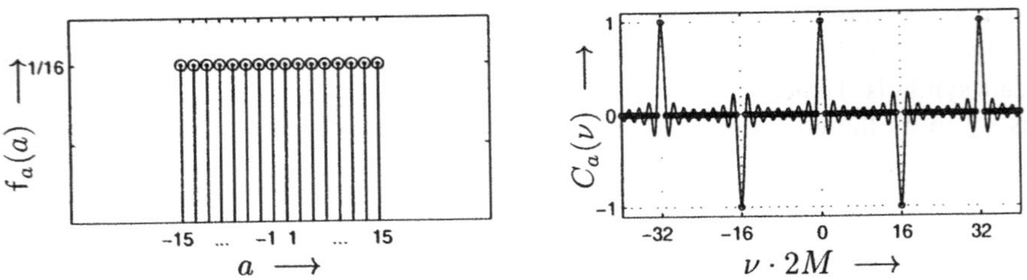

Fig. 3.7 Probability density function of 16-ary ASK and corresponding characteristic function.

Combining (3.2.10), (3.2.12), and (3.2.16), the pdf of the THP transmit symbols for M-ary baseband signaling, reads:

$$
\begin{aligned}
\mathrm{f}_x(x) &= \frac{1}{2M} + \frac{1}{2M}\sum_{l,\, l\neq 0} C_a(l/(2M)) \cdot C_f^*(l/(2M)) \cdot \mathrm{e}^{-\mathrm{j}\frac{2\pi}{2M}lx} \\
&= \frac{1}{2M} + \frac{1}{2M}\sum_{l,\, l\neq 0} C_f^*(l) \cdot \mathrm{e}^{-\mathrm{j}2\pi l x} \\
&\quad -\frac{1}{2M}\sum_{l} C_f^*(l+1/2) \cdot \mathrm{e}^{-\mathrm{j}2\pi(l+1/2)x} , \quad x \in [-M,\, M) . \qquad (3.2.17)
\end{aligned}
$$

From Figure 3.1 we see that $f[k]$ is obtained through filtering with the transfer function $H(z) - 1$. Regarding the central limit theorem[5] [Pap91], the pdf of $f[k]$ is thus well approximated by a zero-mean Gaussian probability density function with variance σ_f^2. The corresponding characteristic function is also Gaussian [Pap77]

$$\mathsf{f}_f(f) \approx \frac{1}{\sqrt{2\pi\sigma_f^2}} \cdot \mathrm{e}^{-f^2/(2\sigma_f^2)} \quad \bullet\!\!-\!\!\circ \quad C_f(\nu) \approx \mathrm{e}^{-(2\pi)^2\nu^2\sigma_f^2/2} \,. \tag{3.2.18}$$

Since the characteristic function decreases significantly over ν, the pdf of the channel symbols $x[k]$ is well approximated by

$$\begin{aligned} \mathsf{f}_x(x) &\approx \frac{1}{2M} - \frac{1}{2M}\left(C_f^*(1/2)\cdot \mathrm{e}^{-\mathrm{j}\pi x} + C_f^*(-1/2)\cdot \mathrm{e}^{+\mathrm{j}\pi x}\right) \\ &= \frac{1}{2M}\cdot\left[1 - 2\cdot \mathrm{e}^{-(\pi^2\sigma_f^2)/2}\cdot\cos(\pi x)\right] \,. \end{aligned} \tag{3.2.19}$$

This equation states that under moderate requirements the channel symbols predicted by THP are almost uniformly distributed over the region $\mathcal{R} = [-M, M)$; see also [MS76, FE91, TF92, EF92]. This approximation becomes more exact, if (a) the number M of signal levels increases, or (b) the length of the channel impulse response—to be precise, $\sum_k |h[k]|^2$—increases, which also increases σ_f^2.

The derivation is no longer valid if the channel only has integer coefficients, in which case the transmit signal is also discrete [FE91, TF92]. For the applications of interest in this book, this special case can be excluded. However, nonlinear precoding with integer coefficients leads to the well-known field of *partial-response* encoding [Pro01, And99, Ber96]. Such techniques are known as *line coding* or *data translation codes*, and are frequently used for shaping the transmit spectrum appropriately. The most prominent example of these is the *alternate mark inversion (AMI)* code. For details, the reader is referred to the literature, e.g., [Bla90, Imm91, Ber96, GG98, And99].

Finally, assuming the pdf of the channel symbols $x[k]$ is well approximated by a uniform density, the transmit power for THP reads

$$\sigma_x^2 = \mathrm{E}\{|x[k]|^2\} = \int_{-M}^{M} \frac{1}{2M}\, x^2 \,\mathrm{d}x = \frac{M^2}{3} \,. \tag{3.2.20}$$

In practice, this approximation to the transmit power is very tight. If upper and lower bounds are desired, the reader is referred to [MS76, PC93b], where a more detailed discussion on the transmit power of THP can be found. [MS76] shows that

$$\frac{M^2-1}{3} \le \sigma_x^2 \le \frac{M^2+2}{3} \tag{3.2.21}$$

holds, but the upper bound is very loose.

[5]We assume the prerequisites for the central limit theorem to be met.

Power Spectral Density of the THP Transmit Signal Now, after having derived the pdf of the channel symbols, we address the autocorrelation sequence $\phi_{xx}[\kappa] = \mathsf{E}\{x[k+\kappa] \cdot x[k]\}$. We therefore consider the samples $x_1 \triangleq x[k_1]$ and $x_2 \triangleq x[k_2]$ at discrete time k_1 and k_2, respectively, and we are interested in the joint distribution $\mathsf{f}_{x_1x_2}(x_1, x_2)$ of these samples. Since the sawtooth device is memoryless, i.e., each sample is modulo reduced separately, the pdf is given in terms of that of $z_1 \triangleq z[k_1]$ and $z_2 \triangleq z[k_2]$ as

$$\mathsf{f}_{x_1x_2}(x_1, x_2) = \left(\sum_{i_1} \sum_{i_2} \mathsf{f}_{z_1z_2}(x_1 - i_1 \cdot \Delta, x_2 - i_2 \cdot \Delta) \right) \cdot \operatorname{rect}(x_1/\Delta) \operatorname{rect}(x_2/\Delta) \,. \tag{3.2.22}$$

Following the steps leading from (3.2.6) to (3.2.10)—now using two-dimensional quantities—this joint pdf can be expressed in terms of the joint characteristic function

$$C_{z_1z_2}(\nu_1, \nu_2) \triangleq \mathcal{F}^{-1}\{\mathsf{f}_{z_1z_2}(z_1, z_2)\} \,. \tag{3.2.23}$$

Regarding $C_{z_1z_2}(0,0) = 1$, we arrive at

$$\mathsf{f}_{x_1x_2}(x_1, x_2) = \begin{cases} \frac{1}{\Delta^2} + \frac{1}{\Delta^2} \sum\limits_{\substack{l_1 \; l_2 \\ (l_1,l_2)\neq(0,0)}}\!\!\sum C_{z_1z_2}(l_1/\Delta, l_2/\Delta) \cdot \mathrm{e}^{-\mathrm{j}\frac{2\pi}{\Delta}(l_1x_1 + l_2x_2)} \,, & -\Delta/2 \le x_1, x_2 < \Delta/2 \\ 0 \,, & \text{else} \end{cases} \,. \tag{3.2.24}$$

From the unambiguousness of the Fourier transform, the condition for a uniform joint pdf follows:

$$\mathsf{f}_{x_1x_2}(x_1, x_2) = \begin{cases} \frac{1}{\Delta^2} \,, & -\Delta/2 \le x_1, x_2 < \Delta/2 \\ 0 \,, & \text{else} \end{cases} \,, \tag{3.2.25a}$$

iff

$$C_{z_1z_2}(l_1/\Delta, l_2/\Delta) = \begin{cases} 1 \,, & (l_1, l_2) = (0,0) \\ 0 \,, & (l_1, l_2) \neq (0,0) \end{cases} \,. \tag{3.2.25b}$$

With the same arguments as in the previous paragraph, we can state that this condition is met in good approximation. Either $C_{a_1a_2}(\nu_1, \nu_2)$ guarantees this property if the i.i.d. data sequence $\langle a[k] \rangle$ is uniformly distributed, or the feedback sequence $\langle f[k] \rangle$ is near Gaussian with sufficient variance.

If (3.2.25a) holds, the samples $x[k_1]$ and $x[k_2]$ are statistically independent and thus uncorrelated, because the joint density can be written as a product of the marginal densities [Pap91]

$$\mathsf{f}_{x_1x_2}(x_1, x_2) = \frac{1}{\Delta^2} = \frac{1}{\Delta} \cdot \frac{1}{\Delta} = \mathsf{f}_{x_1}(x_1) \cdot \mathsf{f}_{x_2}(x_2) \,. \tag{3.2.26}$$

Since this property holds for all k_1, $k_2 \in \mathbb{Z}$, $k_1 \neq k_2$, we have

$$\phi_{xx}[\kappa] = \begin{cases} \sigma_x^2 = \frac{M^2}{3} & , \kappa = 0 \\ 0 & , \text{else} \end{cases} \,, \tag{3.2.27}$$

i.e., the power spectral density (PSD) of the channel symbols is white and reads $\Phi_{xx}(e^{j2\pi fT}) = \frac{M^2}{3}$.

Tomlinson [Tom71] describes the precoder to have the effect of scrambling or randomizing the input data. Even for a monofrequent input it generates a pseudorandom transmit sequence and spreads the frequency spectrum of the input to a uniform one.

The statistical properties of the channel symbols generated by a Tomlinson-Harashima precoder are summarized in the following theorem [FE91, TF92]:

Theorem 3.1: *Statistical Properties of THP Transmit Signal*

The sequence $\langle x[k]\rangle$ of channel symbols generated by Tomlinson-Harashima precoding is almost i.i.d., and uniformly distributed within the region $\mathcal{R} = [-M, M)$. The approximation becomes more exact as M, the number of signal levels, increases.

3.2.3 Tomlinson-Harashima Precoding for Complex Channels

Up to now, Tomlinson-Harashima precoding has been described using baseband transmission and one-dimensional signal sets. The generalization of the "Tomlinson principle" to square QAM constellations, and channels given in the equivalent complex baseband domain, first appeared in [MS76]. Here, the feedback filter is complex-valued, i.e., consists of four cross-coupled real-valued filters, and modulo reduction is applied independently to in-phase and quadrature component (real and imaginary part).

Two-dimensional Tomlinson-Harashima precoding for square QAM constellations is shown in Figure 3.8. The impulse response $\langle h[k]\rangle$, $h[0] = 1$, and the corresponding transfer function of the complex-valued channel (equivalent complex baseband) is decomposed as follows [Fra80, Hub93]

$$\begin{array}{ccccc} h[k] & = & h_{\mathrm{I}}[k] & + & \mathrm{j}\, h_{\mathrm{Q}}[k]\,, \qquad h_{\mathrm{I}}[k],\, h_{\mathrm{Q}}[k] \in \mathbb{R} \\ \circ\!\!-\!\!\bullet & & \circ\!\!-\!\!\bullet & & \circ\!\!-\!\!\bullet \\ H(z) & = & H_{\mathrm{I}}(z) & + & \mathrm{j}\, H_{\mathrm{Q}}(z)\,. \end{array} \tag{3.2.28}$$

Additionally, the compact complex representation is given on the right-hand side.

To obtain a linearized description of Tomlinson-Harashima precoding (cf. Figure 3.4), we first introduce a beneficial one-to-one correspondence between complex numbers and two-dimensional real vectors (group isomorphism $\mathbb{R}^2 \to \mathbb{C}$)

$$a = b + \mathrm{j}c \leftrightarrow \boldsymbol{a} = \begin{bmatrix} b \\ c \end{bmatrix}, \qquad b, c \in \mathbb{R}\,. \tag{3.2.29}$$

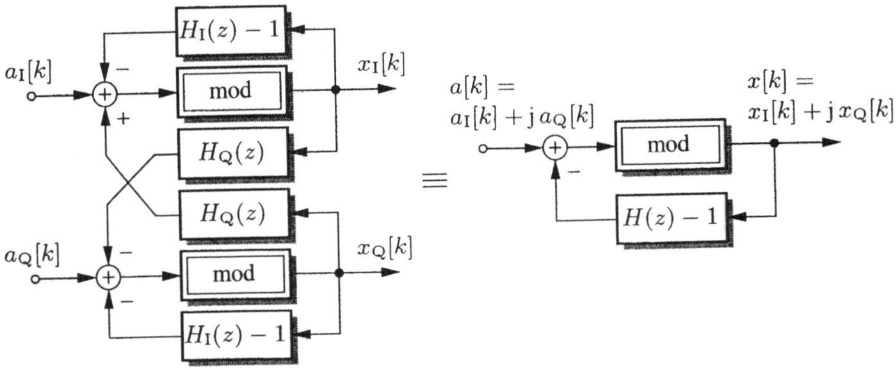

Fig. 3.8 Two-dimensional Tomlinson-Harashima precoding for square QAM constellations and complex-valued channels.

Later, we will use both notations—especially in the context of sets and regions—interchangeably. For the square QAM constellation

$$\begin{aligned} \mathcal{A} &= \left\{a_I + j\, a_Q \mid a_I, a_Q \in \left\{\pm 1, \ldots, \pm(\sqrt{M}-1)\right\}\right\} \\ &\updownarrow \\ \mathcal{A} &= \left\{\pm 1, \ldots, \pm(\sqrt{M}-1)\right\}^2 \end{aligned} , \qquad (3.2.30)$$

with M points, M a square number, the operation of Tomlinson-Harashima precoding is as follows: In both dimensions, over the entire two-dimensional plane, the initial QAM signal constellation is extended periodically. The actual representative is drawn from this set. This selection is done symbol-by-symbol, so that after linear preequalization, the transmit symbols lie in the region $\mathcal{R} = [-\sqrt{M}, +\sqrt{M})^2$, which coincides with the *boundary region* of the constellation $\mathcal{A}$. Let $\mathbf{\Lambda}_a = 2\mathbb{Z}^2$ be the *signal-point lattice*, then the constallation is given by $\mathcal{A} = \left(\mathbf{\Lambda}_a + \left[\begin{smallmatrix}1\\1\end{smallmatrix}\right]\right) \cap \mathcal{R}$. Consequently, the elements $d[k]$ of the precoding sequence are taken from the set $2\sqrt{M}\mathbb{Z}^2$, and the expanded signal set reads $\mathcal{V} = \mathcal{A} + 2\sqrt{M}\mathbb{Z}^2$.

Figure 3.9 shows the expansion of the 16-ary QAM constellation, and shows the signal points, which are congruent to $3 + j \leftrightarrow \left[\begin{smallmatrix}3\\1\end{smallmatrix}\right]$.

The above approach of decomposing the modulo device into a pair of one-dimensional modulo devices is only appropriate for square constellations. In particular, odd numbers of bits per two-dimensional symbol are not supported. Nevertheless, this strategy is used in [ACZ91] in combination with a 128-ary cross constellation, carrying seven information bits. Here, the periodic extension exhibits "holes" due to the missing vertices of the constellation, which increases the power penalty (see below) over square constellations.

Fortunately, the extension of Tomlinson-Harashima precoding to an odd number of bits per two dimensions is possible by simply rotating the boundary region $\mathcal{R}$ by 45° [EF92]. Let M be the number of signal points, $\log_2(M)$ odd, the square has sides $2\sqrt{M}$. Modulo reduction into this region is done jointly for real and imaginary

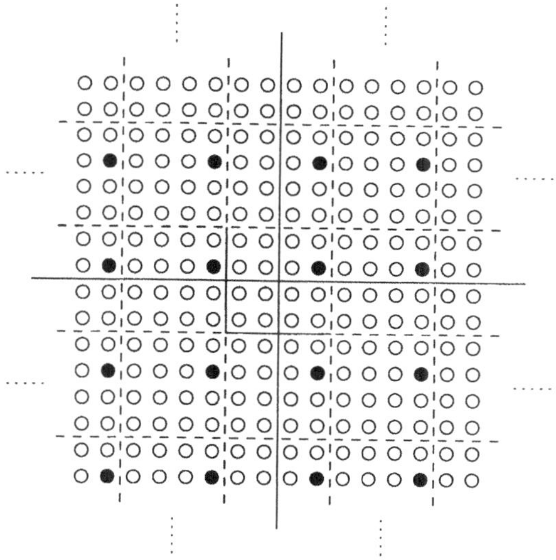

Fig. 3.9 Extended signal set $\mathcal{V}$ for $(M = 16)$-QAM. The coordinates of the points are taken from the set of odd integers. For illustration, the points congruent to the initial point $\left[\begin{smallmatrix}3\\1\end{smallmatrix}\right]$ are marked.

parts. Here, the elements of the precoding sequence $\langle d[k]\rangle$ are drawn from the set $\sqrt{2M}\boldsymbol{R}\mathbb{Z}^2$, where $\boldsymbol{R} = \begin{bmatrix}1 & -1\\ 1 & 1\end{bmatrix}$ is the rotation and scaling operator (for details, see Appendix C). Figure 3.10 shows the boundary regions $\mathcal{R}$ and the set of symbols $d[k]$ ("o") for both, $\log_2(M)$ even and odd.

Instead of rotating the boundary region, the grid of the signal points can be rotated, too, cf. [PE91, Figure 8] or [Wei94, Figure 2]. The resulting transmission schemes are equivalent—all signals are simply turned by $45°$. The advantage is that modulo reduction can again be done independently for real and imaginary part. In this book, we prefer the signal points taken from the grid $2\mathbb{Z}^2$ (or a translate thereof), and thus resort to rotated boundary regions.

Up to now, only constellations with (possibly rotated) square boundaries have been considered—we call it *square constellations*. But in QAM, a much more flexible choice of the signal constellation is possible. With regard to elementary geometry, the design of the signal set $\mathcal{A}$ and the choice of the elements $d[k]$ for Tomlinson-Harashima precoding are equivalent to the problem of tiling the two-dimensional plane without any gap. Hence, unfortunately, neither conventional cross constellations nor circular constellations, which offer a (small) shaping gain, can be used.

A systematic construction of signal sets and the associated modulo operation for generalized Tomlinson-Harashima precoding can be obtained from *lattice theory*. For an introduction to lattices, see Appendix C. In the language of lattice theory, key element for the design of the signal set is the *precoding lattice* Λ_p, from which the samples of the precoding sequence $\langle d[k]\rangle$ are taken. Modulo reduction is done with

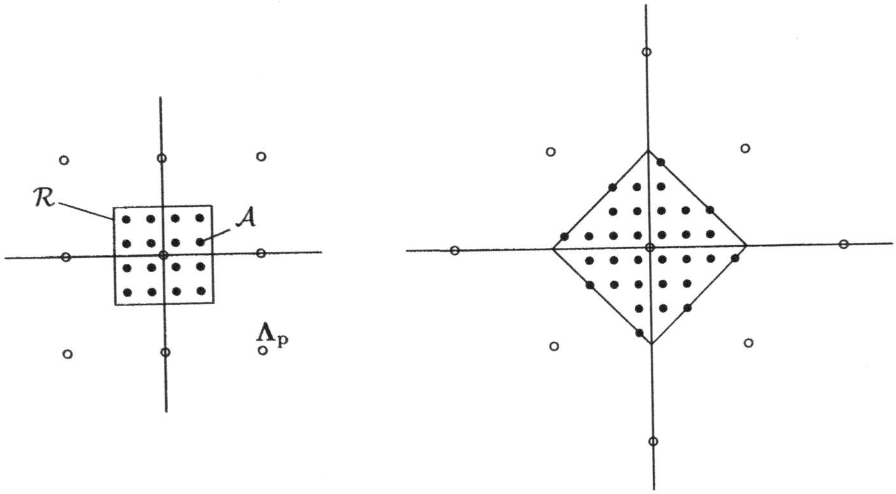

Fig. 3.10 Two-dimensional boundary regions of square constellations for Tomlinson-Harashima precoding.

respect to the precoding lattice $\mathbf{\Lambda}_\mathrm{p}$ into a *fundamental region* $\mathcal{R}(\mathbf{\Lambda}_\mathrm{p})$ of the precoding lattice $\mathbf{\Lambda}_\mathrm{p}$.

The signal constellation $\mathcal{A}$ is then the intersection of the odd-integer grid (translated version of the signal-point lattice $\mathbf{\Lambda}_\mathrm{a} = 2\mathbb{Z}^2$) and $\mathcal{R} = \mathcal{R}(\mathbf{\Lambda}_\mathrm{p})$. Since the channel symbols $x[k]$ fall into the region $\mathcal{R}$, its shape determines the transmit power, and hence it is preferable to choose the *Voronoi region*[6] $\mathcal{R}_\mathrm{V}(\mathbf{\Lambda}_\mathrm{p})$ as the boundary region. Mathematically, assuming channel symbols $x[k]$ uniformly distributed over the region $\mathcal{R}$, the transmit power is given by the *second moment* of the region $\mathcal{R}$ (regions are always assumed to have zero mean). Now, the extended signal set equals the grid of odd-integers, since

$$\begin{aligned}\mathcal{V} &= \mathcal{A} + \mathbf{\Lambda}_\mathrm{p} \\ &= \{a + d \,|\, a \in (2\mathbb{Z}^2 + \left[\begin{smallmatrix}1\\1\end{smallmatrix}\right]) \cap \mathcal{R}(\mathbf{\Lambda}_\mathrm{p}),\ d \in \mathbf{\Lambda}_\mathrm{p}\} \\ &= 2\mathbb{Z}^2 + \left[\begin{smallmatrix}1\\1\end{smallmatrix}\right] .\end{aligned} \tag{3.2.31}$$

All points in $\mathcal{V}$, which are *equivalent modulo* $\mathbf{\Lambda}_\mathrm{p}$, i.e., whose difference is in $\mathbf{\Lambda}_\mathrm{p}$, represent the same data.

Figure 3.11 shows the general structure of a transmission system using Tomlinson-Harashima precoding. The data symbols $a[k]$ are drawn from the signal constellation $\mathcal{A}$. By adding suitably selected elements $d[k]$ of the precoding lattice $\mathbf{\Lambda}_\mathrm{p}$, the effective data sequence with samples taken from the whole grid of odd integers is generated. This sequence is filtered with the (formal) inverse of the channel transfer function

[6]Here, the Voronoi region of a lattice denotes that region, whose points are at least as close to the origin as to any other point of the lattice.

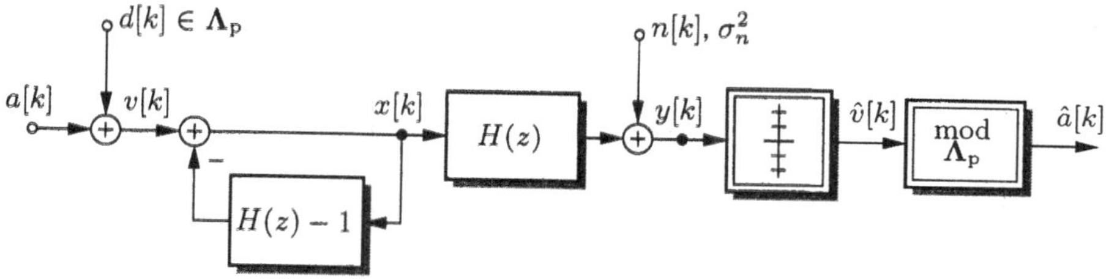

Fig. 3.11 General transmission scheme using Tomlinson-Harashima precoding.

$H(z)$, in order to obtain the channel symbols $x[k]$, which have to lie exclusively within a fundamental region $\mathcal{R}(\Lambda_p)$. These symbols are transmitted over the actual channel, given by its T-spaced model $H(z)$. At the receiver, the noisy observation $y[k]$ is quantized to the translated signal-point lattice $\Lambda_a + \left[\begin{smallmatrix}1\\1\end{smallmatrix}\right]$, resulting in an estimate $\hat{v}[k]$ for the effective data sequence. A reduction modulo Λ_p, i.e., to $\mathcal{R}(\Lambda_p)$, gives the final estimates $\hat{a}[k]$ for the data symbols.

From Figure 3.11 it is again obvious that Tomlinson-Harashima precoding is an extension of *linear preequalization*. Here, the signal set $\mathcal{V}$ is used instead of $\mathcal{A}$.

In [Wei94] a large number of signal constellations and boundary regions suited for Tomlinson-Harashima precoding are given. In particular, "generalized square" constellations, resembling QAM cross constellations, and "generalized hexagonal" constellations are discussed. Using these signal sets, transmission at a fractional number of bits per symbols is possible. The hexagonal constellations (which differ slightly from a regular hexagon) are of particular interest, since they achieve some *shaping gain*, i.e., their second moment is less than that of a square constellation of equal cardinality. Moreover, as the vertices of the square are omitted, a reduction in peak power is also possible. Figure 3.12 shows the boundary regions and the respective tiling of the two-dimensional plane for examples of generalized square and generalized hexagonal constellations.

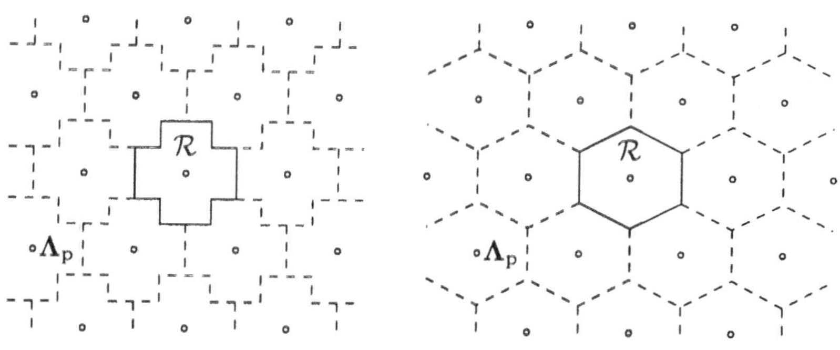

Fig. 3.12 Examples for boundary regions and respective tiling of the two-dimensional plane. Left: Generalized square constellations; Right: Generalized hexagonal constellations.

3.2.4 Precoding for Arbitrary Signal Constellations

The description of Tomlinson-Harashima precoding given above is based on a conventional (one- or two-dimensional) PAM signal constellation $\mathcal{A}$ which is extended periodically. Hence, the set $\mathcal{V}$ of effective data symbols is also a regular PAM constellation. In particular, the signal points are taken from a regular grid (translate of a lattice), and thus are uniformly spaced.

There are (rare) situations where such a constellation $\mathcal{V}$ cannot be used. Recall, the effective data symbols $v[k]$—disturbed by additive noise—are present at the decision device. If the slicer is fixed and does not fit to a uniformly spaced constellation, precoding as described up to now is impossible. The solution to overcome this problem is to adapt the set $\mathcal{V}$ of effective data symbols to the slicer and to slightly change the operation of the precoder.

The best example for such a situation is the up-stream transmission in the latest version of voice-band modems for *pulse code modulation (PCM)* telephone lines (ITU recommendation V.92 [ITU00]). Here, the A/D converter at the central office performs nonuniform, in particular logarithmic, quantization according to A-law companding (Europe) or μ-law companding (North America and Japan) [ITU93, JN84]. Signals with small amplitudes are thereby quantized with a much smaller step size than signals with large amplitudes. In turn, for PAM data transmission over the analog section from the customer to the central office, a suitable nonuniformly spaced signal constellation has to be used. Signal points with small amplitudes are spaced more closely to each other than signal points with large amplitudes. Moreover, due to the dispersive nature of the telephone line, some kind of equalization has to be applied. Since the receiver side (central office) is fixed, precoding is the only way for combating intersymbol interference.

Tomlinson-Harashima-type precoding for arbitrary signal constellations and a desired transmission rate $R_{\mathrm{m}} = \log_2(M)$, $M \in \mathbb{N}$, can be done in the following way [ITU00]. The given signal set

$$\mathcal{V} = \left\{v_0, v_1, \ldots, v_{|\mathcal{V}|-1}\right\} \tag{3.2.32}$$

with $|\mathcal{V}| > M$ signal points is partitioned into M disjoint *equivalent classes* of signal points. This is done such that an equivalent class $\mathcal{E}_i$ contains all points v_l whose *indices* l are congruent modulo M. The M equivalent classes are thus defined as

$$\mathcal{E}_i \triangleq \{v_l \mid l = i + M\mathbb{Z},\, 0 \le l < |\mathcal{V}|\}\ , \quad i = 0, 1, \ldots, M-1\ . \tag{3.2.33}$$

This procedure requires some specific sorting of the signal points v_l. Generally, the equivalent classes have to be interlaced—in particular, $v_l < v_{l+1}$ has to be met for one-dimensional constellations.

At the transmitter, information is first mapped to the number i of the equivalent class. Given the previously transmitted symbols, the postcursor sum $f[k]$ is calculated at the precoder. With this knowledge, the symbol v_l is selected from $\mathcal{E}_i$ as effective data symbol $v[k]$, which—after linear preequalization via $1/H(z)$—minimizes the amplitude of the channel symbol $x[k]$, i.e., for which $|v_l - f[k]|$ is minimum. Other selection strategies, e.g., a long-term selection for minimizing average transmit power, are also possible. (Compare Section 5.2 on shaping without scrambling, a principle which is applicable to the present scenario, as well.) The block diagram of precoding for arbitrary signal constellations is shown in Figure 3.13.

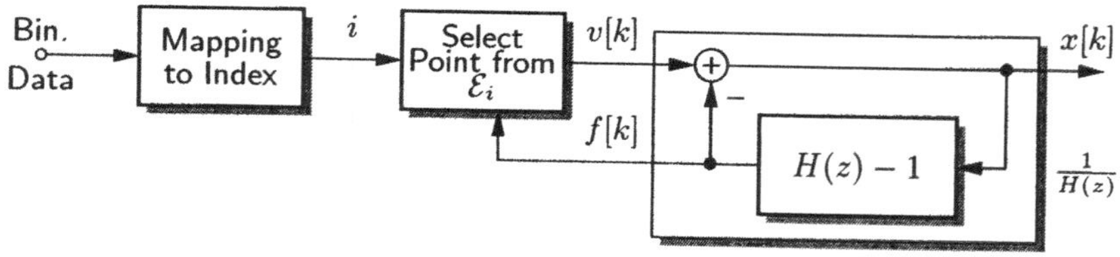

Fig. 3.13 Generalization of Tomlinson-Harashima precoding for arbitrary signal constellations.

Tomlinson-Harashima precoding for arbitrary signal constellations can therefore be done in almost the same way as for uniformly spaced constellations. Instead of treating absolute *signal levels* modulo $2M$, *indices* are treated modulo M. As a consequence of the nonuniform spacing of the effective data symbols $v[k]$, the channel symbols $x[k]$ will not be uniformly distributed over some region.

Later, we will restrict the exposition to one- and two-dimensional regular constellations and will not consider arbitrary signal constellations any further.

3.2.5 Multidimensional Generalization of Tomlinson-Harashima Precoding

The principle of Tomlinson-Harashima precoding can easily be extended to channels with more than one input and one output. Such *matrix channels* or *multiple-input/multiple-output (MIMO) channels* arise, e.g., when regarding parallel, but perhaps mutually interfering transmission of, say, D users. A second example is mobile communications with multiple transmit and receive antennas.

As in the single-user case, each channel may exhibit intersymbol interference and may be distorted by additive noise. But now, additional *interchannel interference* can occur. Mathematically, the transmission characteristics are described by the channel matrix of the $D \times D$ transfer functions (or impulse responses) from each of the D inputs to each of the D outputs. W. van Etten [Ett76] showed that analog to the scalar case, here a *multiple whitened matched filter* constitutes the optimum receiver front-end.[7] The T-spaced discrete-time end-to-end description[8] can then be given as

$$\boldsymbol{H}(z) = \sum_{k=0}^{p} \boldsymbol{h}[k]\, z^{-k} \,, \tag{3.2.34}$$

[7] The solution for linear equalization is given in the preceding paper [Ett75].

[8] Note, a 2×2 vector channel is more general than a complex-valued channel, where the transfer functions between the quadrature components have to meet specific symmetry properties.

where, in each case, $\boldsymbol{h}[0]$ can be chosen as lower left triangular matrix, with diagonal elements all equal to one. Note, $\boldsymbol{h}[k]$ is a constant (not dependent on z) $D \times D$ matrix, which describes the mutual impulse responses at instant k. Similarly as explained in Section 2.3.2, $\boldsymbol{H}(z)$ can be obtained via factorization of a spectral matrix; for the problem of factorizing a matrix, see, e.g., [You61, Dav63].

Based on the discrete-time channel model, in [Ett76], two strategies for maximum-likelihood sequence estimation using the Viterbi algorithm ("vector Viterbi algorithm") are derived. The first approach is a generalization of Forney's solution to MLSE [For72], whereas the second one is a vector version of Ungerböck's algorithm [Ung74], cf. also Section 2.5. Unfortunately, the number of states in the Viterbi algorithm tends to be extremely high, which prohibits practical use.

This problem can be overcome by applying *vector* or *multidimensional decision-feedback equalization*, see, e.g., [Due95]. The feedback filter is given by $\boldsymbol{H}(z) - \boldsymbol{I}$. Since $\boldsymbol{h}[0]$ is forced to be a lower left triangular matrix, decisions can be generated successively, without conflicting causality. Thereby, for decision in channel i the estimates of channels $1, \ldots, i-1$ are already incorporated. In [FHK94, Fis96] it is proved that for optimal power and rate allocation among the channels—they are coordinated—vector DFE in combination with powerful channel codes, designed for the AWGN channel, can approach channel capacity. This generalizes Price's result [Pri72], see also Section 2.3.2. In summary, the application of a multiple whitened matched filter and equalization using vector DFE is a *canonical* receiver concept, as it is in the one-dimensional, scalar case.

Moving the feedback filter of the DFE into the transmitter leads directly to *multidimensional Tomlinson-Harashima precoding*. Because of the lower left triangular structure of the matrix $\boldsymbol{h}[0]$, the components can be processed successively. If the data symbols in the channels are selected independently of each other, each component can be modulo reduced individually, as for square QAM constellations. We have seen, that in QAM the choice of the boundary region of the signal constellation is much more multifarious than in one dimension. Going to D dimensions provides even more flexibility in signal design. Here, again, *lattice theory* is very helpful. As described in the last section, the key point is the (D-dimensional) precoding lattice Λ_{p}. The data signal set is the intersection of a regular grid (e.g., $(2\mathbb{Z}+1)^D$) and a fundamental region (preferably the Voronoi region) $\mathcal{R}(\Lambda_{\mathrm{p}})$ of the precoding lattice. Due to the reduction modulo Λ_{p}, the transmit vectors lie in the region $\mathcal{R}(\Lambda_{\mathrm{p}})$.

Figure 3.14 shows the transmission scheme employing multidimensional Tomlinson-Harashima precoding.

In the remainder of this chapter, we will not discuss multidimensional schemes further, and will return to conventional one- and two-dimensional PAM. Some further aspects of precoding for MIMO channels are discussed in Appendix E.

3.2.6 Signal-to-Noise Ratio

In an earlier section we have shown that the original Tomlinson-Harashima precoder produces a near-i.i.d. transmit sequence $\langle x[k] \rangle$, uniformly distributed over the interval $\mathcal{R} = [-M, M)$. Following the same line of the derivation, it can be deduced that also

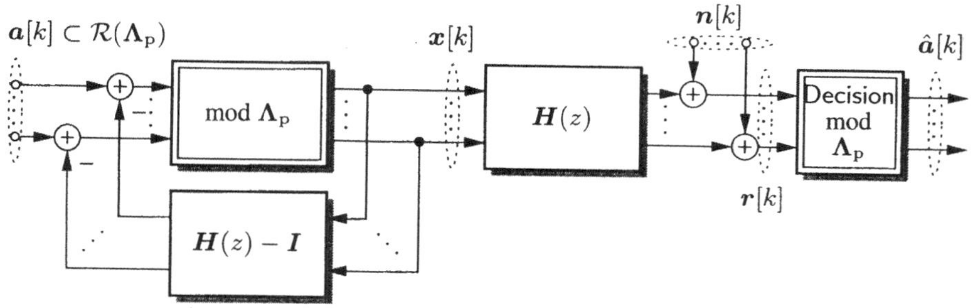

Fig. 3.14 Transmission scheme using multidimensional Tomlinson-Harashima precoding.

for the generalization of Tomlinson-Harashima precoding the discrete-time transmit signal is near i.i.d. and uniformly distributed within the fundamental region $\mathcal{R}$. Thus the transmit signal resembles an ordinary PAM transmit sequence. The only difference is that the variance of the transmit symbols $x[k]$ is increased compared to conventional PAM transmission and discrete signal points $a[k]$.

At the receiver side, the spacing of the signal points and the noise variance are unchanged compared to DFE. Hence, neglecting the slight increase in the number of nearest-neighbor signal points, the decision is as reliable as for DFE. Consequently, except for the transmit power penalty, called *precoding loss*,

$$\gamma_p^2 \triangleq \frac{\mathrm{E}\{|x[k]|^2\}}{\mathrm{E}\{|a[k]|^2\}}, \tag{3.2.35}$$

Tomlinson-Harashima precoding performs as well as the ideal (i.e., error-free) zero-forcing DFE.

For a one-dimensional signal set $\mathcal{A} = \{\pm1, \pm3, \dots, \pm(M-1)\}$, the variance of the data symbols $a[k]$, uniformly distributed over $\mathcal{A}$,

$$\sigma_a^2 = \mathrm{E}\{|a[k]|^2\} = \frac{M^2-1}{3}, \tag{3.2.36}$$

is increased to the variance of the channel symbols $x[k]$, uniformly distributed over $\mathcal{R} = [-M, M)$,

$$\sigma_x^2 = \mathrm{E}\{|x[k]|^2\} = \frac{M^2}{3}. \tag{3.2.37}$$

Hence, for one-dimensional M-ary signaling the precoding loss reads

$$\gamma_{p,1D}^2 = \frac{M^2}{M^2-1}. \tag{3.2.38}$$

Applying two-dimensional (possibly rotated) square constellations with M signal points, the variance of the data symbols $a[k]$, and the variance of the channel symbols $x[k]$, uniformly distributed over the square region $\mathcal{R}$, respectively, are given as

$$\sigma_a^2 = 2\frac{M-1}{3}, \qquad \sigma_x^2 = 2\frac{M}{3}. \tag{3.2.39}$$

Thus, for two-dimensional M-ary square constellations, the precoding loss calculates to

$$\gamma_{\text{p,2D}}^2 = \frac{M}{M-1} . \tag{3.2.40}$$

Table 3.1 shows the precoding loss γ_{p} (in dB) for various constellation sizes. Even

Table 3.1 Precoding loss γ_{p} (in dB) of Tomlinson-Harashima precoding.

$M =$	2	4	8	16	32	64
$10\log_{10}(\gamma_{\text{p,1D}}^2)$ [dB]	1.25	0.28	0.07	0.02	0.004	0.001
$10\log_{10}(\gamma_{\text{p,2D}}^2)$ [dB]	—	1.25	0.58	0.28	0.14	0.07

for moderate sizes M the precoding loss γ_{p}^2 is negligible and vanishes completely as M goes to infinity. We summarize:

Theorem 3.2: *Signal-to-Noise Ratio of Tomlinson-Harashima Precoding*

When using Tomlinson-Harashima precoding, the signal-to-noise ratio is given by

$$\text{SNR}^{(\text{THP})} = \text{SNR}^{(\text{ZF-DFE})} / \gamma_{\text{p}}^2 , \tag{3.2.41}$$

where the signal-to-noise ratio $\text{SNR}^{(\text{ZF-DFE})}$ of ideal ZF-DFE is given in (2.3.55) and

$$\gamma_{\text{p}}^2 = \mathsf{E}\{|x[k]|^2\} / \mathsf{E}\{|a[k]|^2\} \tag{3.2.42}$$

is the *precoding loss*, i.e., the increase in average transmit power relative to conventional PAM signaling.

For DFE, we have seen that an optimization according to the MMSE criterion provides some gain over zero-forcing. The modification of Tomlinson-Harashima precoding for the application of MMSE filters and the resulting consequences will be discussed in Section 3.7.

3.2.7 Combination with Coded Modulation

Usually, precoding does not end in itself, but establishes the basis for coded modulation. Price's important result of digital transmission theory [Pri72, FE91, CDEF95], derived in the last chapter (pages 77 and 95), states that capacity of ISI-producing channels can (asymptotically) be achieved by coding techniques developed for the AWGN channel, if ideal DFE, or equivalently, precoding, is applied. Consequently, the combination of precoding and channel coding is a good choice to closely approach capacity with moderate expenditure.

In this book, we mainly concentrate on PAM transmission with high spectral efficiency using constellations with $M > 2$ signal points. For such nonbinary signaling, channel coding and modulation (mapping of binary vectors to signal points) have to be optimized jointly [Mas74]. The aim is to optimize the code in Euclidean space rather than dealing with Hamming distance, as in classical coding schemes. One representative of *coded modulation* schemes is *trellis-coded modulation (TCM)*, introduced by Ungerböck [UC76, Ung82, Ung87a, Ung87b], and later generalized to signal constellations derived from lattice theory, establishing the so-called *coset codes*, see, e.g., [CS87, For88a, For88b, PC93a]. A further coded modulation scheme is *multilevel coding (MLC)* proposed by Imai [IH77], which, in retrospect, can be seen as a particular lattice construction [CS88]. For a tutorial review of MLC, see [WFH99].

Trellis-coded modulation and multilevel codes have in common that they are based on a repeated partitioning of the signal constellation into subsets. MLC protects each address bit, representing a binary partition step, by its own code, whereas TCM directly treats the K-fold partitioning (K subsets, K a power of two). For the present application, it is sufficient to think of coding based on a partitioning of depth K. In the denomination of lattices (see Appendix C for detais), the subsets are translates of the *"coset lattice"* or *"coding lattice"* Λ_c, which obviously is a sublattice of the signal lattice Λ_a. In TCM, the subsets are selected via a single convolutional coder, and all possible sequences of subsets constitute the trellis code. The actual signal point within the current subset is addressed memoryless, symbol by symbol. The corresponding addressing bits are often called the *uncoded levels*. Figure 3.15 shows the essential parts of the TCM encoder. Multilevel coding is done in the same way, replacing the convolutional encoder by individual binary channel encoders for each address symbol (encoding level).

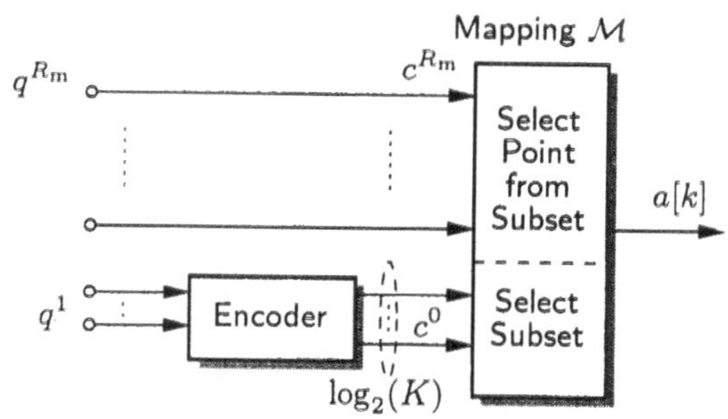

Fig. 3.15 General structure of the TCM encoder.

As already explained, except for the modulo congruence, Tomlinson-Harashima precoding produces an end-to-end AWGN channel. Hence, TCM, designed for the AWGN channel, can be used in the usual way. The periodic continuation of the signal set can be viewed as an extension of the mapping by an infinite number of additional uncoded levels. The only restriction on the signal constellation is that the periodic extension (the additional levels) must not reduce the intrasubset distance.

Mathematically, we demand that

$$\min_{\substack{\lambda_1,\lambda_2 \in \Lambda_c \cap \mathcal{R} \\ \lambda_1 \neq \lambda_2}} |\lambda_1 - \lambda_2|^2 \overset{!}{=} \min_{\substack{\lambda_1,\lambda_2 \in ((\Lambda_c \cap \mathcal{R}) + \Lambda_p) \\ \lambda_1 \neq \lambda_2}} |\lambda_1 - \lambda_2|^2 . \tag{3.2.43}$$

Figure 3.16 visualizes the effect of periodic extension on the distances.

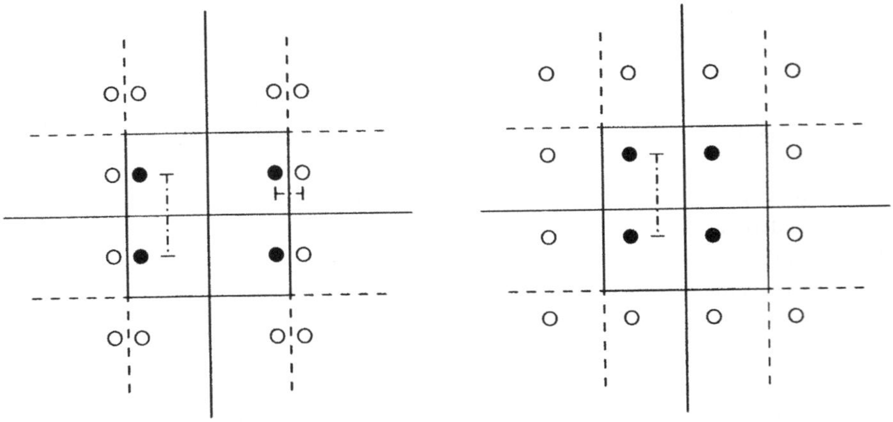

Fig. 3.16 Effect of periodic extension on the intrasubset distance. Solid dots: original subset. Left: decrease of the intrasubset distance due to the periodic extension (open dots); Right: the intrasubset distance is preserved.

The above demand is preferably met if the continuation of each subset entirely falls in a translate of Λ_c. This calls for the precoding lattice Λ_p to be a subset of the coding lattice Λ_c. Since, on the other hand, the coding lattice is a sublattice of the signal lattice Λ_a, a partitioning chain

$$\Lambda_a/\Lambda_c/\Lambda_p \tag{3.2.44}$$

should be constituted. Whether this requirement can be fulfilled depends on the number M of signal points, the boundary region $\mathcal{R}$ of the constellation, and the depth K of the partition. For one-dimensional signal sets and two-dimensional square constellations it is very likely that the desired partitioning can be found. The compatibility of general square or hexagonal constellations (cf. Figure 3.12) for coded modulation with Tomlinson-Harashima precoding is discussed in [Wei94].

Example 3.1: *Signal Lattice, Coset Lattice, and Precoding Lattice*

This example gives the signal lattice Λ_a, the coset lattice Λ_c, and the precoding lattice Λ_p. We start from the signal lattice $\Lambda_a = 2\mathbb{Z}^2$, plotted on the left-hand side. Additionally, the 16-ary QAM constellation derived from the translate $2\mathbb{Z}^2 + \left[\begin{smallmatrix}1\\1\end{smallmatrix}\right]$ is sketched. For two-dimensional signal sets, coded modulation is preferably based on an eight-way partitioning [Ung82]. This results in the coset lattice $\Lambda_c = 4\boldsymbol{R}\mathbb{Z}^2$, displayed in the middle. From the figure, it is easy to verify that this subset of $2\mathbb{Z}^2$ induces eight cosets. Last, a further two-way partitioning leads to the precoding lattice $\Lambda_p = 8\mathbb{Z}^2$ (right-hand side of the figure), which is suited for the 16-ary QAM signal set.

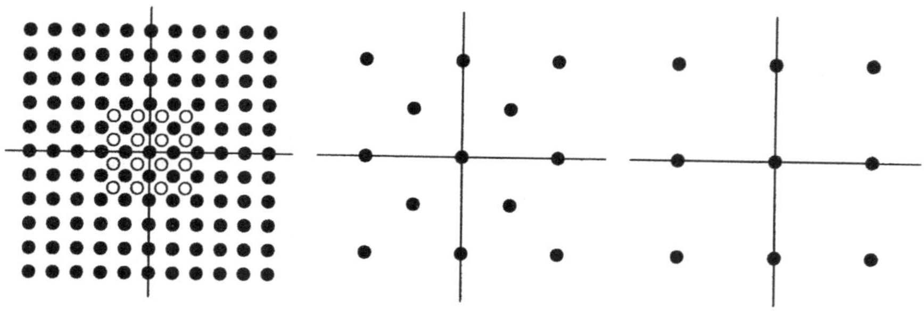

Fig. 3.17 From left to right: signal lattice $\Lambda_a = 2\mathbb{Z}^2$ and 16-ary QAM constellation (open dots); coset lattice $\Lambda_c = 4\boldsymbol{R}\mathbb{Z}^2$; precoding lattice $\Lambda_p = 8\mathbb{Z}^2$.

Now let us turn to decoding. The task of maximum-likelihood (ML) decoding is to find that valid code sequence which produces the received sequence with highest probability. This can be done by setting up the *conditional probability function*—also called *likelihood function*—of the received sequence given the transmit sequence, and finding that transmit sequence which maximizes this function. In trellis-coded modulation, the decoder has first to determine the "best" sequence of subsets. Then, the "best" point within each subset can be detected symbol by symbol.

Regarding the subsets, each subset is composed of a number of actual signal points. Due to the periodic extension in Tomlinson-Harashima precoding, at the receiver each subset is given as a translate of the coding lattice Λ_c, i.e., extends infinitely. Thus, information conveyed in the number of the subset is multiply represented. Strictly speaking, maximum-likelihood decoding has to take this multiple symbol representation into account. The likelihood function for subset $\mathcal{S}$ is hence given as

$$\mathsf{f}_y(y \mid \mathcal{S}) = \sum_{d \in \Lambda_c} \mathsf{f}_n(y - (s + d)) \,, \qquad \text{for some } s \in \mathcal{S} \,, \tag{3.2.45}$$

since the received point y may potentially emerge for any of the subset points. Here, $\mathsf{f}_n(n)$ is the Gaussian density function of the additive noise.

Yet at moderate, and especially at high signal-to-noise ratios, only the signal point s closest to the received point y in Euclidean space contributes significantly to the above sum. Thus ignoring the contributions of the other representatives, taking the logarithm, and ignoring constant terms irrelevant for maximization, decoding can be based on the metric

$$\log \mathsf{f}_y(y \mid s) \stackrel{\approx}{\sim} \min_{d \in \Lambda_c} |y - (s + d)|^2 = \lambda^{(s)} \,. \tag{3.2.46}$$

The metric increment $\lambda^{(s)}$, when regarding the subset specified by its member point s, is simply the squared Euclidean distance between the received point y and the nearest point in the subset.

In summary, a practical approach to ML decoding in Tomlinson-Harashima precoding is to first reduce the received signal $y[k]$ to the fundamental region $\mathcal{R}$, thus ignoring the ambiguity modulo $\boldsymbol{\Lambda}_{\mathrm{p}}$. Then, the "best" representative $a \in \mathcal{A}$ of each subset s is determined, and $\lambda^{(s)}$ is calculated. The uncoded levels are thereby extracted and the ambiguity modulo $\boldsymbol{\Lambda}_{\mathrm{c}}$ is resolved. But, because of the initial mod-$\boldsymbol{\Lambda}_{\mathrm{p}}$ reduction, Euclidean distances have to be calculated modulo the precoding lattice, i.e., also considering points generated by periodic extension. Consequently, the number of nearest-neighbor signal points is (slightly) increased.

Given the metric increments $\lambda^{(s)}[k]$ for all instants k, a Viterbi decoder will produce an estimated sequence $\langle \hat{a}[k] \rangle$. Compared to conventional TCM over the ISI free channel, only an additional modulo reduction of the receiver input signal is present, and Euclidean distances are measured modulo the precoding lattice. Using appropriate fixed-point arithmetic, the metric reduction is done automatically.

Strictly speaking, metrics derived from (3.2.45) are not totally optimal, because the signal points of the periodic extension are not uniformly distributed and exhibit correlation in time—the PSD of the effective data sequence is a scaled version of $|H(e^{j2\pi fT})|^2$. However, even the simplified metric calculation (3.2.46) comes along with almost no loss in performance.

Neglecting the small increase in the number of nearest neighbors, the same coding gain is expected to be realized, as over the AWGN channel. Since TCM expands the signal set, as a side effect, the precoding loss is reduced by a small amount. By doubling the signal constellation from M to $2M$ points (using one redundant bit per symbol), the precoding loss of the Tomlinson-Harashima system now reads

$$\gamma_{\mathrm{p,1D}}^2 = \frac{(2M)^2}{(2M)^2 - 1}, \qquad \gamma_{\mathrm{p,2D}}^2 = \frac{2M}{2M - 1} \tag{3.2.47}$$

for one- and two-dimensional constellations, respectively. Note that the loss does not depend on the actual code being used.

3.2.8 Tomlinson-Harashima Precoding and Feedback Trellis Encoding

Note: Conceptually this section belongs to Tomlinson-Harashima precoding. But it can only be understood if the reader is familiar with combined flexible precoding and trellis-coded modulation. Thus, Section 3.3.5 should have been read first. Please skip this section at first reading, and return after having worked through Section 3.3 on flexible precoding.

As shown above, Tomlinson-Harashima precoding and trellis-coded modulation can be combined in a straightforward manner if a partition chain $\boldsymbol{\Lambda}_{\mathrm{a}}/\boldsymbol{\Lambda}_{\mathrm{c}}/\boldsymbol{\Lambda}_{\mathrm{p}}$ exists. Although this requirement can be fulfilled for a great number of applications, there are situations where signal lattice, coset lattice, and precoding lattice cannot be chosen to be sublattices of each other. For example, consider a 6×6 QAM constellation. Using this signal set, an eight-way partitioning, which is required for two-dimensional TCM, is impossible, since 36 is not an integer multiple of 8. Interest in such a 6×6 constellation stems from the requirements in Fast Ethernet systems (100BASE-T2: 100 Mbit/s Ethernet transmission over voice-grade cables), where besides 4-bit data nibbles, control information must be conveyed [CÖU97a, CÖU97b].

The solution to the aforementioned problem is to combine Tomlinson-Harashima precoding with feedback trellis encoding. Such a scheme was first introduced in [CÖU97b] in the context of channels with spectral zeros, where the authors called it *"trellis-augmented precoding."* The requirements on the signal constellation are relaxed, so that only an efficient periodic continuation with respect to the first partitioning level has to exist. In terms of lattice, we now require $\Lambda_\mathrm{a}/\Lambda_\mathrm{f}/\Lambda_\mathrm{p}$ to be a lattice partition chain.

The transmitter for combined Tomlinson-Harashima precoding and feedback trellis encoding is depicted in Figure 3.18. As the main block, the conventional

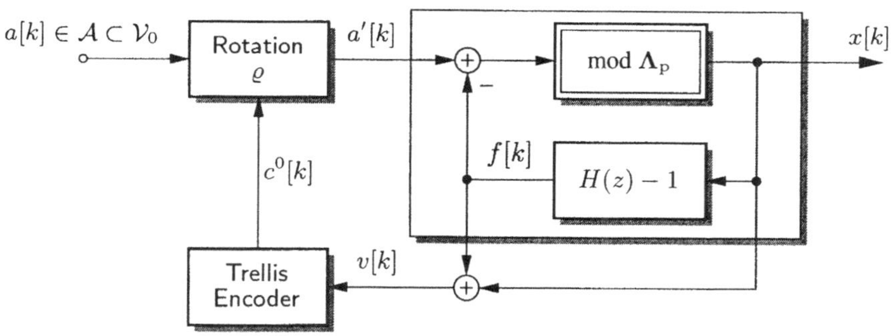

Fig. 3.18 Block diagram of combined Tomlinson-Harashima precoding and feedback trellis encoding [CÖU97b].

Tomlinson-Harashima precoder is recognizable (gray box). Contrary to a straightforward combination, the initial signal points are drawn from a signal constellation $\mathcal{A}$, which is a subset of $\mathcal{V}_0$. This reflects the one-bit redundancy introduced by coded modulation. Note that the extended signal constellation $\mathcal{V}$ may be partitioned into two equal-sized, disjoint subsets according to the set-partitioning principle [Ung82], i.e., $\mathcal{V} = \mathcal{V}_0 \cup \mathcal{V}_1$. Both subsets are translates of the first-partitioning-level lattice Λ_f.

In order to obtain a valid code sequence $\langle v[k] \rangle$ at the channel output, the respective samples $v[k]$ are also generated at the transmitter. Feeding them into the feedback trellis encoder (cf. Figure 3.30), the next sample of the sequence $\langle c^0[k] \rangle$ of parity bits can be calculated. Depending on this binary flag, the initial signal point is either left unchanged ($c^0[k] = 0$), or rotated if $c^0[k] = 1$. Rotation by by 90° (multiplication with $\varrho = \mathrm{j}$) is done for two-dimensional constellations, and by 180° ($\varrho = -1$) for one-dimensional constellations. Via this rotation, signal points drawn from $\mathcal{V}_0$ may be changed into $\mathcal{V}_1$ points. Since we require Λ_f to be a sublattice of Λ_p, $a'[k]$ and its periodic extension $a'[k] + d[k]$, with $d[k] \in \Lambda_{\rm p}$, lie in the same coset of $\Lambda_{\rm p}$. Hence, the subset of the channel output symbols can be determined by rotation. This guarantees that $\langle v[k] \rangle$ is in agreement with the parity condition of the trellis code. In turn, the channel output is ensured to be a code sequence that agrees with the trellis code employed.

Finally, Figure 3.19 shows the receiver corresponding to the above transmitter. First, a Viterbi decoder produces estimates $\hat{v}[k]$ for the extended data symbols $v[k]$.

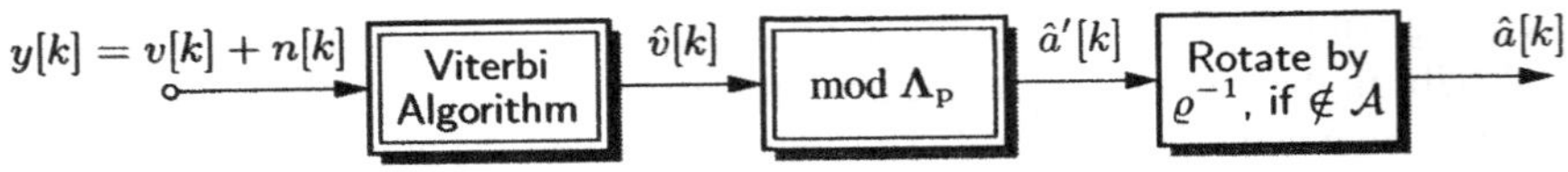

Fig. 3.19 Block diagram of the receiver for combined Tomlinson-Harashima precoding and feedback trellis encoding.

By reducing $\hat{v}[k]$ modulo $\Lambda_{\rm p}$, the symbols $\hat{a}'[k]$ are recovered. Finally, if $\hat{a}'[k]$ lies in $\mathcal{V}_1$, and hence is not an element of $\mathcal{A}$, the symbol is rotated back (by $-90°$ or $-180°$, respectively). This results in estimates $\hat{a}[k]$ for the data carrying symbols.

Please note, here no modulo reduction is possible prior to Viterbi decoding. Metric calculation, of course, has to be based on the coding lattice $\Lambda_{\rm c}$, which (virtually) only exists for the extended signal set. Since $\Lambda_{\rm c}$ is not a sublattice of $\Lambda_{\rm p}$, modulo reduction will cause irreversible aliasing, and hence circumvent recovery of the data symbols $a[k]$.

Just as was described in the last section, albeit a small increase in the number of nearest neighbors, the coding gain of the trellis code over the AWGN channel may be achieved. Moreover, the precoding loss is the same as above; it again does not depend on the actual code used.

3.2.9 Combination with Signal Shaping

The objective of *signal shaping* is to generate a (transmit) signal which exhibits some desired characteristics. The most important purpose is to reduce the average transmit power, which is possible by replacing a uniformly distributed signal by a Gaussian one. In, e.g., [FW89]—see also Chapter 4—it is shown, that the *shaping gain*, the reduction in average power, is limited to $\pi e/6$ or 1.53 dB, and, at least for large constellations (i.e., high rates), is essentially decoupled from the coding gain. Hence, it is common to address coding and shaping separately. Often it is easier to achieve an additional gain by shaping than by going to more powerful channel codes [EFDL93, FE91]. To achieve capacity of the Gaussian channel it is indispensable to use shaping.

Roughly, shaping techniques can be divided into two groups, reflecting the differences between block and convolutional codes. In block shaping techniques, a number of input bits are mapped to a block of the transmit signal. This can basically be done inversely to source coding, i.e., a source decoder can be used as shaping encoder (an example is given in [FGL+84]). An attractive method (derived from vector quantization) is *shell mapping* [LL89, FRC92, EFDL93, KK93, KP93, LFT94], which has been standardized for fast telephone line modems (ITU standard V.34). The only shaping technique—to the knowledge of the author—which uses a convolutional decoder to determine the best sequence is *trellis shaping* [For92]. The principal item is a Viterbi algorithm, which selects the minimum weight sequence. Modifying the branch metrics will adapt the transmit signal to the specific demands. For an in-depth discussion of signal shaping the reader is referred to Chapter 4.

Unfortunately, a straightforward combination of Tomlinson-Harashima precoding and shaping techniques designed for the AWGN channel is not possible, because of the nonlinear device in the feedforward path of the precoder. Signal characteristics prior to precoding may be completely changed by this nonlinearity. The only solution is to combine shaping and precoding into one unit. This, in turn, is only possible for trellis shaping, since only here can the influence of the precoder be incorporated into the shaping algorithm. The combination of trellis shaping and Tomlinson-Harashima precoding is called *trellis precoding* [EF92, FE91]. We will describe combined precoding/shaping techniques in detail in Chapter 5.

3.3 FLEXIBLE PRECODING

During the standards activities for the ITU voice-band modem standard V.34 in the early 1990s, a request for a more flexible precoding scheme arose. The V.34 standard contains a "modulation toolbox," from which after line probing during start-up, an appropriate combination of "tools" is selected. This requires that precoding, channel coding, and signal shaping can be combined in a seamless fashion. In the last section, we saw that this is not possible if Tomlinson-Harashima precoding is used, since it does not preserve the distribution of its input signal points, and, moreover, imposes constraints on the boundary of the signal constellation.

In [EFDL93, LTF93, Lar94] the result of the activities, called *flexible precoding (FLP)* or *distribution-preserving precoding*, is presented. In this section, we describe the operation of flexible precoding and discuss its advantages and disadvantages compared to Tomlinson-Harashima precoding.

3.3.1 Precoder and Inverse Precoder

We anticipate that in contrast to Tomlinson-Harashima precoding which is derived from linear preequalization at the transmitter, flexible precoding resembles linear equalization at the *receiver side*. The disadvantage of linear equalization is that the channel noise is filtered with $1/H(z)$, too, and thus the desired prediction gain is lost. This effect can be avoided, if a decision can be made prior to the inverse channel filter. We now show how this can be accomplished.

First consider one-dimensional signal sets $\mathcal{A}$ taken from $2\mathbb{Z}+1$, a translate of the signal lattice $\Lambda_a = 2\mathbb{Z}$, and uncoded transmission. Assume that $x[k] = a[k] \in \mathcal{A}$ is transmitted over the channel $H(z) = 1 + \sum_{k=1}^{p} h[k] \cdot z^{-k}$ without precoding. Here, the channel output $v[k]$ is given by $a[k]$ plus $f[k] = \sum_{\kappa=1}^{p} h[\kappa] \cdot x[k-\kappa]$, the sum of postcursors of prior transmitted symbols, see Figure 3.20. If $f[k]$ were in Λ_a, i.e.,

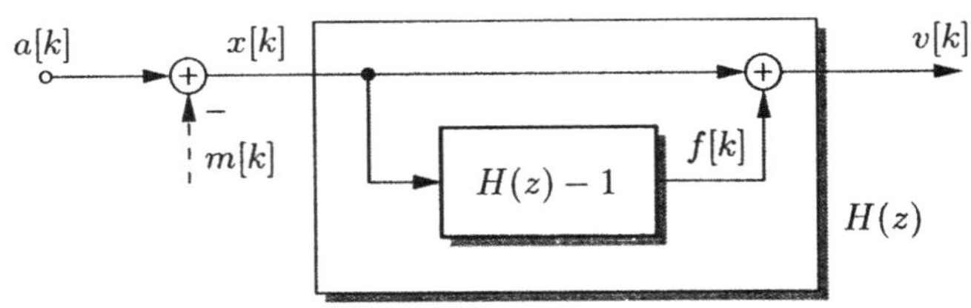

Fig. 3.20 Visualization of the transmit and receive signals.

the signal lattice, then the channel output $v[k]$ would be an odd integer, as is $a[k]$, since $\mathcal{A} + \Lambda_a = 2\mathbb{Z}+1$. Hence, a slicer could eliminate the noise, and $1/H(z)$ could recover data $a[k]$ without any noise enhancement.

Unfortunately, in general, $f[k] \notin \Lambda_a$. Thus we address the other addend, $x[k]$, in order to make $v[k]$ an odd integer. The idea is to modify $x[k]$, i.e., to subtract a "small" sequence $m[k]$ from $a[k]$, such that $v[k]$ is an odd integer, cf. Figure 3.20.

For that, by using $H(z) - 1$, the sequence $\langle f[k] \rangle$ is generated at the transmitter. In contrast to Tomlinson-Harashima precoding, $f[k]$ is not subtracted completely from the data sequence. Rather it is quantized to the nearest point $d[k] \in \Lambda_a$, and only the quantization error $m[k] = f[k] - d[k]$, sometimes called *dither sequence*, is subtracted. Note that due to this construction, $m[k]$ lies in the interval $[-1, 1)$, the Voronoi region on Λ_a. This procedure ensures

$$\begin{aligned} v[k] &= x[k] + f[k] \\ &= a[k] - m[k] + d[k] + m[k] \\ &= a[k] + d[k] \in 2\mathbb{Z} + 1 \end{aligned} \tag{3.3.1}$$

to be taken from the set of odd integers. The block diagram of the flexible precoder is sketched in Figure 3.21, where, on the left-hand side, the calculation of the quantization error $m[k]$ is described as a modulo operation. On the right-hand side the modulo operation is decomposed into a slicer and a direct signal path.

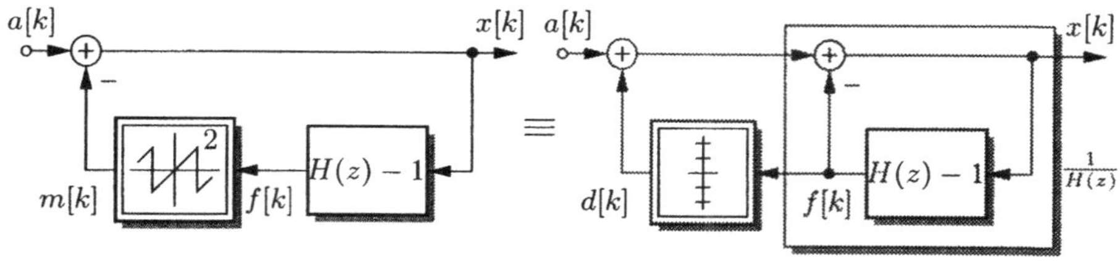

Fig. 3.21 Flexible precoder.

Using basic lattice theory, the extension to two-dimensional signal sets is obvious. The signal constellation is taken from a translate of the signal lattice Λ_a (usually $\Lambda_a = 2\mathbb{Z}^2$), and the samples of the precoding sequence $d[k]$ are drawn from Λ_a, too. The elements of the dither sequence $m[k]$ then lie in a fundamental region $\mathcal{R}(\Lambda_a)$ of the signal lattice.

At this point, an important difference between flexible precoding and Tomlinson-Harashima precoding is evident: Tomlinson-Harashima precoding needs knowledge of the boundary region of the constellation $\mathcal{A}$—or, equivalently, the precoding lattice Λ_p—, but disregards the internal arrangement of the signal points. Conversely, flexible precoding only has to be adapted to the signal lattice Λ_a. Here, the boundary region or the number of signal points is irrelevant. Hence, all constellations based on a regular lattice, e.g., the $\mathbb{Z}^2$ lattice or the hexagonal lattice A_2, can be used. In particular cross constellations or circular constellations are supported. Moreover, the number of signal points is irrelevant, offering high flexibility for the transmission rate.

But, as in Tomlinson-Harashima precoding, the effective data signal $v[k]$ entering the inverse of the channel filter at the precoder, and that is active at the input of the slicer, has high dynamics. For flexible precoding and uncoded one-dimensional transmission the magnitude of the channel output can be as high as already given in (3.2.5). It is noteworthy that the same is true for decision-feedback equalization at the output of the feedforward filter, i.e., prior to subtraction of the postcursor by the feedback part.

At the receiver, first, an estimate $\hat{v}[k]$ of the effective data signal is generated by using a threshold device. In order to recover the actual data $a[k]$, $\hat{v}[k]$ is filtered with the inverse channel filter $1/H(z)$, which results in an estimate $\hat{x}[k]$ of the transmit symbols. Because $x[k] = a[k] - m[k]$, $\hat{f}[k]$ could be generated via $H(z) - 1$, and then be modulo reduced to $\hat{m}[k]$. Finally, the estimate of the data symbols is obtained as $\hat{a}[k] = \hat{x}[k] + \hat{m}[k]$. But the "inverse precoder" can be built much easier. Since $m[k]$ lies in a fundamental region of $\boldsymbol{\Lambda}_a$ (we will soon see that the Voronoi region should be used) and $a[k]$ comes from a translate of $\boldsymbol{\Lambda}_a$, each value of $x[k]$ uniquely stems from a particular signal point $a[k] \in \mathcal{A}$. Hence, data can be recovered by quantizing $\hat{x}[k]$ to the nearest point in $\mathcal{A}$. This is again performed by an ordinary slicer. The block diagram of the transmission system using flexible precoding is shown in Figure 3.22.

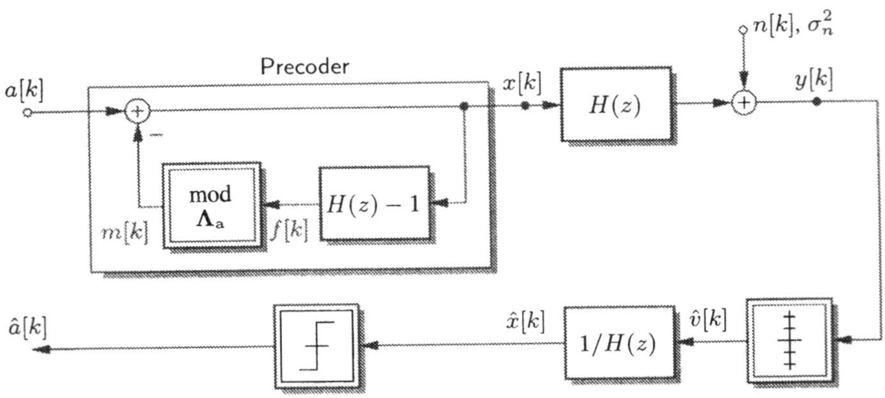

Fig. 3.22 Transmission scheme using flexible precoding.

From Figure 3.22 it can be seen that the inverse of the channel filter has to be applied at the receiver. Hence, it is obvious that $H(z)$ has to be strictly minimum-phase. Zeros on the unit circle, i.e., spectral nulls of $H(z)$, lead to poles in $1/H(z)$, and the inverse channel filter no longer is guaranteed to be stable. As for Tomlinson-Harashima precoding, the application of filters optimized according to the MMSE criterion will be discussed at a later point.

As shown above, Tomlinson-Harashima precoding converts the ISI channel $H(z)$ into a (near) memoryless AWGN channel. For flexible precoding the overall description is neither memoryless, nor is the additive noise Gaussian distributed! Because the dispersive system $1/H(z)$ has to be implemented at the receiver, every decision error leads to a burst of errors in $\hat{x}[k]$. Since $1/H(z)$ has to be stable, the error events at least die out. If the impulse response of $1/H(z)$ is sufficiently long, the error signal at its output is approximately Gaussian. Now, the data signal to be recovered by the second slicer is superimposed by this error signal and, as in the case of error-free transmission, by the uniform density of $m[k]$. In summary, mainly blocks of errors occur—hence the channel has memory—and the additive "noise" is not Gaussian, but its density is approximately given by the convolution of a Gaussian density with a uniform density over the fundamental region of the signal lattice $\boldsymbol{\Lambda}_\mathrm{a}$.

Finally, it should be noted that a multidimensional generalization of flexible precoding—as for Tomlinson-Harashima precoding; see Section 3.2.5—is straightforward. Resorting to lattice theory, for a $D \times D$ matrix channel, modulo reduction is simply done with respect to the D-dimensional signal lattice $\boldsymbol{\Lambda}_\mathrm{a}$. The inverse $\boldsymbol{H}^{-1}(z)$ of the channel matrix has to be implemented at the receiver, hence the determinant $\det(\boldsymbol{H}(z))$ of the channel transfer matrix is required to be nonzero.

3.3.2 Transmit Power and Signal-to-Noise Ratio

We always assume that $\langle a[k] \rangle$ is an i.i.d. data sequence. Then, since $m[k]$ depends only on the past symbols $a[k-\kappa]$, $\kappa = 1, 2, \ldots$, the samples $a[k]$ and $m[k]$ are statistically independent. Hence, the probability density function of the transmit symbols $x[k]$ is given as the convolution of the densities of $a[k]$ and $-m[k]$. In particular, the variance of $x[k]$ is then the sum of the variances of $a[k]$ and $m[k]$, respectively,

$$\sigma_x^2 = \sigma_a^2 + \sigma_m^2 \,. \tag{3.3.2}$$

As σ_x^2 should always be as low as possible, the variance of $m[k]$ should be minimum. This is achieved if $m[k]$ lies in the Voronoi region of $\boldsymbol{\Lambda}_\mathrm{a}$. Thus, quantization and modulo reduction should always be done with respect to this special fundamental region.

Following the derivations in Section 3.2.2, the quantization error $m[k]$ is white and has a near-uniform continuous distribution. Since $a[k]$ is taken from a translate of the signal lattice $\boldsymbol{\Lambda}_\mathrm{a}$ and $m[k]$ lies in a fundamental region of this lattice, the transmit symbols $x[k]$ are uniformly continuously distributed, provided the data symbols $a[k]$ are equiprobable. Otherwise, a stairstep density results, reflecting the probabilities of the signal points. In each case, the support region $\mathcal{R}$ for the transmit symbols is given as $\mathcal{R} = \mathcal{A} + \mathcal{R}(\boldsymbol{\Lambda}_\mathrm{a})$. Moreover, since the signals $\langle a[k] \rangle$ and $\langle m[k] \rangle$ are white and current samples are statistically independent of each other, the transmit sequence $\langle x[k] \rangle$ is also white.

In summary, assuming uncoded transmission and uniform signaling, the statistics of $\langle x[k] \rangle$ are the same as for Tomlinson-Harashima precoding. Hence, the average transmit power is again given by $\mathrm{E}\{|x[k]|^2\} \approx M^2/3$ for one-dimensional M-ary ASK, and $\mathrm{E}\{|x[k]|^2\} \approx 2M/3$ for two-dimensional square QAM constellations with M signal points. The precoding loss is thus equal to that of Tomlinson-Harashima precoding

$$\gamma_{\mathrm{p,FLP}}^2 = \gamma_{\mathrm{p,THP}}^2 \,. \tag{3.3.3}$$

For the actual numbers see Table 3.1, page 144.

Example 3.2: *Density of the Transmit Signal*

The density of the transmit signal is visualized for a quaternary one-dimensional signal constellation. First, the signal points $\pm 1, \pm 3$ are uniformly distributed and the quantization error lies in the Voronoi region $\mathcal{R}_V(2\mathbb{Z}) = [-1, 1)$; see Figure 3.23.

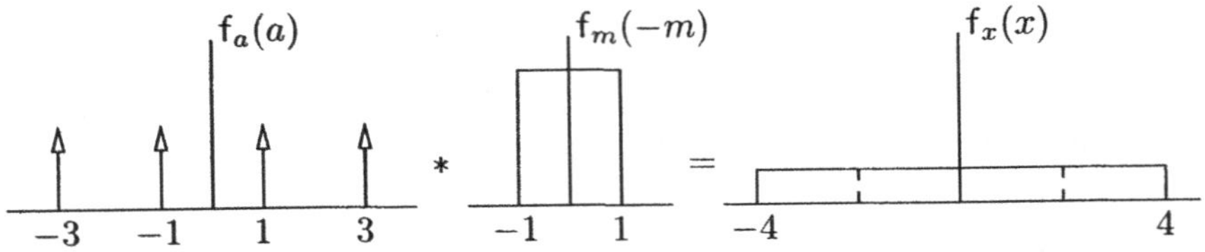

Fig. 3.23 Sketch of the probability density functions of the signals $a[k]$, $-m[k]$, $x[k]$.

In Figure 3.24, the signal points have different probabilities. The corresponding density of the transmit signal is a stairstep function.

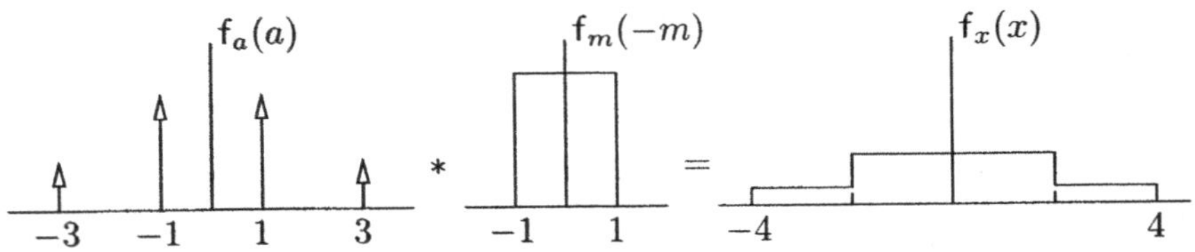

Fig. 3.24 Sketch of the probability density functions of the signals $a[k]$, $-m[k]$, $x[k]$.

Finally, the signal points $\pm 1, \pm 3$ are uniformly distributed; however, quantization is not done with respect to the Voronoi region, but with respect to the fundamental region $\mathcal{R} = (-2, 0]$. The negative of $m[k]$ then lies in $[0, 2)$; see Figure 3.25. The transmit signal now no longer has zero average, which increases the transmit power.

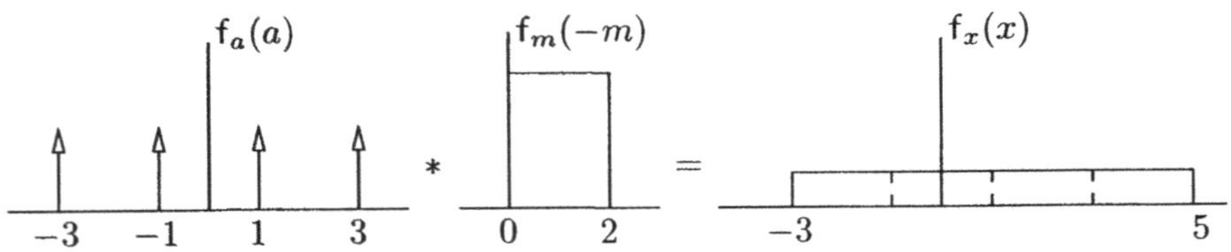

Fig. 3.25 Sketch of the probability density functions of the signals $a[k]$, $-m[k]$, $x[k]$.

3.3.3 Combination with Signal Shaping

Flexible precoding emerged from the desire to support constellation shaping on ISI channels. Because the data signal $a[k]$ is modified only slightly at the precoder by subtracting the signal $m[k]$, the characteristics of the distribution are preserved. Hence the alternative name "distribution-preserving precoding." This separation of shaping and precoding provides some advantages for implementation.

If shaping is performed prior to precoding, the generated near-discrete Gaussian distribution is converted into a stairstep density. For large constellations, i.e., at high rates, this pdf resembles a continuous Gaussian density. However for small constellations, separate shaping and precoding are unfavorable. On the one hand, only low shaping gains can be achieved, cf. the coarse stairstep function in Figure 3.24. The "ultimate shaping gain" of 1.53 dB is only valid for continuous distributions, cf. Chapter 4. On the other hand, this gain is reduced by the precoding loss. Here combined precoding/shaping techniques such as trellis precoding (see Chapter 5) are preferable, since they directly try to generate a continuous (near) Gaussian density.

Moreover, when using flexible precoding, shaping is restricted to "power shaping," i.e., the reduction of the average transmit power, which results in a Gaussian distribution. The generation of other desirable properties, e.g., influencing the shape of the spectral density (spectral shaping), is not possible. Such properties created for the data sequence $\langle a[k] \rangle$ will usually be destroyed due to the additive sequence $-m[k]$.

3.3.4 Straightforward Combination with Coded Modulation

In [LTF93, EFDL93] it is shown how flexible precoding can be combined with coded modulation in a straightforward manner. Since encoding is done prior to precoding, a valid code sequence $\langle a[k] \rangle$ is present at the input of the precoder. Given this, the precoding procedure has to ensure that the channel output sequence $\langle v[k] \rangle$ is also a valid code sequence.

For this, we observe that the code is only specified by the sequences of subsets, but not by the specific signal point. Thus, given a code sequence, any sequence of samples taken from the coset lattice Λ_c can be added without destroying the code property. Referring to Figure 3.15, the uncoded levels can be chosen arbitrarily. In terms of lattices, any sequence of elements equivalent modulo Λ_c to a code sequence is also consistent with the code.

Hence, keeping Figure 3.22 in mind, if $\langle a[k] \rangle$ is a code sequence, the modulo operation has to be performed based on the coset lattice Λ_c rather than the signal lattice Λ_a. Then, $d[k] \in \Lambda_c$ (see Figure 3.21), and $\langle a[k] \rangle$ and $\langle v[k] \rangle$ are the same sequence of subsets—only the specific signal points are different.

At the receiver, first a Viterbi decoder—instead of a slicer, which is used in uncoded transmission—produces an estimate $\langle \hat{v}[k] \rangle$ of the effective data signal. But in contrast to Tomlinson-Harashima precoding the algorithm has to work on the whole dynamic range of $v[k]$. No modulo reduction is possible since here no clear separation into data point and periodic extension is possible. Since $v[k]$ is approximately truncated discrete Gaussian and the boundary is ignored in metrics calculations, as in Tomlinson-Harashima precoding the decoder performs near ML decoding.

In the next step, via $1/H(z)$, an estimate $\hat{x}[k]$ of the channel symbols is generated. We observe that $m[k]$ lies in the Voronoi region of Λ_c, and $v[k]$ and $a[k]$ are taken from the same coset of Λ_c. Hence, the code sequence $\langle \hat{a}[k] \rangle$ is obtained by quantizing $\hat{x}[k]$ to the closest point in the subset given by $\hat{v}[k]$. Figure 3.26 shows the receiver structure.

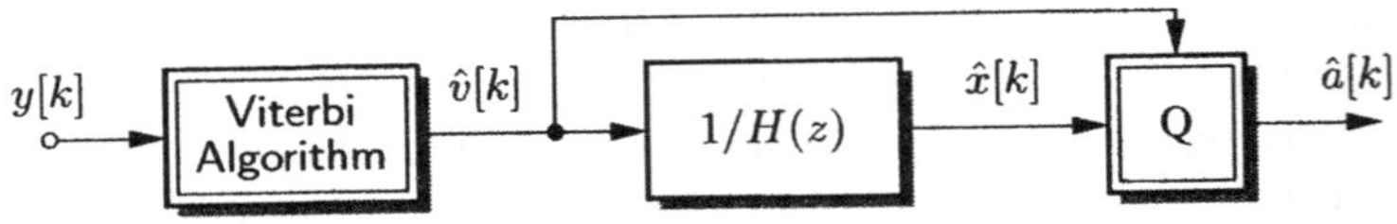

Fig. 3.26 Receiver structure for flexible precoding with channel coding.

The use of a time-variant slicer can be avoided if $\hat{v}[k]$ is subtracted from $\hat{x}[k]$. The current coset is thereby shifted to the lattice Λ_c. Then, $\hat{x}[k] - \hat{v}[k]$ can be quantized to the nearest point in the coset lattice Λ_c. Finally, $\hat{v}[k]$ is again added to get the estimate $\hat{a}[k]$.

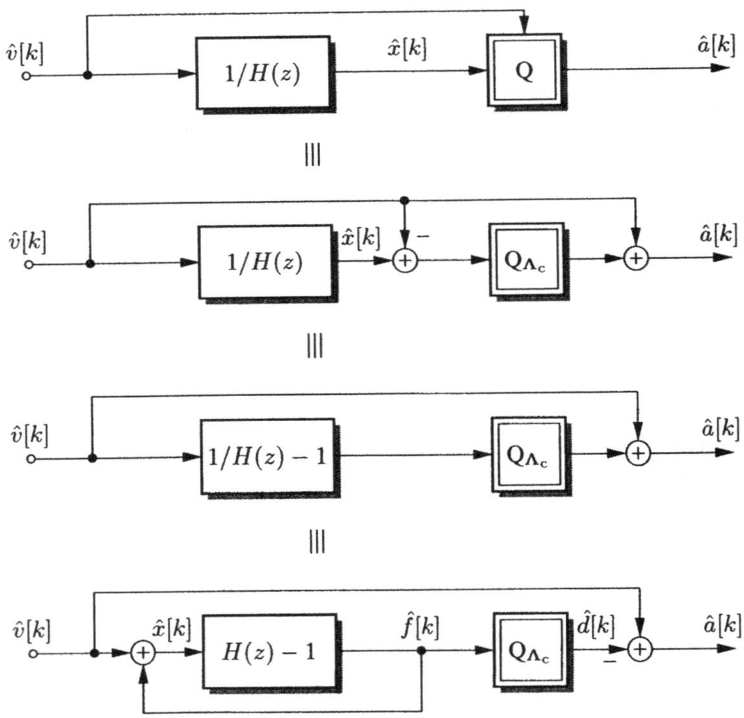

Fig. 3.27 Alternative receiver implementation for flexible precoding with channel coding.

Figure 3.27 shows some basic manipulations of the receiver structure, which finally results in the *reconstruction circuit*proposed in [EFDL93]. With respect to ideal performance the receivers are identical. But for an implementation with finite-precision arithmetics, the last structure is preferable. For a perfect reconstruction, the precoding filter in the transmitter and the ISI removing filter $1/H(z)$ in the receiver should exactly match. In the receiver at the bottom of Figure 3.27, the same operations are performed as in the transmitter—the same filter has to be implemented and the same signals are generated. Hence, identical structures can be used, which guarantees that no mismatch due to finite-precision calculations occurs.

Transmit PDF and Signal-to-Noise Ratio As for uncoded transmission, the transmit pdf is given by the convolution of the densities of $a[k]$ and $-m[k]$, respectively. Now, $m[k]$ lies in a fundamental region—preferably the Voronoi region—of Λ_c, which is larger than the Voronoi region of Λ_a. Thus, the contributions of the signal points to the total transmit pdf overlap, leading to a nonuniform density.

Example 3.3: *Density of the Transmit Signal*

For a coded 8-ary one-dimensional signal constellation, the density of the transmit signal is visualized. The trellis code is based on the four-way partition $2\mathbb{Z}/8\mathbb{Z}$, which is standard for one-dimensional codes [Ung82]. The dither sequence $m[k]$ then lies in the interval $[-4, 4)$. Figure 3.28 shows the probability density functions of the signals $a[k]$, $-m[k]$, and $x[k]$.

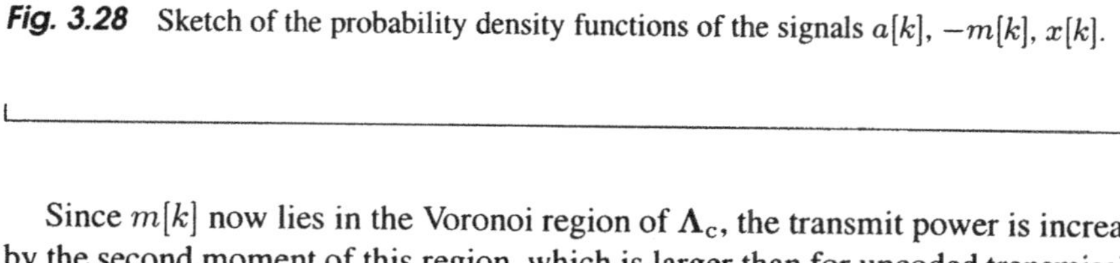

Fig. 3.28 Sketch of the probability density functions of the signals $a[k]$, $-m[k]$, $x[k]$.

Since $m[k]$ now lies in the Voronoi region of Λ_c, the transmit power is increased by the second moment of this region, which is larger than for uncoded transmission. Again we assume that for coded transmission, the signal constellation has $2M$ points, and that the trellis code is based on $K = |\Lambda_a/\Lambda_c|$ subsets ($K \leq \log_2(2M)$).

Using one-dimensional signaling, $\mathcal{A} = \{\pm1, \pm3, \ldots, \pm(2M-1)\}$, and one-dimensional trellis coding, the Voronoi region of Λ_c is $[-K, K)$. The average transmit power $\sigma_a^2 = \frac{(2M)^2-1}{3}$ is increased by $\sigma_m^2 = \frac{K^2}{3}$. Hence the precoding loss $\gamma^2_{\text{p,FLP,c}} = \frac{\sigma_a^2+\sigma_m^2}{\sigma_a^2}$ reads

$$\gamma^2_{\text{p,FLP,c}} = \frac{(2M)^2 - 1 + K^2}{(2M)^2 - 1} . \tag{3.3.4}$$

For two-dimensional square constellations $\mathcal{A} = \{\pm 1, \pm 3, \ldots, \pm(\sqrt{2M}-1)\}^2$, and two-dimensional trellis coding the Voronoi region is given by $[-\sqrt{K}, \sqrt{K})^2$. Here, we have $\sigma_a^2 = 2\frac{2M-1}{3}$ and $\sigma_m^2 = 2\frac{K}{3}$, resulting in the precoding loss

$$\gamma^2_{\mathrm{p,FLP,c}} = \frac{2M-1+K}{2M-1} \,. \tag{3.3.5}$$

Tables 3.2 and 3.3 show the precoding loss of flexible precoding for one- and two-dimensional signal constellations, respectively. The trellis code is based on $K = 4$, 8, 16, and 32 subsets. These results are also valid for multilevel coding (MLC) where $\log_2(K)$ levels are channel encoded and the remaining are uncoded. For comparison the precoding loss of Tomlinson-Harashima precoding is also shown.

Table 3.2 Precoding loss (in dB) of flexible precoding and Tomlinson-Harashima precoding in combination with coded modulation (one-dimensional signaling).

	$10\log_{10}(\gamma^2_{\mathrm{p,FLP,c}})$ [dB]			$10\log_{10}(\gamma^2_{\mathrm{p,THP}})$ [dB]
K	4	8	16	
$2M = 4$	3.15	—	—	0.28
8	0.98	3.04	—	0.07
16	0.26	0.97	3.02	0.02

Table 3.3 Precoding loss (in dB) of flexible precoding and Tomlinson-Harashima precoding in combination with coded modulation (two-dimensional signaling).

	$10\log_{10}(\gamma^2_{\mathrm{p,FLP,c}})$ [dB]			$10\log_{10}(\gamma^2_{\mathrm{p,THP}})$ [dB]
K	4	8	16	
$2M = 16$	1.03	1.86	3.15	0.28
32	0.53	1.00	1.81	0.14
64	0.27	0.52	0.98	0.07

As one can see, the straightforward combination of flexible precoding with coded modulation leads to a significantly higher precoding loss compared to Tomlinson-Harashima precoding. In particular, increasing the number of cosets also increases the precoding loss. In order to approach channel capacity, very powerful codes are required. Since for increased coding gain the number of subsets has to be increased, capacity cannot be achieved with this version of flexible precoding [LTF93, Lar96]. Asymptotically, for $M \to \infty$ and $K \to 2M$, a loss of 3 dB results.

3.3.5 Combined Coding and Precoding

Since separate channel coding and precoding leads to a huge loss in power efficiency, an obvious strategy is to combine both operations into one unit. The concept of generating a coded sequence and preserving its properties in precoding, which can be restated as "*working against the channel*", is dropped. Now, it is only required that the noiseless channel output $\langle v[k] \rangle$ is a valid code sequence. This enables the system to generate the code sequence which is closest to the ISI-corrupted signal produced by the channel. Hence, only a small correction by the precoder is necessary, because now it is "*working in synergy with the channel*". Referring to Figure 3.20, channel coding and precoding shall now be done by choosing $m[k]$ appropriately. The resulting precoding schemes were introduced by Laroia [Lar94, Lar96].

In order to achieve the desired combination, we first have to recall some important characteristics of trellis codes, as introduced by Ungerböck [UC76, Ung82, Ung87a, Ung87b]. For that, let the (extended) signal constellation at the channel output be partitioned into two equal-sized, disjoint subsets, i.e., $\mathcal{V} = \mathcal{V}_0 \cup \mathcal{V}_1$. Both sets are translates of the lattice Λ_{f}, which is given by the first binary partitioning step of the signal lattice Λ_{a}. In summary, we have the partition chain $\Lambda_{\mathrm{a}}/\Lambda_{\mathrm{f}}/\Lambda_{\mathrm{c}}$.

Statement A:

One design rule of Ungerböck-type trellis codes [Ung82, For88a] is that the state of the encoder uniquely determines the subset of the first partitioning step. This is because the current data bits have no immediate influence on the LSB of the binary address vector. Thus, the encoder states can be divided into two types: all outgoing branches of the first type address points in $\mathcal{V}_0$, and all outgoing branches of the second type address points in $\mathcal{V}_1$. The specific coset of the coding lattice Λ_{c} is not known until the data bits are given. Then, the branch is initiated and the next state is determined.

Consequently, the task of the precoder is to track the channel output $v[k]$ and to ensure—by subtracting the dither symbol $m[k]$ from the channel input—that the next symbol falls into the proper subset $\mathcal{V}_b$, $b \in \{0, 1\}$. The actual point of $\mathcal{V}_b$ then gives the current coset of Λ_{c}. Since the code sequences are produced by an unifiliar Markov source, given the current state and the coset which corresponds to a specific branch, the next encoder state can be given. To some respect, this procedure is the opposite of the usual (feedforward) encoder. Now, data are not represented in an obvious way, but as long as the encoding procedure is bijective, no ambiguity occurs.

Statement B:

Any point taken from the lattice Λ_{f} can be added to points drawn from $\mathcal{V}_b$ without changing the subset. Contrary, by adding a point from the complementary lattice $\overline{\Lambda}_{\mathrm{f}} \triangleq \Lambda_{\mathrm{a}} \setminus \Lambda_{\mathrm{f}}$, points from $\mathcal{V}_0$ are translated to points from $\mathcal{V}_1$ and vice versa. It is noteworthy that this reflects the algebraic properties of the quotient group $\Lambda_{\mathrm{a}}/\Lambda_{\mathrm{f}}$.

Combining the above two statements, any uncoded sequence $\langle a[k] \rangle$, with $a[k] \in \mathcal{V}_0$, can be converted into a coded one by adding elements taken from Λ_{f} or $\overline{\Lambda}_{\mathrm{f}}$, respectively, according to the state of the encoder. Since from Figure 3.20 we have

$$v[k] = a[k] + \underbrace{(f[k] - m[k])}_{\in \mathbf{\Lambda}_\mathrm{f} \text{ or } \overline{\mathbf{\Lambda}}_\mathrm{f}} , \tag{3.3.6}$$

the precoder has to generate the small correction symbol $m[k]$, which, together with the postcursors due to prior transmitted symbols $f[k]$, falls into the appropriate subset. Note that the initial signal constellation $\mathcal{A}$ is now a (proper) subset of $\mathcal{V}_0$ rather than of $\mathcal{V}$. The use of only every other point for data transmission reflects the single bit of coding redundancy.

Figure 3.29 is a block diagram of the combined channel encoder/precoder which we subsequently will call *ISI coder* [Lar94, Lar96]. Here, the current state of the

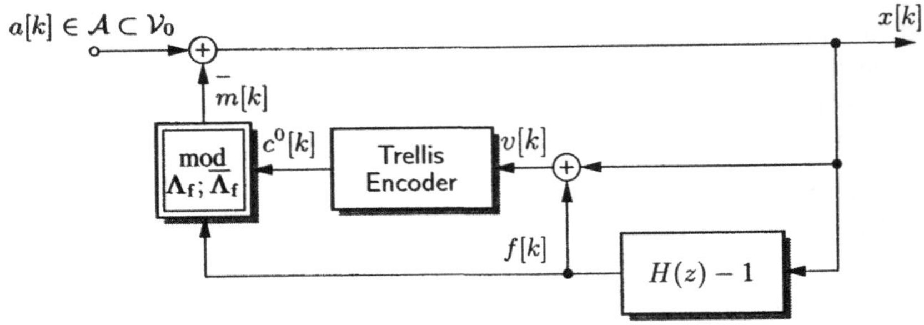

Fig. 3.29 Block diagram of the ISI coder [Lar96].

trellis encoder is represented by the binary number $c^0[k]$, which is simply the parity bit, i.e., the LSB of the binary address label, cf. Figure 3.15. First, in the precoder the postcursor sum $f[k]$ is generated via $H(z) - 1$. Then, depending on $c^0[k]$, $f[k]$ is quantized differently in order to obtain the internal signal $d[k]$. If $c^0[k] = 0$, quantization is done with respect to $\mathbf{\Lambda}_\mathrm{f}$, and if $c^0[k] = 1$, it is done with respect to $\overline{\mathbf{\Lambda}}_\mathrm{f}$. The quantization error $m[k] = f[k] - d[k]$ is subtracted from the data symbol $a[k]$, which is taken from the subset $\mathcal{V}_0$. Again, the calculation of $m[k]$ is displayed as a modulo operation. Finally, in order to track the state of the encoder, the channel output symbol $v[k]$ is also generated at the precoder. From $v[k]$, the next encoder state can be determined.

One possible way, the block labeled "Trellis Encoder" can be implemented is to perform an inverse mapping and to extract the bits involved in the convolutional encoding ("coded levels"). The bit $c^0[k]$ is determined via the usual systematic encoder,[9] which gives the subset $\mathcal{V}_b$ in the next step. Hence, the encoder structure can be exactly the same as in classical TCM encoding, and the same properties of the coded sequence are achieved. Figure 3.30 shows the trellis encoder.

[9]Considerations based on nonrecursive, nonsystematic encoders lead to exactly the same recursive structure.

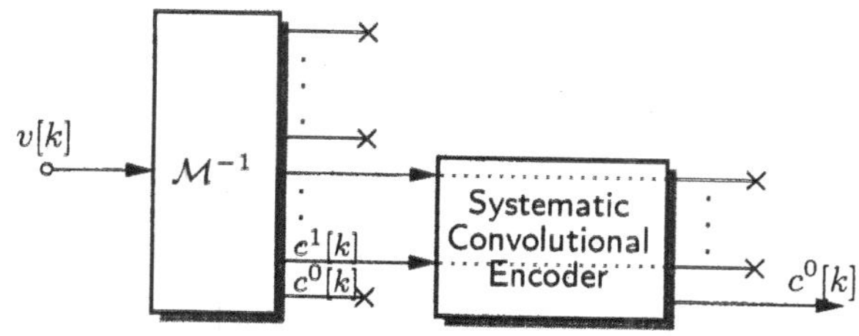

Fig. 3.30 Block diagram of the trellis encoder.

At the receiver, first a trellis decoder provides estimates $\hat{v}[k]$. Filtering $\langle\hat{v}[k]\rangle$ with the inverse $1/H(z)$ of the channel filter gives estimates $\hat{x}[k]$ of the transmit symbols. Since $a[k] \in \mathcal{A} \subset \mathcal{V}_0$, $\mathcal{V}_0$ a translate of Λ_f, and $m[k]$ lie in the Voronoi region of Λ_f, an estimate $\hat{a}[k]$ of the data symbol can be derived from $\hat{x}[k]$ by quantizing this sample to the nearest point in $\mathcal{A}$. The receiver structure is sketched in Figure 3.31.

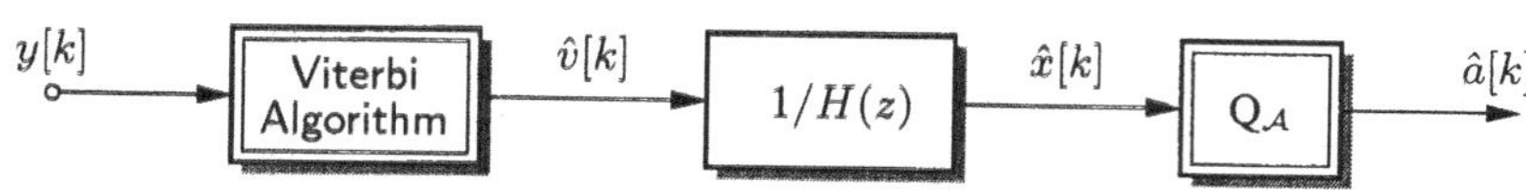

Fig. 3.31 Receiver structure corresponding to the ISI coder.

Transmit PDF and Signal-to-Noise Ratio The transmit pdf is again given by the convolution of the densities of $a[k]$ and $-m[k]$, respectively. Since $m[k]$ lies in the Voronoi region of Λ_f and $\mathcal{A}$ is a subset of a translate of this lattice, the contributions of the individual signal points to the total transmit pdf do not overlap. For one-dimensional signaling and constellations derived from $2\mathbb{Z}+1$, the transmit signal will exhibit a DC offset. To reduce transmit power, this offset should be compensated by an appropriate shift. Alternatively, quantization can be done with respect to some suitable fundamental region instead of the Voronoi region.

Example 3.4: *Density of the Transmit Signal*

For coded one-dimensional transmission with 2 bits per symbol, the density of the transmit signal is shown in Figure 3.32. The uncoded data symbols are taken from $\mathcal{A} = \{-7, -3, 1, 5\}$. The dither sequence $m[k]$ lies in the interval $[-2, 2)$. Please compare these densities with those of Example 3.3, which are valid for the same transmission scenario.

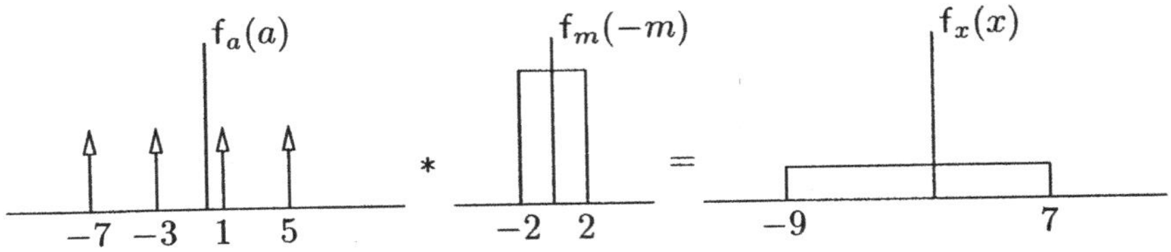

Fig. 3.32 Sketch of the probability density functions of the signals $a[k]$, $-m[k]$, $x[k]$.

Next, two-dimensional transmission with rate 3 bit/symbol is considered. In Figure 3.33 the support of the respective signals is sketched. The signal constellation is a subset of a translate of $2\boldsymbol{R}\mathbb{Z}^2$, and the dither sequence $m[k]$ falls into the region $\boldsymbol{R}[-1,\,1)^2$ (cf. Appendix C).

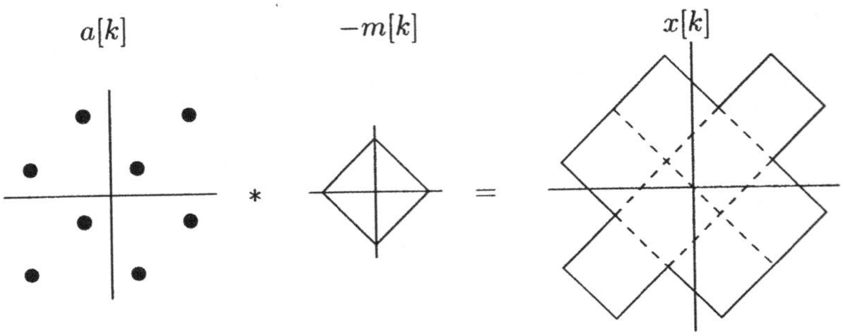

Fig. 3.33 Sketch of the support of the probability density functions of the signals $a[k]$, $-m[k]$, $x[k]$.

In one-dimensional signaling, the M-ary signal constellation $\mathcal{A} = (4\mathbb{Z}+1) \cap [-2M,\, 2M]$ is used. Since $\sigma_m^2 = \frac{4}{3}$ holds, the transmit power is increased from $\mathsf{E}\{a[k]\} = \frac{(2M)^2-1}{3}$ to $\mathsf{E}\{x[k]\} = \frac{(2M)^2+3}{3}$. Hence the precoding loss (without compensating the DC offset) reads

$$\gamma^2_{\text{p,ISI coder}} = \frac{(2M)^2+3}{(2M)^2-1} \,. \tag{3.3.7}$$

For two-dimensional M-ary constellations $\mathcal{A} = (2\boldsymbol{R}\mathbb{Z}^2 + \left[\begin{smallmatrix}1\\1\end{smallmatrix}\right]) \cap [-\sqrt{2M}, \sqrt{2M}]^2$, the dither sequence falls into $\boldsymbol{R}[-1,\,1)^2$. Here, the transmit power $\mathsf{E}\{a[k]\} = 2\frac{2M-1}{3}$ is increased by $\sigma_m^2 = \frac{4}{3}$ to $\mathsf{E}\{x[k]\} = 2\frac{2M+1}{3}$, and the precoding loss calculates to

$$\gamma^2_{\text{p,ISI coder}} = \frac{2M+1}{2M-1} \,. \tag{3.3.8}$$

Tables 3.4 and 3.5 show the precoding loss of the ISI coder for one- and two-dimensional signal constellations, respectively. For comparison the precoding loss of Tomlinson-Harashima precoding is shown as well (equation (3.2.47)). Obviously, the precoding loss is still larger than that of coded transmission using Tomlinson-Harashima precoding.

Table 3.4 Precoding loss (in dB) of the ISI coder and Tomlinson-Harashima precoding and coded modulation (one-dimensional signaling).

$M =$	4	8	16	32	64	128
$10\log_{10}(\gamma^2_{\mathrm{p,ISI\,coder}})$ [dB]	0.27	0.07	0.02	0.004	0.001	0.0003
$10\log_{10}(\gamma^2_{\mathrm{p,THP}})$ [dB]	0.07	0.02	0.004	0.001	0.0003	0.0001

Table 3.5 Precoding loss (in dB) of the ISI coder and Tomlinson-Harashima precoding and coded modulation (two-dimensional signaling).

$M =$	4	8	16	32	64	128
$10\log_{10}(\gamma^2_{\mathrm{p,ISI\,coder}})$ [dB]	1.09	0.54	0.27	0.14	0.07	0.03
$10\log_{10}(\gamma^2_{\mathrm{p,THP}})$ [dB]	0.58	0.28	0.14	0.07	0.03	0.02

Further Reduction of the Precoding Loss The precoding loss can be decreased once again, if an additional degree of freedom for modifying the signal is admitted. In addition to Statement B on page 161, signal points taken from $\mathcal{V}_0$ can also be changed into $\mathcal{V}_1$ points (and vice versa) by an appropriate rotation ϱ. For one-dimensional constellations a rotation by 180° (inversion, multiplication by $\varrho = -1$), and for two-dimensional constellations a rotation by 90° ($\varrho = \mathrm{j}$), has to be performed. The *modified ISI coder* [Lar96] is shown in Figure 3.34.

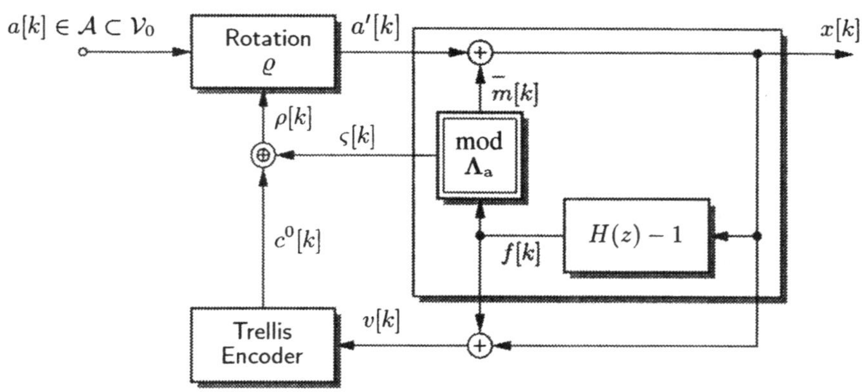

Fig. 3.34 Block diagram of the modified ISI coder.

The key element of the precoder is now identical to that for uncoded transmission, cf. Figure 3.22. In particular, modulo reduction is performed with respect to the densest lattice $\boldsymbol{\Lambda}_a$. Again, as for the ISI coder, the initial signal points are drawn from a subset of $\mathcal{V}_0$. From the linear description in Figure 3.21 we see that depending on $f[k]$, the quantized signal $d[k]$ lies either in $\boldsymbol{\Lambda}_f$ or in $\overline{\boldsymbol{\Lambda}}_f$. Thus the subset of the channel output signal $v[k]$ may be changed to $\mathcal{V}_1$. But for a valid code sequence, the subset has to be chosen depending on the state of the channel encoder (expressed by $c^0[k]$). In order to have an explicit influence on the subset, the initial signal constellation may be rotated in order to guarantee the desired subset. Table 3.6 lists all possible combinations and the resulting action of the rotation unit.

Table 3.6 Combinations of subsets.

Trellis Encoder Requires the Subset ...	Modulo Operation ... the Subset	Rotation Has ... the Subset
not to change	does not change	not to change
not to change	changes	to change
to change	does not change	to change
to change	changes	not to change

Let us define a binary flag $\varsigma[k]$, which is 0 if $d[k] \in \boldsymbol{\Lambda}_f$, and 1 if $d[k] \notin \boldsymbol{\Lambda}_f$. Then, considering the above truth table, a flag $\rho[k]$, which indicates rotation if $\rho[k] = 1$, can be given. It is simply the XOR combination of $c^0[k]$ and $\varsigma[k]$, i.e., $\rho[k] = c^0[k] \oplus \varsigma[k]$ ($\oplus$: modulo-2 addition, XOR).

The modulo-$\boldsymbol{\Lambda}_a$ device quantizes $f[k]$ to the nearest point $d[k]$ in $\boldsymbol{\Lambda}_a$ and outputs the quantization error $m[k] = f[k] - d[k]$. Additionally, it checks whether $d[k]$ is in $\boldsymbol{\Lambda}_f$ (in two dimensions, e.g., by calculating the sum of real and imaginary part and testing whether it is even) and, if so, also outputs $\varsigma[k] = 0$. Otherwise $\varsigma[k] = 1$ is delivered.

The receiver again consists of a trellis decoder (e.g., Viterbi algorithm), which provides $\hat{v}[k]$. After filtering $\langle \hat{v}[k] \rangle$ with $1/H(z)$ the estimates $\hat{x}[k]$ of the transmit symbols are quantized with respect to the extended transmit constellation $\mathcal{A}'$, given by the union of $\mathcal{A}$ and its rotated version: $\mathcal{A}' = \mathcal{A} \cup \varrho\mathcal{A}$. If $\hat{a}'[k]$ does not lie in $\mathcal{A}$, in the last step the point is rotated back (ϱ^{-1}) and constitutes the estimate $\hat{a}[k]$ of the data symbol. The receiver structure for the modified ISI coder is sketched in Figure 3.35.

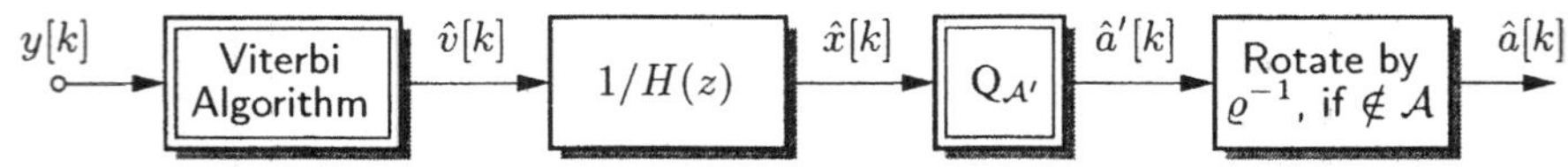

Fig. 3.35 Receiver structure corresponding to the modified ISI coder.

Transmit PDF and Signal-to-Noise Ratio The transmit pdf is now given by the convolution of the densities of $a'[k]$ and $-m[k]$, respectively. However, $f_{a'}(a')$ is the average (assuming that rotation and no rotation occur equally likely) of the pdf of $a[k]$ and its rotated version. The dither $m[k]$ lies in the Voronoi region of Λ_a, and the contributions of the signal points to the transmit pdf do not overlap.

Example 3.5: *Density of the Transmit Signal*

The density of the transmit signal using the modified ISI coder is shown in Figure 3.36 for coded one-dimensional transmission with 2 bits per symbol. The uncoded data symbols are taken from $\mathcal{A} = \{-7, -3, 1, 5\}$. The dither sequence $m[k]$ lies in the interval $[-1, 1)$. These densities have to be compared with those of Examples 3.3 and 3.4.

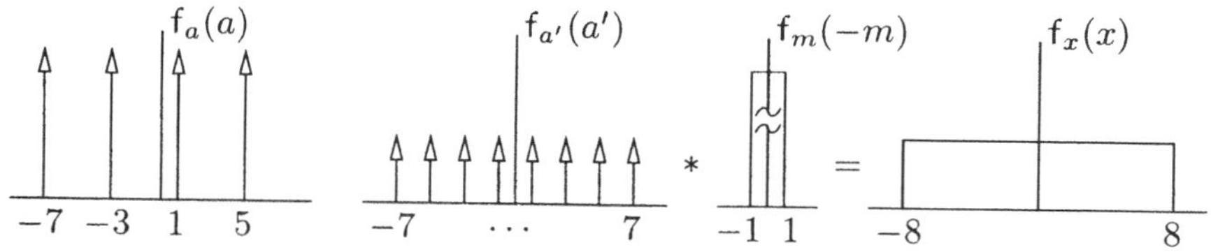

Fig. 3.36 Sketch of the probability density functions of the signals $a[k]$, $a'[k]$, $-m[k]$, $x[k]$.

Next, two-dimensional transmission with a rate of 3 bits/symbol is considered. The support of the respective signals is shown in Figure 3.37. The signal constellation is a subset of a translate of $2R\mathbb{Z}^2$, and the dither sequence $m[k]$ falls in the region $[-1, 1)^2$. For reference, compare this example with Figure 3.33.

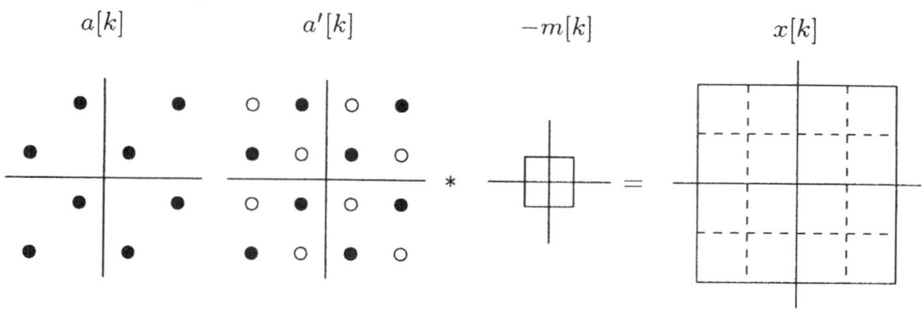

Fig. 3.37 Sketch of the support of the probability density functions of the signals $a[k]$, $a'[k]$, $-m[k]$, $x[k]$.

Compared to the ISI coder, the dither sequence of the modified ISI coder lies in the Voronoi region of Λ_a. Note that rotation does not change the signal power. Hence, following the derivations above, the precoding loss can be calculated by setting $\sigma_m^2 = \frac{1}{3}$ for one-dimensional signaling and $\sigma_m^2 = \frac{2}{3}$ for two-dimensional constellations. The precoding loss now reads

$$\gamma^2_{\text{p,mISI coder}} = \frac{(2M)^2}{(2M)^2 - 1}, \tag{3.3.9}$$

for one-dimensional M-ary constellations, and

$$\gamma^2_{\text{p,mISI coder}} = \frac{2M}{2M - 1}, \tag{3.3.10}$$

for two-dimensional M-ary constellations, respectively.

A comparison of (3.3.9), (3.3.10), and (3.2.47) reveals that the precoding loss of the modified ISI coder is identical to that of Tomlinson-Harashima precoding.

Extension to Other Types of Codes So far, the codes used in the (modified) ISI coder have been matched to the dimensionality of the signal constellation. But in some situations, higher dimensional codes, e.g., four-dimensional trellis codes, are preferred because of their lower constellation expansion ratio or the possibility of achieving rotational invariance. For example, in the ITU voice-band modem standard V.34 four-dimensional trellis coding is specified [ITU94]. First we discuss the combination of 2δ-dimensional codes with one redundant bit per 2δ dimensions, i.e., based on the partition chain $\mathbb{Z}^{2\delta}/\boldsymbol{D}_{2\delta}/\boldsymbol{\Lambda}_c$, and two-dimensional signaling.

Now, δ QAM symbols constitute one code symbol. In order to generate a valid code sequence this hypersymbol has to fall into the right 2δ-dimensional subset, a translate of $\boldsymbol{D}_{2\delta}$. Regardless of the first $\delta - 1$ QAM symbols, a selection of the subset can always be done by choosing the last symbol from an appropriate two-dimensional subset, which is a translate of $\boldsymbol{R}\mathbb{Z}^2$. Hence, for the first $\delta - 1$ symbols, quantization in the ISI coder can always be done with respect to $\mathbb{Z}^2$. Depending on the $\delta - 1$ channel output symbols $v[k]$, quantization is done in the δth step either with respect to $\boldsymbol{R}\mathbb{Z}^2$ or $\boldsymbol{R}\mathbb{Z}^2 + \left[\begin{smallmatrix}1\\0\end{smallmatrix}\right]$. Note that the first $\delta - 1$ QAM symbols are taken from the signal set $\mathcal{A}' = \mathcal{A} \cup \varrho\mathcal{A}$, whereas the last symbol can only be drawn from $\mathcal{A}$. This reflects the single redundant bit. Similarly, during the first $\delta - 1$ steps, the modified ISI coder does not perform any rotation. The two-dimensional and hence 2δ-dimensional subset is determined by an appropriate rotation of the δth QAM symbol.

Note, that using the ISI coder in combination with higher dimensional codes results in a reduced precoding loss compared to two-dimensional codes. For the modified ISI coder, nothing changes.

Analogous to QAM signaling, δ-dimensional trellis codes can be combined with baseband transmission. Again, due to proper quantization/rotation of the last symbol within one coding frame, the desired code property of the channel output signal can be achieved.

Finally, the ISI coders can also be modified in order to work with code with higher redundancy, i.e., $r > 1$ bits per 2δ (δ) dimensions. This is achieved by using 2^r different modulo operations in the ISI coder and 2^r different rotation angles in the modified ISI coder, respectively, for the last symbol within the coding frame. This restricts the modified ISI coder to $r\delta$-dimensional codes with a maximum number of r redundant bits, when using r-dimensional ($r = 1, 2$) signaling. However, in practice, this is not really a disadvantage. For details and some discussion on practical the aspects, such as rotational invariance, the reader is referred to the original work [Lar96].

3.3.6 Spectral Zeros

Since flexible precoding and its evolutions need to implement the inverse of the channel filter at their receiver, it is obvious that $H(z)$ has to be strictly minimum-phase. Zeros at the unit circle, i.e., spectral nulls, lead to spectral poles in $1/H(z)$, and the inverse channel filter no longer is stable. In [Fis95] a method to overcome this problem has been proposed. By modifying only the receiver, these precoding techniques can be used for the broad class of channels which exhibit zeros at $z = 1$ (DC) and/or $z = -1$ (Nyquist frequency). Here, we will recapitulate this modification.

First, we note that if there is no channel noise the spectral poles have no effect, since zeros of $H(z)$ and poles of $1/H(z)$ cancel each other. Hence, because we deal with linear filters, the effect of decision errors may be studied separately.

Decision errors are characterized by the corresponding error sequence $\langle e_v[k]\rangle$, with $e_v[k] = \hat{v}[k] - v[k]$, which is assumed to be time-limited. Since the samples $e_v[k]$ are differences between valid signal points, $e_v[k] \in \mathbf{\Lambda}_\mathrm{a}$ holds. Next, suppose $H(z)$ has a zero at $z = 1$. Filtering $\langle e_v[k]\rangle$ by $1/H(z)$, due to the integrating part $\frac{1}{1-z^{-1}}$, in the steady state, a constant sequence $\langle e_x[k]\rangle$, with $e_x[k] = \hat{x}[k] - x[k]$, of infinite duration results. Likewise, if $H(z)$ has a zero at $z = -1$, $\langle e_x[k]\rangle$ is an alternating sequence with a constant envelope.

Knowing that the transmit symbols $x[k]$ are restricted to a well-specified region $\mathcal{X}$ (see the preceding sections), these error sequences can be detected. If the recovered symbol $\hat{x}[k]$ lies outside the region $\mathcal{X}$, a decision error must have occurred. Then, the error sequence $\langle e_x[k]\rangle$ can be compensated by feeding an additional impulse into the filter $1/H(z)$. Because in the steady state the response must be the negative of $\langle e_x[k]\rangle$, the "correction impulse" also has to have the properties of an error sequence, i.e., its samples have to be taken from $\mathbf{\Lambda}_\mathrm{a}$. The actual impulse should be determined from $\hat{x}[k]$; preferably such that the corrected version lies inside $\mathcal{X}$, and that, in order not to overshoot, the correction impulse has minimum magnitude.

For one-dimensional signaling and for two-dimensional square constellations, the stabilization of $1/H(z)$ can be described easily using a nonlinear device. Figure 3.38 shows the situation for a one-dimensional M-ary signal set $\mathcal{A} = \{\pm1, \pm3, \ldots, \pm(M-1)\}$ and either uncoded transmission or for use with the modified ISI coder. It is noteworthy that to some extent the stabilization of $1/H(z)$ resembles the Tomlinson-Harashima precoder. The region, where the channel symbols $x[k]$ lie is given as $\mathcal{X} = [-M, M)$. For amplitudes within this region, the nonlinearity has a linear course. Signals whose amplitudes lie outside $\mathcal{X}$ are reduced to this range using a sawtooth characteristic with a step size of 2, i.e., equal to the spacing of the signal points. The nonlinear device ensures that $\hat{x}[k]$ is bounded to $\mathcal{X}$—the inverse

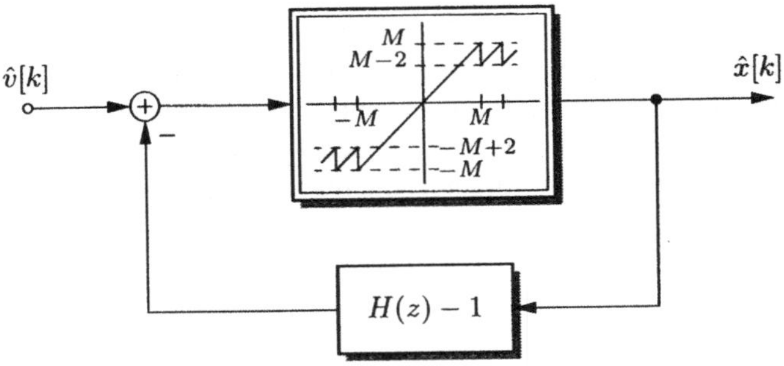

Fig. 3.38 Stable version of the inverse channel filter.

channel filter is BIBO-stable. Moreover, since the step size is 2, the magnitude of the correction impulse is a multiple of 2, and thus has the property of an error sequence. This is a prerequisite for an error event to completely die out in the steady state.

When using two-dimensional square QAM constellations, the above nonlinearity is simply applied independently to real and imaginary part of $\hat{x}[k]$. For other boundary regions which may be desirable (e.g., circular regions), an appropriate two-dimensional "modulo" reduction onto the region $\mathcal{X}$ is always possible.

A similar, almost identical, method for using flexible precoding for channels with spectral nulls was given in [CÖU97b]. Again, stable operation is achieved by projecting samples outside the support of the transmit signal back into this region.

3.4 SUMMARY AND COMPARISON OF PRECODING SCHEMES

In the last two sections, Tomlinson-Harashima precoding and flexible precoding were introduced in detail. As a summary, we now elucidate the general concepts of both precoding schemes and show their differences, cf. [FH97].

Table 3.7 compiles the characteristics of Tomlinson-Harashima precoding and flexible precoding, respectively.

Table 3.7 Characteristics of precoding schemes.

	Tomlinson-Harashima Precoding	Flexible Precoding
Derived from	linear preequalization at the *transmitter*	linear equalization at the *receiver*
Constraints on signal constellation	dependent on the *boundary region* (periodic extension required)	dependent on the *signal lattice* (boundary region immaterial)
Application of coded modulation	straightforward	high precoding loss, unless *precoding and coding* are combined
Application of signal shaping	properties are destroyed, unless *precoding and shaping* are combined	straightforward
Channels with spectral zeros	no restrictions	*not stable* unless receiver is modified
Performance	moderate precoding loss	higher precoding loss and *error propagation*
Implementation	simple *modulo arithmetic* suffices	more complex receiver requires linear systems with *high dynamic range*
No coding, no signal shaping	white transmit sequence $\langle x[k] \rangle$, uniformly distributed over $\mathcal{R}$	

From the above table it is evident that Tomlinson-Harashima precoding and flexible precoding are dual to each other in essential points. First, THP is based on linear preequalization at the *transmitter* side, while FLP can be viewed as being derived from linear equalization at the *receiver*.

A second duality can be noticed if the dependency on the signal constellation is considered. In this case, flexible precoding is much more flexible (*nomen est omen*). Cross or circular constellations can be used and the distribution, e.g., imposed by signal shaping algorithms, is preserved. FLP only has to be adapted to the underlying signal lattice—the boundary region is insignificant. Conversely, Tomlinson-Harashima precoding puts constraints on the support of the signal set. To be power efficient, we require that by repeating the constellation periodically, the two-dimensional plane can be tiled without any gap. Hence, compared to circular constellations here a higher peak-to-average power ratio may result.

Moreover, FLP offers a much simpler way to support a fractional number of bits per symbol. For strictly band-limited channels, transmission at fractional rates is essential for optimum performance. Flexible precoding in combination with *shell mapping* (cf. Chapter 4) can support fractional data rates in a direct way. The shell-mapping algorithm is the stage where the rates can be chosen in a wide range and with small granularity. In contrast to this, Tomlinson-Harashima precoding does not support fractional rates. In [EF92] it is proposed to extend *trellis precoding* (combined precoding and shaping, see Chapter 5) by *constellation switching*, where the sizes of the constellations are changed periodically. The disadvantage of this technique is a (moderate) increase in peak power, as well as in average transmit power. For both precoding procedures signal shaping is essential for transmission at noninteger rates, but FLP does this in a much more flexible way.

But the advantages of flexible precoding with respect to the choice of the constellation are bought at the price of error propagation. In applications, such as voice-band modems, where the channel is strictly band-limited, FLP offers the possibility to adapt the transmission scheme (e.g., center frequency and symbol rate) very tightly to the channel. In contrast to this, for use with digital subscriber lines the loss due to a restriction to integer rates is negligible compared to the loss caused by error propagation. Here, in spite of the restrictions on the signal set, Tomlinson-Harashima precoding is preferable.

The most important duality of THP and FLP arises when a combination with coded modulation or signal shaping, respectively, is considered. The goal of completely separating the three operations of precoding, channel coding, and signal shaping cannot be achieved at all or only with serious disadvantages. In THP it is indispensable to combine precoding and shaping into one unit. Here, trellis-coded modulation can be done separately. Conversely, in order to avoid a huge loss in power efficiency, FLP has to do precoding and channel coding together. Now, signal shaping can be performed prior to the ISI coder.

Lastly, implementation is very simple for THP (cf. also Section 3.5). Using appropriate fixed-point arithmetic, modulo reduction is done automatically. This is especially true for one-dimensional signal sets or two-dimensional square constellations. Modulo reduction is not done at one stage, but at each multiplication and addition. This reduction due to overflow, dreaded in filter design, is the desired property of the Tomlinson-Harashima precoder. The same is true for FLP in combination with constellations based on a rectangular grid. Here, the feedback part can work with fewer digits, resulting in the necessary modulo reduction. The subtraction of $a[k]$ and $m[k]$ (see Figure 3.20) has to be carried out with higher precision to cover the full range of the signals $a[k]$ and $x[k]$, respectively.

Unfortunately, receiver implementation is more complicated. This is because for THP as well as for FLP the channel output has a near discrete Gaussian distribution, and may have a wide dynamic range (cf. equation (3.2.5)). This effect is even increased for channels $H(z)$, which corresponds to prediction-error filters offering high noise prediction gain. Hence, implementation of the receiver is intricate because all arithmetics have to be carried out with large word length. Furthermore, timing recovery and adaptive residual equalization are complicated significantly or even impossible (for details on blind adaptive equalization for precoding schemes see, e.g., [FGH95, Ger98]).

Even worse, the receiver in flexible precoding needs to operate linearly over this large dynamic range, because in contrast to Tomlinson-Harashima precoding no modulo reduction of the receive samples $y[k]$ (see Figure 3.22) is possible. Both, decoder (slicer or Viterbi decoder) and inverse channel filter $1/H(z)$ have to work on the whole dynamic range. It is essential that this filter works linearly, i.e., overflows must not occur. Hence, long word length is required.

In [FGH95, FH95] a combined precoding/shaping technique is proposed through which the dynamic range of $v[k]$ can be reduced by a large amount with almost no loss. We will discuss such techniques in Chapter 5.

Finally, for reference, in Tables 3.8 and 3.9 the precoding loss of the various schemes discussed in the last sections is summarized. The signal constellation $\mathcal{A}$ of the data symbols $a[k]$ and the support of the transmit signal $x[k]$ are listed therein. For FLP the region of the quantization error (dither sequence) $m[k]$ is given. Additionally, the precoding loss $\gamma_{\mathrm{p}}^2 = \sigma_x^2/\sigma_a^2$ is noted. Table 3.8 is valid for one-dimensional transmission, whereas Table 3.9 is valid for two-dimensional square constellations.

Table 3.8 Summary of precoding loss (one-dimensional signal constellations).

		$a[k] \in \dots$ $\sigma_a^2 = \dots$	$m[k] \in \dots$ $\sigma_m^2 = \dots$	$x[k] \in \dots$ $\sigma_x^2 = \dots$	$\gamma_p^2 = \frac{\sigma_x^2}{\sigma_a^2}$
THP	uncoded	$\mathcal{A}_1$ $\frac{M^2-1}{3}$	—	$[-M, M)$ $\frac{M^2}{3}$	$\frac{M^2}{M^2-1}$
	coded	$\mathcal{A}_2$ $\frac{(2M)^2-1}{3}$	—	$[-2M, 2M)$ $\frac{(2M)^2}{3}$	$\frac{(2M)^2}{(2M)^2-1}$
FLP	uncoded	$\mathcal{A}_1$ $\frac{M^2-1}{3}$	$[-1, 1)$ $\frac{1}{3}$	$\sigma_x^2 =$ $\sigma_a^2 + \sigma_m^2$	$\frac{M^2}{M^2-1}$
	coded	$\mathcal{A}_2$ $\frac{(2M)^2-1}{3}$	$[-K, K)$ $\frac{K^2}{3}$		$\frac{(2M)^2-1+K^2}{(2M)^2-1}$
ISI-coder		$\mathcal{A}_3$ $\frac{(2M)^2-1}{3}$	$[-2, 2)$ $\frac{2^2}{3}$		$\frac{(2M)^2+3}{(2M)^2-1}$
mISI-coder		$\mathcal{A}_3$ $\frac{(2M)^2-1}{3}$	$[-1, 1)$ $\frac{1}{3}$		$\frac{(2M)^2}{(2M)^2-1}$

with $\mathcal{A}_1 = \{\pm 1, \pm 3, \dots, \pm(M-1)\}$, $\mathcal{A}_2 = \{\pm 1, \pm 3, \dots, \pm(2M-1)\}$,
$\mathcal{A}_3 = \{a \in \mathcal{A}_2 | a \in 4\mathbb{Z} + 1\}$, K: number of subsets.

Table 3.9 Summary of precoding loss (two-dimensional square constellations).

		$a[k] \in \dots$ $\sigma_a^2 = \dots$	$m[k] \in \dots$ $\sigma_m^2 = \dots$	$x[k] \in \dots$ $\sigma_x^2 = \dots$	$\gamma_p^2 = \frac{\sigma_x^2}{\sigma_a^2}$
THP	uncoded	$\mathcal{A}_1$ $2\frac{M-1}{3}$	—	$[-\sqrt{M}, \sqrt{M})^2$ $2\frac{M}{3}$	$\frac{M}{M-1}$
	coded	$\mathcal{A}_2$ $2\frac{2M-1}{3}$	—	$[-\sqrt{2M}, \sqrt{2M})^2$ $2\frac{2M}{3}$	$\frac{2M}{2M-1}$
FLP	uncoded	$\mathcal{A}_1$ $2\frac{M-1}{3}$	$[-1, 1)^2$ $2\frac{1}{3}$	$\sigma_x^2 =$ $\sigma_a^2 + \sigma_m^2$	$\frac{M}{M-1}$
	coded	$\mathcal{A}_2$ $2\frac{2M-1}{3}$	$[-\sqrt{K}, \sqrt{K})^2$ $2\frac{K}{3}$		$\frac{2M-1+K}{2M-1}$
ISI-coder		$\mathcal{A}_3$ $2\frac{2M-1}{3}$	$[-\sqrt{2}, \sqrt{2})^2$ $2\frac{2}{3}$		$\frac{2M+1}{2M-1}$
mISI-coder		$\mathcal{A}_3$ $2\frac{2M-1}{3}$	$[-1, 1)^2$ $2\frac{1}{3}$		$\frac{2M}{2M-1}$

with $\mathcal{A}_1 = \{\pm 1, \pm 3, \dots, \pm(\sqrt{M}-1)\}^2$, $\mathcal{A}_2 = \{\pm 1, \pm 3, \dots, \pm(\sqrt{2M}-1)\}^2$,
$\mathcal{A}_3 = \{a \in \mathcal{A}_2 | \mathrm{Re}\{a\} + \mathrm{Im}\{a\} \in 4\mathbb{Z} + 2\}$, K: number of subsets.

In order to conclude this section which summarizes and compares precoding schemes, numerical simulations are presented. They clearly visualize the predicted effects and differences between Tomlinson-Harashima precoding and flexible precoding.

Example 3.6: *Numerical Simulations of Precoding Schemes*

In this example we consider two scenarios: First, SDSL, which employs baseband (one-dimensional constellations) with three information bits per symbol. The simplified DSL up-stream example (self-NEXT dominated environment) with cable length of 3 km is studied, cf. Appendix B. The T-spaced discrete-time end-to-end channel model $H(z)$ with monic impulse response of order $p = 10$ is calculated via the Yule-Walker equations.

Second, passband signaling using two-dimensional QAM constellations is considered. The scenario is close to ADSL down-stream transmission (white channel noise), where, as an alternative to discrete multitone modulation, *carrierless AM/PM (CAP)* [Wer93], a variant of QAM, may also be used. Details on the ADSL example can be found in Appendix B as well. Here, a cable length of 5 km is assumed. An optimization identical to that of Example 2.8 reveals that in this situation five information bits per (complex) symbol is optimum. The complex discrete-time channel model $H(z)$, given in the equivalent low-pass domain, is of order $p = 10$, too.

Note that all results will be displayed over the *transmit energy* per information bits E_b at the output of the precoder, divided by the normalized (one-sided) noise power spectral density N_0'. Since the channel impulse response is forced to be monic, and precoding eliminates its tail, in effect a unit-gain discrete-time AWGN channel with noise variance $\sigma_n^2 = N_0'/T$ is present. Hence, a constant channel attenuation is eliminated due to the current normalization, and the simulation results—at least for Tomlinson-Harashima precoding—are valid for all (except some degenerate cases) monic impulse responses.

Uncoded Transmission Figure 3.39 shows the *symbol error rates (SER)* over the signal-to-noise ratio E_b/N_0' in dB for uncoded transmission employing THP and FLP, respectively. The section at the top shows the results for the SDSL scenario (baseband transmission) using a one-dimensional PAM signal constellation $\mathcal{A} = \{\pm1, \pm3, \pm5, \pm7\}$. The error rates of the ADSL scheme are shown at the bottom. Since 5 bits per symbol are transmitted, flexible precoding operates on a cross constellation with 32 signal points (cf., e.g., [Pro01]). Unfortunately, this constellation is not suited for Tomlinson-Harashima precoding, hence here a rotated square constellation (cf. Figure 3.10) is employed. Because of the different signal sets, flexible precoding has some advantage with respect to average transmit power.

For comparison, the theoretic symbol error rate of uncoded 8-ary ASK and 32-ary QAM are plotted, too. In order to account for the periodic extension of the signal set when using precoding, the number of nearest neighbors is increased to 2 and 4. Thus the curves are calculated according to $\mathrm{SER} = 2 \cdot \mathrm{Q}\left(\sqrt{\frac{6\cdot 3}{63} \cdot \frac{E_b}{N_0'}}\right)$ and $\mathrm{SER} = 4 \cdot \mathrm{Q}\left(\sqrt{\frac{3\cdot 5}{31} \cdot \frac{E_b}{N_0'}}\right)$, respectively.

As one can see, the curves for Tomlinson-Harashima precoding are in good agreement with the theoretical statements. In ADSL the precoding loss of $\frac{32}{31} \mathrel{\hat{=}} 0.14$ dB is visible, whereas it vanishes almost completely for SDSL, since $\frac{64}{63} \mathrel{\hat{=}} 0.07$ dB.

Flexible precoding performs worse than Tomlinson-Harashima precoding. Because of the inverse channel filter $1/H(z)$ at the receiver, decision errors propagate through the receiver

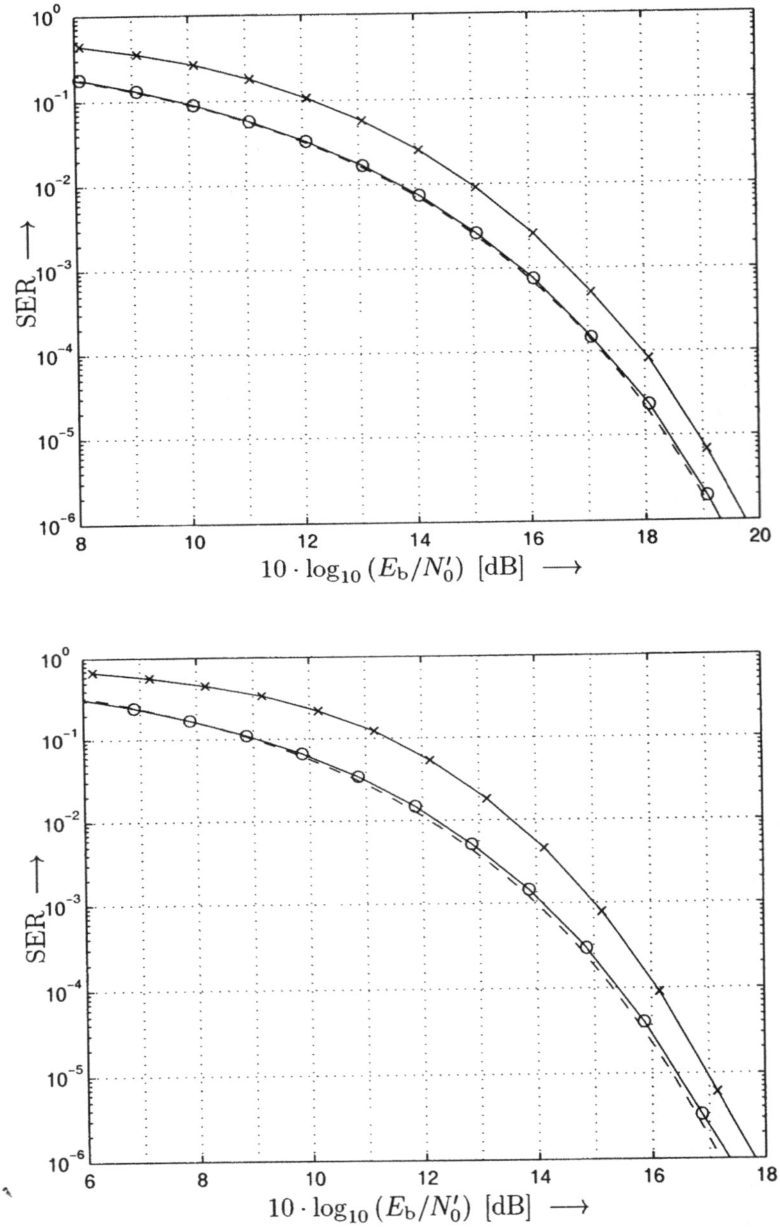

Fig. 3.39 Symbol error rates versus the signal-to-noise ratio. Uncoded transmission using Tomlinson-Harashima precoding (o) and flexible precoding (×). Top: SDSL (baseband) scenario; bottom: ADSL (passband) scenario. Dashed lines: theoretic error rates.

for some time, leading to error multiplication. This effect is visible for baseband, as well as for passband transmission.

Trellis-Coded Transmission We now turn to coded transmission. The symbol error rates over the signal-to-noise ratio E_b/N_0' in dB for the various precoding schemes are plotted in Figure 3.40. For reference, the results for uncoded transmission are repeated. As an example we use the 16-state Ungerböck code [Ung82]. For one-dimensional signaling this code provides an asymptotic coding gain of about 4.3 dB, whereas the two-dimensional 16-state code achieves asymptotically 4.7 dB. In each case, the path register length of the Viterbi decoder is chosen to be 40 symbols.

Note that Tomlinson-Harashima precoding again performs best. This is due to the higher precoding loss of flexible precoding and its enhanced versions, as well as due to error propagation in the inverse precoder. Compared to uncoded transmission, error propagation in the coded flexible precoding scheme is lower. The Viterbi decoder mainly produces bursts of errors, which are only slightly prolonged by the reconstruction filter $1/H(z)$ at the receiver. All schemes fall short of the asymptotic bound, which is given by shifting the error curve for uncoded transmission to the left by the asymptotic coding gain.

A comparison of the straightforward combination of precoding and channel coding and the enhanced schemes reveals some gain. But compared to the ISI coder, the modified versions thereof are almost not beneficial. The reduction in average transmit power can only be exploited asymptotically. Going from flexible precoding to the ISI coder and finally to its modified version, quantization to recover data at the inverse precoder has to be done with respect to an even more dense lattice ($\Lambda_c \to \Lambda_f \to \Lambda_a$). As a result, the error rate is increased, which in turn absorbs the gain in average transmit power. For smaller Voronoi regions, the error event has to ware off much more before the decision is made correctly.

All phenomena described above are valid for the SDSL scenario, as well as for the ADSL example.

In order to further assess coded transmission in combination with precoding, Figure 3.41 compares SDSL transmission (three information bits per symbol) and transmission over the AWGN channel using the same signal constellations and trellis code.

For uncoded transmission, except for the precoding loss of 0.07 dB, almost no difference in performance is visible. The increase in number of nearest neighbors from $\frac{14}{8}$ to 2 is negligible. Regarding trellis-coded transmission, a somewhat higher error rate for the precoding scheme is visible. The trellis code is more affected by the periodic extension of the signal constellation than a simple slicer. This is because signal points at the perimeter of the constellation, which may by highly reliable, are no longer present. But in sequence decoding, such reliable spots are dispersed over the entire series. Asymptotically, this effect can be neglected and the loss disappears.

To summarize, except for a small degradation, especially for symbol error rate below, say, $10^{-5} \dots 10^{-6}$, almost the same coding gain as over the AWGN channel can be achieved for precoding schemes. Since the periodic extension is the same for all versions of precoding, this statement is true for flexible precoding as well as its modifications.

Channels with Spectral Zeros We now study the effect of zeros in the end-to-end channel transfer function more closely. For brevity, we restrict ourselves to fourth-order channels with $H(z) = (1 - \rho z^{-1})(1 + \rho z^{-1})(1 - 0.9e^{j2\pi/3} z^{-1})(1 - 0.9e^{-j2\pi/3} z^{-1})$, i.e., zeros at $\pm\rho$, $|\rho| \leq 1$, on the real axis and at $z = 0.9e^{\pm j2\pi/3}$. Figure 3.42 shows the error rate using flexible precoding for different ρ. Here, again uncoded baseband transmission with an 8-ary ASK signal constellation is considered.

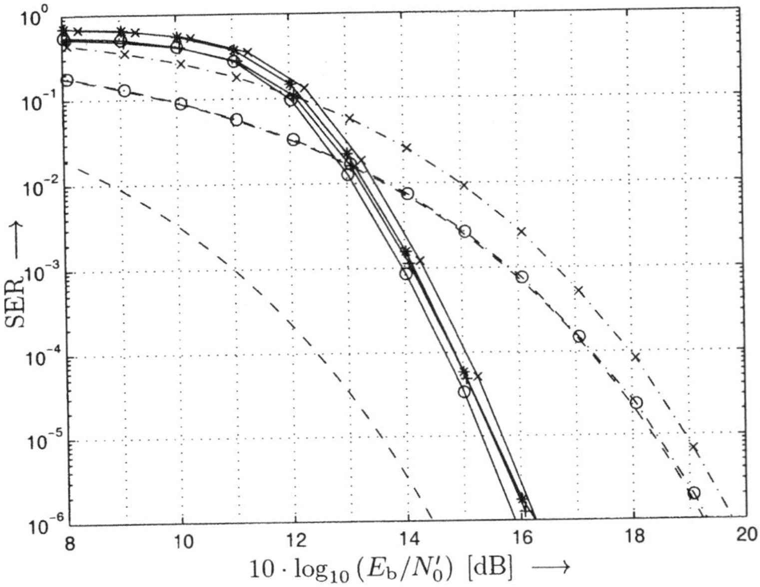

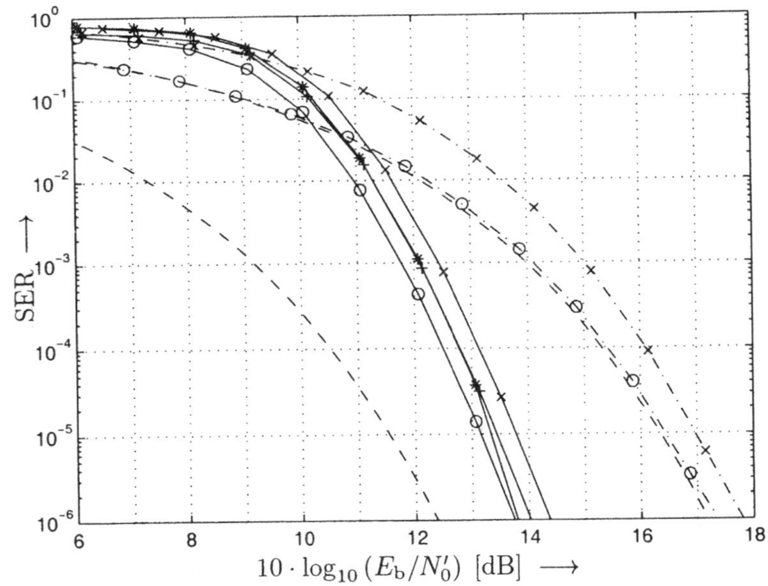

Fig. 3.40 Symbol error rates versus the signal-to-noise ratio. Trellis-coded transmission using Tomlinson-Harashima precoding (○), flexible precoding (×), and the modified versions thereof (ISI coder (+), modified ISI coder (∗)). Top: SDSL (baseband) scenario; bottom: ADSL (passband) scenario. Dashed lines: theoretic error rates.

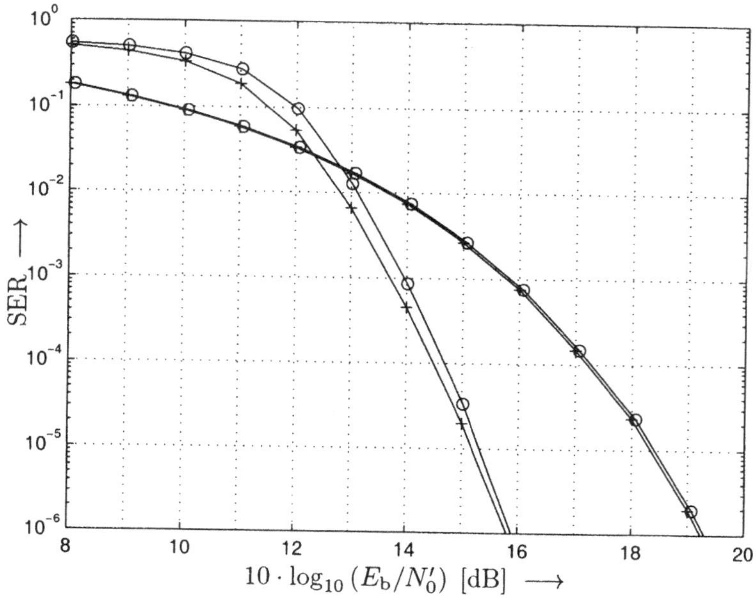

Fig. 3.41 Symbol error rates versus the signal-to-noise ratio. Trellis-coded transmission over AWGN channel (+) and SDSL transmission using Tomlinson-Harashima precoding (o).

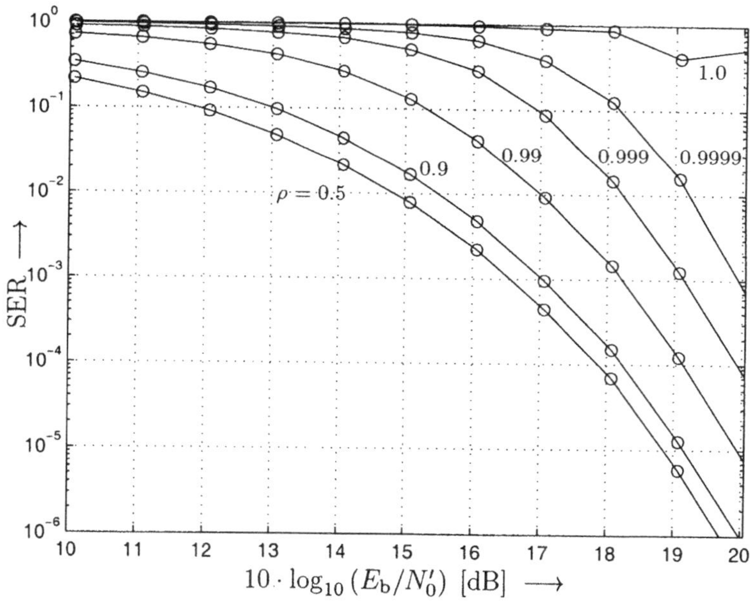

Fig. 3.42 Symbol error rates versus the signal-to-noise ratio. Uncoded baseband transmission using flexible precoding over $H(z) = (1 - \rho z^{-1})(1 + \rho z^{-1})(1 - 0.9e^{j2\pi/3}z^{-1})(1 - 0.9e^{-j2\pi/3}z^{-1})$ Bottom to top: $\rho = 0.5$, 0.9, 0.99, 0.999, 0.9999, and 1.0.

Increasing ρ shifts the zeros of $H(z)$ more closely toward the unit circle. Consequently, the inverse channel filter at the inverse precoder has poles that are in increasingly closer proximity to the unit circle. Hence, its impulse response becomes longer and longer. This enhances error propagation, and the symbol error rate is increased. For $\rho = 1$, i.e., spectral zeros, transmission becomes impossible.

We now concentrate on $\rho = 1$—the channel $H(z)$ has spectral zeros. In Figure 3.43 the performance of Tomlinson-Harashima precoding and flexible precoding is compared. Here, uncoded, as well as trellis-coded transmission using a 16-state Ungerböck, code are studied.

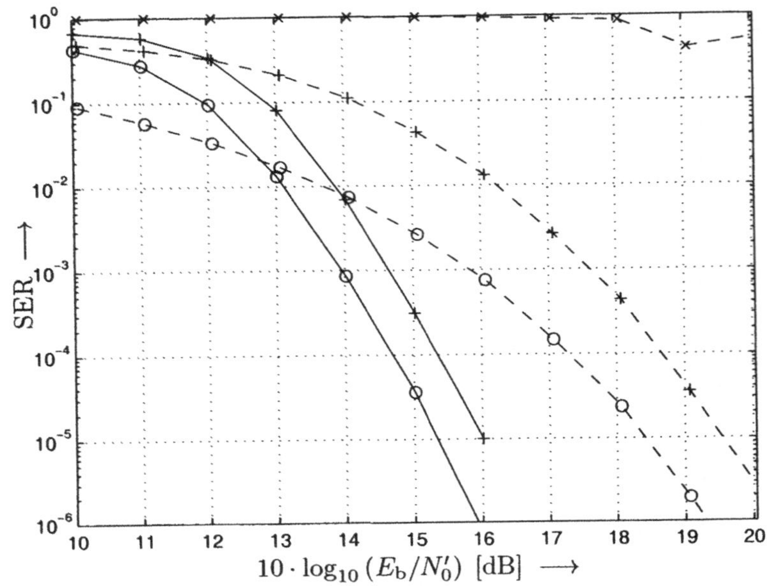

Fig. 3.43 Symbol error rates versus the signal-to-noise ratio. Uncoded (dashed lines) and trellis-coded (16 states, solid lines) transmission using Tomlinson-Harashima precoding (o), flexible precoding ($\times$), and flexible precoding with the proposed receiver modification (+). $H(z) = (1 - z^{-1})(1 + z^{-1})(1 - 0.9e^{j2\pi/3}z^{-1})(1 - 0.9e^{-j2\pi/3}z^{-1})$.

The curve at the very top of the figure is again valid for flexible precoding. Due to infinite error propagation, no reliable transmission is possible. Applying the proposed modified receiver (nonlinear version of the inverse channel filter, cf. Section 3.3.6), stable operation is assured in spite of the spectral nulls. But compared to Tomlinson-Harashima precoding, due to error propagation, the symbol error rate is approximately 10 times higher. If the stable version of the inverse channel filter is also used in the modified ISI code, trellis-coded transmission over this channel with spectral zeros is enabled. Here, since TCM produces bursts of error, the error multiplication in the inverse precoder is somewhat lower than for uncoded transmission.

To summarize, by using the proposed modified receiver, flexible precoding can be extended to the wide class of channels with zeros at DC and/or the Nyquist frequency. The additional complexity is negligible, and all desirable properties of flexible precoding are not affected.

3.5 FINITE-WORD-LENGTH IMPLEMENTATION OF PRECODING SCHEMES

In the previous sections, the theoretical performance of precoding schemes was discussed. However, it is also important how efficiently the system can be implemented in practice. In high-speed communication, the precoder still cannot be realized using a general-purpose *digital signal processor (DSP)*, but *field-programmable gate arrays (FPGAs)* or *application-specific integrated circuits (ASICs)* have to be utilized. Here, the costs are determined by the number of gates, and hence, restricted word length is still an important issue in implementation.

ASICs are of special interest when the precoder is fixed to a typical reference application. This avoids the necessity of transmitting back channel data to the transmitter. Using adaptive residual linear equalization at the receiver, a fixed precoder causes only minor degradation, even if actual and reference situation exhibit noticeable differences. For details, the reader is referred to [FGH95, Ger98].

Now, some aspects of finite-word-length effects in precoding schemes are addressed. For brevity, we restrict the discussions to baseband signaling, i.e., one-dimensional constellations, as used, e.g., in SDSL. Moreover, we are concerned with a full customer fixed-point implementation. Quantization effects at the precoder are investigated in two steps: on the one hand, finite-word-length restriction on the precoder coefficients is studied, and on the other hand, quantization of data at the precoder is considered. The effect of quantization noise taking effect at the decoder input is described analytically.

3.5.1 Two's Complement Representation

For fast hardware implementation, a suitable representation of numbers and an efficient implementation of arithmetic has to be available. Here, we focus on fixed-point arithmetics with a word length of w binary digits. Because of its special properties, it is a natural choice to consider exclusively *two's-complement representation*.

In contrast to most of the literature, where numbers are usually normalized to the interval $[-1, 1)$, because of the range of the data symbols, we have to represent numbers in the range of $[-2^{w_\mathrm{I}}, 2^{w_\mathrm{I}})$, $w_\mathrm{I} \in \mathbb{N}$. Here, w_I denotes the number of binary digits representing the integer part. Providing one digit of the total word length w as the sign bit, the fraction is represented with $w_\mathrm{F} = w - 1 - w_\mathrm{I}$ bits. Hence, the quantization step size is given by $Q = 2^{-w_\mathrm{F}}$.

With the above definitions, the two's-complement representation of the decimal number x reads (e.g., [Bos85, PM88]):

$$\begin{aligned} x &= 2^{w_\mathrm{I}} \cdot \left(-b_0 2^0 + \sum_{i=1}^{w-1} b_i 2^{-i} \right), \quad b_i \in \{0, 1\}, \quad i = 0, 1, \ldots, w-1 \\ &= \underbrace{-b_0 2^{w_\mathrm{I}}}_{\text{sign bit}} + \underbrace{\sum_{i=1}^{w_\mathrm{I}} b_i 2^{(w_\mathrm{I}-i)}}_{\text{integer}} + \underbrace{\sum_{i=w_\mathrm{I}+1}^{w-1} b_i 2^{(w_\mathrm{I}-i)}}_{\text{fraction}} \end{aligned} \qquad (3.5.1)$$

We also write compactly

$$x = [\underbrace{b_0}_{\text{sign bit}} \mid \underbrace{b_1 b_2 \cdots b_{w_\mathrm{I}}}_{\text{integer}} \mid \underbrace{b_{w_\mathrm{I}+1} \cdots b_{w-1}}_{\text{fraction}}]_2 \,. \tag{3.5.2}$$

Since $-x$ is obtained by complementing the whole binary representation of the number and incrementing this binary word by one [Bos85], i.e., written in real numbers

$$-x = 2^{w_\mathrm{I}} \cdot \left(-\overline{b_0}\, 2^0 + \sum_{i=1}^{w-1} \overline{b_i}\, 2^{-i} \right) + 2^{-w_\mathrm{F}} \,, \tag{3.5.3}$$

in two's complement, the range of representable numbers is asymmetric. The largest number is given by

$$x_{\max} \;=\; 2^{w_\mathrm{I}} - 2^{w_\mathrm{F}} \;=\; [\,0\,|\,1\,\cdots\,1\,|\,1\,\cdots\,1\,]_2 \tag{3.5.4a}$$

and the smallest number (toward minus infinity) reads

$$x_{\min} \;=\; -2^{w_\mathrm{I}} \qquad\;=\; [\,1\,|\,0\,\cdots\,0\,|\,0\,\cdots\,0\,]_2 \,. \tag{3.5.4b}$$

We will now show how arithmetic is done in two's complement. After each arithmetic operation with two w-bit numbers, the result has to be represented again by w bits. It is then that overflow and/or round-off errors occur.

Binary Fixed-Point Arithmetic: Addition When two numbers are added, say x_1 and x_2, each represented by w digits, the sum $x_1 + x_2$ has to be represented by $w + 1$ bits in general. Quantization step size is unaffected, but the integer part has to be extended. In two's complement such overflows are treated by simply deleting the surplus digit. This corresponds to a repeated addition or subtraction of $2^{(w_\mathrm{I}+1)}$, so that the final result falls in a range which can be expressed with w bits. Thus, each addition in two's complement—symbolized by "$\boxplus$"—can be written as

$$\begin{aligned} y &= x_1 \boxplus x_2 \\ &= x_1 + x_2 + d \cdot 2^{(w_\mathrm{I}+1)} \,, \qquad \text{with } d \in \mathbb{Z}, \text{ so that } y \in [-2^{w_\mathrm{I}},\, 2^{w_\mathrm{I}}) \,. \end{aligned} \tag{3.5.5}$$

This procedure is equivalent to a modulo reduction of the sum to the range $[-2^{w_\mathrm{I}},\, 2^{w_\mathrm{I}})$ which, as we will see shortly, directly leads to the desired property in Tomlinson-Harashima precoding. Figure 3.44 shows the addition in two's complement.

Binary Fixed-Point Arithmetic: Multiplication The multiplication of two fixed point numbers, each with w binary digits, results in a product of word length $2w$ bits. Overflow is again treated as above by adding an integer multiple of $2^{(w_\mathrm{I}+1)}$, i.e., by modulo reduction. Having two numbers, each with a quantization step size Q, the product is of precision Q^2, corresponding to $2w_\mathrm{F}$ bits representing the fraction.

For shortening the number to w_F bits again, two methods are feasible: *rounding* and *two's-complement truncation*, cf. [Bos85, PM88]. The respective quantization characteristics are shown in Figure 3.45.

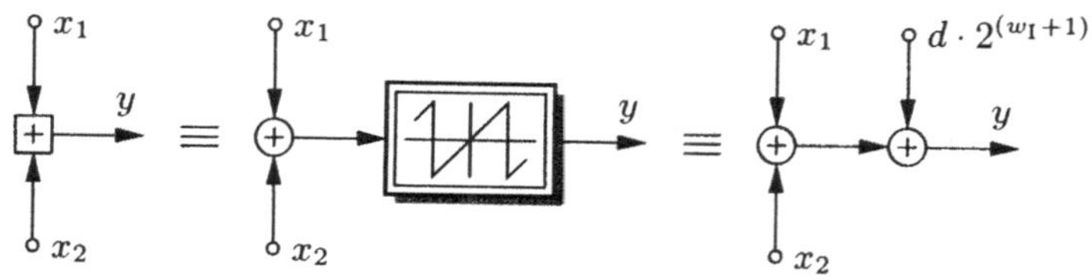

Fig. 3.44 Addition in two's complement. The boxed plus "⊞" denotes operation in two's complement, whereas operation symbolized by circles are performed ideally (real numbers). The integer d has to be chosen, so that $y \in [-2^{w_I}, 2^{w_I})$.

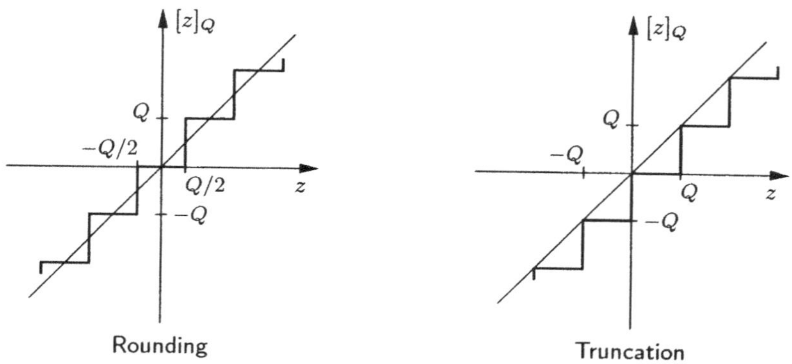

Fig. 3.45 Rounding (left) and truncation (right) of two's-complement numbers. $[z]_Q$ symbolizes the quantization of z.

Denoting the quantization error by ε, the multiplication in two's complement—symbolized by "⊠"—can be written as

$$\begin{aligned} y &= x_1 \boxtimes x_2 \\ &= x_1 \cdot x_2 + \varepsilon + d \cdot 2^{(w_I+1)} \quad \text{with } d \in \mathbb{Z}, \text{so that } y \in [-2^{w_I}, 2^{w_I}) \; . \end{aligned} \tag{3.5.6}$$

Figure 3.46 shows the multiplication in two's complement.

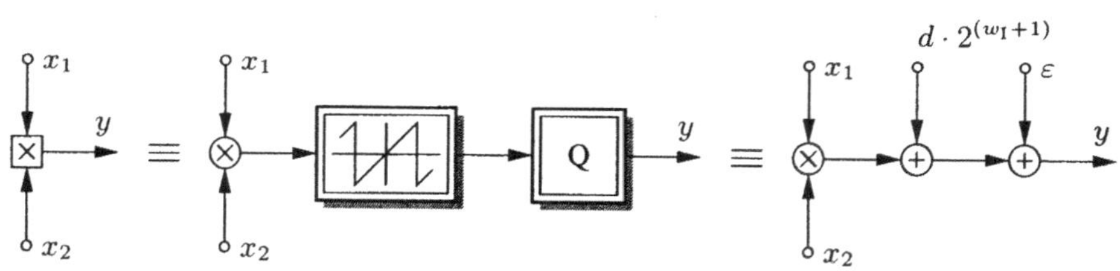

Fig. 3.46 Multiplication in two's complement. The boxed multiplication sign "⊠" denotes operation in two's complement, whereas operation symbolized by circles are performed ideally (real numbers). The integer d has to be chosen, so that $y \in [-2^{w_I}, 2^{w_I})$.

Because in digital systems at least one factor is a random variable, the quantization error ε is also random. Thus, it has to be characterized by its probability density function (pdf) and its autocorrelation sequence. Here, we omit a detailed derivation of the statistics of quantization errors and only apply the results. For a detailed discussion on the topic, [HM71, Sjo73, ES76, SS77, BTL85, Bos85] may serve as starting points.

Let $\langle x[k] \rangle$ be an i.i.d. uniformly distributed data sequence with $w_\mathrm{D} \triangleq w_{\mathrm{F},x}$ digits representing the fraction, and c a fixed coefficient with an actual number (not counting least significant zeros) of $w_\mathrm{C} \triangleq w_{\mathrm{F},c}$ bits. The product $y[k] = c \cdot x[k]$ should be represented using w_D bits for the fraction, too. Since the signal to be quantized is of finite precision, the quantization error has a discrete distribution. Moreover, if the signal range covers many quantization intervals, the quantization error will be uniformly distributed. Hence, the pdf $\mathsf{f}_\varepsilon(\varepsilon)$ of the quantization error ε is well approximated by

$$\mathsf{f}_\varepsilon(\varepsilon) = 2^{-w_\mathrm{C}} \cdot \sum_{\nu \in \mathcal{L}} \delta(\varepsilon - \nu \cdot 2^{-(w_\mathrm{D}+w_\mathrm{C})}) \,, \tag{3.5.7a}$$

where

$$\mathcal{L} = \mathbb{Z} \cap \begin{cases} (-2^{(w_\mathrm{C}-1)},\, 2^{(w_\mathrm{C}-1)}], & \text{rounding} \\ (-2^{w_\mathrm{C}},\, 0], & \text{truncation} \end{cases} \,, \tag{3.5.7b}$$

and $\delta(\cdot)$ is the delta function. Figure 3.47 displays the densities $\mathsf{f}_\varepsilon(\varepsilon)$ of the quantization error ε for rounding and truncation in two's complement, respectively.

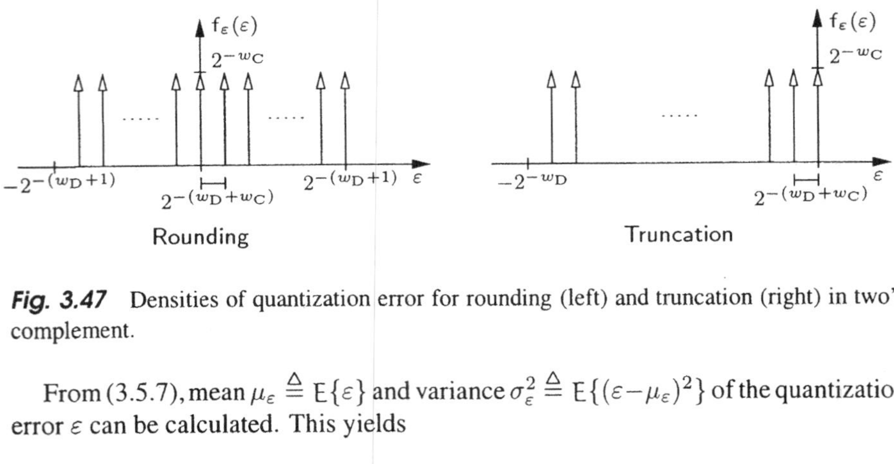

Fig. 3.47 Densities of quantization error for rounding (left) and truncation (right) in two's complement.

From (3.5.7), mean $\mu_\varepsilon \triangleq \mathsf{E}\{\varepsilon\}$ and variance $\sigma_\varepsilon^2 \triangleq \mathsf{E}\{(\varepsilon-\mu_\varepsilon)^2\}$ of the quantization error ε can be calculated. This yields

$$\mu_\varepsilon = \int_{-\infty}^{+\infty} \varepsilon \cdot \mathsf{f}_\varepsilon(\varepsilon)\, d\varepsilon = \begin{cases} \begin{cases} 2^{-(w_\mathrm{C}+w_\mathrm{D}+1)}, & w_\mathrm{C} \neq 0 \\ 0, & w_\mathrm{C} = 0 \end{cases} & \text{rounding} \\ 2^{-(w_\mathrm{D}+1)} \cdot (2^{-w_\mathrm{C}} - 1), & \text{truncation} \end{cases} \,, \tag{3.5.8a}$$

and for both cases

$$\sigma_\varepsilon^2 = \int_{-\infty}^{+\infty} (\varepsilon - \mu_\varepsilon)^2 \cdot \mathsf{f}_\varepsilon(\varepsilon)\, d\varepsilon = 2^{-2w_\mathrm{D}}/12 \cdot (1 - 2^{-2w_\mathrm{C}}) \,. \tag{3.5.8b}$$

It is noteworthy that the number of bits representing the integer part does not appear in the above formulae. Moreover, for $w_C \to \infty$, the variance of the error sequence tends to $2^{-2w_D}/12 = Q^2/12$ and the mean values become zero and $-2^{-(w_D+1)}$, respectively. These are the parameters of the well-known "$Q^2/12$" model, which is not precise enough for our analysis.

Finally, as usual, $\varepsilon[k]$ and $\varepsilon[k+\kappa]$, $\kappa \in \mathbb{Z} \setminus \{0\}$, are assumed to be statistically independent. Hence, the autocorrelation sequence of the quantization error shall be given as

$$\phi_{\varepsilon\varepsilon}[\kappa] \triangleq \mathrm{E}\{\varepsilon[k+\kappa] \cdot \varepsilon^*[k]\} = \begin{cases} \mu_\varepsilon^2 + \sigma_\varepsilon^2, & \kappa = 0 \\ \mu_\varepsilon^2, & \kappa \neq 0 \end{cases} . \tag{3.5.9}$$

3.5.2 Fixed-Point Realization of Tomlinson-Harashima Precoding

Remember again the derivation of decision-feedback equalization via noise prediction in Section 2.3.1, page 49. Starting from optimal linear zero-forcing equalization (ZF-LE), based on the knowledge of the autocorrelation sequence $\phi_{nn}^{(\mathrm{ZF\text{-}LE})}[\kappa]$ of the T-spaced discrete-time noise, a noise whitening filter has been introduced and optimized. The optimum tap weights of this monic finite impulse response (FIR) filter are given as the solution of the *Yule-Walker equations*, see (2.3.14). By varying p, the order of the whitening filter, an exchange between complexity and prediction gain (see (2.3.16)) is possible. In turn, the whitening filter gives the end-to-end channel model to which a precoder has to be adapted.

Optimization of the Noise Whitening Filter Applying the Yule-Walker equations gives a perfect whitening filter, but generally coefficients result which cannot be represented with a finite number of binary digits. Quantizing these coefficients will lead to a loss in prediction gain. Thus, an obvious task is to determine the optimal tap weights under the additional constraint of a finite word length w. Especially for short word lengths w an optimization may provide significant improvements compared to a straightforward quantization of the coefficients. Unfortunately, discrete optimization is difficult and for the present situation no closed-form solution is available. A possible solution to overcome this problem is to apply *simulated annealing* (e.g. [PTVF92]) or other numerical methods for combinatorial optimization. The following example (Example 3.7) shows the procedure for a typical scenario and compares the results to the optimum solution.

Example 3.7: *Optimization of Fixed-Point Coefficients*

As an example we again consider the simplified DSL up-stream example (self-NEXT dominated environment) according to Appendix B. The field length of the cable is chosen to be 3.0 km. In Figure 3.48 the prediction gain G_p (in dB) is plotted over the order p of the whitening filter. The total word length $w = 1 + w_I + w_F$ is chosen to be 3, 4, 6, and 8, respectively. The partitioning of the total word length into the integer part w_I and fractional part w_F is therefore left to the optimization procedure. Interestingly, in each case the optimization will result in a minimum-phase impulse response. For reference, the exchange between prediction order and gain for an infinite word length is shown as well (solid line; cf. also Figure 2.18).

Note that for short word lengths increasing the order p is not rewarding, because the optimization results in trailing zeros of the impulse response $\langle h[k] \rangle$, and hence the actual order is lower than the designed one (visible as horizontal asymptotes in the figure).

From Figure 3.48 it can be seen that for $w \geq 6$ the loss in prediction gain compared to the optimum one is negligible. Thus, it is useless to use a higher precision for the coefficients of the precoder.

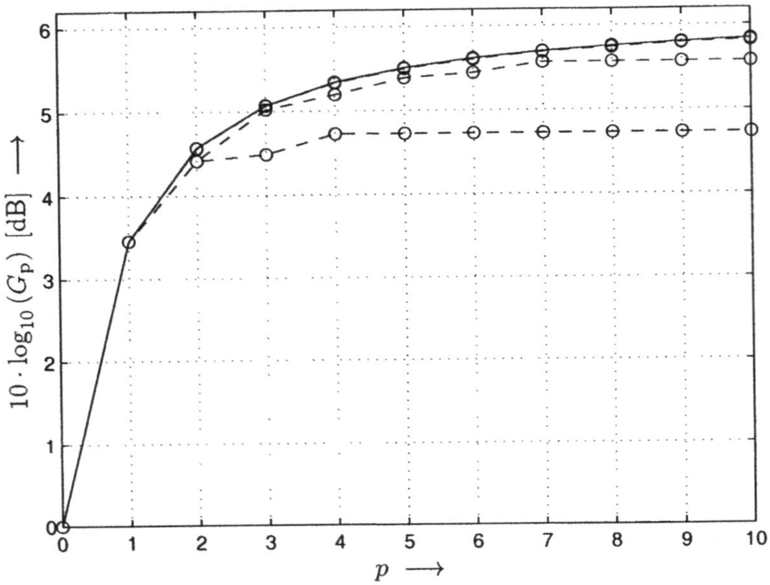

Fig. 3.48 Prediction gain over the order of the whitening filter for finite word lengths w of the coefficients. Bottom to top: $w = 3, 4, 6, 8$. Solid line: infinite word length.

Table 3.10 Example for coefficients of the whitening filter: $p = 5$, $w_\mathrm{I} = 1$, $w_\mathrm{F} = 4$.

i	$h[i]$ Decimal	$h[i]$ Two's Compl.	$w_{\mathrm{C},i}$
0	1.0000	01.0000	0
1	1.3750	01.0110	3
2	1.1250	01.0010	3
3	0.8125	00.1101	4
4	0.5000	00.1000	1
5	0.1875	00.0011	4

Table 3.10 exemplary shows one set of coefficients for $w_\mathrm{I} = 1$, $w_\mathrm{F} = 4$, i.e., $w = 6$, and $p = 5$, which is used in later examples. In addition, the actual word length $w_{\mathrm{C},i}$ (not counting trailing zeros) of tap $h[i]$ is given. Because all coefficients $h[i]$ in this example are positive, the sign bit is always zero and does not contribute to the implementation effort.

For implementation, the analog receiver front-end filter and the discrete-time whitening filter are usually combined into a single system followed by T-spaced sampling. Now, the task of the total receiver input filter is to limit noise bandwidth and to force the end-to-end impulse response to the desired, optimized one. In the following we assume that this is done by an analog front-end filter, so that at this stage no additional quantization noise emerges.

Quantization Effects at the Precoder Quantizing the coefficients of the end-to-end impulse response $\langle h[k] \rangle$ for use in Tomlinson-Harashima precoding does not change the operation of the precoder at all. Merely a (slightly) different impulse response has to be equalized. A much more crucial point is the finite-word-length restriction on the data signals within the precoder. It is well known that in recursive digital systems this may lead to oscillations, called *limit cycles* and/or additive *quantization noise*. Later on, these effects are analyzed in detail.

Two's-complement representation is very well suited for implementation of Tomlinson-Harashima precoding. In his original work [Tom71], Tomlinson called this "*modulo arithmetic.*" If the number M of signal points is a power of two, an explicit modulo reduction is not required. Choosing the number of bits representing the integer part to be $w_\mathrm{I} = \mathrm{ld}(M)$, the overflow handling in two's-complement—i.e., neglecting the surplus digits (carry bits)—directly carries out the desired modulo reduction. This is not performed at one stage, as shown in Figure 3.4, but at every single addition and multiplication. It should be emphasized that this overflow reduction, dreaded in the context of linear filter design because of the resulting limit cycles, is the basic and desired property of Tomlinson-Harashima precoding. Moreover, the half-open interval for the channel symbols $x[k]$ is a consequence of two's-complement representation with its asymmetric range of representable numbers.

For mathematical analysis, Figure 3.49 shows the implementation of Tomlinson-Harashima precoding using two's-complement arithmetic. All operations represented by square boxes ("⊞" and "⊠") are performed in two's complement. Conversely, as usual, operations for real numbers will later be symbolized by circles.

In order to calculate the effects due to quantization, all additions and multiplications are replaced by their descriptions (3.5.5) and (3.5.6) given above. With regard to Figures 3.44 and 3.46, additions are substituted by an ideal addition plus an integer multiple of $2M$, and multiplications by an ideal multiplier with subsequent addition of an integer multiple of $2M$ and of the quantization error ε. As the transmit signal $\langle x[k] \rangle$ is an almost white, uniformly distributed sequence (cf. Theorem 3.1), the statistical description (3.5.8) and (3.5.9) for the quantization error ε is assumed to be valid.

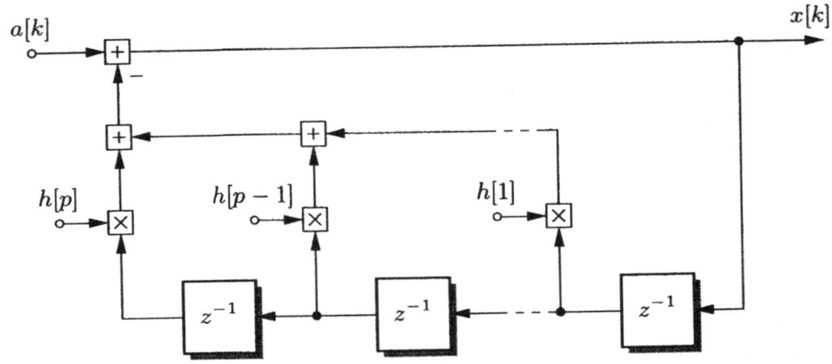

Fig. 3.49 Implementation of Tomlinson-Harashima precoding using two's-complement arithmetic.

With these considerations a linearized model of Tomlinson-Harashima precoding can be set up, see Figure 3.50. Here, all additive terms are moved to the input of the remaining ideal and linear system $1/H(z)$. Please note, the quantization error at time index k stemming from the multiplication with coefficient $h[i]$ is denoted by $\varepsilon_i[k]$. It is evident that, due to negative feedback, the total error signal $\varepsilon[k]$, which is the superposition of the individual quantization errors, reads

$$\varepsilon[k] \triangleq -\sum_{i=1}^{p} \varepsilon_i[k] \,. \tag{3.5.10}$$

From Figure 3.50 it is obvious that neglecting the channel noise $n[k]$, the slicer at the receiver has to work on $a[k] + 2M \cdot d[k] + \varepsilon[k]$. Hence, after modulo reduction into

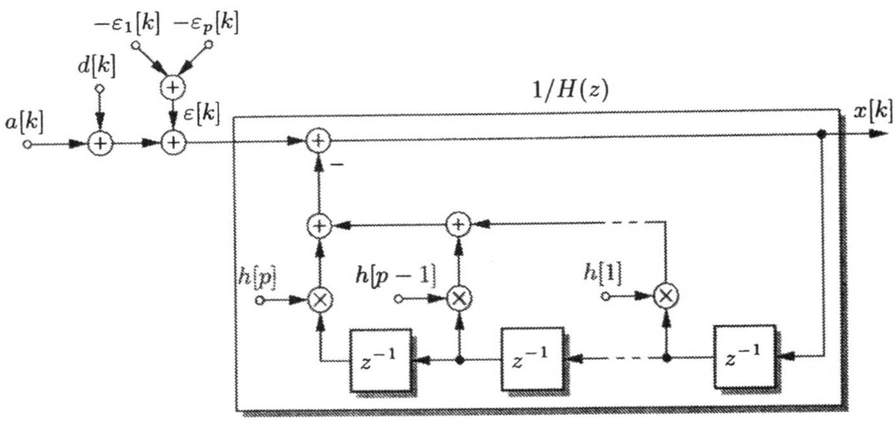

Fig. 3.50 Linearized model of Tomlinson-Harashima precoding using two's-complement arithmetic.

the interval $[-M, +M)$, the data signal $a[k]$ is disturbed by the quantization noise[10] $\varepsilon[k]$.

If, for simplicity and lacking a more accurate model, we assume the individual quantization errors $\varepsilon_i[k]$, $i = 1, 2, \ldots, p$, to be i.i.d. and statistically independent of each other, the probability density function $\mathsf{f}_\varepsilon(\varepsilon)$ of the effective quantization error ε calculates to (see [Pap91]; $*$: convolution)

$$\mathsf{f}_\varepsilon(\varepsilon) = \mathsf{f}_{\varepsilon_1}(-\varepsilon) * \mathsf{f}_{\varepsilon_2}(-\varepsilon) * \cdots * \mathsf{f}_{\varepsilon_p}(-\varepsilon) \triangleq \mathop{\Huge *}_{i=1}^{p} \mathsf{f}_{\varepsilon_i}(-\varepsilon)\,. \quad (3.5.11)$$

Inserting the pdfs $\mathsf{f}_{\varepsilon_i}(\varepsilon_i)$ of the individual quantization errors and taking the definition of the sets $\mathcal{L}$ (equations (3.5.7a) and (3.5.7b)) into consideration, finally yields

$$\mathsf{f}_\varepsilon(\varepsilon) = \mathop{\Huge *}_{i=1}^{p} 2^{-w_{\mathrm{C},i}} \sum_{\nu\in\mathcal{L}} \delta\left(-\varepsilon - \nu \cdot 2^{-(w_\mathrm{D}+w_{\mathrm{C},i})}\right)\,. \quad (3.5.12)$$

From (3.5.12) it can be seen that the granularity of the total quantization error ε is $2^{-(w_\mathrm{D}+w_\mathrm{C})}$. Here, $w_\mathrm{C} = \max_{i=0,\ldots,p}\{w_{\mathrm{C},i}\}$ is the maximum actual number of bits representing the fraction of the coefficients $h[i]$, and w_D denotes the number of bits representing the fraction of the data signals. Hence, using the definition $p_\nu \triangleq \Pr\left\{\varepsilon = \nu \cdot 2^{-(w_\mathrm{D}+w_\mathrm{C})}\right\}$, the pdf $\mathsf{f}_\varepsilon(\varepsilon)$ can be written as

$$\mathsf{f}_\varepsilon(\varepsilon) = \sum_{\nu=-\infty}^{\infty} p_\nu \cdot \delta\left(\varepsilon - \nu \cdot 2^{-(w_\mathrm{D}+w_\mathrm{C})}\right)\,. \quad (3.5.13)$$

Finally, using (3.5.8a) and (3.5.8b), the mean μ_ε and variance σ_ε^2 of the total quantization error ε can be easily calculated, resulting in

$$\mu_\varepsilon = -\sum_{i=1}^{p} \mu_{\varepsilon_i} = \begin{cases} -\sum\limits_{\substack{i=1\\ w_{\mathrm{C},i}\neq 0}}^{p} 2^{-(w_\mathrm{D}+w_{\mathrm{C},i}+1)}, & \text{rounding} \\ -2^{-(w_\mathrm{D}+1)} \cdot \sum\limits_{i=1}^{p} (2^{-w_{\mathrm{C},i}} - 1), & \text{truncation} \end{cases} \quad (3,5.14a)$$

and

$$\sigma_\varepsilon^2 = \sum_{i=1}^{p} \sigma_{\varepsilon_i}^2 = 2^{-2w_\mathrm{D}}/12 \cdot \sum_{i=1}^{p} (1 - 2^{-2w_{\mathrm{C},i}})\,. \quad (3.5.14b)$$

Observe that only coefficients $h[i]$, $i = 0, 1, \ldots, p$, with $w_{\mathrm{C},i} > 0$, contribute to mean μ_ε and variance σ_ε^2, because multiplications with integers are performed perfectly.

[10]Because $|\varepsilon[k]| < 1$ is valid in almost all cases (see Example 3.10), signal-dependent folding of the pdf of $\varepsilon[k]$ can be neglected.

Example 3.8: Quantization Error in Fixed-Point Precoders

Figure 3.51 shows the pdf of the total quantization error ε for the exemplary impulse response, the coefficients of which are given in Table 3.10. For the present figure, the data signals are represented by $w_D = 4$ bits for the fractional part. The integer part does not influence the results, and hence they are valid for all M-ary baseband transmission schemes. At the precoder two's-complement rounding is employed for quantization.

The theoretical result $f_\varepsilon(\varepsilon)$ according to (3.5.13) is displayed on the left-hand side—to be precise, the weights p_ν of the delta pulses are given. Conversely, on the right hand side simulation results $\hat{f}_\varepsilon(\varepsilon)$ are shown. It can be seen that the theoretical derivation and simulation result match very closely. Since we have $w_C = 4$, the error signal has a granularity of $2^{-(w_D+w_C)} = 2^{-8} = 0.0039$. Moreover, the mean value of $\varepsilon[k]$ is given by $\mu_\varepsilon = -0.0273$, and the error variance is $\sigma_\varepsilon^2 = 0.0015$.

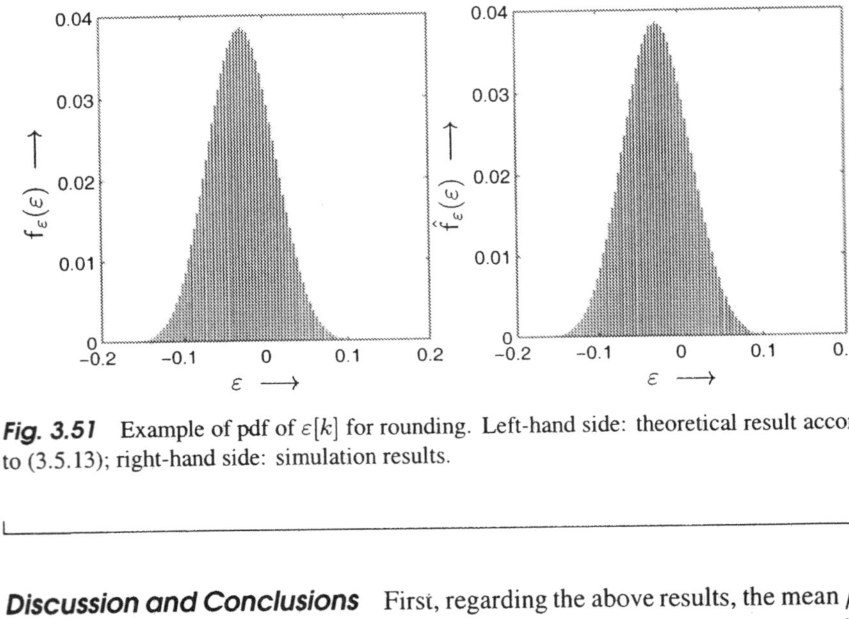

Fig. 3.51 Example of pdf of $\varepsilon[k]$ for rounding. Left-hand side: theoretical result according to (3.5.13); right-hand side: simulation results.

Discussion and Conclusions First, regarding the above results, the mean μ_ε of the total error signal does not contribute to the disturbance. This constant can easily be balanced by a shift of the decision levels; only a residual zero-mean error $\varepsilon' \triangleq \varepsilon - \mu_\varepsilon$ remains. Thus, as ε' has the same statistical properties whether rounding or truncation is used in quantization, it is insignificant which type of word length reduction is done. Because of its somewhat lower complexity, for hardware implementation two's-complement truncation is the preferable method.

Next, reviewing (3.5.14b) two effects are recognizable. On the one hand, the variance σ_ε^2 increases (almost linearly) with the order p of the whitening filter. On the other hand, the well-known gain (decrease of error variance) of 6 dB per bit word length for the data signal is evident. An upper bound on σ_ε^2 is simply given by the "$Q^2/12$" model, for which $\sigma_\varepsilon^2 = p \cdot 2^{-2w_D}/12$ results, if the word length of the coefficients is sufficiently large ($w_{C,i} \to \infty$, $i = 1, \ldots, p$). In order to achieve best performance of Tomlinson-Harashima precoding with the lowest complexity, prediction order p, coefficient word length w_C, and data word length w_D have to be optimized jointly. Of course, increasing p does provide a higher prediction gain G_p (lower variance σ_n^2 of the effective channel noise, see Section 2.3.1), but it also increases σ_ε^2 linearly with p. Since both disturbances are statistically independent, and hence their variances add up, an optimum order p for the discrete-time noise whitening filter—an optimum exchange between σ_n^2 and σ_ε^2—exists. The following example shows this trade-off.

Example 3.9: *Optimum Order of the Whitening Filter*

Continuing the above examples, Figure 3.52 sketches the gain over linear zero-forcing equalization, now taking the additional quantization error into account. Denoting the channel noise variance for linear ZF equalization by $\sigma^2_{n,\mathrm{LE-ZF}}$, and the channel noise variance when using a whitening filter of order p by $\sigma_n^2(p)$, the displayed gain is calculated as

$$G_\mathrm{p}(p) = \frac{\sigma^2_{n,\mathrm{LE-ZF}}}{\sigma_n^2(p) + p \cdot 2^{-2w_\mathrm{D}}} \,. \tag{3.5.15}$$

The symbol error rate for ideal Tomlinson-Harashima precoding and $p \to \infty$ is chosen to be 10^{-7}, which is the usual requirement in DSL applications. The word length of the fractional part for the data signal in the precoder is fixed to $w_\mathrm{D} = 3$. For reference, the ideal exchange ($w_\mathrm{D} \to \infty$, cf. Figure 3.48) is given, too.

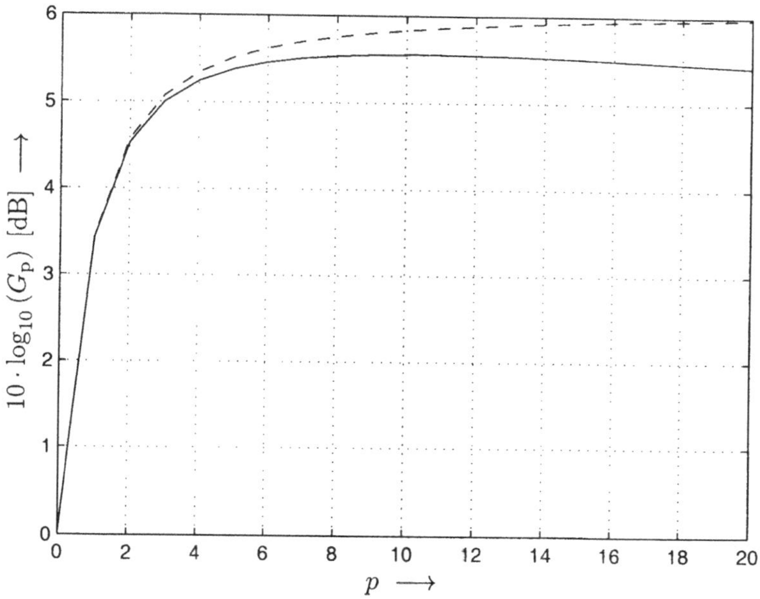

Fig. 3.52 Prediction gain over the order of the whitening filter for finite word lengths, taking the quantization error into account.

Increasing p first leads to a significant gain. Then, under ideal conditions ($w_\mathrm{D} \to \infty$), the gain flattens out, but the quantization error variance increases linearly. In turn, this leads to an optimum order of the whitening filter. The optimum is relatively broad and here lies in the region of $p = 10$. In conclusion, one can state that very high orders of the filters are counterproductive and do not provide further gain.

Finally, the increase in error probability of the transmission system due to quantization errors will be calculated. For that, we denote the total noise at the slicer input by $n_\mathrm{T}[k] \triangleq n[k]+\varepsilon'[k]$. Here, as usual, channel noise $n[k]$ is assumed to be Gaussian with zero mean and variance σ_n^2. Because of the statistical independence of channel noise and quantization error, considering (3.5.13) the pdf of the total disturbance $n_\mathrm{T}[k]$ reads

$$\begin{aligned} \mathrm{f}_{n_\mathrm{T}}(n_\mathrm{T}) &= \mathrm{f}_n(n_\mathrm{T}) * \mathrm{f}_{\varepsilon'}(n_\mathrm{T}) \\ &= \frac{1}{\sqrt{2\pi}\sigma_n}\mathrm{e}^{-\frac{n_\mathrm{T}^2}{2\sigma_n^2}} * \sum_{\nu=-\infty}^{\infty} p_\nu \cdot \delta\left(n_\mathrm{T} - \nu \cdot 2^{-(w_\mathrm{D}+w_\mathrm{C})} + \mu_\varepsilon\right) \\ &= \frac{1}{\sqrt{2\pi}\sigma_n} \sum_{\nu=-\infty}^{+\infty} p_\nu \cdot \mathrm{e}^{-\frac{(n_\mathrm{T}-\nu\cdot 2^{-(w_\mathrm{D}+w_\mathrm{C})}+\mu_\varepsilon)^2}{2\sigma_n^2}} \; . \end{aligned} \tag{3.5.16}$$

Assuming, as usual, that signal points and decision threshold are spaced by 1, the probability of error is proportional to $\Pr\{n_\mathrm{T} > 1\}$ (the multiplicity is given by the number of nearest neighbors) given as

$$\Pr\{n_\mathrm{T} > 1\} = \int_1^\infty \mathrm{f}_{n_\mathrm{T}}(n_\mathrm{T}) \, \mathrm{d}n_\mathrm{T} \; , \tag{3.5.17a}$$

and which, taking (3.5.16) into account, calculates to

$$= \sum_{\nu=-\infty}^{+\infty} p_\nu \cdot \mathrm{Q}\left(\frac{1 - \nu \cdot 2^{-(w_\mathrm{D}+w_\mathrm{C})} + \mu_\varepsilon}{\sigma_n}\right) \; . \tag{3.5.17b}$$

Here, $\mathrm{Q}(x) = 1/\sqrt{2\pi} \cdot \int_x^\infty \mathrm{e}^{-t^2/2} \mathrm{d}t$ is again the complementary Gaussian integral function.

If, additionally to the channel noise, the quantization noise $\varepsilon'[k]$ also is expected to be Gaussian (variance σ_ε^2), the total error is Gaussian, too, and a simple approximation for (3.5.17b) can be given as

$$\Pr\{n_\mathrm{T} > 1\} \approx \mathrm{Q}\left(\frac{1}{\sqrt{\sigma_n^2 + \sigma_\varepsilon^2}}\right) \; . \tag{3.5.18}$$

Example 3.10: *Increase in Error Probability*

Figure 3.53 sketches the probability of a decision error for the example impulse response given in Table 3.10 over the word length w_D of the fractional part. Here, the variance σ_n^2 of the channel noise is fixed so that ideal Tomlinson-Harashima precoding again performs at an error rate 10^{-7}.

First, it is recognizable that the Gaussian approximation (3.5.18) is very tight compared to the more detailed analysis. Second, almost no loss in performance occurs for word length $w_D > 4$. Thus, for, e.g., 16-ary transmission ($w_{I,x} = 4$) as in SDSL, it is not necessary to use higher precision than 10 bit ($1 + w_{I,x} + w_D = 1 + 4 + 5$) at the precoder.

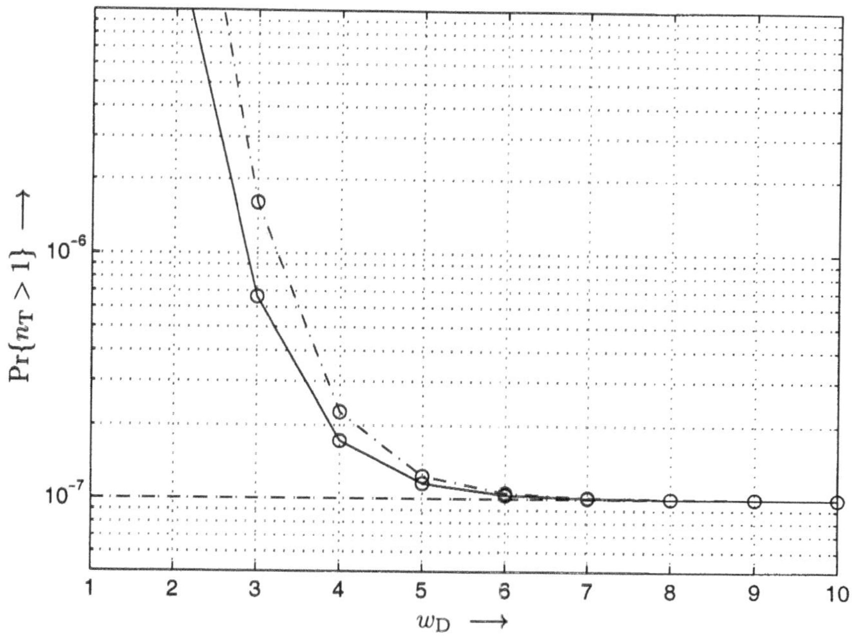

Fig. 3.53 Error probability versus the word length w_D of the fractional part. Solid line: exact calculations; dashed line: Gaussian approximation.

A lower bound for the word length can be derived from the fact that by inspection $|\varepsilon'[k]| < p \cdot Q/2 = p \cdot 2^{-(w_D+1)}$ holds. Hence, for $w_D > \log_2(p) - 1$ the amplitude of the total quantization error (after removal of the mean value) is limited to $|\varepsilon[k]| < 1$, and under noise-free conditions ($\sigma_n^2 = 0$), the system operates without errors. For the present example $w_D > \log_2(5) - 1 \approx 1.3219$, i.e., at least $w_D = 2$ should be chosen.

To summarize, finite-word-length realization of Tomlinson-Harashima precoding can be done at little expense; especially for one-dimensional signal sets or two-dimensional square constellations. When applying a suitable two's-complement representation of numbers, the desired modulo reduction is done automatically, not at one stage, but at each multiplication and addition. Here, this overflow handling, dreaded in filter design, is the desired property. In spite of its recursive structure, in each case, the precoder operates stably in the BIBO sense. This is true even if the minimum-phase property of $H(z)$ may be violated. Limit cycles do not occur, but the precoder produces additional noise, which takes effect at the decision point.

The additional noise, effective at the decision point, due to data quantization at the precoder (which is preferably done by truncation) can usually be neglected compared to the channel noise. Moreover, the analysis shows the important result that the order of the whitening filter should not be chosen as high as possible, but there always exists an optimum order.

The same considerations are true for flexible precoding in combination with constellations drawn from the integer lattice. Here, the feedback part of the precoder can be implemented with even a smaller number of digits resulting in the necessary modulo reduction. But the subtraction of $a[k]$ and $m[k]$ (see Figure 3.20) has to be carried out with a word length comparable to that in Tomlinson-Harashima precoding to cover the full range of the data signal $a[k]$ and the transmit signal $x[k]$, respectively. As already discussed in the last section, the receiver always has to work linearly over the full dynamic range, and hence requires much more complexity.

3.6 NONRECURSIVE STRUCTURE FOR TOMLINSON-HARASHIMA PRECODING

In the last section, we saw that finite-word-length implementation of precoding schemes is rather uncritical. Nevertheless, we now present an alternative structure which, moreover, completely avoids any quantization noise.

3.6.1 Precoding for IIR Channels

Most of the time, the T-spaced discrete-time end-to-end channel model is described as an FIR filter. Consequently, the precoder is derived from an all-pole IIR filter, and hence is recursive. In Section 2.4 we saw alternatively, that the discrete-time noise whitening filter can be implemented as an all-pole filter. Moreover, in typical DSL applications this provides better performance when comparing filters of the same order. Hence, for such all-pole end-to-end impulse responses (including transmit filter, actual channel, and receiver front-end filter), the precoder will become a nonrecursive structure.

Let $H(z) = 1/C(z) = \frac{1}{1+\sum_{k=1}^{p} c[k]z^{-k}}$ be the all-pole end-to-end impulse response. For implementation, again, the word lengths of the coefficients $c[k]$, $k = 1, 2, \ldots, p$, have to be restricted. As above, this leads to some small loss in prediction gain. But, contrary to FIR whitening filters, here it has to be assured that the quantized version of $C(z)$ remains strictly minimum phase, i.e., that $C(z)$ has a stable inverse.

Having derived a suitable discrete-time IIR channel model, Tomlinson-Harashima precoding can be implemented by a nonrecursive structure. In practice, an adaptive equalizer will ensure that the end-to-end impulse response equals the desired one. The direct approach of replacing the feedback filter in the conventional precoder (cf. Figure 3.4) by $1/C(z) - 1$ would be possible, but does not lead to the desired result. Figure 3.54 shows a nonrecursive structure based on an input-delay-line realization of $C(z)$, which is suited for the Tomlinson-Harashima type of precoding. The subsequent explanation is for one-dimensional baseband signaling, but the generalization is straightforward.

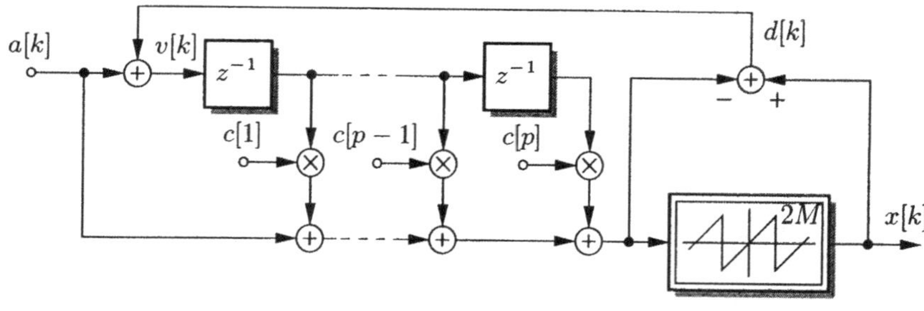

Fig. 3.54 "Nonrecursive" structure for Tomlinson-Harashima precoding.

First, in order to preequalize the channel, the data sequence $\langle a[k] \rangle$ is filtered with the finite impulse response $\langle c[k] \rangle \circ\!\!-\!\!\bullet\, C(z)$. Then, the output of this system is modulo reduced into the interval $[-M,\, M)$. In contrast to conventional TH precoding, the precoding sequence $\langle d[k] \rangle$ is now calculated explicitly and added to $a[k]$. The resulting effective data symbol $v[k] = a[k] + d[k]$ is finally stored in the delay line. Note that even though it may seem so at first sight, there is no delay-free (and hence nonrealizable) loop at this precoder. Because of the feedback of the $d[k]$'s, strictly speaking this structure is not purely nonrecursive. But since, neglecting the multiple symbol representation of the data, filtering is done by $C(z) = 1 + \sum_{k=1}^{p} c[k] z^{-k}$, we denote this structure as "nonrecursive" to distinguish it from conventional Tomlinson-Harashima precoding.

This structure has some advantages for implementation. Because the effective data symbols $v[k]$ are integers (in fact, odd integers: $v[k] \in 2\mathbb{Z} + 1$), multiplication with the coefficients $c[k]$, $k = 1, 2, \ldots, p$, can be performed without any quantization error! Moreover, if the number M of signal points is a power of two, the calculation of the signals $x[k]$ and $d[k]$ is trivial, i.e., modulo reduction and calculation of the difference. It is easily done by splitting the binary representation into the least significant bits ($x[k]$) and most significant bits ($d[k]$).

Unfortunately, the effective data symbols $v[k]$ can assume very large values, cf. Section 3.2.1. All arithmetics have to be performed without any inherent modulo reduction in order to ensure proper performance. Thus, a larger number of bits representing the integer part is necessary. Since $|v[k]| \leq V_{\max}$, $\forall k$, holds, at least $w_{\mathrm{D}} \geq \log_2(V_{\max})$ has to be chosen. However, we will see in Chapter 5 that this nonrecursive structure has special advantages when we lower $V_{\max}$ by means of signal shaping.

Because in DSL applications, when applying the whitened matched filter, the end-to-end impulse response typically has IIR characteristic, nonrecursive Tomlinson-Harashima precoding is more suitable. For achieving the same prediction gain, a lower order p is sufficient compared to an FIR whitening filter (cf. Example 2.11). Hence in implementation, a smaller number of arithmetic operations, but with somewhat higher word length, has to be carried out.

Example 3.11: *Optimization of Fixed-Point Coefficients*

Continuing Example 3.7, we now study the optimization of fixed-point all-pole whitening filters.

In Figure 3.55 the prediction gain G_{p} (in dB) is plotted over the order p of the whitening filter. The total word length $w = 1 + w_{\mathrm{I}} + w_{\mathrm{F}}$ is chosen to be 3, 4, 6, and 8, respectively. Partitioning of the total word length into the integer part w_{I} and fractional part w_{F} is again left to the optimization procedure. For reference, the exchange between prediction order and gain for an infinite word length is shown as well (solid line). For it, the calculation was based on an FIR whitening filter of order $p = 100$. Note that the numerical optimization is not based on any auxiliary filter (see $W(z)$ in Section 2.4.2).

The same phenomena as in Example 3.7 are visible. Increasing the order p is not rewarding for short word lengths, since the optimization results in trailing zeros of the impulse response.

For word length $w \geq 6$ the prediction gain almost equals that of the optimum, i.e., infinite-order, whitening filter.

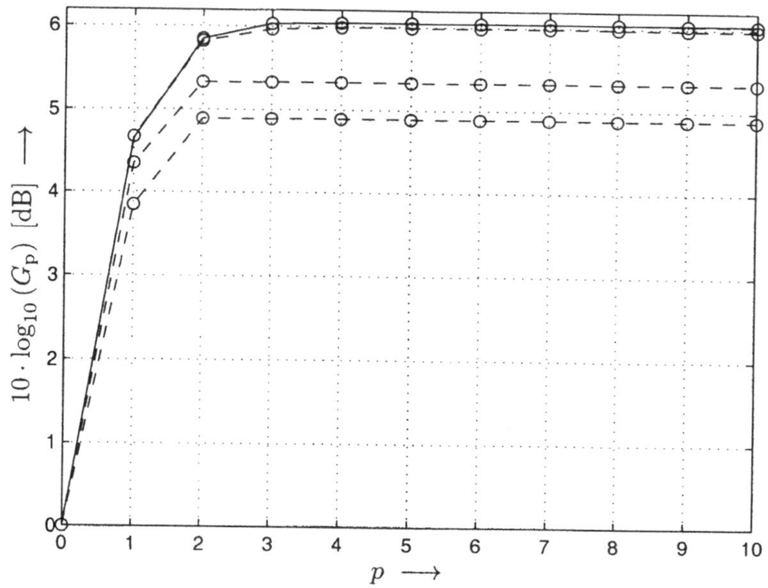

Fig. 3.55 Prediction gain over the order of the all-pole whitening filter for finite word lengths w of the coefficients. Bottom to top: $w = 3, 4, 6, 8$. Solid line: infinite word length.

3.6.2 Extension to DC-free Channels

As explained in Section 2.4.3, sometimes the discrete-time channel should exhibit a spectral zero at DC. For example, this may model transformer coupling. For DC-free impulse responses $h[k] = \mathcal{Z}^{-1}\{H(z)\}$ the operation of Tomlinson-Harashima precoding does not change at all. All statements of the preceding sections remain valid, and the examples and phenomena described above are also representative in this case.

For FIR whitening filters the optimal coefficients can be calculated in two steps (see page 104 and [Hub92b]). First, an FIR filter $H_0(z) = 1 + \sum_{k=1}^{p} h_0[k] \cdot z^{-k}$ is determined via the Yule-Walker equations applied to the modified autocorrelation sequence $\phi_{nn}^{(\text{ZF-LE})}[\kappa] * (-\delta[\kappa+1] + 2\delta[\kappa] - \delta[\kappa-1])$. Then, the optimal whitening filter of order $p+1$ under the additional constraint $H(z=1) = 0$ is given by $H(z) = (1 - z^{-1}) \cdot H_0(z)$.

If the nonrecursive precoding structure should be extended to DC-free channels, the all-pole restriction on the channel transfer function $H(z)$ has to be dropped. This is because spectral zeros can only be achieved via zeros of the numerator of the transfer function. Hence, we drop the restriction and resort to a pole-zero model (cf. also page 102), in particular we choose $H(z) = (1 - z^{-1})/C(z)$. For that, following the steps in Section 2.4.2, an all-pole whitening filter $C(z)$ fitted to the above modified autocorrelation sequence is calculated (twofold application of the Yule-Walker equations). Finally, $H(z) = (1 - z^{-1})/C(z)$ is the desired pole-zero whitening filter.

Since the precoder now has to implement $C(z)/(1-z^{-1})$, an accumulator has to be included in the nonrecursive structure of Figure 3.54. After some basic manipulations, we arrive at the precoding structure shown in Figure 3.56. As $a[k]$ and $v[k]$ are still integers, all points concerning the nonrecursive structure discussed recently apply here, too. Moreover, since the effective data symbols $v[k]$ are limited in amplitude, no overrun problems at the accumulator occur.

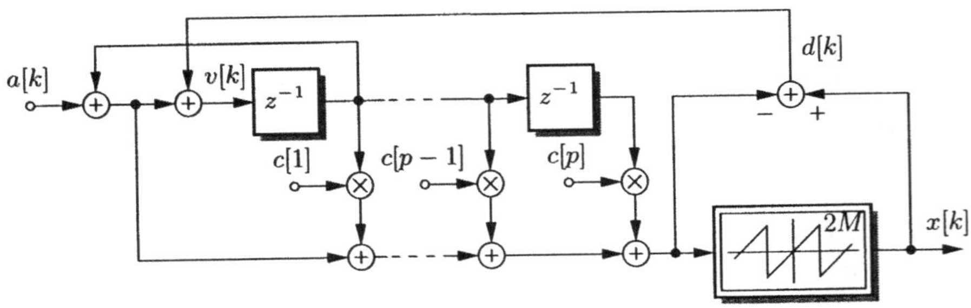

Fig. 3.56 "Nonrecursive" structure for Tomlinson-Harashima precoding and DC-free channels.

Again it should be emphasized that the proposed alternative, nonrecursive structure for Tomlinson-Harashima precoding can be implemented without producing any quantization noise. Furthermore, nonrecursive precoding and recursive end-to-end discrete-time channel descriptions are better suited for typical DSL scenarios.

3.7 INFORMATION-THEORETICAL ASPECTS OF PRECODING

Before concluding this chapter on precoding schemes, we address some information-theoretical aspects of precoding. This includes the question of a precoding scheme which is optimal in the MMSE sense, and we study the capacity achievable therewith.

3.7.1 Precoding Designed According to MMSE Criterion

In Chapter 2, we saw that optimizing the system with respect to the MMSE criterion leads to a gain over the ZF solution. This is especially true for low signal-to-noise ratios. Up to now, precoding has only been addressed as the counterpart to ZF-DFE, where—except for the leading coefficient "1"—feedforward filter $F(z)$ and feedback filter $B(z) - 1$ are identical, cf. Figures 2.17 and 2.33. Hence a natural question is whether precoding, optimized according to the MMSE criterion—we call it *MMSE precoding*—can simply be obtained by transferring the feedback part of MMSE-DFE into the transmitter. To answer this question, we first regard finite-length filters. After having derived the basic result, the extension to infinite filter orders is addressed.

As with the derivation of MMSE-DFE, we start from the T-spaced discrete-time channel model, when applying the matched-filter front-end. All quantities are again expressed in terms of the PSD $\Phi_{hh}(z) \bullet\!\!-\!\!\circ\, \varphi_{hh}[k]$, which is defined in (2.3.30) and (2.2.38b) as

$$\Phi_{hh}(e^{j2\pi fT}) = \sigma_a^2 \sum_{\mu} \left| H_T(f - \tfrac{\mu}{T}) H_C(f - \tfrac{\mu}{T}) \right|^2 . \qquad (3.7.1)$$

Remember, for the matched-filter front-end, both signal transfer function and noise PSD are proportional to this quantity.

Finite-Length Results Taking the linearized description (valid for Tomlinson-Harashima precoding, as well as for flexible precoding) of precoding into account, Figure 3.57 sketches the configuration, the components of which have to be optimized. The precoder now employs the feedback filter $B(z) - 1$, which is causal and has a

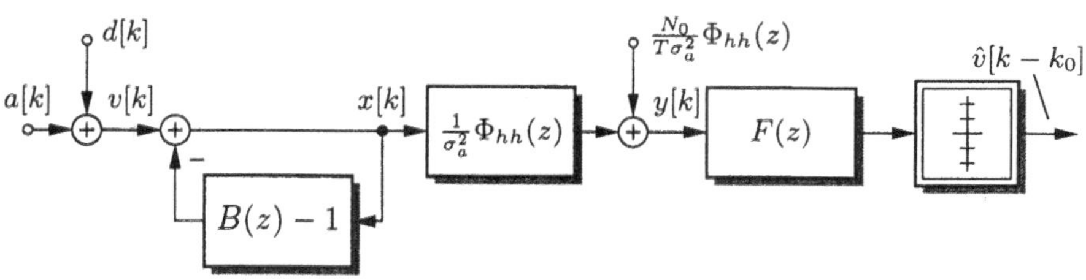

Fig. 3.57 Structure of the transmission scheme for MMSE precoding.

monic polynomial $B(z) = 1 + \sum_{k=1}^{q_b} b[k] z^{-k}$ of order q_b. At the receiver, the causal feedforward filter $F(z)$ of order q_f, i.e., $F(z) = \sum_{k=0}^{q_f} f[k] z^{-k}$, is present. Again, a delay k_0 for producing the estimates $\hat{v}[k]$—now with respect to the extended signal set—is admitted.

Using the above definitions, the error signal at the input of the slicer is given as

$$e[k] = \sum_{\kappa=0}^{q_f} f[\kappa] y[k-\kappa] - v[k-k_0] \,. \tag{3.7.2}$$

Furthermore, due to preequalization at the transmitter, we have

$$x[k] = v[k] - \sum_{\kappa=0}^{q_b} b[\kappa] x[k-\kappa] \,. \tag{3.7.3}$$

Solving (3.7.3) for $v[k]$ and plugging it into (3.7.2) yields for the error signal

$$e[k] = \sum_{\kappa=0}^{q_f} f[\kappa] y[k-\kappa] - \sum_{\kappa=0}^{q_b} b[\kappa] x[k-k_0-\kappa] - x[k-k_0] \,. \tag{3.7.4}$$

Using the following definitions for the vectors

$$\begin{aligned} \boldsymbol{f} &= \begin{bmatrix} f^*[0] \\ f^*[1] \\ \vdots \\ f^*[q_f] \end{bmatrix}, & \boldsymbol{b} &= \begin{bmatrix} b^*[1] \\ b^*[2] \\ \vdots \\ b^*[q_b] \end{bmatrix}, \\ \boldsymbol{y}[k] &= \begin{bmatrix} y[k] \\ y[k-1] \\ \vdots \\ y[k-q_f] \end{bmatrix}, & \boldsymbol{x}[k] &= \begin{bmatrix} x[k-k_0-1] \\ x[k-k_0-2] \\ \vdots \\ x[k-k_0-q_b] \end{bmatrix}. \end{aligned} \tag{3.7.5}$$

the error can finally be written compactly as

$$e[k] = \boldsymbol{f}^{\mathsf{H}} \boldsymbol{y}[k] - \boldsymbol{b}^{\mathsf{H}} \boldsymbol{x}[k] - x[k-k_0] \,. \tag{3.7.6}$$

Comparing equation (3.7.6) with its corresponding counterpart for MMSE-DFE, equation (2.3.83), shows that the problem of determining the filters $F(z)$ and $B(z)$ for minimum error variance, i.e., $\mathsf{E}\{|e[k]|^2\} \to \min$, is almost identical for MMSE-DFE and MMSE precoding. Only the data signal $\langle a[k] \rangle$ has to be replaced by the precoded channel signal $\langle x[k] \rangle$. Both signals are white (cf. Section 3.2.2), but, because of the precoding loss, $x[k]$ has a (slightly) increased variance compared to $a[k]$. Hence, the optimum filters for MMSE precoding can be derived as for MMSE-DFE, but replacing the variance σ_a^2 by σ_x^2. In particular, this holds for the correlation matrices and vectors according to (2.3.85a) through (2.3.85e).

Consequently, the filters in MMSE-DFE cannot be used directly for precoding. Transferring the feedback filter to the transmitter does not give optimum performance. This fact, usually ignored in the literature, was first observed in [Ger98].

As for large constellations, where the precoding loss vanishes asymptotically and σ_x^2 approaches σ_a^2, the mismatch is only relevant for small signal sets. Moreover, since for large signal-to-noise ratios zero-forcing and minimum mean-squared error solution coincide, significant differences will only be noticeable at low SNR.

Infinite-Length Results Now we turn to asymptotic results, employing infinite-length filters. In order to eliminate the decision delay, as in Section 2.3.4 the feedforward filter $F(z)$ is assumed to be two-sided and IIR, but the feedback part $B(z) - 1$, of course, is strictly causal.

Regarding again Figure 3.57, the error at the decision device, using the z-transform, is now given as

$$E(z) = Y(z)F(z) - V(z) = Y(z)F(z) - X(z)B(z)\,. \tag{3.7.7}$$

A comparison with the respective result of infinite-length MMSE-DFE (equation (2.3.97)) reveals that the only difference is that the z-transform of the sequence $\langle a[k]\rangle$ has to be replaced by that of the precoded sequence $\langle x[k]\rangle$. Recapitulating the derivations in Section 2.3.4, and simply replacing $a[k]$ by $x[k]$ and σ_a^2 by σ_x^2, the key point in the design of infinite-length MMSE precoding is the factorization problem

$$\Phi_{ff}(z) \triangleq \sigma_x^2 H^{(\mathsf{MF})}(z) + \frac{N_0}{T} \overset{!}{=} \sigma_g^2 \cdot G(z) \cdot G^*(z^{-*})\,. \tag{3.7.8}$$

The polynomial $G(z)$ is again forced to be causal, monic, and minimum-phase; $G^*(z^{-*})$ is hence anticausal, monic, and maximum-phase.

Finally, from the factorization, the feedforward and feedback filter should be chosen to be

$$B(z) = G(z)\,, \qquad F(z) = \frac{\sigma_x^2}{\sigma_g^2}\frac{1}{G^*(z^{-*})}\,. \tag{3.7.9}$$

Since now the PSD of the error sequence calculates to

$$\Phi_{nn}^{(\mathsf{MMSE\text{-}Prec})}(z) = \frac{\sigma_x^2}{\sigma_g^2}\frac{N_0}{T}\,, \tag{3.7.10}$$

the signal-to-noise ratio is given by

$$\begin{aligned}
\mathrm{SNR}^{(\mathsf{MMSE\text{-}Prec})} &= \frac{\sigma_a^2}{\mathrm{E}\{|e[k]|^2\}} \\
&= \frac{\sigma_a^2}{\sigma_x^2}\cdot\frac{\sigma_g^2}{N_0/T} \\
&= \frac{1}{\gamma_{\mathrm{p}}^2}\cdot \exp\left\{ T \int_{-\frac{1}{2T}}^{\frac{1}{2T}} \log\left(\gamma_{\mathrm{p}}^2\, \widetilde{\mathrm{SNR}}(\mathrm{e}^{\mathrm{j}2\pi fT}) + 1\right)\,\mathrm{d}f \right\} \\
&= \exp\left\{ T \int_{-\frac{1}{2T}}^{\frac{1}{2T}} \log\left(\widetilde{\mathrm{SNR}}(\mathrm{e}^{\mathrm{j}2\pi fT}) + \frac{1}{\gamma_{\mathrm{p}}^2}\right)\,\mathrm{d}f \right\}\,.
\end{aligned} \tag{3.7.11}$$

Here, γ_{p}^2 denotes the precoding loss, and $\widetilde{\mathrm{SNR}}(\mathrm{e}^{\mathrm{j}2\pi fT})$ is the folded spectral signal-to-noise ratio, cf. (2.2.18), (2.2.19). Note that, contrary to the zero-forcing case (Theorem 3.2), the SNR is not simply given by dividing the signal-to-noise ratio obtained for MMSE decision-feedback equalization by the precoding loss. But as γ_{p}^2 tends to one, the SNR approaches that of MMSE-DFE. Finally, anticipating later results, since $\frac{1}{\gamma_{\mathrm{p}}^2} < 1$, the SNR expression (3.7.11) already gives a hint that at low SNR the capacity of the underlying channel cannot be approached by precoding.

Applying the above results, the end-to-end transfer function seen by the channel symbols $\langle x[k] \rangle$, i.e., the input to the transmit filter $H_{\mathrm{T}}(f)$, can be derived as follows

$$
\begin{aligned}
H^{(\text{MMSE-Prec})}(z) &= \frac{1}{\sigma_a^2}\Phi_{hh}(z)\cdot F(z) \\
&= \frac{1}{\sigma_a^2}\frac{\sigma_a^2}{\sigma_x^2}\left(\Phi_{ff}(z) - \frac{N_0}{T}\right)\cdot\frac{\sigma_x^2}{\sigma_g^2}\cdot\frac{1}{G^*(z^{-*})} \\
&= \frac{\sigma_g^2 G(z)G^*(z^{-*}) - N_0/T}{\sigma_g^2 G^*(z^{-*})} \\
&= G(z) - \frac{N_0/T}{\sigma_g^2 G^*(z^{-*})}\,.
\end{aligned}
\tag{3.7.12}
$$

The transfer function is thus composed of two parts: first, the causal part $G(z)$. Since $B(z) = G(z)$ these postcursors are treated in the precoder. Second, an anticausal, i.e., precursor-producing, part which cannot be processed by a causal precoder.

Starting from (3.7.12), the overall transfer function for the effective data sequence $\langle v[k] \rangle$ reads

$$
\begin{aligned}
H(z) &= \frac{1}{B(z)}\cdot H^{(\text{MMSE-Prec})}(z) \\
&= \frac{1}{G(z)}\cdot\left(G(z) - \frac{N_0/T}{\sigma_g^2 G^*(z^{-*})}\right) \\
&= 1 - \frac{N_0/T}{\Phi_{ff}(z)}\,.
\end{aligned}
\tag{3.7.13}
$$

Since $\Phi_{ff}(z) = \sigma_g^2 G(z)G^*(z^{-*})$, with $G(z) = 1 + g[1]z^{-1} + g[2]z^{-2} + \cdots$, is a linear phase polynomial, this property also holds for its inverse. Moreover, we can identify the coefficient at instant zero to be $\frac{N_0/T}{\sigma_g^2}$. Hence, the end-to-end impulse response can be written in the form

$$
H(z) = 1 - \left(\frac{N_0/T}{\sigma_g^2} + H_+(z) + H_+^*(z^{-*})\right)\,, \tag{3.7.14}
$$

with the strictly causal polynomial $H_+(z) = \sum_{k=1}^{\infty} h[k]z^{-k}$.

From (3.7.14) we see the following: first, the MMSE precoding solution is biased, i.e., part of the data signal is falsely apportioned to the noise. To compensate for this bias, the receive signal should be scaled prior to threshold decision by

$$\left(1 - \frac{N_0/T}{\sigma_g^2}\right)^{-1} = \frac{\sigma_g^2}{\sigma_g^2 - N_0/T} ; \qquad (3.7.15)$$

this term again coincides with that for MMSE-DFE; equation (2.3.114). This correction in turn decreases the signal-to-noise ratio by one, but improves performance.

Second, in contrast to MMSE-DFE where the feedback filter eliminates the postcursors completely, here precursors as well as postcursors contribute to the residual intersymbol interference. This fact has already been observed in [SL96, Ger98]. Thus, the overall channel for MMSE precoding is no longer AWGN, which makes the application of standard coding techniques, developed for the ISI-free channel, doubtful.

A possible solution to overcome residual, data-dependent correlations is interleaving. In Tomlinson-Harashima precoding and the initial version of flexible precoding, where channel coding is separated from precoding, interleaving can be done without any problems. The only disadvantage is the introduction of additional delay. But for the combined coding/precoding schemes (ISI coder and its modified version), interleaving is very intricate or even impossible.

Finally a remark on flexible precoding (and its enhanced versions) and MMSE filtering: Here, $B(z)$ has to be used at the transmitter, but at the receiver, in the inverse precoder which reconstructs the sent data, the end-to-end transfer function $H^{(\text{MMSE-Prec})}(z)$ has to be inverted. Hence, the system $H^{(\text{MMSE-Prec})}(z)$ now is required to be minimum-phase, but not $B(z)$.

3.7.2 MMSE Precoding and Channel Capacity

In Chapter 2, we have shown that, in principle, ideal (error-free) MMSE-DFE, in combination with powerful channel coding schemes, is able to approach channel capacity. Unfortunately, this proposition is only of minor practical benefit, because error-free decisions cannot be generated; in particular not at zero delay. In [SL96] it is shown that the assumption of an error-free (i.e., genie-aided) feedback in MMSE-DFE leads to contradictions: canceling of the tail of the impulse response leads to an increase in capacity rather than a decrease. Examples can be given where an optimization leads to the strange situation of a feedforward filter $F(z) \equiv 0$. All the information is then "transmitted" via the feedback filter rather than over the actual channel. Since the feedback is supposed to be error-free, the capacity will be infinite, which shows the inherent paradox.

The question which now arises is whether channel capacity can be approached by MMSE precoding, where no zero-delay decisions are required. The first obstacle for a straightforward correspondence to DFE—as we have just seen—is that the optimal filters for MMSE-DFE are not the optimal choice for MMSE precoding. But more critical, for DFE (page 95), Gaussian transmit symbols have been assumed which are necessary for approaching Shannon capacity. Unfortunately, precoding produces channel symbols uniformly distributed over some boundary region. Moreover, at the receiver, the modulo congruence of the effective data symbols is resolved by performing a modulo reduction of the receive signal into the above-mentioned boundary region. Hence the additive noise becomes folded. Following the exposition in [WC98], and the much more general proceedings in [FTC00], we now derive the capacity utilizable when applying Tomlinson-Harashima precoding. We conjecture that the results can also be applied to other types of precoding.

Zero-Forcing Precoding Let us start with zero-forcing Tomlinson-Harashima precoding. The linearized model of the communication system is again depicted in Figure 3.58. All signals are complex. The precoding lattice is designated by $\mathbf{\Lambda}_{\mathrm{p}}$,

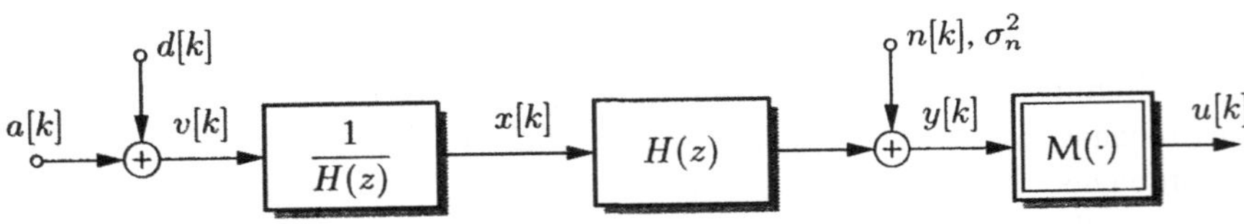

Fig. 3.58 Transmission scheme using zero-forcing Tomlinson-Harashima precoding.

and the respective fundamental region (preferably the Voronoi region) is $\mathcal{R}(\mathbf{\Lambda}_{\mathrm{p}})$. At the receiver front-end a modulo operation $\mathsf{M}(y) \triangleq y \bmod \mathbf{\Lambda}_{\mathrm{p}}$ is performed (for details on lattices, see Appendix A). The present channel, including the modulo operation, is called a *mod-$\mathbf{\Lambda}_{\mathrm{p}}$ channel* or, in the case of discrete distributed input, a $\mathbf{\Lambda}_{\mathrm{a}}/\mathbf{\Lambda}_{\mathrm{p}}$ channel [FTC00]. Such types of channels, assuming arbitrary lattices, are treated exhaustively in [FTC00] in connection with multilevel codes. (Lower levels in such coding schemes experience the same multiple symbol representation (cf., e.g., [WFH99]) as the symbols in precoding schemes.)

First, the output of the modulo device reads (omitting the time index for brevity)

$$u = \mathsf{M}(a + d + n) = \mathsf{M}(a + n) \,. \tag{3.7.16}$$

With this, the mutual information[11] $I(A;U)$ (e.g. [Gal68, CT91]) of the overall channel calculates to ($h(\cdot)$ denotes differential entropy)

$$I(A;U) = h(U) - h(U \mid A) \,. \tag{3.7.17}$$

The symbol u is restricted to the fundamental region $\mathcal{R}(\mathbf{\Lambda}_{\mathrm{p}})$. It is easy to show that $h(U)$ is maximum, if and anly if u is uniformly distributed over $\mathcal{R}(\mathbf{\Lambda}_{\mathrm{p}})$. Then

$$h(U) = -\int_{\mathcal{R}(\mathbf{\Lambda}_{\mathrm{p}})} \frac{1}{V(\mathbf{\Lambda}_{\mathrm{p}})} \log_2\left(\frac{1}{V(\mathbf{\Lambda}_{\mathrm{p}})}\right) \, \mathrm{d}u = \log_2\left(V(\mathbf{\Lambda}_{\mathrm{p}})\right) \tag{3.7.18}$$

[11]Random variables are denoted by the corresponding capital letters.

holds, where $V(\mathbf{\Lambda}_\text{p})$ is the (fundamental) volume of the precoding lattice. For the second term, we regard the conditional pdf $\mathsf{f}_u(u\,|\,a)$. Taking the Gaussian density of the channel noise (variance σ_n^2) into account, we have

$$\mathsf{f}_u(u\,|\,a) = \sum_{d\in\mathbf{\Lambda}_\text{p}} \frac{1}{\pi\sigma_n^2} \mathrm{e}^{-|u-a+d|^2/\sigma_n^2} = \mathsf{f}_{\tilde{n}}(u-a)\,, \quad u \in \mathcal{R}(\mathbf{\Lambda}_\text{p})\,, \tag{3.7.19}$$

where the $\mathbf{\Lambda}_\text{p}$-aliased Gaussian noise $\tilde{n} = \mathsf{M}(n)$ with

$$\mathsf{f}_{\tilde{n}}(\tilde{n}) \triangleq \sum_{d\in\mathbf{\Lambda}_\text{p}} \frac{1}{\pi\sigma_n^2} \mathrm{e}^{-|\tilde{n}+d|^2/\sigma_n^2} \tag{3.7.20}$$

has been introduced. Since $\mathsf{f}_{\tilde{n}}(\tilde{n})$ is $\mathbf{\Lambda}_\text{p}$-periodic, independent of a, we have

$$\begin{aligned} h(U\,|\,A) &= -\int_{\mathcal{R}(\mathbf{\Lambda}_\text{p})} \mathsf{f}_{\tilde{n}}(u-A) \log_2\left(\mathsf{f}_{\tilde{n}}(u-A)\right)\,\mathrm{d}u \\ &= -\int_{\mathcal{R}(\mathbf{\Lambda}_\text{p})} \mathsf{f}_{\tilde{n}}(\tilde{n}) \log_2\left(\mathsf{f}_{\tilde{n}}(\tilde{n})\right)\,\mathrm{d}\tilde{n} \\ &= h(\tilde{N}) \\ &= h(\mathsf{M}(N))\,. \end{aligned} \tag{3.7.21}$$

In summary, the maximum mutual information, i.e., the capacity, of zero-forcing Tomlinson-Harashima precoding reads

$$C_\text{ZF-THP} = \log_2\left(V(\mathbf{\Lambda}_\text{p})\right) - h(\mathsf{M}(N))\,, \tag{3.7.22}$$

which is achieved for i.i.d. symbols $a[k]$, uniformly distributed over $\mathcal{R}(\mathbf{\Lambda}_\text{p})$, since then $u[k]$ is uniformly distributed, too.

Minimum Mean-Squared Error Precoding Unlike in zero-forcing precoding, in MMSE precoding, residual ISI (both precursors and postcursors) remains. Moreover, the filtered channel noise is no longer white. In order to apply standard coding techniques and to enable analysis, we now assume interleaving of sufficient depth. Because exploitation of the correlation would improve performance, the derived capacity may be somewhat lower than the true capacity of the channel with memory.

The situation we now have to deal with is shown in Figure 3.59. Here, the unbiased MMSE solution is considered. In addition to the desired signal $a[k] + d[k]$ and the noise sample $n[k]$, the intersymbol interference term $i[k]$ is present.

For MMSE precoding, we have

$$u = \mathsf{M}(a+d+i+n) = \mathsf{M}(a+i+n)\,. \tag{3.7.23}$$

The respective differential entropies now calculate to

$$h(U) = h(\mathsf{M}(A+I+N))\,, \tag{3.7.24a}$$

and

$$h(U \mid A) \;=\; h(\mathrm{M}(A+I+N) \mid A)\,. \tag{3.7.24b}$$

Since these differential entropies cannot be calculated analytically, we resort to upper and lower bounds. An upper bound on the channel capacity can be established by taking $h(U) \leq \log_2\left(V(\mathbf{\Lambda}_\mathrm{p})\right)$ into account and

$$h(\mathrm{M}(A+I+N) \mid A) \geq h(\mathrm{M}(A+I+N) \mid A, I) = h(\mathrm{M}(N))\,. \tag{3.7.25}$$

Hence, we have

$$C_{\mathrm{MMSE-THP}} \leq \log_2\left(V(\mathbf{\Lambda}_\mathrm{p})\right) - h(\mathrm{M}(N))\,. \tag{3.7.26}$$

The equation resembles (3.7.22), but please keep in mind that the noise samples $n[k]$ for ZF and MMSE precoding have different variances.

Now, let us assume that $a[k]$ is uniformly distributed over $\mathcal{R}(\mathbf{\Lambda}_\mathrm{p})$. Then, the capacity reads

$$C_{\mathrm{MMSE-THP}} = \log_2\left(V(\mathbf{\Lambda}_\mathrm{p})\right) - h(\mathrm{M}(A+I+N) \mid A)\,. \tag{3.7.27}$$

To establish a lower bound on $C_{\mathrm{MMSE-THP}}$, we note that

$$h(\mathrm{M}(A+I+N) \mid A) = h(\mathrm{M}(I+N) \mid A) \leq h(\mathrm{M}(I+N))\,. \tag{3.7.28}$$

From the derivation of the MMSE filters, $e[k] = n[k] + i[k]$ holds, and hence the variance of $n[k] + i[k]$ is given by $\sigma_e^2 = \mathrm{E}\{|e[k]|^2\}$. Now, an upper bound on the differential entropy $h(\mathrm{M}(I+N))$ can be given by that of a truncated Gaussian distribution $G(\sigma_e^2, \mathcal{R}(\mathbf{\Lambda}_\mathrm{p}))$, having the same variance σ_e^2 (after truncation to $\mathcal{R}(\mathbf{\Lambda}_\mathrm{p})$) as $\langle n[k] + i[k] \rangle$ [WC98]. We denote this entropy by $h(G(\sigma_e^2, \mathcal{R}(\mathbf{\Lambda}_\mathrm{p})))$. For one-dimensional signaling this entropy is explicitly given in [SD94, WC98]. In summary, a lower bound on the capacity of MMSE precoding reads

$$C_{\mathrm{MMSE-THP}} \geq \log_2\left(V(\mathbf{\Lambda}_\mathrm{p})\right) - h(G(\sigma_e^2, \mathcal{R}(\mathbf{\Lambda}_\mathrm{p})))\,. \tag{3.7.29}$$

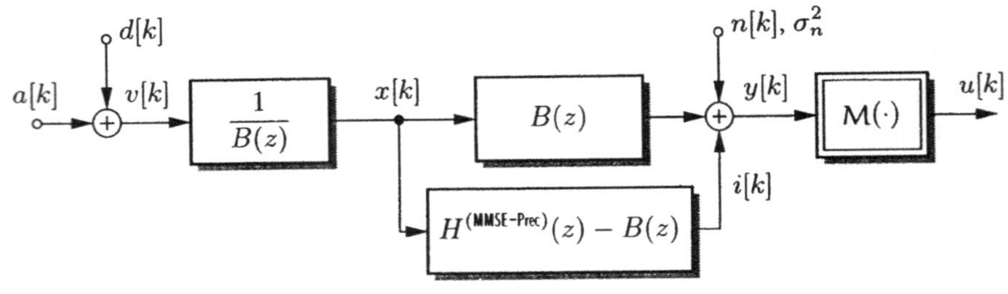

Fig. 3.59 Transmission scheme using MMSE Tomlinson-Harashima precoding.

Discussion The first observation to be made is that for high signal-to-noise ratio equations (3.7.22), (3.7.26), and (3.7.29) coincide and converge to

$$C_{\text{THP}} \to \log_2\left(V(\mathbf{\Lambda}_{\text{p}})\right) - \log_2(\pi \text{e}\, \mathsf{E}\{|e[k]|^2\}) \tag{3.7.30}$$

for two-dimensional signaling. Assuming square constellations with support equal to $[-\sqrt{M},\, \sqrt{M}]^2$, we have $V(\mathbf{\Lambda}_{\text{p}}) = 4M$, and

$$C_{\text{THP}} \to \log_2\left(\frac{4M}{\pi \text{e}\, \mathsf{E}\{|e[k]|^2\}}\right) . \tag{3.7.31}$$

This has to be compared with the capacity of ideal, unbiased MMSE-DFE. Combining (2.3.127), (2.3.117), (2.3.108) with $\sigma_a^2 = 2\frac{M}{3}$ for the square constellation, the capacity reads

$$C_{\text{MMSE-DFE}} = \log_2\left(\frac{2M}{3\, \mathsf{E}\{|e[k]|^2\}}\right) . \tag{3.7.32}$$

A comparison of (3.7.31) and (3.7.32) yields a difference of

$$\begin{aligned} \Delta C &\triangleq C_{\text{MMSE-DFE}} - C_{\text{THP}} \\ &= \log_2\left(\frac{\pi \text{e}}{6}\right) \approx 0.51 \text{ bit} \end{aligned} \tag{3.7.33}$$

or

$$\Delta\text{SNR} = 10 \log_{10}\left(\frac{\pi \text{e}}{6}\right) \approx 1.53 \text{ dB} \tag{3.7.34}$$

in favor of MMSE-DFE. But, as will be shown in detail in Chapter 4, this is exactly the ultimate shaping gain—the difference between a Gaussian distribution and a uniform one having the same variance. Hence, at high signal-to-noise ratios, the only loss associated with precoding is due to the uniformly distributed channel symbols. In Chapter 5 on combined precoding and signal shaping, we show how to overcome this gap.

In order to elucidate the effect of the modulo receiver front-end, the capacities for the nondispersive one-dimensional AWGN channel (without the need for precoding) are evaluated. Figure 3.60 displays the Shannon capacity (Gaussian symbols), the capacity for uniformly distributed transmit symbols, and for uniformly distributed transmit symbols but modulo receiver front-end. Note, the latter capacity equals that for ZF Tomlinson-Harashima precoding. Similar curves can be found in [WC98, FTC00]. Moreover, tight bounds on the asymptotic behavior of the mod-$\mathbf{\Lambda}$ channel capacity are derived in [FTC00].

Of course, Shannon capacity is superior over the whole range of signal-to-noise ratios. For high SNR, the modulo front-end is ineffective and the curves for uniformly distributed symbols, both with and without modulo reduction, converge. Asymptotically, a gap of 1.53 dB compared to Shannon capacity remains. It is well known that shaping does not provide any gains for low SNR—the capacity for uniformly distributed symbols approaches that for Gaussian symbols. But in this low-SNR region, the capacity when applying a modulo device at the receiver clearly stays behind

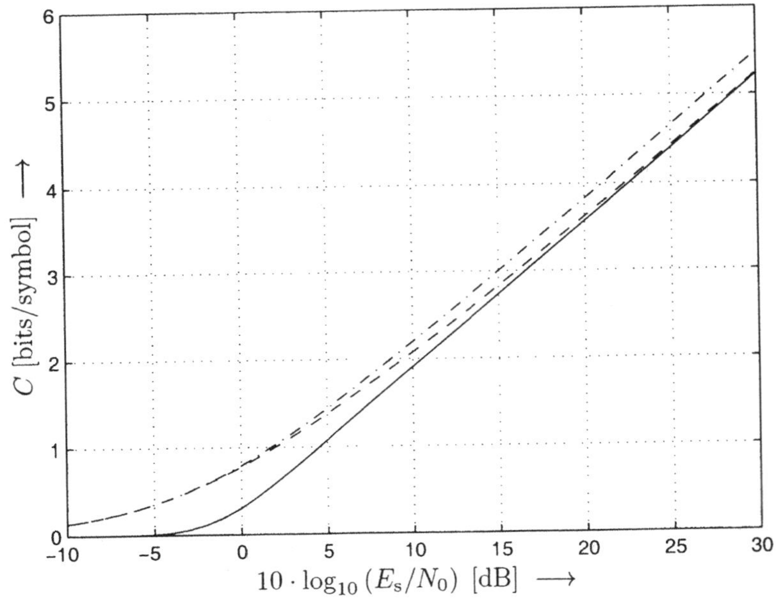

Fig. 3.60 Capacities for the AWGN channel. Top to bottom: Shannon capacity; uniform distributed channel input; uniform distributed channel input and modulo front-end.

capacity without this modulo reduction. This clearly indicates that the modulo device is responsible for the enormous loss at low SNR.

Finally, to conclude this section, we calculate the achievable capacity for our standard DSL example.

Example 3.12: *Achievable Capacity of Precoding*

Again the simplified DSL down-stream (white noise) example using one-dimensional signaling is considered. Figures 3.61, 3.62, and 3.63 show the capacity achievable by Tomlinson-Harashima precoding. The dashed line corresponds to zero-forcing precoding, the solid line is the MMSE lower bound, and the dash-dotted curve represents the MMSE upper bound. Additionally, the water-pouring capacity of the channel is given (dotted). The three figures are valid for cable lengths of 1, 3, and 5 km.

All curves are plotted over the transmit energy per symbol ($E_s = \sigma_x^2 T$), divided by the virtual noise power spectral density N_0'. For ZF precoding (cf. Figure 3.58), $\sigma_n^2 = N_0'/(2T)$ holds, since we regard baseband transmission. The actual PSD of the underlying channel reads $N_0 = N_0' \cdot \sigma_g^2/\sigma_x^2$, where σ_g^2 is obtained from spectral factorization of $\sigma_x^2 H^{(\mathrm{MF})}(z)$. The present normalization makes the results comparable with that of the ISI-free AWGN channel.

For high signal-to-noise ratios, the capacity of ZF precoding and the bounds for MMSE precoding merge. Compared to the optimum, which is given for a water-pouring transmit PSD, only the shaping gap of 1.53 dB (or 0.255 bit for one-dimensional signaling) remains. For low SNR and large cable length, MMSE precoding can provide gains over the zero-forcing

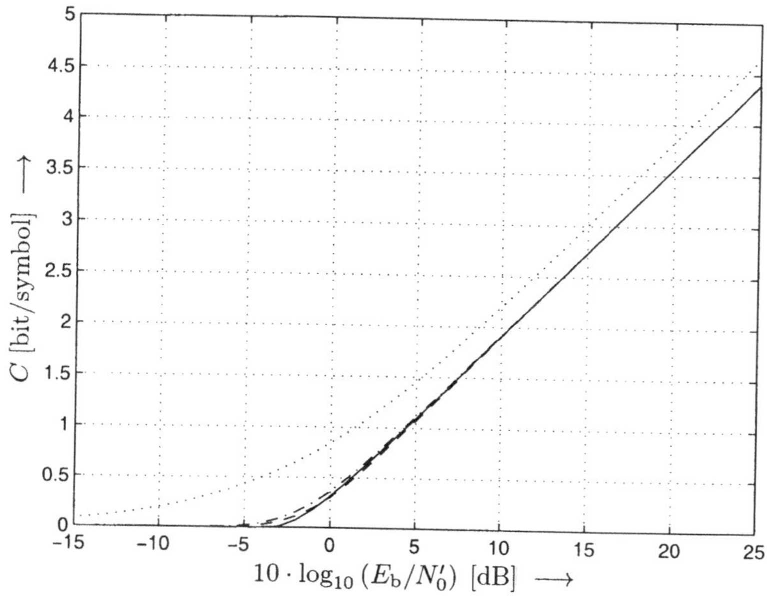

Fig. 3.61 Capacity achievable by Tomlinson-Harashima precoding. DSL down-stream example, cable length 1 km. Dashed line: zero-forcing precoding; solid line: MMSE lower bound; dash-dotted line: MMSE upper bound; dotted line: water-pouring capacity.

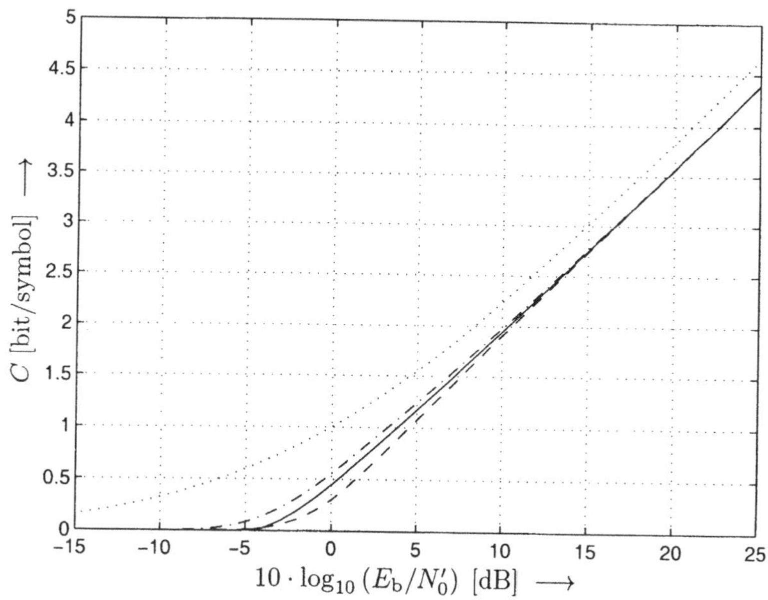

Fig. 3.62 Capacity achievable by Tomlinson-Harashima precoding. DSL down-stream example, cable length 3 km. Dashed line: zero-forcing precoding; solid line: MMSE lower bound; dash-dotted line: MMSE upper bound; dotted line: water-pouring capacity.

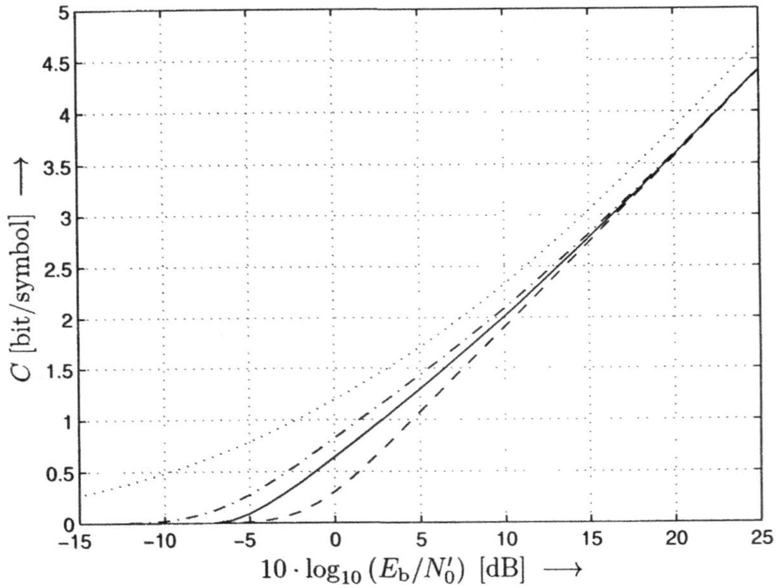

Fig. 3.63 Capacity achievable by Tomlinson-Harashima precoding. DSL down-stream example, cable length 5 km. Dashed line: zero-forcing precoding; solid line: MMSE lower bound; dash-dotted line: MMSE upper bound; dotted line: water-pouring capacity.

solution. The gap between ZF precoding and actual channel capacity is bridged to some extent. For increasing cable length, MMSE filtering clearly outperforms the ZF approach.

Note that even when using MMSE precoding, the capacity of the underlying channel can not be utilized entirely.

REFERENCES

[ACZ91] A. K. Aman, R. L. Cupo, and N. A. Zervos. Combined Trellis Coding and DFE through Tomlinson Precoding. *IEEE Journal on Selected Areas in Communications*, JSAC-9, pp. 876–884, August 1991.

[And99] J. B. Anderson. *Digital Transmission Engineering*. IEEE Press, Piscataway, NJ, 1999.

[Ber96] J. W. M. Bergmans. *Digital Baseband Transmission and Recording*. Kluwer Academic Publishers, Dordrecht, The Netherlands, 1996.

[Bla90] R. E. Blahut. *Digital Transmission of Information*. Addison-Wesley Publishing Company, Reading, MA, 1990.

[Bos85] N. K. Bose. *Digital Filters — Theory and Applications*. North-Holland, Amsterdam, 1985.

[BTL85] C. W. Barnes, B. N. Tran, and S. H. Leung. On the Statistics of Fixed-Point Roundoff Error. *IEEE Transactions on Acoustics, Speech, and Signal Processing*, ASSP-33, pp. 595–606, June 1985.

[CDEF95] J. M. Cioffi, G. P. Dudevoir, M. V. Eyuboğlu, and G. D. Forney. MMSE Decision-Feedback Equalizers and Coding — Part I: Equalization Results, Part II: Coding Results. *IEEE Transactions on Communications*, COM-43, pp. 2582–2604, October 1995.

[CÖU97a] G. Cherubini, S. Ölçer, and G. Ungerböck. 100BASE-T2: A New Standard for 100 Mb/s Ethernet Transmission over Voice-Grade Cables. *IEEE Communications Magazine*, Vol. 35, pp. 115–122, November 1997.

[CÖU97b] G. Cherubini, S. Ölçer, and G. Ungerböck. Trellis Precoding for Channels with Spectral Nulls. In *Proceedings of the IEEE International Symposium on Information Theory*, p. 464, Ulm, Germany, June/July 1997.

[CS87] A. R. Calderbank and N. J. A. Sloane. New Trellis Codes Based on Lattices and Cosets. *IEEE Transactions on Information Theory*, IT-33, pp. 177–195, 1987.

[CS88] J. H. Conway and N. J. A. Sloane. *Sphere Packings, Lattices and Groups*. Springer Verlag, New York, Berlin, 1988.

[CT91] T. M. Cover and J. A. Thomas. *Elements of Information Theory*. John Wiley & Sons, Inc., New York, 1991.

[Dav63] M. C. Davis. Factoring the Spectral Matrix. *IEEE Transactions on Automatic Control*, AC-7, pp. 296-305, October 1963.

[Due95] A. Duel-Hallen. A Family of Multiuser Decision-Feedback Detectors for Asynchronous Code-Division Multiple-Access Channels. *IEEE Transactions on Communications*, COM-43, pp. 421–434, February/March/April 1995.

[EF92] M. V. Eyuboğlu and G. D. Forney. Trellis Precoding: Combined Coding, Precoding and Shaping for Intersymbol Interference Channels. *IEEE Transactions on Information Theory*, IT-38, pp. 301–314, March 1992.

[EFDL93] M. V. Eyuboğlu, G. D. Forney, P. Dong, and G. Long. Advanced Modulation Techniques for V.fast. *European Transactions on Telecommunications*, ETT-4, pp. 243–256, May/June 1993.

[ES76] B. Eckhardt and H. W. Schüßler. On the Quantization Error of a Multiplier. In *Proceedings of the International Symposium on Circuits and Systems*, pp. 634–637, München, April 1976.

[Ett75] W. van Etten. An Optimum Linear Receiver for Multiple Channel Digital Transmission Systems. *IEEE Transactions on Communications*, COM-23, pp. 828–834, August 1975.

[Ett76] W. van Etten. Maximum Likelihood Receiver for Multiple Channel Transmission Systems. *IEEE Transactions on Communications*, COM-24, pp. 276–283, February 1976.

[Eyu88] M. V. Eyuboğlu. Detection of Coded Modulation Signals on Linear, Severely Distorted Channels Using Decision-Feedback Noise Prediction with Interleaving. *IEEE Transactions on Communications*, COM-36, pp. 401–409, April 1988.

[FE91] G. D. Forney and M. V. Eyuboğlu. Combined Equalization and Coding Using Precoding. *IEEE Communications Magazine*, Vol. 29, pp. 25–34, December 1991.

[FGH95] R. Fischer, W. Gerstacker, and J. Huber. Dynamics Limited Precoding, Shaping, and Blind Equalization for Fast Digital Transmission over Twisted Pair Lines. *IEEE Journal on Selected Areas in Communications*, JSAC-13, pp. 1622–1633, December 1995.

[FGL+84] G. D. Forney, R. G. Gallager, G. R. Lang, F. M. Longstaff, and S. U. H. Qureshi. Efficient Modulation for Band-Limited Channels. *IEEE Journal on Selected Areas in Communications*, JSAC-2, pp. 632–647, September 1984.

[FH95] R. Fischer and J. Huber. Dynamics Limited Shaping for Fast Digital Transmission. In *Proceedings of the IEEE International Conference on Communications (ICC'95)*, pp. 22–26, Seattle, WA, June 1995.

[FH97] R. Fischer and J. Huber. Comparison of Precoding Schemes for Digital Subscriber Lines. *IEEE Transactions on Communications*, COM-45, pp. 334–343, March 1997.

[FHK94] R. Fischer, J. Huber, and G. Komp. Coordinated Digital Transmission: Theory and Examples. *Archiv für Elektronik und Übertragungstechnik (International Journal of Electronics and Communications)*, Vol. 48, pp. 289–300, November/December 1994.

[Fis95] R. Fischer. Using Flexible Precoding for Channels with Spectral Nulls. *Electronics Letters*, Vol. 31, pp. 356–358, March 1995.

[Fis96] R. Fischer. *Mehrkanal- und Mehrträgerverfahren für die schnelle digitale Übertragung im Ortsanschlußleitungsnetz.* PhD Thesis, Technische Fakultät der Universität Erlangen–Nürnberg, Erlangen, Germany, October 1996. (In German.)

[For72] G. D. Forney. Maximum Likelihood Sequence Estimation of Digital Sequences in the Presence of Intersymbol Interference. *IEEE Transactions on Information Theory*, IT-18, pp. 363–378, May 1972.

[For88a] G. D. Forney. Coset Codes — Part I: Introduction and Geometrical Classification. *IEEE Transactions on Information Theory*, IT-34, pp. 1123–1151, September 1988.

[For88b] G. D. Forney. Coset Codes — Part II: Binary Lattices and Related Codes. *IEEE Transactions on Information Theory*, IT-34, pp. 1152–1187, September 1988.

[For92] G. D. Forney. Trellis Shaping. *IEEE Transactions on Information Theory*, IT-38, pp. 281–300, March 1992.

[FRC92] P. Fortier, A. Ruiz, and J. M. Cioffi. Multidimensional Signal Sets Through the Shell Construction for Parallel Channels. *IEEE Transactions on Communications*, COM-40, pp. 500–512, March 1992.

[FTC00] G. D. Forney, M. D. Trott, and S.-Y. Chung. Sphere-Bound-Achieving Coset Codes and Multilevel Coset Codes. *IEEE Transactions on Information Theory*, IT-46, pp. 820–850, May 2000.

[FW89] G. D. Forney and L.-F. Wei. Multidimensional Constellations — Part I: Introduction, Figures of Merit, and Generalized Cross Constellations. *IEEE Journal on Selected Areas in Communications*, JSAC-7, pp. 877–892, August 1989.

[Fra80] L. E. Franks. Carrier and Bit Synchronization in Data Communication — A Tutorial Review. *IEEE Transactions on Communications*, COM-28, pp. 1107–1121, August 1980.

[Gal68] R. G. Gallager. *Information Theory and Reliable Communication*. John Wiley & Sons, Inc., New York, London, 1968.

[Ger98] W. Gerstacker. Entzerrverfahren für die schnelle digitale Übertragung über symmetrische Leitungen. PhD Thesis, Technische Fakultät der Universität Erlangen–Nürnberg, Erlangen, Germany, December 1998. (In German.)

[GG98] I. A. Glover and P. M. Grant. *Digital Communications*. Prentice-Hall, Inc., Englewood Cliffs, NJ, 1998.

[HM69] H. Harashima and H. Miyakawa. A Method of Code Conversion for Digital Communication Channels with Intersymbol Interference. *Transactions of the Institute of Electronics and Communincations Engineers of Japan.*, 52-A, pp. 272–273, June 1969. (In Japanese.)

[HM71] N. Halyo and G. A. McAlpine. A Discrete Model for Product Quantization Errors in Digital Filters. *IEEE Transactions on Audio and Electroacoustics*, AU-19, pp. 255–256, September 1971.

[HM72] H. Harashima and H. Miyakawa. Matched-Transmission Technique for Channels with Intersymbol Interference. *IEEE Transactions on Communications*, COM-20, pp. 774–780, August 1972.

[Hub92a] J. Huber. Personal Communications. Erlangen, March 1992.

[Hub92b] J. Huber. Reichweitenabschätzung durch Kanalcodierung bei der digitalen Übertragung über symmetrische Leitungen. Internal Report, Lehrstuhl für Nachrichtentechnik, Universität Erlangen–Nürnberg, Erlangen, Germany, 1992. (In German.)

[Hub93] J. Huber. Signal- und Systemteoretische Grundlagen zur Vorlesung Nachrichtenübertragung. Skriptum, Lehrstuhl für Nachrichtentechnik II, Universität Erlangen–Nürnberg, Erlangen, Germany, 1993. (In German.)

[IH77] H. Imai and S. Hirakawa. A New Multilevel Coding Method Using Error Correcting Codes. *IEEE Transactions on Information Theory*, IT-23, pp. 371–377, May 1977.

[Imm91] K. A. S. Immink. *Coding Techniques for Digital Recorders*. Prentice-Hall, Inc., Hertfordshire, UK, 1991.

[ITU00] ITU-T Recommendation V.92. *Enhancements to Recommendation V.90*. International Telecommunication Union (ITU), Geneva, Switzerland, November 2000.

[ITU93] ITU-T Recommendation G.711. *Pulse Code Modulation (PCM) of Voice Frequencies*. International Telecommunication Union (ITU), Geneva, Switzerland, 1994.

[ITU94] ITU-T Recommendation V.34. *A Modem Operating at Data Signalling Rates of up to 28800 bit/s for Use on the General Switched Telephone Network and on Leased Point-to-Point 2-Wire Telephone-Type Circuits.* International Telecommunication Union (ITU), Geneva, Switzerland, September 1994.

[JN84] N. S. Jayant and P. Noll. *Digital Coding of Waveforms.* Prentice-Hall, Inc., Englewood Cliffs, NJ, 1984.

[KK93] A. K. Khandani and P. Kabal. Shaping Multidimensional Signal Spaces — Part I: Optimum Shaping, Shell Mapping, Part II: Shell-Addressed Constellations. *IEEE Transactions on Information Theory*, IT-39, pp. 1799–1819, November 1993.

[KP93] F. R. Kschischang and S. Pasupathy. Optimal Nonuniform Signaling for Gaussian Channels. *IEEE Transactions on Information Theory*, IT-39, pp. 913–929, May 1993.

[Kre66] E. R. Kretzmer. Generalization of a Technique for Binary Data Communication. *IEEE Transactions on Communication Technology*, COM-14, pp. 67–68, February 1966.

[Lar94] R. Laroia. Coding for Intersymbol Interference Channels — Combined Coding and Precoding. In *Proceedings of the IEEE International Symposium on Information Theory*, p. 328, Trondheim, Norway, June 1994.

[Lar96] R. Laroia. Coding for Intersymbol Interference Channels — Combined Coding and Precoding. *IEEE Transactions on Information Theory*, IT-42, pp. 1053–1061, July 1996.

[Len64] A. Lender. Correlative Digital Communication Techniques. *IEEE Transactions on Communication Technology*, COM-12, pp. 128–135, December 1964.

[LFT94] R. Laroia, N. Farvardin, and S. A. Tretter. On Optimal Shaping of Multidimensional Constellations. *IEEE Transactions on Information Theory*, IT-40, pp. 1044–1056, July 1994.

[LL89] G. R. Lang and F. M. Longstaff. A Leech Lattice Modem. *IEEE Journal on Selected Areas in Communications*, JSAC-7, pp. 986–973, August 1989.

[LTF93] R. Laroia, S. A. Tretter, and N. Farvardin. A Simple and Effective Precoding Scheme for Noise Whitening in Intersymbol Interference Channels. *IEEE Transactions on Communications*, COM-41, pp. 1460–1463, October 1993.

[Mas74] J. L. Massey. Coding and Modulation in Digital Communications. In *Proceedings of the 1974 Intern. Zürich Seminar on Digital Communications*, Zürich, Switzerland, March 1974.

[MS76] J. E. Mazo and J. Salz. On the Transmitted Power in Generalized Partial Response. *IEEE Transactions on Communications*, COM-24, pp. 348–351, March 1976.

[Pap77] A. Papoulis. *Signal Analysis*. McGraw-Hill, New York, 1977.

[Pap91] A. Papoulis. *Probability, Random Variables, and Stochastic Processes*. McGraw-Hill, New York, 3rd edition, 1991.

[PC93a] S. S. Pietrobon and D. J. Costello, Jr. Trellis Coding with Multidimensional QAM Signal Sets. *IEEE Transactions on Information Theory*, IT-39, pp. 325–336, March 1993.

[PC93b] G. M. Pitstick and J. R. Cruz. An Efficient Algorithm for Computing Bounds on the Average Transmitted Power in Generalized Partial Response. In *Proceedings of the IEEE Global Telecommunications Conference'93*, pp. 2006-2010, Houston, TX, December 1993.

[PE91] G. J. Pottie and M. V. Eyuboğlu. Combined Coding and Precoding for PAM and QAM HDSL Systems. *IEEE Journal on Selected Areas in Communications*, JSAC-9, pp. 861–870, August 1991.

[PM88] J. G. Proakis and D. G. Manolakis. *Introduction to Digital Signal Processing*. Macmillan Publishing Company, New York, 1988.

[Pri72] R. Price. Nonlinear Feedback Equalized PAM versus Capacity for Noisy Filter Channels. In *Proceedings of the IEEE International Conference on Communications (ICC'72)*, pp. 22.12–22.17, 1972.

[Pro01] J. G. Proakis. *Digital Communications*. McGraw-Hill, New York, 4th edition, 2001.

[PTVF92] W. H. Press, S. A. Teukolsky, W. T. Vetterling, and B. P. Flannery. *Numerical Recipes in C—The Art of Scientific Computing*. Cambridge University Press, Cambridge, 2nd edition, 1992.

[Sch94] H. W. Schüßler. *Digitale Signalverarbeitung, Band I*. Springer Verlag, Berlin, Heidelberg, 4th edition, 1994. (In German.)

[SD94] S. Shamai (Shitz) and A. Dembo. Bounds on the Symmetric Binary Cutoff Rate for Dispersive Gaussian Channels. *IEEE Transactions on Communications*, COM-42, pp. 39–53, January 1994.

[Sjo73] T. W. Sjoding. Noise Variance for Rounded Two's Complement Product Quantization. *IEEE Transactions on Audio and Electroacoustics*, AU-21, pp. 378–380, August 1973.

[SL96] S. Shamai (Shitz) and R. Laroia. The Intersymbol Interference Channel: Lower Bounds on Capacity and Channel Precoding Loss. *IEEE*

Transactions on Information Theory, IT-42, pp. 1388–1404, September 1996.

[SS77] A. B. Sripad and D. L. Snyder. A Necessary and Sufficient Condition for Quantization Errors to be Uniform and White. *IEEE Transactions on Acoustics, Speech, and Signal Processing*, ASSP-52, pp. 442–448, October 1977.

[TF92] T. Trump and U. Forssén. On the Statistical Properties of Tomlinson Filters. Telecommunication Theory, Royal Institute of Technology, Stockholm, Sweden, March 1992.

[Tom71] M. Tomlinson. New Automatic Equaliser Employing Modulo Arithmetic. *Electronics Letters*, Vol. 7, pp. 138–139, March 1971.

[UC76] G. Ungerböck and I. Csajka. On Improving Data-Link Performance by Increasing Channel Alphabet and Introducing Sequence Coding. In *Proceedings of the IEEE International Symposium on Information Theory*, Ronneby, Sweden, June 1976.

[Ung74] G. Ungerböck. Adaptive Maximum-Likelihood Receiver for Carrier-Modulated Data-Transmission Systems. *IEEE Transactions on Communications*, COM-22, pp. 624–636, May 1974.

[Ung82] G. Ungerböck. Channel Coding with Multilevel/Phase Signals. *IEEE Transactions on Information Theory*, IT-28, pp. 55–67, January 1982.

[Ung87a] G. Ungerböck. Trellis-Coded Modulation with Redundant Signal Sets, Part I: Introduction. *IEEE Communications Magazine*, Vol. 25, pp. 5–11, February 1987.

[Ung87b] G. Ungerböck. Trellis-Coded Modulation with Redundant Signal Sets, Part II: State ot the Art. *IEEE Communications Magazine*, Vol. 25, pp. 12–21, February 1987.

[WC98] R. D. Wesel and J. M. Cioffi. Achievable Rates for Tomlinson-Harashima Precoding. *IEEE Transactions on Information Theory*, IT-44, pp. 824–831, March 1998.

[Wei94] L.-F. Wei. Generalized Square and Hexagonal Constellations for Intersymbol-Interference Channels with Gereralized Tomlinson-Harashima Precoders. *IEEE Transactions on Communications*, COM-42, pp. 2713–2721, September 1994.

[Wer93] J. J. Werner. Tutorial on Carrierless AM/PM — Part I: Fundamentals and Digital CAP Transmitter; Part II: Performance of Bandwith-Efficient Line Codes. AT&T Bell Laboratories, Middletown, NJ, 1992/1993.

[WFH99] U. Wachsmann, R. F. H. Fischer, and J. B. Huber. Multilevel Codes: Theoretical Concepts and Practical Design Rules. *IEEE Transactions on Information Theory*, IT-45, pp. 1361–1391, July 1999.

[You61] D. C. Youla. On the Factorization of Rational Matrices. *IEEE Transactions on Information Theory*, IT-7, pp. 172–189, July 1961.

4

Signal Shaping

Each communications scenario has its specific demands—hence, for best performance, the transmission system should be tailored to the actual situation as close as possible. This implies that the transmit signal should match the requirements stipulated by the communication link. In its broadest definition, the task of *signal shaping* is to generate signals which meet specific demands. Shaping aims can be as multifarious as the transmission scenarios can be.

The most popular aim of signal shaping is to generate signals with least average power, without sacrificing performance. Especially in crosstalk-limited transmission scenarios, *average transmit power* is of major interest. Here, transmit power of one link directly translates to noise power experienced by the other lines. Hence, transmission with least average power is desired. By simply scaling the signal on one line a reduction of the power level would be possible, but at the same time performance of this link is reduced, as well.

Another signal property which often has to be controlled is the power spectral density (PSD). In some situations, a specific shape of the PSD is advantageous or even necessary. For example, in magnetic recording, when using transformer coupling, the power contents at low frequencies should be as small as possible.

With respect to the general definition of shaping, even the precoding schemes of the last chapter can be viewed as a special form of signal shaping. Here, transmit signals are generated which result in equalized data signals after transmission over an ISI channel.

In this chapter, signal shaping schemes are discussed and their performance is analyzed. Because of its importance, we primarily focus on signal shaping for minimum average transmit power. Moreover, only transmission over the intersymbol-interference-free AWGN channel is considered in this chapter. The combination of signal shaping and equalization of ISI channels via precoding is the subject of Chapter 5.

First, an intuitive explanation is given of how a reduction of average transmit power is possible. The differences and similarities between shaping and source and channel coding are studied. Then, performance bounds on shaping are derived and the various effects of signal shaping are discussed. *Shell mapping*, a specific shaping algorithm is explained in detail and the statistical properties of the transmit symbols are calculated. Thereafter, a second important scheme, *trellis shaping* is studied and its performance is assessed. In the context of trellis shaping we show that other shaping aims than reducing average transmit power can also be met easily. The chapter closes with a discussion on how performance can be improved even if we restrict ourselves to equiprobable signaling.

4.1 INTRODUCTION TO SHAPING

Undoubtedly, one of the most important parameters for signal design is average transmit power. Low transmit power is, e.g., desirable in mobile terminals because of limited battery capacity, or for remotely supplied modems where power is provided over the same line as the transmission takes place. Moreover, in multiple-access situations transmit power of one user directly translates into noise for the other users. For instance, consider fast digital transmission over subscriber lines where binder groups with hundreds of lines emerge at the central office. Mainly due to capacitive coupling, crosstalk occurs among the various lines. In such situations it is beneficial to reduce transmit power. This of course should be done without sacrificing performance. We will now give an intuitive explanation of how this can be achieved.

Consider a PAM transmit signal $s(t) = \sum_k a[k] g_{\mathrm{T}}(t - kT)$ with i.i.d. zero-mean data sequence $\langle a[k] \rangle$ as introduced in Chapter 2. It is well known that for such signals the average transmit power calculates to[1] [Pro01]

$$S \triangleq \frac{1}{T} \int_0^T \mathsf{E}\{|s(t)|^2\}\, dt = \frac{\mathsf{E}\{|a[k]|^2\}}{T} E_{\mathrm{T}} , \tag{4.1.1}$$

where $E_{\mathrm{T}} = \int |g_{\mathrm{T}}(t)|^2 \, dt$ is the energy of the transmit pulse. Since we fix pulse shape $g_{\mathrm{T}}(t)$ and symbol spacing T, the average $\sigma_a^2 \triangleq \mathsf{E}\{|a[k]|^2\}$ over the squared magnitudes of the zero-mean PAM data symbols $a[k]$, i.e., the variance of $a[k]$, directly relates to average transmit power S. Hence, subsequently we only treat the discrete-time sequence of PAM data symbols.

[1] Since PAM data signals are samples of a *cyclostationary* process with period equal to the symbol interval T, expectation is carried out over the process and, additionally, over one (arbitrary) period of duration T.

A very helpful geometric interpretation of signals is to represent blocks of symbols as a point in a higher dimensional signal space, e.g., [Sha49]. Because successive symbols are mapped independently and are assigned to mutually orthogonal pulses, each time index constitutes a separate coordinate, orthogonal to all other dimensions. Recall, in digital signal processing, the sum over squared magnitudes of symbols is called *energy* [OS75]. Consequently, the energy within one block of symbols is now given as the squared Euclidean distance from the origin, i.e., the *Euclidean norm*. Note that the energy simply adds up along the different dimensions. When dealing with the discrete-time sequence $\langle a[k] \rangle$, we thus speak of (average or peak) *energy*, whereas when studying the continuous-time transmit signal $s(t)$ we talk of (average or peak) *power*.

For the following, imagine *baseband signaling* employing one-dimensional PAM data symbols $a[k]$. Furthermore, we assume a uniform i.i.d. binary data sequence, i.e., ideally source-coded, to be transmitted. Traditionally, in transmission schemes the signal points are selected equiprobable. If the number of signal points is a power of two, this property directly results from mapping blocks of data bits to the signal points.

Then, regarding blocks of two consecutive symbols (first symbol: x-axis, second symbol: y-axis), all pairs of signal points are arranged within a square. This is visualized for 16-ary one-dimensional symbols on the left-hand side of Figure 4.1. In general, for independently, uniformly mapping of N successive one-dimensional symbols, the signal point is enclosed in an N-*cube*.

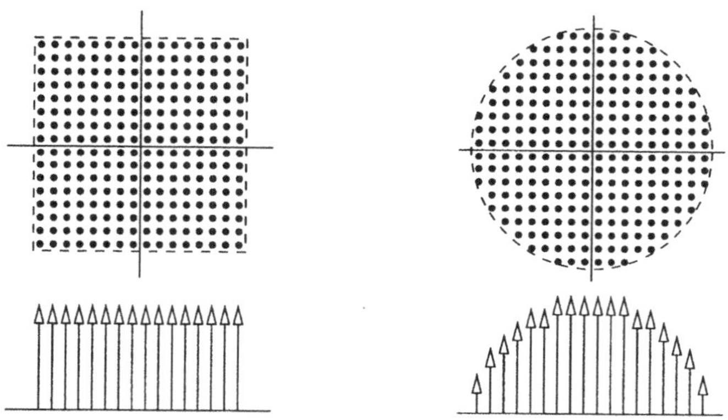

Fig. 4.1 Independent mapping of two consecutive 16-ary symbols (left) and joint mapping (right) for minimum average energy. Bottom: Probability density, i.e., projection to one dimension.

By jointly mapping two time intervals, the average energy of the two-dimensional arrangement can be lowered. This is achieved by moving points with high energy (especially the vertices) to positions (near the coordinate axis) with lower energy. The underlying regular grid, and hence the minimum distance between the points, is thereby preserved. It is intuitively clear that the boundary for lowest average energy is a circle. The optimal two-dimensional arrangement of 16^2 signal points and its boundary is shown on the right-hand side of Figure 4.1. Considering again N dimensions, the signal points should be enclosed in an N-*sphere* rather than an N-*cube*. Of course, it is more difficult to select the points from an N-sphere than individually selecting the coordinates in each dimension. Hence, shaping has to be payed for with an increased addressing complexity.

From this simple example we can conclude that signal shaping is responsible for the design of the *shape* of the signal constellation in N-dimensional space. This moreover explains the term "*shaping*," or sometimes "*constellation shaping*". To the contrary, the task of channel coding is to arrange the points *within* the signal set—classically in order to achieve large distances. For example, in two dimensions, a hexagonal grid would be preferable over the rectangular one. Hence, coding and shaping are in a way dual operations, and—at least for large constellations—are separable, i.e., both tasks can be performed individually, and the respective gains (in dB) add up.

Now, in signal shaping, instead of addressing the points equiprobably in one dimension, the points are selected equiprobably from an N-dimensional sphere. Going back again to one dimension by regarding the respective projection,[2] we see that the signal points in one dimension are no longer uniformly distributed. This projection is of some importance, because the transmitter still has to work on sequences of one-dimensional symbols. The one-dimensional projections of the points along the axis are also shown on the bottom of Figure 4.1. Clearly, the projection of the square is a uniform density, whereas the projection of the circle induces a one-dimensional density, where points with low energy occur more often than points at the perimeter.

This observation leads us to a second, different approach to signal shaping. Instead of generating a high-dimensional, uniformly distributed constellation, one can also try to directly generate an appropriate nonuniform low-dimensional distribution. Since this is usually done by means of some kind of block coding over a number of consecutive symbols, both approaches are closely related to each other.

It is noteworthy that generating a nonuniform distribution from a redundancy-free data sequence is the dual operation to *source coding*. In source coding, nonequiprobable (in general, redundant, if the source has memory) input is converted into a (often binary) redundancy-free output; hence, the symbols are equiprobable. This duality allows the use of source decoders as encoders in shaping schemes. As we have seen above, shaping and channel coding are dual operations, too. Hence, it is possible to use channel decoders for signal shaping, as well. We will return to these points often in this chapter. In summary, the three items, *source coding*, *channel coding*, and *signal shaping*, are mutually dual. Figure 4.2 shows the relations.

Furthermore, we note from the above example that the constituent one-dimensional constellation is *expanded*; in this example 18 signal levels per dimension are visible compared to 16. In regard to the high-dimensional constellation, no expansion takes place, i.e., the number of signal points is the same. This expansion of the low-

[2]We assume that the projections into any dimension are identical.

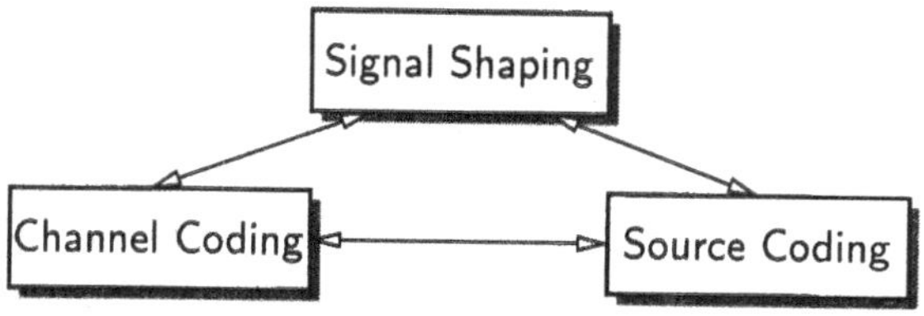

Fig. 4.2 Duality of source coding, channel coding, and signal shaping.

dimensional signal constellations is a general principle in signal shaping. Information theory tells us that for fixed entropy, a nonuniform distribution requires more symbols than a uniform distribution.

Finally, signal shaping is intended to *decrease average transmit power*. But due to the constellation expansion, it *increases peak power* of the (one-dimensional) transmit signal. Constellation expansion and increased peak energy are the price to be paid for a gain in average energy.

Please note that the above considerations are likewise valid if passband signaling with two-dimensional signal sets is regarded. Here, two real-valued dimensions are combined into a complex-valued one. In QAM modems the properties of the two-dimensional constituent constellation are of major importance, rather than those of the one-dimensional projection.

4.1.1 Measures of Performance

In the above discussion, we have seen the advantages and briefly discussed the drawbacks of signal shaping. For performance evaluation of signal shaping, the following specific quantities are of special interest [FW89]:

- **Shaping Gain:**
  The shaping gain (sometimes also called shape gain [FW89]) is defined as the ratio of average signal energy for equiprobable signaling and average signal energy when applying signal shaping and transmitting the same rate. Usually, this gain is expressed in dB.

- **Constellation Expansion Ratio:**
  The constellation expansion ratio gives the number of signal points in the low-dimensional constituent constellation relative to the number of points required to transmit the same rate by equiprobable signaling. The constellation expansion ratio is always greater than or equal to unity.

- **Peak-to-Average Energy Ratio:**
  The peak-to-average energy ratio relates the peak energy of the low-dimensional constituent constellation to its average energy. This quantity is also often expressed in dB and is always greater than 0 dB.

Later on, we will calculate these quantities for specific scenarios and give the general relation between them.

4.1.2 Optimal Distribution for Given Constellation

After having discussed the basic principles of signal shaping, we now turn to the problem of finding the optimal distribution of the signal points for a given constellation. This knowledge is required if, starting from a low-dimensional constellation, shaping gain should be achieved by directly imposing a suitable distribution.

Let $\mathcal{A} = \{a_i\}$ be a given D-dimensional signal constellation with $|\mathcal{A}|$ points a_1, $a_2, \ldots, a_{|\mathcal{A}|}$, and $R^{(D)}$ be the rate per D dimensions to be transmitted using $\mathcal{A}$. This of course requires that $\mathcal{A}$ is capable to support this rate; mathematically $R^{(D)} \leq \log_2(|\mathcal{A}|)$ has to hold. The aim is to minimize average energy[3] $E(\mathcal{A}) = \mathrm{E}\{|a_i|^2\}$ of the constellation by adjusting the probabilities $p_i \triangleq \Pr\{a_i\}$ of the signal points. Mathematically, the optimization problem is given as[4]

- Minimize average energy $E(\mathcal{A}) = \sum_i p_i\,|a_i|^2$
- under the additional constraints that
  (i) $\{p_i\}$ is a probability distribution, $\sum_i p_i = 1$, $p_i \geq 0$, and
  (ii) the entropy of the constellation equals the desired rate:
  $H(\mathcal{A}) = -\sum_i p_i \log_2(p_i) = R^{(D)}$.

Using the method of Lagrange multipliers, we can set up the following Lagrange function with Lagrange multipliers μ and ν

$$L(\{p_i\}) = \sum_i p_i\,|a_i|^2 - \mu \sum_i p_i + \nu \sum_i p_i \log(p_i)\,. \tag{4.1.2}$$

Here for the moment it is convenient to use the natural logarithm (base e) rather than the logarithm dualis, i.e., to think in nats rather than bits.

The optimal solution for the probabilities p_i is a stationary point of the Lagrange function. Hence, derivation of $L(\{p_i\})$ with respect to p_i leads to

$$\frac{\mathrm{d}L(\{p_i\})}{\mathrm{d}p_i} = |a_i|^2 - \mu + \nu \cdot (\log(p_i) + 1) \stackrel{!}{=} 0\,, \qquad \forall i\,. \tag{4.1.3}$$

Solving for p_i gives

$$p_i = \mathrm{e}^{-\frac{|a_i|^2}{\nu}} \cdot \mathrm{e}^{\frac{\mu}{\nu} - 1} \tag{4.1.4}$$

or, when substituting the reciprocal of the multiplier ν by a new variable λ,

$$p_i = K(\lambda) \cdot \mathrm{e}^{-\lambda |a_i|^2}\,, \qquad \lambda \geq 0\,. \tag{4.1.5}$$

The factor $K(\lambda) = \left(\sum_{a \in \mathcal{A}} \mathrm{e}^{-\lambda |a|^2}\right)^{-1}$ normalizes the distribution, and the parameter λ governs the trade-off between average energy $E(\mathcal{A})$ and entropy $H(\mathcal{A})$ of

[3]Unless otherwise stated, we assume symmetric constellations with zero mean value, i.e., $\mathrm{E}\{a_i\} = 0$. Here, average energy $E(\mathcal{A})$ over the D-dimensional signal points equals their variance σ_a^2.

[4]Here, $\sum_i(\cdot)$ stands for $\sum_{i=1}^{|\mathcal{A}|}(\cdot)$.

signal points. For $\lambda = 0$ a uniform distribution results, whereas for $\lambda \to \infty$ only the signal points closest to the origin remain. Since low-energy signal points should always be at least as likely as high-energy points, λ is nonnegative.

With regard to (4.1.5), it is obvious that the optimal distribution is *discrete* or *sampled Gaussian*. This distribution, which maximizes the entropy under an average energy constraint, is sometimes also called a *Maxwell-Boltzmann distribution* [KP93].

To conclude this derivation, we note that the factor $K(\lambda)$, which is called *partition function* [KP93] has some interesting properties. First, $K(\lambda)$ may be obtained from the *theta series* or *Euclidean weight enumerator* [CS88] $\Theta(x) = \sum_{a \in \mathcal{A}} x^{|a|^2}$ of the constellation as $K(\lambda) = \Theta(\mathrm{e}^{-\lambda})$ (Note: This relation is analogous to the union bound for error probability and the distance profile.) Furthermore, it is easy to verify that from $K(\lambda)$ average energy is obtained as [KP93]

$$E(\lambda) = \frac{\mathrm{d}}{\mathrm{d}\lambda} \log(K(\lambda)) , \tag{4.1.6}$$

and entropy per D dimensions equals

$$H(\lambda) = \lambda^2 \frac{\mathrm{d}}{\mathrm{d}\lambda} \left(\frac{\log_2(K(\lambda))}{\lambda} \right) . \tag{4.1.7}$$

The following example—inspired by one given in [FGL$^+$84]—shows how the optimal probability distribution can be approximated by using a simple source decoder.

Example 4.1: *Shaping using Huffman Decoder*

As explained above, since signal shaping is the dual operation to source coding, a source decoder can be used as shaping encoder. Here, we employ a simple Huffman code with 21 codewords, whose code tree is depicted in Figure 4.3.

In the transmitter, the binary data sequence is parsed and partitioned into valid codewords. Each codeword corresponds to one signal point. Since the Huffman code satisfies the prefix condition (no codeword is prefix of any other codeword) that it is a self-punctuating code, a unique mapping from the data stream to codewords is possible. Note that this procedure requires that each node in the tree be either a leaf or has two children; mathematically speaking, Kraft's inequality [CT91] has to be met with equality.

Let l_i denote the length of the ith codeword. Assuming an i.i.d. uniform data sequence, the probability for this codeword is then $p_i = 2^{-l_i}$. From these probabilities, the entropy, and hence the average transmission rate, can be given as

$$H = -\sum_i p_i \log_2(p_i) = \sum_i l_i \, 2^{-l_i} , \tag{4.1.8}$$

which in our example calculates to $H = 3 \cdot 3 \cdot 2^{-3} + 6 \cdot 4 \cdot 2^{-4} + 4 \cdot 5 \cdot 2^{-5} + 8 \cdot 6 \cdot 2^{-6} = 4$. Hence, the baseline system for equiprobable transmission is an 16-ary QAM constellation, where the average energy equals $E_\square = 10$ (real and imaginary part of the signal points may assume values ± 1, ± 3).

Figure 4.4 shows the expanded signal constellation; 21 instead of 16 signal points are used. The points are labeled by their associated code word. Straightforward calculations gives

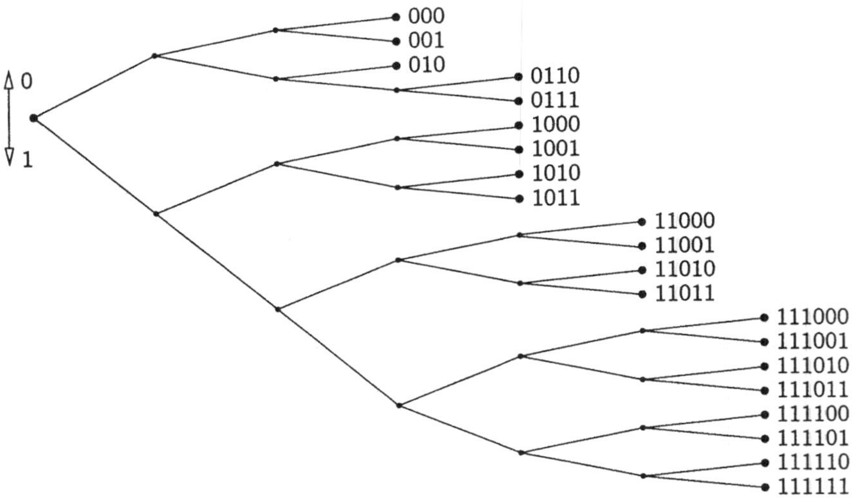

Fig. 4.3 Code tree of the Huffman code.

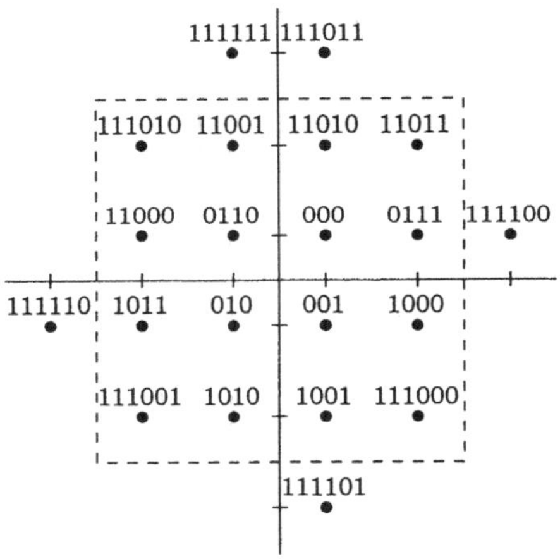

Fig. 4.4 Signal constellation used for shaping. The signal points are labeled by their corresponding codeword. Dashed line: Boundary region for 16QAM.

the average energy of the shaped constellations to be $E = 8.38$. Hence a shaping gain of $10/8.38 = 1.19$ or 0.77 dB is achieved. Unfortunately, the constellation does not have zero mean. Compensating the mean $0.22 - 0.16\mathrm{j}$ by an appropriate shift, average energy is further decreased (now $E = 8.30$) and shaping gain is increased to 0.80 dB.

The price to be paid is the moderate constellation expansion ratio in two dimensions of $21/16 = 1.31$. Furthermore, the peak-to-average energy ratio is increased. For 16QAM, the peak-to-average energy ratio in two dimensions reads $18/10$ or 2.55 dB. Using shaping, the peak-to-average energy ratio calculates to $25/8.38 = 2.99$, equivalent to 4.76 dB; i.e., an increase by more than 2 dB.

Finally, using the same 21-ary constellations, but selecting the probabilities according to the optimal distribution (4.1.5), where the parameter λ is adjusted so that the entropy is 4 bits, the average energy can be lowered to $E = 7.92$. This translates to a shaping gain of 1.02 dB.

The simple Huffman coding scheme is thus able to achieve most of the gain possible. However, the main disadvantage in using a variable-length code for signal shaping is that transmission rate is probabilistic; it varies over time. In turn, sufficiently large buffers at the transmitter and receiver are required to compensate for rate fluctuations. For practice, fixed-rate schemes with only small buffers, and hence a small transmission delay, are clearly preferable.

4.1.3 Ultimate Shaping Gain

In the example above we have seen that gains on the order of 1 dB are possible by simple means. The question that arises now is, what is the maximum shaping gain. Here, we will give a preliminary answer, and return to the question later on, when some additional constraints are imposed. Without loss of generality, the derivations are done for one-dimensional constellations.

First, we note that the baseline system again uses a uniform distribution of the signal points. Following the above derivations, the shaped system should exhibit a (discrete) Gaussian distribution. In order to transmit at the same rate, both distributions have to have the same entropy.

When considering constellations with a large number of signal points, it is more convenient to approximate the distribution by a continuous probability density function (pdf). Hence we compare a continuous uniform pdf with a Gaussian one [FW89]. Instead of fixing the entropy, we now have to compare the *differential entropies* of the distributions.

Letting $E_{\square}$ be the average energy of the reference system, the differential entropy $h(X)$ of its transmit symbols x is given as [CT91]

$$h(X) = \frac{1}{2} \cdot \log_2 (12 E_{\square}) \; . \tag{4.1.9}$$

If x is Gaussian distributed with average energy $E_{\circ}$, its entropy calculates to [CT91]

$$h(X) = \frac{1}{2} \cdot \log_2 (2\pi \mathrm{e} E_{\circ}) \; . \tag{4.1.10}$$

Since the above entropies should be equal, we arrive at

$$\log_2 (12 E_{\square}) \overset{!}{=} \log_2 (2\pi \mathrm{e} E_{\circ}) \; , \tag{4.1.11}$$

which translates to

$$G_{\mathrm{s},\infty} \triangleq \frac{E_{\square}}{E_{\circ}} = \frac{\pi \mathrm{e}}{6} \mathrel{\widehat{=}} 1.53 \text{ dB} \; . \tag{4.1.12}$$

The quantity $G_{\mathrm{s},\infty}$ is called the *ultimate shaping gain* [FGL$^+$84, FW89, FU98], and as we will see later, upper bounds the actual shaping gain. Even though the achievable gain seems to be small, in many situations it is easier to obtain shaping gain than to provide a similar gain by more powerful channel coding. In order to approach the Shannon limit—which requires continuous Gaussian signals—shaping is indispensable. We summarize:

Theorem 4.1: *Ultimate Shaping Gain*

The *shaping gain*, i.e., the gain in reducing average energy compared to a signal with uniform distribution, is limited to the *ultimate shaping gain*

$$G_{s,\infty} = \frac{\pi e}{6} \mathrel{\hat{=}} 1.53\,\text{dB}\,, \tag{4.1.13}$$

which is achieved for a continuous Gaussian probability density function.

To conclude this section, we note that a deeper analysis is still necessary to get more insight into signal shaping. For example, the asymptotic result gives no hint of the number of dimensions on which a shaping algorithm should work. Moreover, there is no statement on how the number of signal points influences the shaping gain. In the course of this chapter, we derive bounds on the shaping gain, taking these facts into account; this moreover gives some advice for practical implementation. But the main drawback of the present study is that only the transmit signal is considered. In Section 4.2.6, we will discuss whether shaping gain, i.e., saving in average energy, is the only parameter of importance.

4.2 BOUNDS ON SHAPING

As we have seen, the maximum shaping gain of 1.53 dB can be readily derived from basic information theory. In this section, we calculate the achievable gain under some additional constraints and limiting factors on shaping. These observations give guidelines for practical systems.

4.2.1 Lattices, Constellations, and Regions

For the analysis of signal shaping, we first have to consider some important properties of and performance measures for signal constellations, the underlying lattice, and the boundary region.

Let Λ be an N-dimensional *lattice*, from which the signal points are drawn. Here, for simplicity, we always assume that N is an even integer, and that the lattice spans N dimensions. Again, for an introduction to lattices, see Appendix C. Furthermore, let $\mathcal{R} \subset \mathbb{R}^N$ be a finite N-dimensional *region*. The N-dimensional *signal constellation*[5] $\mathcal{C}$ is then the finite set of $|\mathcal{C}|$ points from the lattice Λ (or a translate thereof; in the present context a translation is of no importance) that lie within the region $\mathcal{R}$. Mathematically, $\mathcal{C}$ is the intersection of Λ and $\mathcal{R}$, namely

$$\mathcal{C} = \Lambda \cap \mathcal{R} . \tag{4.2.1}$$

The constellation is able to support a maximum number of $\log_2 |\mathcal{C}|$ bits per N dimensions.

Here we assume that the *projection* (along the axis) of $\mathcal{C}$ to any D dimensions (usually, $D = 1$ or 2; D divides N) is the same for all coordinate D tuples, i.e., $\mathcal{C}$ is *D-dimensional symmetric*. The projection, denoted by $\mathcal{C}_D$, is called the *constituent constellation*, and is the set of D-dimensional symbols which occur, as the N-dimensional points range through all values of $\mathcal{C}$ [FW89]. Since actual transmission takes place on the constituent constellation, we have $\mathcal{A} = \mathcal{C}_D$. In practice, the one-dimensional ($D = 1$) constituent constellation and its properties are of interest when considering baseband signaling, whereas for QAM transmission the two-dimensional ($D = 2$) constituent constellation matters.

Like the N-dimensional constellation, the constituent constellation can also be written as $\mathcal{C}_D = \Lambda_D \cap \mathcal{R}_D$, where Λ_D and $\mathcal{R}_D$ are the constituent lattice and constituent boundary region, respectively. Both quantities are the projection of their N-dimensional counterparts. Note that $\mathcal{C}$ is a subset of the N/D-fold Cartesian product of $\mathcal{C}_D$

$$\mathcal{C} \subseteq \mathcal{C}_D^{N/D} , \tag{4.2.2}$$

which implies $|\mathcal{C}| \leq |\mathcal{C}_D^{N/D}| = |\mathcal{C}_D|^{N/D}$.

[5] In order to distinguish the high-dimensional constellation from the one- or two-dimensional PAM signal constellation $\mathcal{A}$, we denote it by $\mathcal{C}$. The signal points are vectors $\boldsymbol{c}$, which also emphasizes that the set $\mathcal{C}$ is some kind of code.

Important Parameters For the design and analysis of signal constellations, the following parameters of the underlying lattice $\mathbf{\Lambda}$ are of interest [FW89], cf. also Appendix C:

The *minimum squared distance* of the lattice $\mathbf{\Lambda}$, $d^2_{\min}(\mathbf{\Lambda}) = \min_{\boldsymbol{\lambda}\in\mathbf{\Lambda}\setminus\{\mathbf{0}\}} |\boldsymbol{\lambda}|^2$, gives the distance of the signal points, and hence is directly related to error performance in uncoded systems.

The *fundamental volume* of the lattice, denoted by $V(\mathbf{\Lambda})$, is the volume of N space corresponding to each lattice point, i.e., the volume of the *Voronoi region* $\mathcal{R}_V(\mathbf{\Lambda})$ of the lattice. Moreover, the *volume* of the boundary region $\mathcal{R}$, $V(\mathcal{R}) = \int_{\mathcal{R}} \mathrm{d}\boldsymbol{r}$, is important.

Using the points of $\mathcal{C}$ equiprobably, the *rate* per D dimensions is given by

$$R^{(D)} \triangleq \frac{D}{N} \cdot \log_2(|\mathcal{C}|) \,. \qquad (4.2.3)$$

Mostly, we deal with the *rate per dimension*, and write $R \triangleq R^{(1)}$.

Under the same assumptions, the *average energy* per D dimensions calculates to

$$E^{(D)}(\mathcal{C}) \triangleq \frac{D}{N}\frac{1}{|\mathcal{C}|}\sum\nolimits_{\boldsymbol{c}\in\mathcal{C}} |\boldsymbol{c}|^2 \,. \qquad (4.2.4)$$

Note that energy of an N-dimensional point $\boldsymbol{c}$ equals its Euclidean norm $|\boldsymbol{c}|^2$, and is additive over dimensions. Likewise, the average energy of a region $\mathcal{R}$ is the average energy of a uniform probability distribution over this region: $E^{(N)}(\mathcal{R}) \triangleq \frac{1}{V(\mathcal{R})}\int_{\mathcal{R}} |\boldsymbol{r}|^2 \, \mathrm{d}\boldsymbol{r}$.

In order to get rid of scaling, energy is often normalized to a given volume. The *normalized second moment* [FW89, CS88] (sometimes also called *dimensionless second moment*) of a signal uniformly distributed within an N-dimensional boundary region $\mathcal{R}$ with volume $V(\mathcal{R})$ is defined as (cf. also Appendix C)

$$\mathsf{G}(\mathcal{R}) \triangleq \frac{E^{(1)}(\mathcal{R})}{V(\mathcal{R})^{2/N}} = \frac{\int_{\mathcal{R}} |\boldsymbol{r}|^2 \, \mathrm{d}\boldsymbol{r}}{N \cdot V(\mathcal{R})^{2/N+1}} \,. \qquad (4.2.5)$$

Note that this parameter is also invariant to the Cartesian product operation.

Continuous Approximation When dealing with signal shaping, signal constellations with a large number of signal points are often considered. Instead of considering each single signal point, it is more convenient to treat the constellation as a whole. The constellation is then approximated by a continuous probability density function uniform over the boundary region $\mathcal{R}$. This principle is called the *continuous approximation* or *integral approximation* [FGL+84, FW89].

In particular, we can derive approximations to the above parameters in a simple way, where, given a function $f(\cdot)$, with $f : \mathbb{R}^N \mapsto \mathbb{R}$, we have to evaluate $\sum_{a\in\mathcal{C}} f(a)$. Going the opposite way as taken in numerical integration, in particular regarding the Riemann sum (numerical integration of degree zero [BS98]), we have the approximation [KP93]

$$\sum\nolimits_{\boldsymbol{c}\in\mathcal{C}} f(\boldsymbol{c}) \approx \frac{1}{V(\mathbf{\Lambda})}\int_{\mathcal{R}} f(\boldsymbol{r}) \, \mathrm{d}\boldsymbol{r} \,. \qquad (4.2.6)$$

As $|\mathcal{C}|$ increases, the approximation becomes more accurate.

Setting $f(\boldsymbol{r}) = 1$, the size of the constellation is approximated by [FW89, Proposition 1]

$$|\mathcal{C}| = \sum\nolimits_{c\in\mathcal{C}} 1 \approx \frac{1}{V(\boldsymbol{\Lambda})}\int_{\mathcal{R}} \mathrm{d}\boldsymbol{r} = \frac{V(\mathcal{R})}{V(\Lambda)} \,. \tag{4.2.7}$$

The interpretation is intuitively clear: The boundary region has volume $V(\mathcal{R})$ and each point takes $V(\Lambda)$. Hence, to the accuracy of the continuous approximation, the quotient is the number of points.

For $f(\boldsymbol{r}) = |\boldsymbol{r}|^2$, the average energy per dimension calculates to [FW89, Proposition 2]

$$E^{(N)}(\mathcal{C}) = \frac{1}{|\mathcal{C}|}\sum\nolimits_{c\in\mathcal{C}} |\boldsymbol{c}|^2 \approx \frac{1}{|\mathcal{C}|}\frac{1}{V(\boldsymbol{\Lambda})}\int_{\mathcal{R}} |\boldsymbol{r}|^2\mathrm{d}\boldsymbol{r} = \frac{1}{V(\mathcal{R})}\int_{\mathcal{R}} |\boldsymbol{r}|^2\mathrm{d}\boldsymbol{r}\,, \tag{4.2.8}$$

or in other words,[6] $E^{(N)}(\mathcal{C}) \approx E^{(N)}(\mathcal{R})$, with the obvious definition of $E(\mathcal{R})$ (right-hand side of (4.2.8)).

Measures of Performance Finally, we want to define important parameters for performance evaluation. Some of them have already been briefly discussed in the last section.

The *shaping gain* of some N-dimensional constellation $\mathcal{C}$ over a baseline hypercube constellation $\mathcal{C}_\square$, is the ratio of the respective average energies

$$G_\mathrm{s}(\mathcal{C}) \triangleq \frac{E(\mathcal{C}_\square)}{E(\mathcal{C})} \approx \frac{E(\mathcal{R}_\square)}{E(\mathcal{R})} = \frac{\mathsf{G}(\mathcal{R}_\square)}{\mathsf{G}(\mathcal{R})}\,, \tag{4.2.9}$$

if $\mathcal{R}_\square$ denotes the boundary region of $\mathcal{C}_\square$. Since, in continuous approximation, the volume is a measure for the number of signal points, which has to be equal for the constellations, we can use (4.2.5) and rewrite the shaping gain in terms of normalized second moments.

The *constellation expansion ratio* (CER) relates the number of signal points in the constituent constellation $\mathcal{C}_D$ to the number of points which is required to transmit the same rate by equiprobable signaling. Applying the continuous approximation we, have

$$\mathrm{CER}^{(D)}(\mathcal{C}) \triangleq \frac{|\mathcal{C}_D|}{|\mathcal{C}|^{D/N}} \approx \frac{V(\mathcal{R}_D)}{V(\mathcal{R})^{D/N}}\,. \tag{4.2.10}$$

Note, the constellation expansion ratio depends on the dimensionality of the constituent constellation, and, from the discussion above, we have $\mathrm{CER}^{(D)}(\mathcal{C}) \geq 1$. Alternatively, the shaping scheme can be characterized by a *shaping redundancy* which is given by the difference of maximum rate supported by the constituent constellation and the actual rate. For D dimensions, it calculates to $\log_2(\mathrm{CER}^{(D)}(\mathcal{C}))$.

[6] Note that in [FTC00] an *improved continuous approximation* is given, which reads $E^{(N)}(\mathcal{C}) = E^{(N)}(\mathcal{R}) - E^{(N)}(\boldsymbol{\Lambda})$, where $E(\boldsymbol{\Lambda})$ is calculated like $E(\mathcal{R})$, but replacing $\mathcal{R}$ by the Voronoi region of $\boldsymbol{\Lambda}$.

For power amplification, the *peak-to-average energy ratio* (PAR) is of importance. It is the ratio of peak energy of $\mathcal{C}_D$ to its average energy. Again using the continuous approximation, we may write

$$\mathrm{PAR}^{(D)}(\mathcal{C}) \triangleq \frac{\max_{\boldsymbol{c}\in\mathcal{C}_D} |\boldsymbol{c}|^2}{E^{(D)}(\mathcal{C})} \approx \frac{\max_{\boldsymbol{r}\in\mathcal{R}_D} |\boldsymbol{r}|^2}{E^{(D)}(\mathcal{R})} . \tag{4.2.11}$$

The peak-to-average energy ratio, usually expressed in dB, is always greater than 0 dB.

4.2.2 Performance of Shaping and Coding

With the above definitions in mind, we are now able to quantify what gains are possible by signal shaping and channel coding, and derive relations between the various parameters.

As a starting point, we consider the *error probability* when transmitting over an ISI-free AWGN channel with noise variance σ_n^2 per dimension. Using union bound techniques and considering only the first term (e.g., [Pro01]), the symbol error rate is readily estimated as

$$\mathrm{SER} \approx K_{\min} \cdot \mathrm{Q}\left(\sqrt{\frac{d^2_{\min}(\mathcal{C})}{2\sigma_n^2}}\right) . \tag{4.2.12}$$

Here, $d^2_{\min}(\mathcal{C}) = d^2_{\min}(\boldsymbol{\Lambda})$ is the minimum squared Euclidean distance between signal points (which is equal to the minimum squared distance of the lattice $\boldsymbol{\Lambda}$), and $K_{\min}$ denotes the (average) number of nearest-neighbor signal points—points at distance $d^2_{\min}(\mathcal{C})$—to any point in $\mathcal{C}$.

For a given constellation $\mathcal{C}$, minimum distance $d^2_{\min}(\mathcal{C})$ and average energy $E(\mathcal{C})$ are the primary quantities which determine the power efficiency in uncoded systems. Since $d^2_{\min}(\boldsymbol{\Lambda})$ should be as large as possible, whereas $E(\mathcal{C})$ should be as small as possible, it is beneficial to define a *constellation figure of merit (CFM)* [FW89]

$$\mathrm{CFM}(\mathcal{C}) \triangleq \frac{d^2_{\min}(\mathcal{C})}{E^{(1)}(\mathcal{C})} \approx \frac{d^2_{\min}(\boldsymbol{\Lambda})}{E^{(1)}(\mathcal{R})} . \tag{4.2.13}$$

Since numerator and denominator are both of type squared distance, CFM is dimensionless. By normalizing CFM to the transmission rate, a normalized minimum distance results, which is also sometimes used in the literature, e.g., [Bla90, page 131].

Baseline System The simplest situation is independent mapping of successive symbols, each drawn from the same regularly spaced one- or two-dimensional constellation. Hence, it is common to regard the integer lattice $\boldsymbol{\Lambda} = \mathbb{Z}^N$ with $d^2_{\min}(\mathbb{Z}^N) = 1$ and $V(\mathbb{Z}^N) = 1$ as baseline; moreover, the boundary region is assumed to be an N-cube.

If R denotes the rate per dimension, for the $\mathbb{Z}$ lattice the constituent boundary region $\mathcal{R}_\square$ is the interval $[-2^R/2,\, 2^R/2]$. Its volume and energy are given by

$V(\mathcal{R}_\Box) = 2^R$ and $E_\Box \triangleq E(\mathcal{R}_\Box) = 2^{-R}\int_{-2^R/2}^{2^R/2} r^2 \, \mathrm{d}r = 2^{2R}/12$, respectively. From (4.2.5), the normalized second moment of $\mathcal{R}_\Box$ is $\mathsf{G}_\Box \triangleq \mathsf{G}(\mathcal{R}_\Box) = 2^{2R}/12/2^{2R} = 1/12$.

Since the energy of the signal constellation $\mathcal{C}_\Box = \mathbb{Z} \cap \mathcal{R}_\Box$ calculates to $E(\mathcal{C}_\Box) = (2^{2R}-1)/12$, the baseline constellation figure of merit reads

$$\mathrm{CFM}_\Box \triangleq \mathrm{CFM}(\mathcal{C}_\Box) = \frac{12}{2^{2R}-1} \,. \tag{4.2.14}$$

The subscript "□" is intended to indicate the rectangular shape of the N-dimensional boundary region.

Gains by Coding and Shaping By using a signal constellation derived from some lattice $\boldsymbol{\Lambda}$ and a boundary region $\mathcal{R}$, the constellation figure of merit is increased relative to the baseline performance. Regarding the continuous approximation, we have [KP93]

$$\begin{aligned}\frac{\mathrm{CFM}(\boldsymbol{\Lambda})}{\mathrm{CFM}_\Box} &= \frac{d^2_{\min}(\mathcal{R})}{E^{(1)}(\mathcal{R})} \cdot \frac{2^{2R}-1}{12} \\ &= \frac{d^2_{\min}(\boldsymbol{\Lambda})}{V(\boldsymbol{\Lambda})^{2/N}} \cdot \frac{V(\boldsymbol{\Lambda})^{2/N}2^{2R}}{12E^{(1)}(\mathcal{R})} \cdot (1-2^{-2R}) \\ &\triangleq G_{\mathrm{c}}(\boldsymbol{\Lambda}) \cdot G_{\mathrm{s}}(\mathcal{R}) \cdot G_{\mathrm{d}}(R) \,.\end{aligned} \tag{4.2.15}$$

Hence, the performance advantage can be expressed in terms of three items:

- The first factor is the *coding gain* of the lattice $\boldsymbol{\Lambda}$ [FW89]

$$G_{\mathrm{c}}(\boldsymbol{\Lambda}) \triangleq \frac{d^2_{\min}(\boldsymbol{\Lambda})}{V(\boldsymbol{\Lambda})^{2/N}} \,. \tag{4.2.16}$$

  This gain is due to the internal arrangement of the signal points and is solely a parameter of the underlying lattice $\boldsymbol{\Lambda}$.

- The second term is the *shaping gain*, cf. (4.2.9), of the region $\mathcal{R}$ over an N-cube with $\mathsf{G}_\Box = 1/12$. With regard to (4.2.7), $V(\mathcal{R}) = 2^{NR}V(\boldsymbol{\Lambda})$, and (4.2.5), the definition of the normalized second moment, we have

$$G_{\mathrm{s}}(\mathcal{R}) \triangleq \frac{V(\boldsymbol{\Lambda})^{2/N}2^{2R}}{12E^{(1)}(\mathcal{R})} = \frac{V(\mathcal{R})^{2/N}}{12E^{(1)}(\mathcal{R})} = \frac{1}{12\mathsf{G}(\mathcal{C})} = \frac{E_\Box}{E(\mathcal{R})} \,. \tag{4.2.17}$$

  Interestingly, this factor only depends on the shape of the boundary region. Inserting the energy of the baseline system $E_\Box$, alternative expressions for the shaping gain read

$$G_{\mathrm{s}}(\mathcal{R}) = \frac{2^{2R}}{12E^{(1)}(\mathcal{R})} = \frac{2^{R^{(2)}}}{6E^{(2)}(\mathcal{R})} \,. \tag{4.2.18}$$

- Finally, the third factor, usually ignored in the literature, is a *discretization factor* [KP93]
$$G_{\mathrm{d}}(R) \triangleq (1 - 2^{-2R}) \,. \tag{4.2.19}$$
It can be considered as a quantization loss due to the approximation of a continuous distribution by a discrete distribution with entropy R per dimension. In terms of one-dimensional signaling, it is the energy ratio of a 2^R-ary equiprobable constellation and a continuous, uniform distribution.

For "large" constellations ($R \to \infty$), the discretization factor can be neglected as it tends to one. Then, from (4.2.15), the total gain by signal design is simply the sum (in dB) of coding and shaping gain. Hence, asymptotically we can regard coding and shaping as separable.

Using the continuous approximation, we express the constellation expansion ratio as

$$\begin{aligned}
\mathrm{CER}^{(D)}(\mathcal{C}) &= \frac{|\mathcal{C}_D|}{|\mathcal{C}|^{D/N}} = \frac{V(\mathcal{R}_D)/V(\mathbf{\Lambda}_D)}{(V(\mathcal{R})/V(\mathbf{\Lambda}))^{D/N}} \\
&= \frac{V(\mathbf{\Lambda})^{D/N}}{V(\mathbf{\Lambda}_D)} \cdot \frac{V(\mathcal{R}_D)}{V(\mathcal{R})^{D/N}} \\
&\triangleq \mathrm{CER}_{\mathrm{c}}^{(D)} \cdot \mathrm{CER}_{\mathrm{s}}^{(D)} \,.
\end{aligned} \tag{4.2.20}$$

The constellation expansion is thus determined by two independent factors. On the one hand, $\mathrm{CER}_{\mathrm{c}}^{(D)}$ is the expansion due to channel coding, and on the other hand, $\mathrm{CER}_{\mathrm{s}}^{(D)}$ is that caused by signal shaping.

The peak-to-average energy ratio in D dimensions can be split into factors as follows

$$\begin{aligned}
\mathrm{PAR}^{(D)}(\mathcal{C}) &= \frac{\max_{\mathbf{c}\in\mathcal{C}_D} |\mathbf{c}|^2}{E^{(D)}(\mathcal{C})} = \frac{\max_{\mathbf{r}\in\mathcal{R}_D} |\mathbf{r}|^2}{E^{(D)}(\mathcal{R}_D)} \frac{E^{(D)}(\mathcal{R}_D)}{E^{(D)}(\mathcal{R})} \\
&= \mathrm{PAR}^{(D)}(\mathcal{R}_D) \cdot \frac{\mathsf{G}(\mathcal{R}_D) V(\mathcal{R}_D)^{2/D}}{\mathsf{G}(\mathcal{R}) V(\mathcal{R})^{2/N}} \\
&= \frac{\mathrm{PAR}^{(D)}(\mathcal{R}_D)}{G_{\mathrm{s}}(\mathcal{R}_D)} \cdot G_{\mathrm{s}}(\mathcal{R}) \cdot \left(\frac{V(\mathcal{R}_D)}{V(\mathcal{R})^{D/N}} \right)^{2/D} \\
&= \frac{\mathrm{PAR}^{(D)}(\mathcal{R}_D)}{G_{\mathrm{s}}(\mathcal{R}_D)} \cdot G_{\mathrm{s}}(\mathcal{R}) \cdot \left(\mathrm{CER}^{(D)}(\mathcal{R}) \right)^{2/D} \,.
\end{aligned} \tag{4.2.21}$$

The peak-to-average energy ratio thus depends on (i) the PAR of the D-dimensional constituent region $\mathcal{R}_D$, lowered by the shaping gain of this region, i.e., a factor dependent only on the constituent region, (ii) the shaping gain achieved by the N-dimensional region $\mathcal{R}$, and (iii) the constellation expansion ratio of the region $\mathcal{R}$ in D dimensions to the power of $2/D$. Since the peak-to-average energy ratio should be as low as possible, this relationship suggests using boundary regions for shaping whose D-dimensional constituent constellation has (a) a low PAR, and (b) a low constellation expansion.

4.2.3 Shaping Properties of Hyperspheres

In the introduction to shaping we have seen that the best choice for the N-dimensional boundary region, from which the signal points are selected equiprobably, is a hypersphere. Again, an N-cube constitutes the baseline, and hence the maximum shaping gain possible in N dimensions reads from (4.2.17)

$$G_{s,\circ}(N) = \frac{1}{12\mathsf{G}_\circ(N)}\,, \tag{4.2.22}$$

where $\mathsf{G}_\circ(N)$ denotes the normalized second moment of the N-sphere and the subscript "○" stands for sphere (circle in two dimensions).

To calculate $\mathsf{G}_\circ(N)$, we first note a useful integral equation [GR80, Equation 4.642]

$$\underset{\sum_{i=1}^N x_i^2 \le r^2}{\int \cdots \int} f(\sqrt{x_1^2 + \cdots + x_N^2}\,)\; \mathrm{d}x_1 \cdots \mathrm{d}x_N = \frac{2\cdot \pi^{N/2}}{\Gamma(N/2)} \int_0^r x^{N-1} f(x)\; \mathrm{d}x\,, \tag{4.2.23}$$

where $\Gamma(x) = \int_0^\infty \mathrm{e}^{-t} t^{x-1}\, \mathrm{d}t$ is the Gamma function [BS98].

Choosing $f(x) = 1$ and considering $x \cdot \Gamma(x) = \Gamma(x+1)$, the volume $V_\circ(N)$ of an N-sphere with radius $r_\circ$ calculates to

$$\begin{aligned} V_\circ(N) &= \underset{\sum_{i=1}^N x_i^2 \le r^2}{\int \cdots \int} \mathrm{d}x_1 \cdots \mathrm{d}x_N = \frac{2\cdot \pi^{N/2}}{\Gamma(N/2)} \int_0^{r_\circ} x^{N-1}\, \mathrm{d}x \\ &= \frac{\pi^{N/2}\cdot r_\circ^N}{\Gamma(N/2+1)}\,. \end{aligned} \tag{4.2.24}$$

The average energy $E_\circ(N)$ of the N-sphere is obtained by setting $f(x) = x^2/V_\circ(N)$, which leads to

$$\begin{aligned} E_\circ(N) &= \frac{1}{V_\circ(N)} \cdot \underset{\sum_{i=1}^N x_i^2 \le r_\circ^2}{\int \cdots \int} (x_1^2 + \cdots + x_N^2)\; \mathrm{d}x_1 \cdots \mathrm{d}x_N \\ &= \frac{1}{V_\circ(N)} \cdot \frac{2\cdot \pi^{N/2}}{\Gamma(N/2)} \int_0^{r_\circ} x^{N+1}\, \mathrm{d}x = \frac{1}{V_\circ(N)} \cdot \frac{N\cdot \pi^{N/2}\cdot r_\circ^{N+2}}{(N+2)\cdot \Gamma(N/2+1)} \\ &= \frac{N \cdot r_\circ^2}{N+2}\,. \end{aligned} \tag{4.2.25}$$

In summary, the normalized second moment of the N-sphere reads

$$\mathsf{G}_\circ(N) = \frac{E_\circ(N)/N}{V_\circ^{2/N}(N)} = \frac{r_\circ^2}{N+2} \cdot \left(\frac{\Gamma(N/2+1)}{\pi^{N/2}\cdot r_\circ^N}\right)^{2/N} = \frac{\Gamma^{2/N}(N/2+1)}{(N+2)\cdot \pi}\,. \tag{4.2.26}$$

Inserting (4.2.26) into (4.2.22), the shaping gain of an N-sphere over an N-cube is readily obtained as

$$G_{\mathrm{s,o}}(N) = \frac{\pi(N+2)}{12 \cdot \Gamma^{2/N}(N/2+1)} \,. \tag{4.2.27}$$

Theorem 4.2: Maximum Shaping Gain in N Dimensions

The shaping gain in N dimensions, i.e., when considering blocks of N ($N/2$) consecutive one-(two-)dimensional symbols, is maximum when bounding the N-dimensional signal constellation by a *hypersphere*, and calculates to

$$G_{\mathrm{s,o}}(N) = \frac{\pi(N+2)}{12 \cdot \Gamma^{2/N}(N/2+1)} \,. \tag{4.2.28}$$

Here, $\Gamma(x) = \int_0^\infty \mathrm{e}^{-t} t^{x-1} \, dt$ is the Gamma function.

Asymptotic Shaping Gain In order to further evaluate the asymptotic shaping gain, i.e., $G_{\mathrm{s,o}}(N)$ as $N \to \infty$, we approximate the Gamma function by applying Stirling's formula [BS98], namely

$$\Gamma(x+1) \approx \left(\frac{x}{\mathrm{e}}\right)^x \,, \tag{4.2.29}$$

which becomes exact as the argument tends to infinity. Using (4.2.29), we arrive at

$$G_{\mathrm{s,o}}(N) \approx \frac{\pi(N+2)}{12\left(\frac{N/2}{\mathrm{e}}\right)} = \frac{\pi\mathrm{e}(N+2)}{6N} \,, \tag{4.2.30}$$

which converges to the *ultimate shaping gain*

$$G_{\mathrm{s,o}}(N) \overset{N\to\infty}{\longrightarrow} \frac{\pi\mathrm{e}}{6} \,. \tag{4.2.31}$$

Figure 4.5 plots the shaping gain of an N-sphere over the dimensionality N. Additionally, the ultimate shaping gain $\frac{\pi\mathrm{e}}{6} \mathrel{\widehat{=}} 1.53$ dB is shown. Note that the shaping gain in two dimensions (circle over square) is $\pi/3 \mathrel{\widehat{=}} 0.2$ dB, and already for $N = 16$ a gain of about 1 dB is possible. However, going to larger dimensions, the ultimate shaping gain is approached rather slowly.

Density Induced on the Constituent Constellation Although the signal points are chosen uniformly in N dimensions, the points of the constituent constellation have different probabilities, see, e.g., Figure 4.1. We now calculate the pdf of the signal in D dimensions, when bounding the N-dimensional constellation within

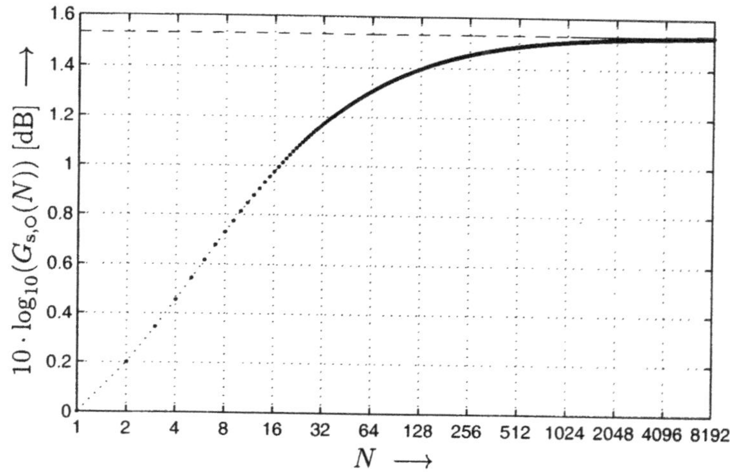

Fig. 4.5 Shaping gain of N-spheres over N-cubes over the dimensionality N. Dashed: ultimate shaping gain.

a hypersphere. Because a sphere is rotationally invariant, any D dimensions can be regarded—here without loss of generality we pick the first D ones.

In order to obtain clear results, we assume the volume to be normalized to one. Generalization is easily possible by scaling. The reference scheme, a hypercube with unit volume, then has a one-dimensional projection, uniformly over the interval $[-1/2, 1/2]$. From (4.2.24), the radius of the sphere thus has to be adjusted as

$$V_{\mathrm{O}}(N) \stackrel{!}{=} 1 \quad \Rightarrow \quad r_{\mathrm{O}}^2(N) = \frac{\Gamma^{2/N}(N/2+1)}{\pi} . \tag{4.2.32}$$

Letting $\boldsymbol{x} \triangleq [x_1, x_2, \ldots, x_D]^{\mathsf{T}}$ be a vector comprising the first D coordinates, the pdf of $\boldsymbol{x}$ is given as the D-dimensional *projection* of the N-sphere with radius $r_{\mathrm{O}}^2(N)$. We arrive at

$$\begin{aligned}
\mathrm{f}_{\boldsymbol{x}}(\boldsymbol{x}) &= \begin{cases} \frac{1}{V_{\mathrm{O}}(N)} \displaystyle\int \cdots \int_{\sum_{i=D+1}^{N} x_i^2 \le r_{\mathrm{O}}^2(N) - |\boldsymbol{x}|^2} \mathrm{d}x_{D+1} \cdots \mathrm{d}x_N, & |\boldsymbol{x}|^2 \le r_{\mathrm{O}}^2(N) \\ 0, & \text{else} \end{cases} \\
&= \frac{2 \cdot \pi^{(N-D)/2}}{\Gamma((N-D)/2)} \cdot \int_0^{\sqrt{r_{\mathrm{O}}^2(N) - |\boldsymbol{x}|^2}} x^{N-D-1} \, \mathrm{d}x \\
&= \frac{2 \cdot \pi^{(N-D)/2}}{\Gamma((N-D)/2)} \cdot \frac{1}{N-D} \left(r_{\mathrm{O}}^2(N) - |\boldsymbol{x}|^2 \right)^{(N-D)/2} \\
&= \frac{\pi^{(N-D)/2}}{(N-D)/2 \cdot \Gamma((N-D)/2)} \cdot \left(\frac{\Gamma^{2/N}(N/2+1)}{\pi} - |\boldsymbol{x}|^2 \right)^{(N-D)/2} .
\end{aligned} \tag{4.2.33}$$

Asymptotic Distribution From (4.2.33) the asymptotic distribution, as $N \to \infty$, can be obtained if the Gamma function is again approximated using Stirling's formula (4.2.29). We have

$$\begin{aligned}
\mathsf{f}_{\boldsymbol{x}}(\boldsymbol{x}) &\approx \frac{\pi^{(N-D)/2}}{((N-D)/(2\mathrm{e}))^{(N-D)/2}} \cdot \left(\frac{1}{\pi} \left((N/(2\mathrm{e}))^{N/2} \right)^{2/N} - |\boldsymbol{x}|^2 \right)^{(N-D)/2} \\
&= \left(\frac{2\pi\mathrm{e}}{N-D} \right)^{(N-D)/2} \cdot \left(\frac{N}{2\pi\mathrm{e}} - |\boldsymbol{x}|^2 \right)^{(N-D)/2} \\
&= \left(\frac{N - 2\pi\mathrm{e}|\boldsymbol{x}|^2}{N-D} \right)^{(N-D)/2} \\
&= \left(1 + \frac{D/2 - \pi\mathrm{e}|\boldsymbol{x}|^2}{(N-D)/2} \right)^{(N-D)/2} .
\end{aligned} \tag{4.2.34}$$

Using $\lim_{N\to\infty} \left(1 + \frac{x}{N}\right)^N = \lim_{N\to\infty} \left(1 + \frac{1}{N}\right)^{Nx} = \mathrm{e}^x$, the definition of the exponential function [BS98], the distribution converges to

$$\begin{aligned}
\mathsf{f}_{\boldsymbol{x}}(\boldsymbol{x}) &\overset{N\to\infty}{\to} \mathrm{e}^{D/2 - \pi\mathrm{e}|\boldsymbol{x}|^2} \\
&= \mathrm{e}^{D/2} \cdot \mathrm{e}^{-\pi\mathrm{e}|\boldsymbol{x}|^2} \\
&= \left(\frac{1}{\sqrt{2\pi\sigma_{\bigcirc}^2}} \right)^D \cdot \mathrm{e}^{-|\boldsymbol{x}|^2/(2\sigma_{\bigcirc}^2)} , \qquad \sigma_{\bigcirc}^2 = \frac{1}{2\pi\mathrm{e}} .
\end{aligned} \tag{4.2.35}$$

As the dimensionality of the N-sphere goes to infinity, any projection to D ($D < N$) dimensions converges to a D-dimensional Gaussian distribution. The variance per dimension is then $\sigma_{\bigcirc}^2 = \frac{1}{2\pi\mathrm{e}}$. Compared to the baseline system with a uniformly distributed signal in $[-1/2, 1/2]$ and variance $\sigma_{\square}^2 = 1/12$ per dimension, the ultimate shaping gain $\frac{1/12}{1/(2\pi\mathrm{e})} = \frac{\pi\mathrm{e}}{6}$ is again visible.

This result, moreover, is in perfect agreement with Section 4.1.3, where we derived the ultimate shaping gain by comparing differential entropies of uniform and Gaussian distributions.

Finally, in Figure 4.6, the evolution of the one-dimensional projection of an N-sphere is plotted. To emphasize the energy reduction, and hence the increasing shaping gain, dashed lines at $\pm\sigma$, the standard deviation of the distribution, are drawn. The dimensionality of the hypersphere ranges from 1 (= interval) to infinity, where a perfect Gaussian distribution results.

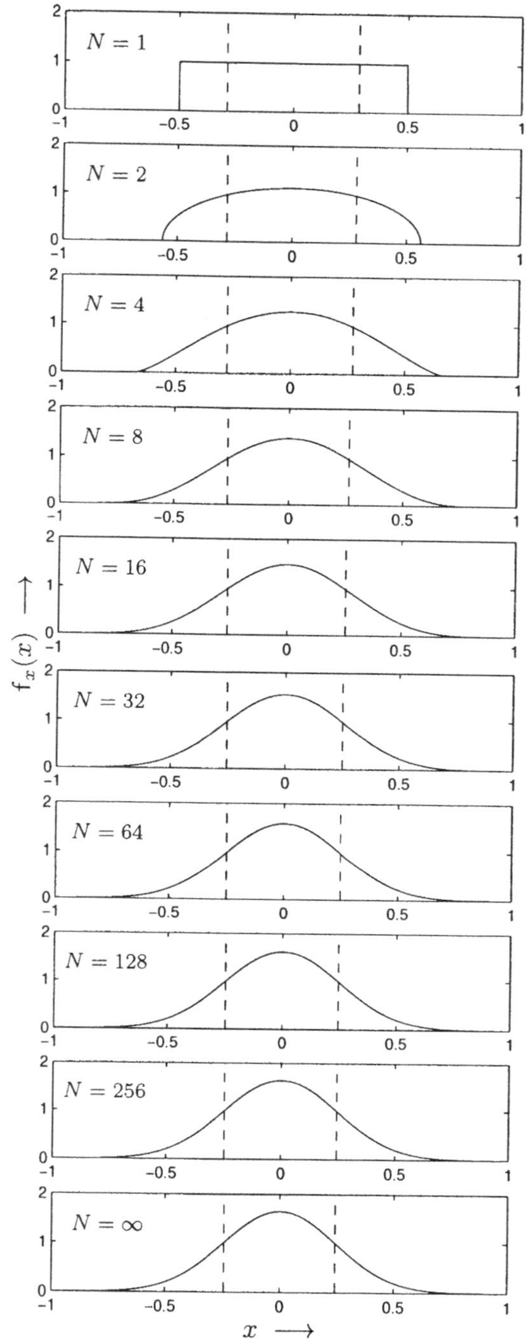

Fig. 4.6 Evolution of the one-dimensional projection of an N-sphere with unit volume. Dashed: $\pm$ standard deviation of the distribution.

Performance Parameters To conclude this section on the properties of hyperspheres, we give their constellation expansion ratio and the peak-to-average energy ratio.

Any projection of an N-sphere with radius $r_{\rm o}$ to any D dimensions again gives a hypersphere of radius $r_{\rm o}$. Since its volume is given by (4.2.24), using (4.2.10), the constellation expansion ratio calculates to

$$\mathrm{CER}_{\rm o}^{(D)}(N) = \frac{\frac{\pi^{D/2} \cdot r_{\rm o}^D}{\Gamma(D/2+1)}}{\left(\frac{\pi^{N/2} \cdot r_{\rm o}^N}{\Gamma(N/2+1)}\right)^{D/N}} = \frac{\Gamma^{D/N}(N/2+1)}{\Gamma(D/2+1)} \,. \tag{4.2.36}$$

In particular for $D = 1$, we have

$$\mathrm{CER}_{\rm o}^{(1)}(N) \;=\; \frac{2}{\sqrt{\pi}} \cdot \Gamma^{1/N}(N/2+1) \;\approx\; \mathrm{const} \cdot \sqrt{N} \,, \tag{4.2.37}$$

where the approximation results when Stirling's formula is applied. Similarly, for $D = 2$, we have

$$\mathrm{CER}_{\rm o}^{(2)}(N) \;=\; \Gamma^{2/N}(N/2+1) \qquad \approx\; \mathrm{const} \cdot N \,. \tag{4.2.38}$$

For achieving large shaping gain, the dimensionality N has to be high. Unfortunately, at the same time, the constellation expansion of the D-dimensional constituent constellations grows according to $N^{D/2}$. To approach the ultimate shaping gain, an infinite expansion results, which is obvious, since the induced pdf is Gaussian.

The peak-to-average energy ratio can be calculated according to (4.2.11), leading to

$$\mathrm{PAR}_{\rm o}^{(D)}(N) = \frac{r_{\rm o}^2}{\frac{D}{N} E_{\rm o}(N)} = \frac{\left(\frac{\Gamma(N/2+1)}{\pi^{N/2}}\right)^{2/N}}{D\frac{(\Gamma(N/2+1))^{2/N}}{\pi(N+2)}} = \frac{N+2}{D} \,. \tag{4.2.39}$$

Here, for $D = 1$ we have $\mathrm{PAR}_{\rm o}^{(1)}(N) = N+2$, and for $D = 2$ the PAR reads $\mathrm{PAR}_{\rm o}^{(2)}(N) = \frac{N}{2}+1$. As with the constellation expansion, for $D < N$, the peak-to-average energy ratio also tends to infinity as the dimensionality N grows. Interestingly, the PAR in N dimensions tends to one, i.e., asymptotically all points are concentrated in close proximity of the surface of the sphere [FW89]; a phenomenon known as *sphere hardening*.

Combing equations (4.2.27), (4.2.36), and (4.2.39) on shaping gain, constellation expansion, and peak-to-average energy ratio of an N-sphere, the exchange between these parameters can be evaluated. Figure 4.7 shows the exchange of constellation expansion and shaping gain. In Figure 4.8, the trade-off between PAR and shaping gain is depicted. Both figures are valid for two-dimensional ($D = 2$) constituent constellations. The circles mark $N = 2, 4, 8, 16, 32, 64, 128$, and 256; the square marks the performance of the reference N-cube.

Already for shaping gains in the region of 1 dB, the constituent 2-dimensional constellation has to be expanded by a factor of 4. Larger gains require enormous

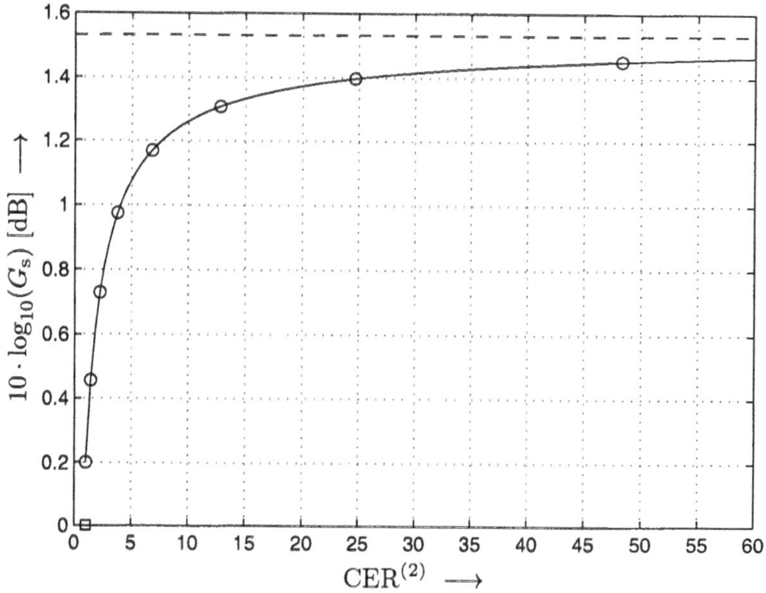

Fig. 4.7 Shaping gain versus two-dimensional constellation expansion ratio of an N-sphere. The circles mark $N = 2, 4, 8, 16, 32, 64, 128$, and 256; the square marks the performance of the reference N-cube.

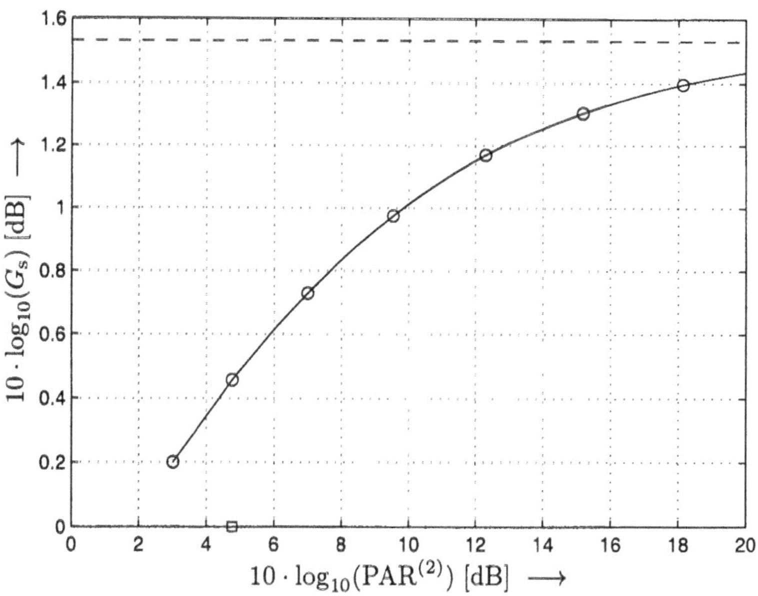

Fig. 4.8 Shaping gain versus two-dimensional peak-to-average energy ratio of an N-sphere. The circles mark $N = 2, 4, 8, 16, 32, 64, 128$, and 256; the square marks the performance of the reference N-cube.

expansions. The same is true for the peak-to-average energy ratio. Shaping gains of 1 dB are accompanied by a PAR of 10 dB, which has to be compared to the PAR of 4.77 dB of an uniform distribution. Note that the choice of $N = 2$, i.e., a two-dimensional circular constellation, offers both shaping gain of 0.2 dB and a gain in PAR (3 dB instead of 4.77 dB) without any two-dimensional constellation expansion, cf. Figure 4.1.

Although an N-sphere offers the maximum shaping gain, it comes along with undesirably large constellation expansion and peak-to-average energy ratio. Hence, we have to think of strategies to lower CER and PAR significantly, while sacrificing only a little shaping gain. This point is addressed in the sequel.

4.2.4 Shaping Under a Peak Constraint

In the above discussion we have seen that spherical boundary regions provide the maximum shaping gain for given dimensionality. Unfortunately, both constellation expansion and peak-to-average energy ratio become undesirably large. The solution to this problem is to impose restrictions on the constituent constellation.

Figure 4.9 sketches two-dimensional boundary regions. On the left-hand side, the

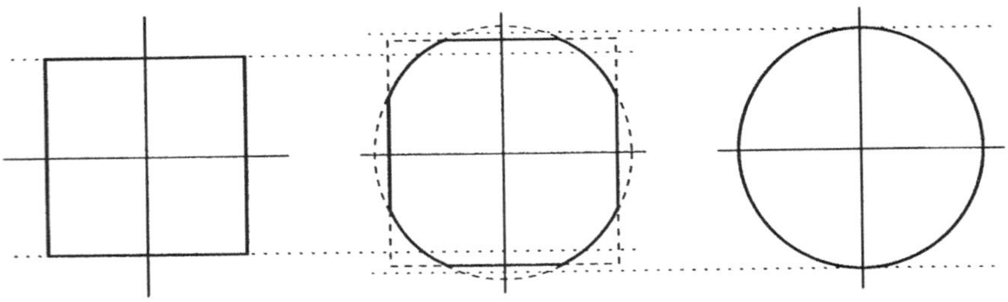

Fig. 4.9 Illustration of shaping under a peak constraint.

two-dimensional square region is the two-fold Cartesian product of a one-dimensional interval. Here, no shaping gain is achieved, but $\text{CER}^{(1)}$ and $\text{PAR}^{(1)}$ are as low as possible. On the right-hand side of Figure 4.9, a spherical boundary region, optimal with respect to shaping gain is shown. If we now restrict the one-dimensional constituent constellation $\mathcal{R}_1$, the signal points in two dimensions are restricted to the region $\mathcal{R}_1^2$. The optimal selection of signal points from $\mathcal{R}_1^2$ obviously falls within a circle, whose radius is adjusted so that all boundary regions have the same area. In the middle of Figure 4.9 such a boundary region, optimal under a peak energy constraint in one dimension is sketched. By varying the peak limit, a trade-off between shaping gain and $\text{CER}^{(1)}$, $\text{PAR}^{(1)}$ over the whole range from that of a square to that of a circle is possible.

This illustration shows the general principle. Here, we again focus on two-dimensional constellations; thus we prescribe the two-dimensional constituent constellation $\mathcal{R}_2$. The signal points in N dimensions, N even, are then restricted to the region $\mathcal{R}_2^{N/2}$. From that region, a subset is selected as the actual constellation. It is intuitively clear that for fixed region $\mathcal{R}_2$, and thus fixed $\mathrm{CER}^{(2)}$, the minimum-energy selection of signal points falls within an N-sphere. The N-dimensional boundary region is thus the intersection of the $N/2$-fold Cartesian product of $\mathcal{R}_2$ and an N-sphere. Moreover, since $\mathrm{PAR}^{(2)}$ should be as low as possible (cf. (4.2.21)), the two-dimensional constituent constellation $\mathcal{R}_2$ itself has to have a PAR as low as possible. Hence, $\mathcal{R}_2$ is chosen to be a circle. A rigorous mathematical proof that this construction leads to the optimal trade-off between shaping gain G_s and $\mathrm{CER}^{(2)}$, $\mathrm{PAR}^{(2)}$ is given, e.g., in [KK93, KP94, LFT94]. Such regions are sometimes called *truncated polydiscs* [KP94].

Performance Parameters Let r_2 be the radius of the two-dimensional constituent circular constellation, and r_N the radius of the N-sphere. The N-dimensional boundary region is then formally given as

$$\mathcal{R}_{\mathbf{o}} = \left\{ \begin{bmatrix} x_1 \\ \vdots \\ x_N \end{bmatrix} \,\middle|\, x_{2p-1}^2 + x_{2p}^2 \le r_2^2,\ p = 1, \ldots, N/2,\ \sum\nolimits_{i=1}^{N} x_i^2 \le r_N^2, \right\} . \tag{4.2.40}$$

The above-mentioned trade-off is now possible via the parameter $\rho = r_N^2/r_2^2$, i.e., the ratio of the radii in N and in two dimensions. For $\rho = N/2$ only the peak constraint in two dimensions is active, i.e., $\mathcal{R}_{\mathbf{o}}$ is the Cartesian product of two-dimensional discs, whereas for $\rho = 1$ only the N-dimensional energy constraint remains and a hypersphere results for $\mathcal{R}_{\mathbf{o}}$.

For calculating the performance parameters, we note a useful integral equation, which supersedes (4.2.23), valid for N-spheres. In [KK93] (see also [KP94]) the following equation is derived

$$\begin{aligned} &\int_{\mathcal{R}_{\mathbf{o}}} \cdots \int f(x_1^2 + \cdots + x_N^2)\, \mathrm{d}x_1 \cdots \mathrm{d}x_N \qquad (4.2.41) \\ &\quad = (\pi r_2^2)^n \sum_{k=0}^{\lfloor \rho \rfloor} (-1)^k \binom{n}{k} \frac{(\rho - k)^n}{(n-1)!} \int_0^1 f(r_2^2((\rho - k)x + k)x^{n-1}\, \mathrm{d}x\ , \end{aligned}$$

where $n \triangleq N/2$. Choosing $f(x) = 1$, the volume $V_{\mathbf{o}}(N, \rho)$ of $\mathcal{R}_{\mathbf{o}}$ calculates to ($\lfloor x \rfloor$: largest integer not exceeding x)

$$\begin{aligned} V_{\mathbf{o}}(N, \rho) &= \int_{\mathcal{R}_{\mathbf{o}}} \cdots \int \mathrm{d}x_1 \cdots \mathrm{d}x_N \\ &= \frac{(\pi r_2^2)^n}{n(n-1)!} \sum_{k=0}^{\lfloor \rho \rfloor} (-1)^k \binom{n}{k} (\rho - k)^n\ , \end{aligned} \tag{4.2.42}$$

and the average energy $E_{\mathbf{o}}(N, \rho)$ per two dimensions is obtained by setting $f(x) = \frac{2}{N} \frac{x}{V_{\mathbf{o}}(N,\rho)}$, which leads to

$$E_{\mathbf{o}}(N, \rho) = \frac{2}{N} \frac{1}{V_{\mathbf{o}}(N, \rho)} \int_{\mathcal{R}_N} (x_1^2 + \cdots + x_N^2)\, \mathrm{d}x_1 \cdots \mathrm{d}x_N$$

$$= \frac{(\pi r_2^2)^n}{n(n-1)!}\frac{1}{V_{\mathbf{o}}(N,\rho)}\sum_{k=0}^{\lfloor\rho\rfloor}(-1)^k\binom{n}{k}(\rho-k)^n r_2^2\left(\frac{\rho-k}{n+1}+\frac{k}{n}\right)$$

$$= \frac{r_2^2\sum_{k=0}^{\lfloor\rho\rfloor}(-1)^k\binom{n}{k}(\rho-k)^n\left(\frac{\rho-k}{n+1}+\frac{k}{n}\right)}{\sum_{k=0}^{\lfloor\rho\rfloor}(-1)^k\binom{n}{k}(\rho-k)^n} . \tag{4.2.43}$$

From that, using (4.2.17), the shaping gain reads

$$G_{\mathrm{s},\mathbf{o}}(N,\rho) \triangleq G_{\mathrm{s}}(\mathcal{R}_{\mathbf{o}}) = \frac{V_{\mathbf{o}}^{2/N}(N,\rho)}{6\,E_{\mathbf{o}}(N,\rho)} ; \tag{4.2.44}$$

with regard to (4.2.10), the constellation expansion ratio is given by

$$\mathrm{CER}_{\mathbf{o}}^{(2)}(N,\rho) \triangleq \mathrm{CER}^{(2)}(\mathcal{R}_{\mathbf{o}}) = \frac{\pi r_2^2}{V_{\mathbf{o}}^{2/N}(N,\rho)} ; \tag{4.2.45}$$

and with (4.2.11), the peak-to-average energy ratio results in

$$\mathrm{PAR}_{\mathbf{o}}^{(2)}(N,\rho) \triangleq \mathrm{PAR}^{(2)}(\mathcal{R}_{\mathbf{o}}) = \frac{r_2^2}{E_{\mathbf{o}}(N,\rho)} . \tag{4.2.46}$$

Asymptotic Density As in the case of hyperspheres above, the probability density induced on the constituent constellation can be derived. Lengthy integration reveals that for $N \to \infty$ a Gaussian distribution, truncated to the constituent circular region results [KK93, FU98]. Asymptotically, this pdf, which enables an optimal trade-off between energy and entropy (via λ), reads for pairs of coordinates $\boldsymbol{x} \triangleq [x_{2p-1},\, x_{2p}]^{\mathsf{T}}$, $p = 1, \ldots, N/2$

$$\mathsf{f}_{\boldsymbol{x}}(\boldsymbol{x}) = \begin{cases} K(\lambda)\cdot \mathrm{e}^{-\lambda|\boldsymbol{x}|^2/r_2^2}, & |\boldsymbol{x}|^2 \le r_2^2 \\ 0, & \text{else} \end{cases} . \tag{4.2.47}$$

Defining [FW89]

$$C_1(\lambda) \triangleq 1-\mathrm{e}^{-\lambda}, \qquad C_2(\lambda) = 1-\mathrm{e}^{-\lambda}-\lambda\mathrm{e}^{-\lambda} , \tag{4.2.48}$$

the normalization factor $K(\lambda)$ calculates to $K(\lambda) = \lambda/(\pi r_2^2 C_1(\lambda))$, and using (4.1.6) and (4.1.7), average energy and entropy per two dimensions are finally obtained as

$$E^{(2)}(\lambda) = \frac{r_2^2 C_2(\lambda)}{\lambda C_1(\lambda)} , \tag{4.2.49}$$

and

$$H^{(2)}(\lambda) = -\log_2\left(\frac{\lambda}{\pi r_2^2 C_1(\lambda)}\right) + \frac{C_2(\lambda)}{C_1(\lambda)}\log_2(\mathrm{e}) . \tag{4.2.50}$$

From (4.2.18) the shaping gain of a truncated two-dimensional Gaussian distribution compared to the reference system transmitting $H^{(2)}(\lambda)$ by uniform signaling calculates to [FW89]

$$G_s(\lambda) = \frac{2^{H^{(2)}(\lambda)}}{6E^{(2)}(\lambda)} = \frac{\pi C_1^2(\lambda)}{6C_2(\lambda)} \cdot \mathrm{e}^{C_2(\lambda)/C_1(\lambda)} , \qquad (4.2.51)$$

the constellation expansion ratio reads

$$\mathrm{CER}^{(2)}(\lambda) = \frac{\pi r_2^2}{2^{H^{(2)}(\lambda)}} = \frac{\lambda}{C_1(\lambda)} \cdot \mathrm{e}^{-C_2(\lambda)/C_1(\lambda)} , \qquad (4.2.52)$$

and the peak-to-average energy ratio is given by

$$\mathrm{PAR}^{(2)}(\lambda) = \frac{r_2^2}{E^{(2)}(\lambda)} = \frac{\lambda C_1(\lambda)}{C_2(\lambda)} . \qquad (4.2.53)$$

Using the quantities derived above, Figures 4.10 and 4.11 plot the exchange between constellation expansion ratio, respectively, peak-to-average energy ratio and shaping gain. From bottom to top, the curves correspond to $N = 4, 8, 16, 32, 64, 128$, and the asymptotic case of a truncated Gaussian distribution.

For $\rho = 1$, i.e., $r_N^2 = r_2^2$, no peak limitation is active and the boundary region is an N-sphere. These points are marked by circles, and the trade-offs for spheres are repeated from Figures 4.7 and 4.8, respectively. Conversely, as $r_N^2 = \frac{N}{2} r_2^2$, the boundary region simply becomes the Cartesian product of two-dimensional discs (called a polydisc in [KP94]), and the gain reduces to $\pi/3$, i.e., that of a 2-sphere. The upper curve is valid as $N \to \infty$, i.e., for a truncated Gaussian distribution. Here, the exchange is governed by the parameter λ (the shaping gain is monotonic increasing with λ). Compared to the N-sphere, larger shaping gains are possible for much lower $\mathrm{CER}^{(2)}$ and $\mathrm{PAR}^{(2)}$.

Of course, for a finite number N of dimensions, the shaping gain is limited to that of an N-sphere; but almost all of this gain is possible with much lower $\mathrm{CER}^{(2)}$ and $\mathrm{PAR}^{(2)}$. It is visible that for $\mathrm{CER}^{(2)} = 2$ (points marked with "+"), the loss in shaping gain is negligible. Even for $\mathrm{CER}^{(2)} = \sqrt{2}$ ("×" points) significant shaping gains can be achieved; here the $\mathrm{PAR}^{(2)}$ is approximately 2 dB lower than for $\mathrm{CER}^{(2)} = 2$. For practice we can summarize that 0.5 bit redundancy per dimension corresponding to $\mathrm{CER}^{(2)} = 2$ is by far sufficient for shaping, cf. also [WFH99].

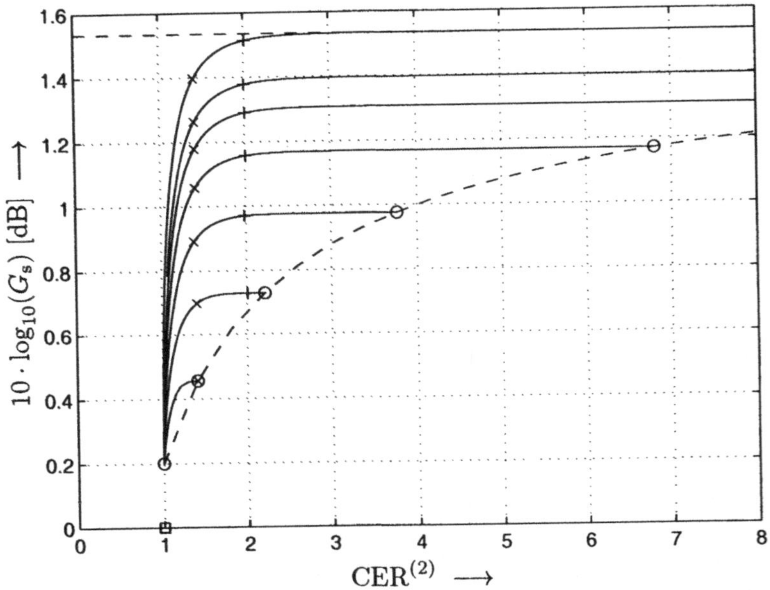

Fig. 4.10 Shaping gain versus two-dimensional constellation expansion ratio of truncated N-spheres. The circles mark $N = 2, 4, 8, 16, \ldots$; the square marks the reference N-cube. The points marked with $\times$ and $+$ correspond to $\mathrm{CER}^{(2)} = \sqrt{2}$ and 2, respectively.

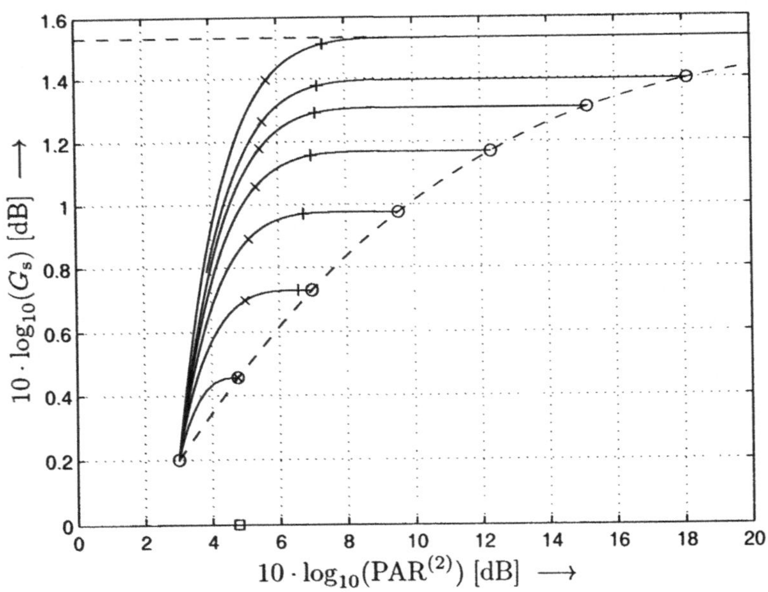

Fig. 4.11 Shaping gain versus two-dimensional peak-to-average energy ratio of truncated N-spheres. The circles mark $N = 2, 4, 8, 16, \ldots$; the square marks the reference N-cube. The points marked with $\times$ and $+$ correspond to $\mathrm{CER}^{(2)} = \sqrt{2}$ and 2, respectively.

4.2.5 Shaping on Regions

We have seen that for achieving shaping gain, an increase in peak-to-average energy ratio and the number of constellation points cannot be avoided. But there is a third drawback: the *addressing complexity* is much higher compared to mapping each dimension separately. Especially for "large" constituent constellations and large block sizes (higher dimensionality) the number of points to be addressed in N-space is tremendous. Consider, e.g., a two-dimensional constituent constellation with only 64 points and shaping in $2 \cdot 16$ dimensions. Then $64^{16} \approx 8 \cdot 10^{28}$ points have to be handled.

A solution to this problem is not to address each point on its own by the shaping algorithm, but to form *groups of signal points* or *regions*. The groups are selected by an algorithm, whereas the actual point from the group is addressed memoryless. Again a duality of signal shaping to channel coding is visible: In *coset coding* [For88], subsets of the underlying lattice are selected by the channel code. The actual point from the subset is addressed by the "uncoded bits." Channel coding deals with the internal arrangement of the points, whereas shaping treats *regions*. Roughly speaking, assuming a suitable labeling of the signal points, i.e., by Ungerböck's set partitioning [Ung82], channel coding operates on the least significant bits and shaping on the most significant ones. Figure 4.12 sketches the general concept, which should be compared with Figure 3.15. Here, $\mathcal{M}$ denotes the mapping into the (low-dimensional) constituent constellation.

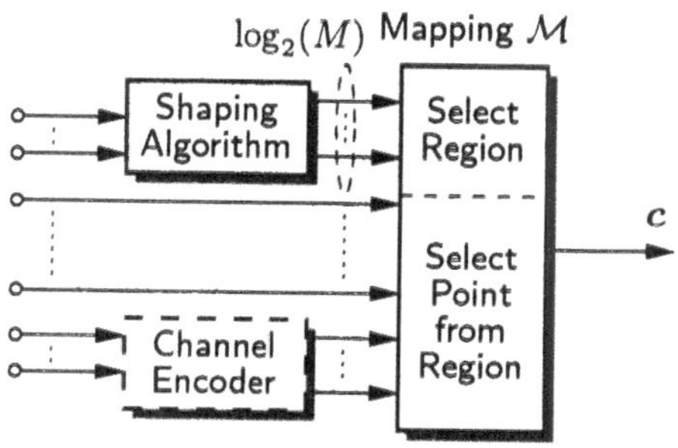

Fig. 4.12 General concept of shaping on regions (combined with channel coding).

An obvious approach is to arrange points with equal, or at least similar, energy into the same region [CO90]. Since energy is proportional to the squared Euclidean norm, the regions are then nested *shells* or *rings*, as is the case for onions. The advantage is that each shell is uniquely characterized by its (average) energy and the shaping algorithm only has to know the associated ring number.

Moreover, we concentrate on *equal-size regions*. If each ring has the same number of signal points, the rate which is transmitted by selecting a point from the ring is constant. Shaping using *variable-size regions* can be found in [Liv92]. Here, buffers are required to average the rate which is probabilistic. Furthermore, in case of transmission errors, bit insertions and bit deletions occur. In practical systems these properties are highly undesirable.

Under minor restrictions, shaping on regions is fully compatible with coded modulation [CO90]. As discussed in Section 3.2.7, channel coding usually affects the least significant address bits and selects sequences of cosets. In order to operate shaping and coding separately, the partitioning into regions has to be done such that each region contains an equal number of signal points from each coset of the coding lattice. Then, regardless of the actually selected region, coded modulation works as usual; in particular the achievable coding gain is not affected.

In the following, we again study two-dimensional constituent constellations and follow the approach of [CO90], where the generalization to an arbitrary number of dimensions also can be found. The starting point is two-dimensional circular constellations, which are partitioned into M equal-sized shells, indexed by their number $s = 0, 1, \ldots, M-1$, starting with the innermost shell. Figure 4.13 shows the partitioning of a circular boundary region into eight annular shells of equal volume.

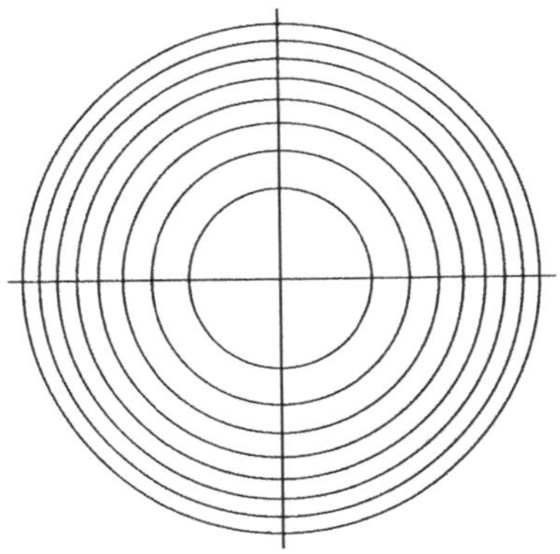

Fig. 4.13 Partitioning of a circular region into 8 shells of equal volume.

Performance of Shaping on Regions The volume (area in two dimensions) of the innermost ring with radius r_0 is $V_0 = r_0^2\pi$. Since all rings should have the same volume, for the sth ring with radius r_s, we require

$$V_s = r_s^2\pi - s \cdot r_0^2\pi \overset{!}{=} r_0^2\pi \,, \tag{4.2.54}$$

which leads to

$$r_s^2 = (s+1) \cdot r_0^2 \,. \tag{4.2.55}$$

For equal-sized regions, the squared radius of the rings thus increases linearly with the shell index s.

Let $E_0 = \frac{1}{r_0^2\pi} \int_{|\boldsymbol{r}|^2 \leq r_0^2} |\boldsymbol{r}|^2 \, \mathrm{d}\boldsymbol{r}$ be the average energy of shell 0. Scaling the size of the region by s increases the integral to $\int_{|\boldsymbol{r}|^2 \leq s r_0^2} |\boldsymbol{r}|^2 \, \mathrm{d}\boldsymbol{r} = s^2 \int_{|\boldsymbol{r}'|^2 \leq r_0^2} |\boldsymbol{r}'|^2 \, \mathrm{d}\boldsymbol{r}'$, i.e., by a factor s due to increased energy, and by the same factor due to increased volume. Since shell s is given by scaling shell 0 by $s+1$ minus the same shell scaled by s, the energy of shell s calculates to

$$\begin{aligned} E_s &= (s+1)^2 E_0 - s^2 E_0 \\ &= (2s+1) E_0 , \quad s = 0, 1, \ldots M-1 . \end{aligned} \tag{4.2.56}$$

If moreover the shells are used with probability $p_s = \Pr\{s\}$, average energy is given by

$$E_{\circledcirc} = \sum_{s=0}^{M-1} p_s E_s = E_0 \sum_{s=0}^{M-1} (2s+1) p_s , \tag{4.2.57}$$

and the rate, transmitted in the selection of the shells, becomes

$$R_{\circledcirc} = H(\{p_s\}) = - \sum_{s=0}^{M-1} p_s \log_2(p_s) . \tag{4.2.58}$$

The subscript "◎" is intended to symbolize the shell arrangement.

If the same rate shall be transmitted by equiprobable signaling on circular regions, the size of the region has to be $2^{R_{\circledcirc}}$ times that of shell 0. Hence, the average energy of the reference system is

$$E_{\bullet} = E_0 \cdot 2^{H(\{p_s\})} , \tag{4.2.59}$$

which translates to a so-called *biasing gain* [CO90], the gain when using the shells with frequencies p_s, of

$$G_{\mathrm{b}}(\{p_s\}) \triangleq \frac{E_{\bullet}}{E_{\circledcirc}} = \frac{2^{H(\{p_s\})}}{\sum_{s=0}^{M-1} (2s+1) p_s} . \tag{4.2.60}$$

Since the constituent circular constellation already exhibits a shaping gain of $G_{\mathrm{s,O}}(2) = \pi/3$ (cf. (4.2.27)), the total *shaping gain* results in

$$G_{\mathrm{s},\circledcirc}(\{p_s\}) = \frac{\pi}{3} \cdot G_{\mathrm{b}}(\{p_s\}) . \tag{4.2.61}$$

Finally, the constellation expansion in two dimensions calculates to

$$\mathrm{CER}_{\circledcirc}^{(2)} = \frac{M}{2^{H(\{p_s\})}} , \tag{4.2.62}$$

and the peak-to-average energy ratio specializes to

$$\begin{aligned} \mathrm{PAR}_{\circledcirc}^{(2)} &= \frac{M \cdot r_0^2}{E_0 \sum_{s=0}^{M-1} (2s+1) p_s} \\ &= \frac{r_0^2}{E_0} \cdot \frac{2^{H(\{p_s\})}}{\sum_{s=0}^{M-1} (2s+1) p_s} \cdot \frac{M}{2^{H(\{p_s\})}} \\ &= \mathrm{PAR}_{\mathrm{O}}^{(2)} \cdot G_{\mathrm{b}}(\{p_s\}) \cdot \mathrm{CER}_{\circledcirc}^{(2)} , \end{aligned} \tag{4.2.63}$$

which is in agreement with the general statement (4.2.21).

Maximum Biasing Gain We now look for the optimal probabilities p_s maximizing the biasing gain, and hence the total shaping gain. Taking the logarithm—which is a strictly monotonic increasing function—of (4.2.60), the optimization task is

- Maximize $H(\{p_s\}) - \log_2\left(\sum_{s=0}^{M-1}(2s+1)p_s\right)$,
- subject to the additional constraint $\sum_{s=0}^{M-1} p_s = 1$.

As in Section 4.1.2, a Lagrange function with Lagrange multiplier μ can be set up:

$$L(\{p_s\}) = -\sum\nolimits_{s=0}^{M-1} p_s \log(p_s) - \log\left(\sum\nolimits_{s=0}^{M-1}(2s+1)p_s\right) - \mu\sum\nolimits_{s=0}^{M-1} p_s \; . \tag{4.2.64}$$

Again, for convenience, we have changed the basis of the logarithm. Derivation of $L(\{p_s\})$ leads to

$$\begin{aligned} \frac{\mathrm{d}L(\{p_s\})}{\mathrm{d}p_s} &= -(1+\log(p_s)) - \frac{2s+1}{\sum_{t=0}^{M-1}(2t+1)p_t} - \mu \\ &\overset{!}{=} 0\,, \qquad \forall\, s = 0,1,\ldots,M-1\,. \end{aligned} \tag{4.2.65}$$

Since the sum in the denominator does not depend on the shell number s, we have

$$\log(p_s) = -\mu - 1 - \frac{2s+1}{\text{const}} \triangleq \text{const} - \lambda\cdot s\,, \tag{4.2.66}$$

or finally

$$p_s = K(\lambda)\cdot \mathrm{e}^{-\lambda s}\,. \tag{4.2.67}$$

Since the ring index s is proportional to the energy of the shell, cf. (4.2.56), the optimal distribution is again discrete Gaussian. The function $K(\lambda)$ once more normalizes the distribution.

For the analysis later in this chapter, it is convenient to define $x \triangleq \mathrm{e}^{-\lambda}$ and rewrite the distribution in terms of x. Now, $\lambda = 0$, i.e., a uniform distribution, corresponds to $x = 1$, and $\lambda \to \infty$ transforms to $x \to 0$. Having the finite sum over a geometric series in mind, the optimal probabilities can be written as

$$p_s = \frac{1-x}{1-x^M}\,x^s\,, \qquad s = 0,1,\ldots,M-1\,. \tag{4.2.68}$$

Asymptotic Behavior If the number of shells goes to infinity, from (4.2.68), and considering $\sum_{s=0}^{\infty} x^s = \frac{1}{1-x}$, the optimal probabilities become [CO90]

$$p_s = (1-x)\,x^s\,, \qquad s = 0,1,\ldots\,. \tag{4.2.69}$$

Then, using $\sum_{s=0}^{\infty} x^s = \frac{1}{1-x}$ and $\sum_{s=0}^{\infty} s x^s = \frac{x}{(1-x)^2}$, after some manipulations, the logarithm of the biasing gain (4.2.60) is given by

$$\log(G_{\mathrm{b}}(x)) = -\sum_{s=0}^{\infty}(1-x)x^s\log\left((1-x)\,x^s\right) - \log\left(\sum_{s=0}^{\infty}(2s+1)(1-x)x^s\right)$$

$$
\begin{aligned}
&= -\left(\log(1-x) + \frac{x\log(x)}{1-x}\right) - \log\left(1 + \frac{2x}{1-x}\right) \\
&= -\frac{x}{1-x}\log(x) - \log(1+x) \; . \qquad (4.2.70)
\end{aligned}
$$

Inspection of this function reveals that it has no local maximum for x in the interval $[0, 1]$; the maximum is approached for $x \to 1$ and, using L'Hopital's rule, reads

$$
\begin{aligned}
\log(G_{\mathrm{b},\infty}) &\triangleq \lim_{x\to 1} \log(G_{\mathrm{b}}(x)) \\
&= \lim_{x\to 1}\left(-\frac{1+\log(x)}{-1}\right) - \log(2) \\
&= \log(\mathrm{e}/2) \; . \qquad (4.2.71)
\end{aligned}
$$

The ultimate biasing gain is thus $G_{\mathrm{b},\infty} = \mathrm{e}/2$ or 1.33 dB, which according to (4.2.61), translates to the ultimate shaping gain of

$$
G_{\mathrm{s},\infty} = \frac{\pi}{3} \cdot G_{\mathrm{b},\infty} = \frac{\pi}{3} \cdot \frac{\mathrm{e}}{2} = \frac{\pi \mathrm{e}}{6} \; . \qquad (4.2.72)
$$

Since the signal points within each shell are used equiprobably, and in the optimum the shells are used according to a sampled Gaussian distribution, the pdf is a circular stairstep function. As the number of shells goes to infinity, the size of the regions goes to zero and the stairstep density converges to a continuous Gaussian distribution. Hence, the ultimate shaping gain of 1.53 dB can be approached by shaping on regions, as the number of shells is chosen sufficiently large.

Discussion Figure 4.14 shows the biasing gain $G_{\mathrm{b}}(x)$ over the parameter x for different numbers M of shells. Additionally, the ultimate biasing gain is marked. As one can see, the possible gain increases with the number M of shells. Given M, the optimum parameter x_{opt}, which directly corresponds to the variance of the underlying Gaussian distribution, has to be selected.

For the optimal choice, Figure 4.15 shows the resulting shaping gain $G_{\mathrm{s},\mathbf{o}}(x_{\mathrm{opt}})$ over the number of shells. In the case of a single shell, only 0.2 dB shaping gain of the circular constellation over the square is possible. However, for $M \geq 16$ the ultimate shaping gain of 1.53 dB is almost reached. Hence, in practice we can conclude that it is sufficient to use 8 to 16 shells.

Finally, Figure 4.16 displays the shaping gain $G_{\mathrm{s},\mathbf{o}}$ over the rate $R_{\mathbf{o}}$ for various numbers M of shells; i.e., the x-axis of Figure 4.14 is transformed to rate. All curves start at a zero rate at 0.2 dB, which corresponds to $x = 0$. Here, only the innermost shell is used and no information can be transmitted via the shell numbers. For $x = 1$, all shells are used with the same frequency, and the maximum rate $\log_2(M)$, but only the minimum shaping gain of 0.2 dB is achieved. Between these two extreme cases, an optimum for the shaping gain exists. The upper envelope, valid for $M \to \infty$, is given by (4.2.70); here the rate readily calculates to $R_{\mathbf{o}}(x) = -\log_2(1-x) - (x \cdot \log_2(x))/(1-x)$. As the number of shells increases, the ultimate shaping gain (1.53 dB) is approached closer and closer.

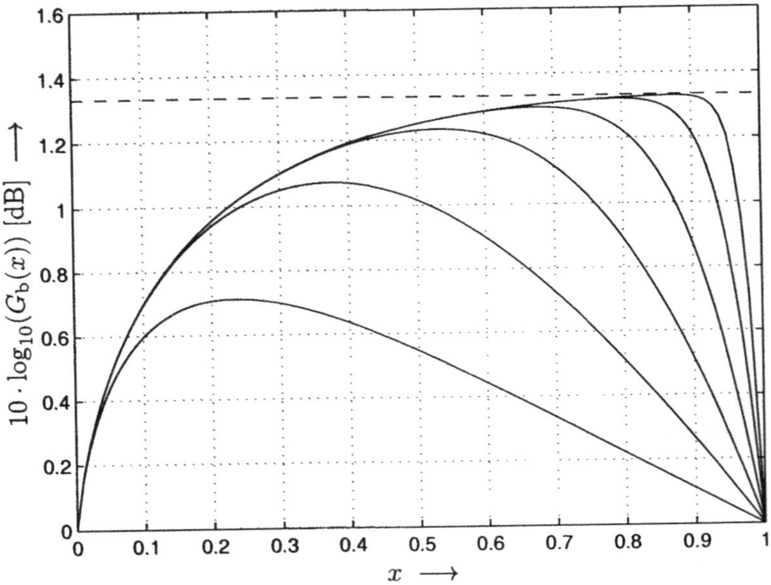

Fig. 4.14 Biasing gain over the parameter x for (bottom to top) $M = 2, 4, 8, 16, 32$, and 64 shells. Dashed: ultimate biasing gain.

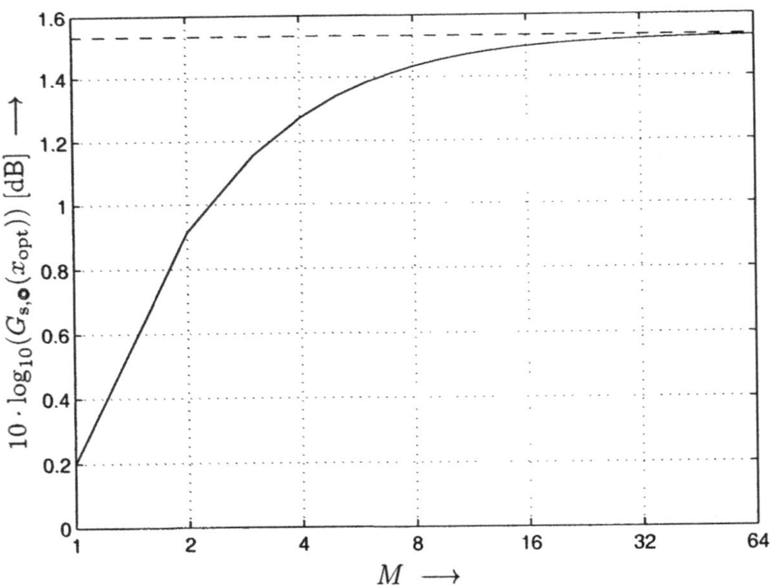

Fig. 4.15 Shaping gain for optimal choice of x over the number M of shells. Dashed: ultimate shaping gain.

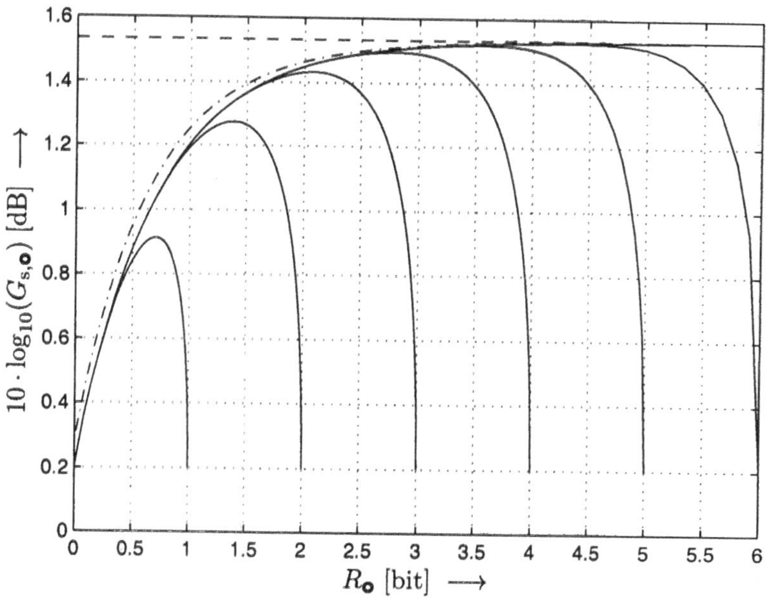

Fig. 4.16 Shaping gain over the rate for (left to right) $M = 2, 4, 8, 16, 32$, and 64 shells. Dashed: ultimate shaping gain. Dash-dotted: approximation derived from discretization factor (4.2.19).

The dashed-dotted line is an approximation to the shaping gain and reads (cf. [KP93])

$$G_s(R) \approx \frac{\pi e}{6} \cdot \left(1 - 2^{2(R+1)}\right) . \tag{4.2.73}$$

The ultimate shaping gain is hence lowered by the discretization factor (4.2.19). Since in each dimension, the left and right part of each shell can be distinguished, the rate in (4.2.73) is the rate transmitted in the selection of the shell indices plus one extra bit. Surprisingly, the approximation is very tight.

Note that the above figure has an intriguing similarity to [For89, Figure 9], where the maximum shaping gain attainable with Voronoi constellations dependent on the partitioning depth of the binary lattice is given. Finally, graphs on constellation expansion and peak-to-average energy ratio can be found in [CO90].

4.2.6 AWGN Channel and Shaping Gain

Up to now, for the design of the signal constellation, we have restricted the discussion to parameters (transmit power, CER, PAR) seen at the transmitter side. The actual transmission over a channel has been disregarded, as is done in the vast majority of the literature. Consequently, we now study the achievable shaping gains when transmitting over an additive white Gaussian noise (AWGN) channel.

In Section 4.1.3, the maximum shaping gain was derived by fixing the (differential) entropy $h(X)$ of the channel input symbols x. Disregarding the channel noise, this entropy directly translates into transmission rate. But when transmitting over an AWGN channel, the maximum rate which can be transmitted reliably is given by the *mutual information* $I(X;Y)$ between channel input X and output Y. Denoting the channel noise by N, the mutual information can be calculated as, e.g., [Gal68, Bla87, CT91]

$$I(X;Y) = h(X) - h(X|Y) \tag{4.2.74a}$$
$$= h(Y) - \underbrace{h(Y|X)}_{=h(N)} . \tag{4.2.74b}$$

Here, $h(Z) \triangleq -\int \mathrm{f}_z(z)\log_2(\mathrm{f}_z(z))\,\mathrm{d}z$ denotes the differential entropy of a random variable Z with pdf $\mathrm{f}_z(z)$.

Unfortunately, fixing the entropy $h(X)$, but changing the shape of the distribution of X, also has an effect on the conditional entropy $h(X|Y)$. In particular, going from a uniform distribution to a Gaussian one increases $h(X|Y)$ for fixed $h(X)$—even given Y, it is harder to predict a Gaussian variable than a uniform one. Hence, from (4.2.74a), the mutual information is decreased. If we aim to transmit a desired data rate, this loss has to be compensated, e.g., by again increasing transmit power, which annihilates parts of the shaping gain.

Since the entropy $h(N)$ of the noise is given, an obvious design rule results from (4.2.74b): Instead of fixing the entropy of the *channel input* X, the entropy of the *channel output* Y should be kept constant. Thus, here we distinguish between *shaping gain*—gain in average transmit power when fixing $h(Y)$—and the gain for fixed mutual information $I(X;Y)$. We denote the latter gain by *capacity gain*. As the signal-to-noise ratio increases, the differences of both approaches vanish, and the whole shaping gain can be utilized. Subsequently, the maximum capacity gain in using a Gaussian distribution—and hence regarding Shannon limit—over an uniform channel input is derived.

The capacity[7] $C_{\circ} = \max I(X;Y)$ (in bits per dimension) of an AWGN channel with input variance σ_x^2 and noise variance σ_n^2 reads [Sha49]

$$C_{\circ} = \frac{1}{2}\log_2\left(1 + \sigma_x^2/\sigma_n^2\right) . \tag{4.2.75}$$

Expressing this quantity by the bit energy E_b and the one-sided noise power spectral density N_0, Shannon capacity is given by [Bla90, Pro01]

$$C_{\circ} = \frac{1}{2}\log_2\left(1 + \frac{RE_\mathrm{b}}{N_0/2}\right) . \tag{4.2.76}$$

Here, R denotes the transmission rate in bits per dimension (one-dimensional symbol). The Shannon capacity $C_{\circ}$ is the upper limit for the transmission rate, and is

[7]The indices "○" for Gaussian distributions and "□" for uniform distributions again remind us of the corresponding high-dimensional sphere and cube.

achieved for Gaussian channel input. For $R = C_{\bigcirc}$, we can rewrite (4.2.76) as

$$\frac{E_b}{N_0} = \frac{2^{2R} - 1}{2R} , \qquad (4.2.77)$$

which gives a connection between rate and signal-to-noise ratio E_b/N_0 for AWGN channel with Gaussian input.

When considering uniformly distributed signals, i.e., $f_x(x) = 1/(2\sqrt{3\sigma_x^2})$ for $|x| \le \sqrt{3\sigma_x^2}$ and $f_x(x) = 0$ otherwise, for calculation of the constellation-constrained capacity $C_{\square}$, we have to resort to (4.2.74b). The differential entropy $h(Y)$ has to be evaluated numerically from the pdf

$$f_y(y) = f_x(y) * f_n(y) = \int_{-\sqrt{3\sigma_x^2}}^{\sqrt{3\sigma_x^2}} \frac{1}{2\sqrt{3\sigma_x^2}} \frac{1}{\sqrt{2\pi\sigma_n^2}} e^{-(y-x)^2/(2\sigma_n^2)} \, dx , \qquad (4.2.78)$$

and the differential entropy of the noise with pdf $f_n(n) = 1/(\sqrt{2\pi\sigma_n^2})e^{-n^2/(2\sigma_n^2)}$ calculates to $h(N) = \frac{1}{2}\log_2(2\pi e\sigma_n^2)$ [CT91]. With regard to the equation $\sigma_x^2/\sigma_n^2 = 2RE_b/N_0$, valid for the baseband AWGN channel, a relation between minimum required SNR E_b/N_0 and rate R, analogous to (4.2.77), can be calculated numerically for uniform channel input.

In Figure 4.17, the capacity curves for the AWGN channel with Gaussian input (Shannon limit) and uniform input are plotted. As one can see, as C goes to zero (for which E_b/N_0 approaches -1.59 dB [Bla90, Pro01]), the gap between Gaussian and uniform signaling vanishes. Here, in each case, $h(Y) = h(X + N) \to h(N)$.

Using the entropy power inequality [Bla87, CT91], the gap between both curves can be bounded to [ST93]

$$\begin{aligned} C_{\bigcirc} - C_{\square} &= h(X_{\bigcirc} + N) - h(X_{\square} + N) \\ &\le h(X_{\bigcirc} + N) - \frac{1}{2}\log_2\left(2^{2h(X_{\bigcirc})} + 2^{2h(N)}\right) \\ &= \frac{1}{2}\log_2\left(2\pi e(\sigma_x^2 + \sigma_n^2)\right) - \frac{1}{2}\log_2\left(12\sigma_x^2 + 2\pi e\sigma_n^2\right) , \end{aligned}$$

and since $2\pi e \approx 17.1 > 12$

$$\begin{aligned} &< \frac{1}{2}\log_2\left(2\pi e(\sigma_x^2 + \sigma_n^2)\right) - \frac{1}{2}\log_2\left(12(\sigma_x^2 + \sigma_n^2)\right) \\ &= \frac{1}{2}\log_2\left(\frac{\pi e}{6}\right) . \end{aligned} \qquad (4.2.79)$$

As the signal-to-noise ratio, and hence the differential entropies, increases, the entropy power inequality becomes exact.

In other words, $C_{\square}$ asymptotically becomes [OW90, ST93]

$$C_{\square} \approx \frac{1}{2}\log_2\left(1 + \frac{6}{\pi e}\frac{RE_b}{N_0/2}\right) , \qquad (4.2.80)$$

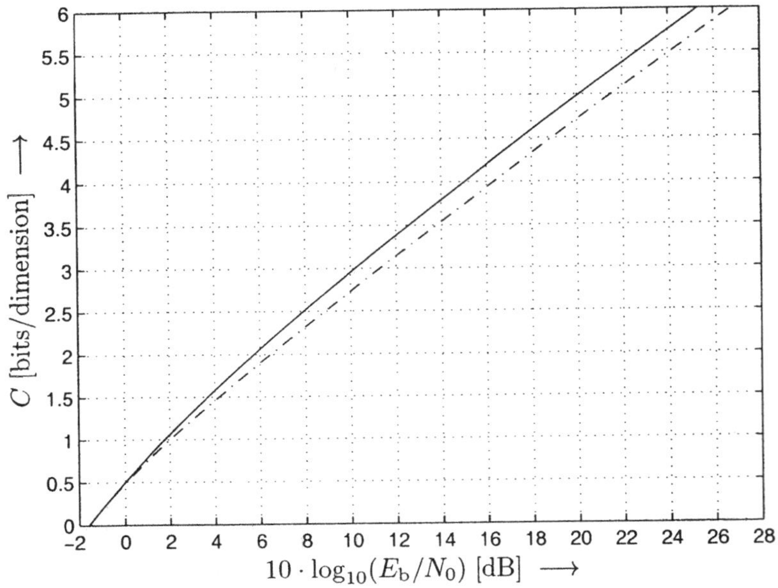

Fig. 4.17 Capacity of the baseband AWGN channel with uniform (dash-dotted) and Gaussian (solid) input over the signal-to-noise ratio E_b/N_0 in dB.

i.e., in terms of SNR, the curves are spaced by the ultimate shaping gain $\frac{\pi e}{6}$. For high SNR, and hence high rates, the energy gain directly translates into a capacity gain. But for moderate rates, the capacity gain stays below this limit.

Figure 4.18 summarizes the capacity gain versus the rate. Additionally, the maximum shaping gain of discrete constellations

$$G_s(R) \approx \frac{\pi e}{6} \cdot \left(1 - 2^{2R}\right) , \tag{4.2.81}$$

which is again a consequence of the discretization factor (4.2.19), is shown. As one can see, over a wide range, the shaping gain is much greater than the gain in capacity. Strictly speaking, the true shaping gain is even greater, because some constellation expansion is necessary to realize coding gain. Hence, the shaping gain of a constellation supporting a certain rate R plus the coding redundancy can be exploited. Thus, the shaping gain curve has to be moved left by the redundancy.[8] Only for asymptotically high rates does the whole shaping gain translate directly to a gain in capacity, approaching the ultimate shaping gain (1.53 dB). In contrast, for $C \to 0$, the capacity gain completely vanishes.

To summarize, when transmitting over a channel, shaping gain, i.e., the reduction in average transmit power, is not the only relevant parameter. If a moderate, fixed rate

[8]Note that an additional loss appears for discrete constellations compared to the corresponding continuous ones. Hence, for discrete constellations, the capacity gain is actually a lower bound.

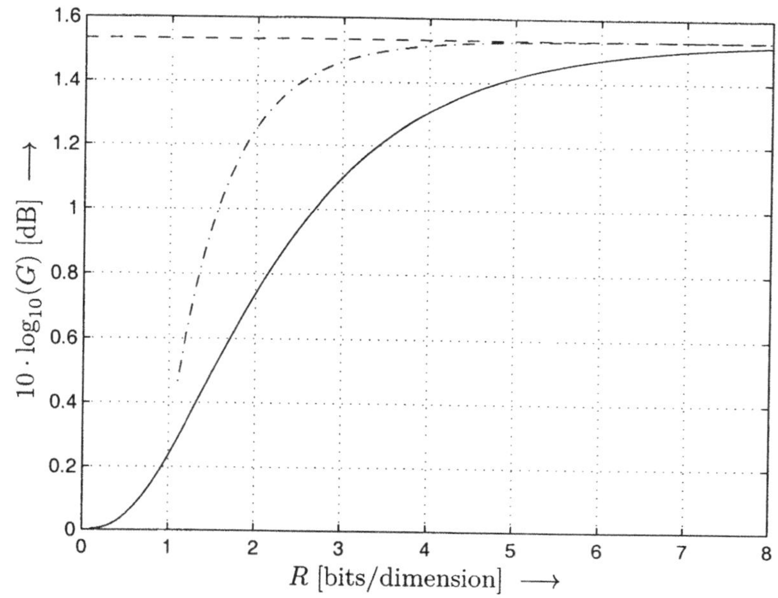

Fig. 4.18 Capacity gain (solid line) and approximation of shaping gain for discrete constellation (dash-dotted line), both in dB, over the rate. Dashed: ultimate shaping gain.

should be transmitted as reliably as possible, i.e., close to capacity limit, the effective gain is much lower than predicted by the ultimate shaping gain. One interpretation of this phenomenon is that for moderate constellation sizes channel coding and signal shaping interact. Here, the gains (in dB) cannot simply be added, and a joint design is required [WFH99].

4.3 SHELL MAPPING

Shell mapping, widely discussed in the literature, e.g., [LL89, EFDL93, LFT94, KK93], is a very efficient algorithm to signal shaping. It is part of the international telephone-line modem standard ITU recommendation V.34 [ITU94] and its successors. Shell mapping has a source-coding counterpart, namely a version of fixed-rate vector quantization [LF93a, LF93b].

In principle, the signal points in an N-dimensional space are labeled in order of increasing signal energy. Shell mapping defines a one-to-one mapping between blocks of, say K, input bits and blocks of N points from the constituent low-dimensional constellation. Since only the 2^K least-energy combinations are used, shaping gain may be achieved. Moreover, since mapping is done on blocks of N (one- or two-)dimensional symbols, a rate granularity of $1/N$ bit/symbol is obtained. On strictly band-limited channels, fractional rates are essential for optimum performance.

In this section, we concentrate on two-dimensional constituent constellations and shell mapping is used for selecting the rings when doing shaping on regions. We first introduce shell mapping for frame sizes N that are a power of two. Thereafter, the generalization to other block sizes N and constituent constellations is discussed. For calculating the average transmit power in shell mapping schemes, a simple algorithm for counting the shell frequencies is presented in Appendix D. Please note, in this section on shell mapping, N denotes the number of *complex* dimensions; the signal space is hence of dimensionality $2N$.

4.3.1 Preliminaries

In what follows, we assume that the constituent two-dimensional signal constellation contains $M \cdot 2^q$ signal points. The signal set is partitioned into M groups ("shells"), denoted by $s = 0, 1, \ldots, M-1$, as in Section 4.2.5, each containing an equal number (2^q) of points. To each shell a *cost* $C(s)$ is assigned, which represents the energy of the shell. We assume that the ordering of shells is such that the costs increase monotonously, i.e., $C(0) \leq C(1) \leq \cdots \leq C(M-1)$. Since energy is additive along the coordinates, the total cost $C^{(N)}$ in N dimensions is the sum of the individual costs: $C^{(N)} \triangleq \sum_{i=1}^{N} C(s_{(i)})$.

In Section 4.2.5 we derived that for two-dimensional constituent constellations the energy of shell s can be expressed as $2s - 1$ times the energy of the innermost shell. Because in the following an additive constant and a scaling of the energy is irrelevant, in QAM signaling we may apply the simple cost function $C(s) = s$, $s = 0, 1, \ldots, M-1$ [EFDL93, LFT94]. Henceforth, the shell index and cost are used identically. Note that this cost function also results from (4.2.55), when we approximate the energy of the shells by their squared radius.

Shell mapping operates on a *frame* of N consecutive symbols, and in each of the N *positions* one particular shell is selected. The points within the shells are addressed memoryless with q "uncoded bits." The task of the shell mapping encoder is thus to map a number (or index) consisting of K bits to an N-tuple of shell indices, shown in

Figure 4.19. Since the constituent constellation is two-dimensional, the signal space of one mapping frame has dimensionality $2N$. Conversely to the encoder, the shell

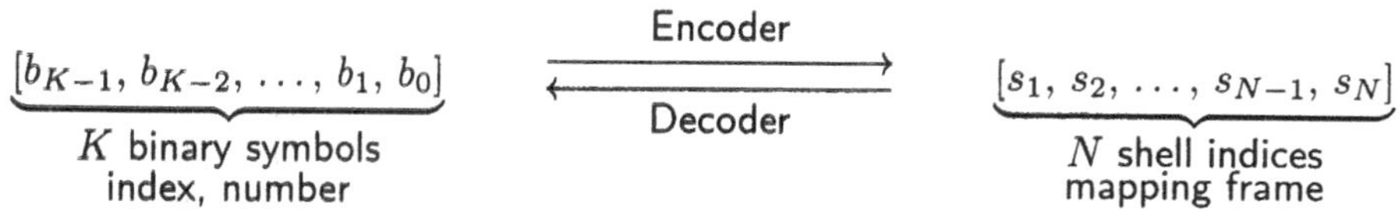

Fig. 4.19 Illustration of the task of shell mapping.

mapping decoder gathers N shell indices and returns the binary K-tuple.

Since K bits are mapped in one frame out of the M^N possible combinations (N-tuples, vectors) of shells, only the $2^K \leq M^N$ vectors with the smallest total cost $\sum_{i=1}^{N} C(s_{(i)})$ are used. Since we assume independent, uniformly distributed data, these 2^K combinations are used with equal probability, i.e., the signal points in $2N$ dimensions are uniformly distributed. As a consequence, in shell mapping schemes shells with lower costs are used more frequently than shells with larger costs, i.e., the signal points in the constituent constellation are nonuniformly distributed. In general, the frequencies of shells are different for each of the N positions.

4.3.2 Sorting and Iteration on Dimensions

The main idea of shell mapping is to *implicitly sort* the vectors of shell indices according to their total cost: tuples of shell indices with lower total cost are associated with a lower index than tuples with larger costs. In order to sort vectors with equal cost, different strategies are possible. For example, if N is even, the cost of the first $N/2$ shell indices can serve as the criterion.

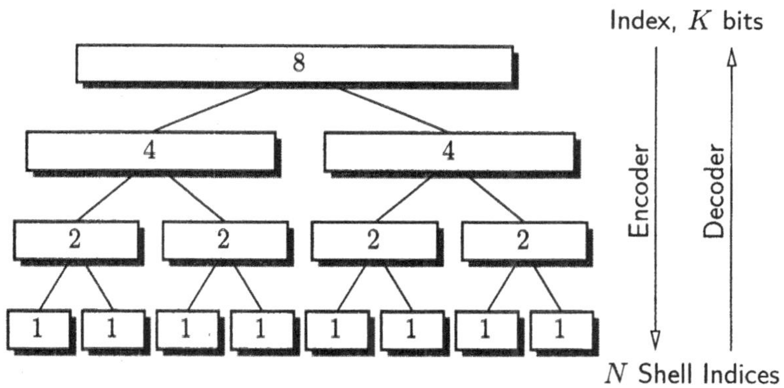

Fig. 4.20 Sketch of recursion on dimensions for $N = 8$. The numbers of dimension are indicated.

Performing a *recursion on dimensions*, the encoding problem is split into two problems of half the number of dimensions. This is iterated until only N scalar, and hence trivial, problems are left. If the frame size is not a power of two, mapping can be done symbol by symbol, but taking the remaining dimensions into account, see Section 4.3.4. Here, for the moment, we assume N to be a power of two, and Figure 4.20 shows the procedure of shell mapping encoder and decoder.

First the building blocks of the encoder and decoder—the splitting of an n-dimensional mapping into two $n/2$-dimensional ones (and the other way round)—are explained. Then the shell mapping encoder and decoder are readily described.

Sorting of 2-Tuples In order to explain the decomposition of one two-dimensional mapping into two one-dimensional ones (and the reverse process), we study the implicit sorting of shell 2-tuples $[s_{(1)}, s_{(2)}]$.

These 2-tuples are first sorted according to their total cost, which is given as

$$C^{(2)} = C^{(1)}_{(1)} + C^{(1)}_{(2)} = s_{(1)} + s_{(2)} \, . \tag{4.3.1a}$$

The superscript n in the denomination $\bullet^{(n)}_{(i)}$ gives the number of dimensions (it is the respective quantity of an n-tuple), and the subscript i, if present, indicates the position of this n-tuple within the $2n$-dimensional mapping frame.

Vectors with equal total cost are sorted according to the cost s_1 of the first shell index. Then, shell 2-tuples are uniquely characterized by (i) the *total cost* $C^{(2)}$, and (ii) an *index* $I^{(2)}$ into the set of 2-tuples with the same total cost. For combinations with total cost less than M, index $I^{(2)}$ equals $s_{(1)}$. Because the largest cost of the second half is $M-1$, the first half has minimum cost $C^{(2)} - M + 1$, if the total cost is larger than or equal to M. Hence, since index zero corresponds to this minimum cost $s_{(1)}$, index $I^{(2)}$ is given by $s_{(1)} - C^{(2)} + M - 1 = M - 1 - s_{(2)}$. In summary, the index is calculated according to

$$I^{(2)} = \begin{cases} s_{(1)} \, , & C^{(2)} < M \\ M - 1 - s_{(2)} \, , & C^{(2)} \geq M \end{cases} = \min\left\{s_{(1)}, M - 1 - s_{(2)}\right\} \, . \tag{4.3.1b}$$

Equations (4.3.1a) and (4.3.1b) establish the step from two one-dimensional problems to one two-dimensional problem required at the shell mapping *decoder*. The reverse procedure needed at the *encoder* is to calculate the final shell indices $s_{(1)}$ and $s_{(2)}$ given $C^{(2)}$ and $I^{(2)}$. Regarding the above equations, the encoding step is

$$C^{(2)} < M: \quad \begin{array}{l} s_{(1)} = I^{(2)} \\ s_{(2)} = C^{(2)} - s_{(2)} \end{array} , \qquad C^{(2)} \geq M: \quad \begin{array}{l} s_{(2)} = M - 1 - I^{(2)} \\ s_{(1)} = C^{(2)} - s_{(2)} \end{array} . \tag{4.3.2}$$

Example 4.2: *Sorting of Shell* 2*-Tuples*

As an example, the sorting of shell 2-tuples is shown for $M = 4$ shells. Table 4.1 summarizes all $4^2 = 16$ possible combinations of two shells and the corresponding total cost $C^{(2)}$ and index $I^{(2)}$.

Table 4.1 Sorting of all possible combinations of two shells and corresponding total cost $C^{(2)}$ and index $I^{(2)}$; $M = 4$.

$[s_{(1)},$	$s_{(2)}]$	$C^{(2)}$	$I^{(2)}$
0	0	0	0
0	1	1	0
1	0		1
0	2		0
1	1	2	1
2	0		2
0	3		0
1	2	3	1
2	1		2
3	0		3
1	3		0
2	2	4	1
3	1		2
2	3	5	0
3	2		1
3	3	6	0

Sorting of n-Tuples In order to get efficient algorithms for encoding and decoding, we define the coefficient $g^{(n)}(c)$ as the number of distinct shell n-tuples having total cost c. Combining these coefficients into a formal power series, *generating functions* of costs

$$G^{(n)}(x) \triangleq \sum_{c=0}^{(M-1)n} g^{(n)}(c) \cdot x^c \tag{4.3.3}$$

are defined. Since $g^{(n)}(c) = \sum_{i=0}^{c} g^{(n/2)}(i) g^{(n/2)}(c-i)$ is valid, which is the convolution of the coefficients $g^{(n/2)}(c)$, we have the equality

$$G^{(n)}(x) = \left(G^{(n/2)}(x)\right)^2 . \tag{4.3.4}$$

Starting from

$$G^{(1)}(x) = 1 + x + x^2 + \cdots + x^{M-1} , \tag{4.3.5}$$

the generating function of costs of a single ring, the generating functions $G^{(2)}(x)$, $G^{(4)}(x)$, $G^{(8)}(x)$, ..., can easily be calculated by repeated squaring, i.e., convolution of the coefficients.

The shell mapping algorithm implicitly assumes the following sorting of shell n-tuples $[s_{(1)}^{(1)}, \ldots, s_{(n)}^{(1)}] \triangleq [s_{(1)}^{(n/2)}, s_{(2)}^{(n/2)}]$:

- First, the n-tuples are sorted according to their total cost $C^{(n)}$.
- Second, n-tuples with equal cost are sorted such that tuples whose first half $s_{(1)}^{(n/2)}$ has lowest cost $C_{(1)}^{(n/2)}$ come first.
- Third, if total cost and cost of the first half of two shell n-tuples are the same, then the n-tuple whose *second half* has a lower index $I_{(2)}^{(n/2)}$ is ranked first.

As in Table 4.1, Table 4.2 contains the sorting of shell n-tuples, starting with the lowest index at the top.

We now turn to the problem of indexing n-tuples. As for 2-tuples, each shell n-tuple is now uniquely characterized by the *total cost* $C^{(n)}$ and the *index* $I^{(n)}$ into the set of n-tuples that have the same total cost. It should be noted that for the index, $0 \le I^{(n)} < g^{(n)}(C^{(n)})$ holds.

Again, we first consider the decoder, i.e., the task of finding $C^{(n)}$ and $I^{(n)}$, given costs $C_{(1)}^{(n/2)}$, $C_{(2)}^{(n/2)}$ and indices $I_{(1)}^{(n/2)}$, $I_{(2)}^{(n/2)}$ of first and second half, respectively. Obviously, the total cost calculates to

$$C^{(n)} = C_{(1)}^{(n/2)} + C_{(2)}^{(n/2)} . \tag{4.3.6a}$$

The index $I^{(n)}$ in the set of shell combinations having total cost $C^{(n)}$ can in principle be obtained by counting the combinations with a lower index. First, the shell combination considered is preceded by all n-tuples whose first half has a smaller cost than $C_{(1)}^{(n/2)}$. Since there are $g^{(n/2)}(c)$ $n/2$-tuples with cost equal to c, the number of combinations with cost of the first half smaller than $C_{(1)}^{(n/2)}$ is

$$\sum_{c=0}^{C_{(1)}^{(n/2)}-1} g^{(n/2)}(c) \cdot g^{(n/2)}(C^{(n)} - c) .$$

Table 4.2 Visualization of the sorting of shell n-tuples.

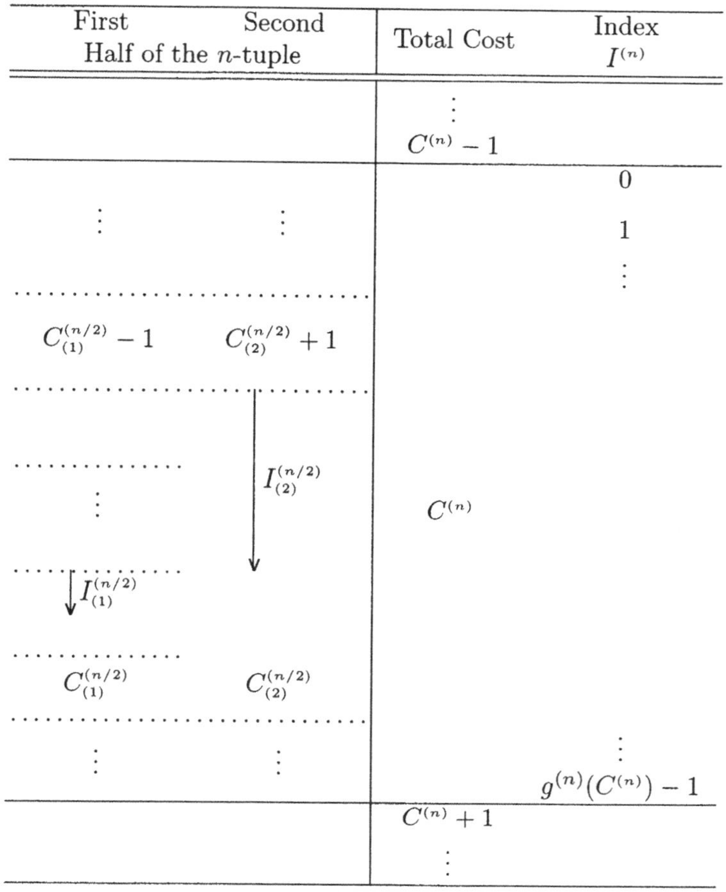

Second, we regard the combinations that have first-half cost $C_{(1)}^{(n/2)}$ and second-half cost $C_{(2)}^{(n/2)}$. Due to the specific sorting, all combinations with a second-half index smaller than $I_{(2)}^{(n/2)}$ are sorted first. Since for each value of $I_{(2)}^{(n/2)}$ there are $g^{(n/2)}(C_{(1)}^{(n/2)})$ possible $n/2$-tuples for the first half, the index has to be increased by

$$I_{(2)}^{(n/2)} \cdot g^{(n/2)}(C_{(1)}^{(n/2)}) \, .$$

Finally, there are

$$I_{(1)}^{(n/2)}$$

tuples with a lower index of the first half. In summary, the requested index $I^{(n)}$ can be computed as

$$I^{(n)} = \sum_{c=0}^{C_{(1)}^{(n/2)}-1} g^{(n/2)}(c) \cdot g^{(n/2)}(C^{(n)}-c) + I_{(2)}^{(n/2)} \cdot g^{(n/2)}(C_{(1)}^{(n/2)}) + I_{(1)}^{(n/2)} \, . \quad (4.3.6b)$$

Equations (4.3.6a) and (4.3.6b) therefore give the desired *decoding* step from two $n/2$-dimensional problems to one n-dimensional problem.

The task of the encoder is to calculate costs $C_{(1)}^{(n/2)}$, $C_{(2)}^{(n/2)}$ and indices $I_{(1)}^{(n/2)}$, $I_{(2)}^{(n/2)}$ of the first and second half, respectively, when given total cost $C^{(n)}$ and index $I^{(n)}$. To this end, we note that $0 \leq I_{(i)}^{(n/2)} < g^{(n/2)}(C_{(i)}^{(n/2)})$ holds; by construction, the index is a nonnegative integer and the maximum number of shell combinations of cost c is given by the coefficient $g^{(n/2)}(c)$. Hence

$$\begin{aligned} I_{(2)}^{(n/2)} \cdot g^{(n/2)}(C_{(1)}^{(n/2)}) + I_{(1)}^{(n/2)} &\leq \left(g^{(n/2)}(C_{(2)}^{(n/2)}) - 1\right) \cdot g^{(n/2)}(C_{(1)}^{(n/2)}) \\ &\quad + \left(g^{(n/2)}(C_{(1)}^{(n/2)}) - 1\right) \\ &= g^{(n/2)}(C_{(1)}^{(n/2)}) \cdot g^{(n/2)}(C^{(n)} - C_{(1)}^{(n/2)}) - 1 \,. \end{aligned} \tag{4.3.7}$$

In words, the sum of the last two terms in (4.3.6b) is smaller than the next term (for $c = C_{(1)}^{(n/2)}$) in the sum would be. Hence, as long as the residue is positive, we subtract the terms $g^{(n/2)}(0) \cdot g^{(n/2)}(C^{(n)}), g^{(n/2)}(1) \cdot g^{(n/2)}(C^{(n)}-1), g^{(n/2)}(2) \cdot g^{(n/2)}(C^{(n)}-2)$, ... in sequence from $I^{(n)}$. Cost $C_{(1)}^{(n/2)}$ of the first half is then the largest integer, for which

$$J \triangleq I^{(n)} - \sum_{c=0}^{C_{(1)}^{(n/2)}-1} g^{(n/2)}(c) \cdot g^{(n/2)}(C^{(n)} - c) \tag{4.3.8a}$$

is positive. The cost of the second half is then, of course, $C^{(n)} - C_{(1)}^{(n/2)}$.

Since, according to (4.3.6b) $J = I_{(2)}^{(n/2)} \cdot g^{(n/2)}(C_{(1)}^{(n/2)}) + I_{(1)}^{(n/2)}$, and $I_{(1)}^{(n/2)} \leq g^{(n/2)}(C_{(1)}^{(n/2)})$, the indices to the first and second half can be obtained by dividing J by $g^{(n/2)}(C_{(1)}^{(n/2)})$. Index $I_{(2)}^{(n/2)}$ of the second half is the quotient, and index $I_{(1)}^{(n/2)}$ of the first half is the remainder. We can write this formally as (mod: modulo operation)

$$\begin{aligned} I_{(1)}^{(n/2)} &= J \bmod g^{(n/2)}(C_{(1)}^{(n/2)}) \\ I_{(2)}^{(n/2)} &= (J - I_{(1)}^{(n/2)})/g^{(n/2)}(C_{(1)}^{(n/2)}) \end{aligned} \,. \tag{4.3.8b}$$

Equations (4.3.8a) and (4.3.6b) constitute the *encoding* step from one n-dimensional problem to two $n/2$-dimensional problems.

Example 4.3: *Sorting of Shell* 4*-Tuples*

We continue Example 4.2 and show the sorting of shell 4-tuples. Again the number of shells is $M = 4$. Table 4.3 shows an excerpt of the ordering of shell 4-tuples. The gray shaded row corresponds to the subsequent numerical example.

Table 4.3 Excerpt of sorting of shell 4-tuples and corresponding total cost $C^{(4)}$ and index $I^{(4)}$.

First Half	Second Half	Total Cost $C^{(4)}$		Index $I^{(4)}$
⋮	⋮	≤ 3		⋮
00	13	4		0
00	22	$C^{(2)}_{(1)} = 0$	$C^{(2)}_{(2)} = 4$	1
00	31			2
01	03			3
10	03			4
01	12			5
10	12			6
01	21	$C^{(2)}_{(1)} = 1$	$C^{(2)}_{(2)} = 3$	7
10	21			8
01	30			9
10	30			10
02	02	$C^{(2)}_{(1)} = 2$	$C^{(2)}_{(2)} = 2$	11
11	02		$I^{(2)}_{(2)} = 0$	12
20	02			13
02	11			14
11	11		1	15
20	11			16
02	20	$I^{(2)}_{(1)} = 0$		17
11	20	1	2	18
20	20	2		19
03	01			20
12	01	$C^{(2)}_{(1)} = 3$	$C^{(2)}_{(2)} = 1$	21
31	01			22
⋮	⋮			⋮
⋮	⋮	≥ 5		⋮

For decoding consider the following example. Let costs and indices be

$$C^{(2)}_{(1)} = 2\,, \quad I^{(2)}_{(1)} = 1$$
$$C^{(2)}_{(2)} = 2\,, \quad I^{(2)}_{(2)} = 2\,,$$

which (cf. Table 4.1) corresponds to the shell 2-tuples $s^{(2)}_{(1)} = [1\ 1]$ and $s^{(2)}_{(2)} = [2\ 0]$, respectively. From (4.3.6a) total cost $C^{(4)}$ calculates to

$$C^{(4)} = C^{(2)}_{(1)} + C^{(2)}_{(2)} = 2 + 2 = 4 .$$

Using (4.3.6b), and taking $G^{(2)}(x) = 1+2x+3x^2+4x^3+3x^4+2x^5+x^6$ into considerations, the index of the 4-tuple is given as

$$I^{(4)} = (1 \cdot 3 + 2 \cdot 4) + 2 \cdot 3 + 1 = 18 .$$

The inverse operation is performed in the encoder. Given $C^{(4)} = 4$ and $I^{(4)} = 18$, according to (4.3.8a), $C^{(2)}_{(1)}$ is determined. Considering

$$\begin{aligned} g^{(2)}(0) \cdot g^{(2)}(4) &= 1 \cdot 3 = 3 \\ g^{(2)}(1) \cdot g^{(2)}(3) &= 2 \cdot 4 = 8 \\ g^{(2)}(2) \cdot g^{(2)}(2) &= 3 \cdot 3 = 9 \\ &\vdots \end{aligned}$$

we find that $3 + 8 < 18 < 20$, hence $C^{(2)}_{(1)} = 2$, thus $C^{(2)}_{(2)} = C^{(4)} - C^{(2)}_{(1)} = 4 - 2 = 2$, and $J = 18 - (3 + 8) = 7$. Using (4.3.8b) the indices are obtained as

$$\begin{aligned} I^{(2)}_{(1)} &= J \bmod g^{(2)}(2) = 7 \bmod 3 = 1 \\ I^{(2)}_{(2)} &= (J - 1)/g^{(2)}(2) = (7 - 1)/3 = 2 , \end{aligned}$$

which agrees with the input of the decoder.

4.3.3 Shell Mapping Encoder and Decoder

We are now in a position to explain the shell mapping encoder and decoder. As noted above, the encoder maps an index, a number I of K binary digits, to N shell indices.

In order to start the recursion on dimensions, in the first step total cost $C^{(N)}$ and index $I^{(N)}$ have to be calculated from I. This can be done in a simple way by searching over coefficients $z^{(N)}(c)$. These coefficients are defined as the number of n-tuples having a total cost *less* than c. Clearly, we have

$$z^{(N)}(c) \triangleq \sum_{i=0}^{c-1} g^{(N)}(i) . \tag{4.3.9}$$

Since there are $z^{(N)}(c)$ shell combinations with cost less than c, and the shell N-tuples are sorted according to their total cost, the initial index lies in the range

$$z^{(N)}(C^{(N)}) \leq I < z^{(N)}(C^{(N)} + 1) . \tag{4.3.10a}$$

Cost $C^{(N)}$ is thus that position for which $z^{(N)}(c)$ is smaller than or at most equal to I. Since the requested shell combination has index I and $z^{(N)}(C^{(N)})$ combinations have smaller total cost, index $I^{(N)}$ in the set of shell tuples with cost $C^{(N)}$ calculates to

$$I^{(N)} = I - z^{(N)}(C^{(N)}) \, . \tag{4.3.10b}$$

Given $C^{(N)}$ and $I^{(N)}$, the recursion on dimensions can be started. Using (4.3.8a) and (4.3.8b), the N-dimensional problem is split into two $N/2$-dimensional ones. Each of the $N/2$-dimensional problems is dissected into two $N/4$-dimensional ones. This procedure continues until $N/2$ two-dimensional tasks remain. The final shell indices are then calculated using (4.3.2), which completes the encoder.

The shell mapping decoder accepts a block of N shell indices as input. Using (4.3.1a) and (4.3.1b), pairs of shells are combined into a two-dimensional entity. Applying (4.3.6a) and (4.3.6b) to $N/2$ pairs of two-dimensional problems, $N/4$ four-dimensional ones result. The recursion on dimensions is continued until a single N-dimensional problem, characterized by $C^{(N)}$ and $I^{(N)}$, remains. In the last step, taking (4.3.10b) into consideration, the final K-bit index I is output as

$$I = I^{(N)} + z^{(N)}(C^{(N)}) \, , \tag{4.3.11}$$

which completes the decoder.

Figure 4.21 shows the costs and indices which occur in shell mapping. Here, $N = 8$ is assumed, which is also used in the V.34 shell mapper.

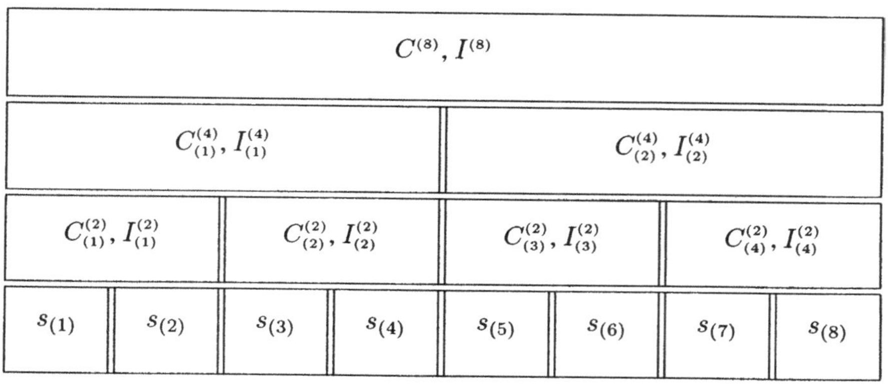

Fig. 4.21 Costs and indices which occur in shell mapping for $N = 8$.

Example 4.4: Distribution on the Constituent Constellation

Having explained the mapping algorithm, we show the distribution which is induced on the two-dimensional constituent constellation.

As an example, we consider the circular two-dimensional constellation defined in the ITU V.34 recommendation [ITU94, Figure 5]. The innermost 512 points are subsequently used. In addition, the shell mapper, according to the same standard, is applied. The frame size is $N = 8$, and we divide the constellation into $M = 8$ concentric shells. Hence, each shell contains 64 points, corresponding to $q = 6$ "uncoded" bits.

Figure 4.22 show the frequencies of the signal points a for $K = 4, 10, 17$, and 24. For each signal point a square bar whose height corresponds to the probability is drawn—this reflects the situation of discrete signal points.

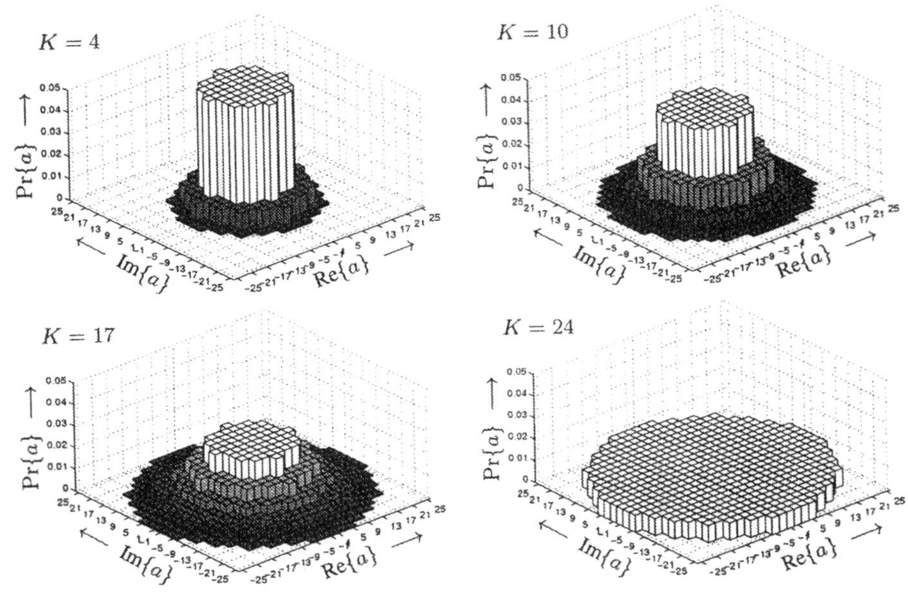

Fig. 4.22 Frequencies of the signal points obtained with shell mapping. $N = 8$, $M = 8$, $K = 4, 10, 17$, and 24.

For $K = 4$, only the $2^4 = 16$ shell combinations with the lowest total cost are used. The 64 innermost signal points (shell $s = 0$) are used most, while shell $s = 1$ and especially shell $s = 2$ are used much less often. By looking at bars with equal height, the shell arrangement is visible. Increasing K results in a distribution which becomes more uniform. Moreover, further shells are used in additionally. When choosing $K = 24$, all $2^{24} = 8^8$ shell 8-tuples are used. Then each shell occurs equally often—a uniform distribution results and no shaping gain is achieved.

Example 4.5: G_s-CER and G_s-PAR Trade-Off

This example assesses the performance of shell mapping. Using the shell mapper given in [ITU94] and the circular constellation specified there, for various numbers M of shells and for different numbers K of input bits, the relevant parameters are calculated. Each shell comprises the maximum number of signal points, such that the total number of points $M \cdot 2^q$ is less than 240. For $M = 4, 8, 16, 32$, this results in $q = 5, 4, 3, 2$.

Figure 4.23 shows the trade-off between shaping gain G_s and constellation expansion ratio $\mathrm{CER}^{(2)}$ (left-hand side) and between shaping G_s and peak-to-average energy ratio $\mathrm{PAR}^{(2)}$ (right-hand side).

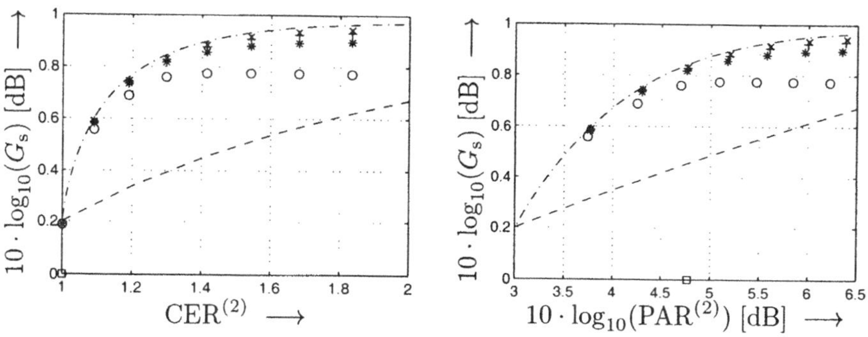

Fig. 4.23 Trade-off between G_s and $\mathrm{CER}^{(2)}$ (left) and G_s and $\mathrm{PAR}^{(2)}$ (right) for shell mapping. $N = 8$, $M = 4$ (○), 8 (∗), 16 (+), 32 (×), $K = 8 \cdot \log_2(M) - 7, \ldots, 8 \cdot \log_2(M)$. Dashed: optimum trade-off under peak constraint (16 dimensions) and for spherical shaping (cf. Figures 4.10, 4.11).

If $K = 8 \cdot \log_2(M)$ is chosen, only the shaping gain in two dimensions is possible. Reducing the number of bits entering the shell mapper, shaping gain is achieved at the price of increased constellations and increased peak-to-average energy ratio.

The dashed lines are the theoretical trade-off for (lower curve) a spherical boundary and for (upper curve) shaping under peak constraints in 16 dimensions. Since here shell mapping uses a frame size of $N = 8$ and each constituent constellation is two-dimensional, a 16-dimensional shaping is present. It can bee seen clearly that shell mapping with a sufficient number of rings enables trade-offs close to the ultimate limits, and hence can be judged as very efficient.

4.3.4 Arbitrary Frame Sizes

In the preceding discussion we have assumed that the frame size N is a power of two, and hence an iterated splitting into two problems of the same dimension is possible. Shell mapping can be straightforwardly generalized to arbitrary frame sizes. The key point is to replace equations (4.3.6a), (4.3.6b), and (4.3.8a), (4.3.8b), which assume parts of equal size, by a more general form.

Assume we want to combine an n_1-dimensional and an n_2-dimensional problem into an $(n_1 + n_2)$-dimensional one (decoding step). The cost and index of the first part are denoted as $C^{(n_1)}$ and $I^{(n_1)}$, those of the second part as $C^{(n_2)}$ and $I^{(n_2)}$. Resorting to the sorting strategy explained above, we can calculate cost $C^{(n_1+n_2)}$ and index $I^{(n_1+n_2)}$ as

$$C^{(n_1+n_2)} = C^{(n_1)} + C^{(n_2)} , \tag{4.3.12a}$$

$$I^{(n_1+n_2)} = \sum_{c=0}^{C^{(n_1)}-1} g^{(n_1)}(c) \cdot g^{(n_2)}(C^{(n_1+n_2)} - c) + I^{(n_2)} \cdot g^{(n_1)}(C^{(n_1)}) + I^{(n_1)} . \tag{4.3.12b}$$

Note that in this step generating functions $G^{(n_1)}(x)$ and $G^{(n_2)}(x)$ are required.

The encoding step from n_1+n_2 dimensions to n_1 and n_2 dimensions, respectively, is readily obtained as follows: cost $C^{(n_1)}$ of the first part is then the largest integer, for which

$$J \triangleq I^{(n_1+n_2)} - \sum_{c=0}^{C^{(n_1)}-1} g^{(n_1)}(c) \cdot g^{(n_2)}(C^{(n_1+n_2)} - c) \tag{4.3.13a}$$

is positive. The cost of the second half is $C^{(n_2)} = C^{(n_1+n_2)} - C^{(n_1)}$, and the indices are obtained as

$$\begin{aligned} I^{(n_1)} &= J \bmod g^{(n_1)}(C^{(n_1)}) \\ I^{(n_2)} &= (J - I^{(n_1)})/g^{(n_1)}(C^{(n_1)}) . \end{aligned} \tag{4.3.13b}$$

When n_1 and n_2 are selected appropriately, any frame size can be used in shell mapping. One of the simplest choices is $n_1 = 1$. Here, in each iteration, one dimension is split off, and the remaining problem has a dimensionality decreased by one. Such a procedure is described in detail in [LFT94].

However, it is more efficient to split the encoding task into problems of approximately equal dimensionality. The following examples show the sorting of shell triples and possible procedures if the frame size is $N = 6$.

Example 4.6: *Sorting of Shell Triples*

The sorting of shell triples is shown in this example. Again $M = 4$ shells are assumed, and $n_1 = 1$ and $n_2 = 2$ are chosen. Table 4.4 shows an excerpt of the $4^3 = 64$ possible combinations. Additionally, the corresponding total cost $C^{(3)}$ and index $I^{(3)}$ are given. Compare the number of entries in the table with the generating function $G^{(3)}(x) = (1 + x + x^2 + x^3)^3 = 1 + 3x + 6x^2 + 10x^3 + 12x^4 + 12x^5 + 10x^6 + 6x^7 + 3x^8 + x^9$.

Table 4.4 Sorting of shell triples, with corresponding total cost $C^{(3)}$ and index $I^{(3)}$; $M = 4$.

$s_{(1)}$	$s_{(2)}, s_{(3)}$	$C^{(3)}$	$I^{(3)}$
0	00	0	0
0	01	1	0
0	10		1
1	00		2
0	02	2	0
0	11		1
0	20		2
1	01		3
1	10		4
2	00		5
0	03	3	0
0	12		1
0	21		2
0	30		3
1	02		4
1	11		5
1	20		6
2	01		7
2	10		8
3	00		9
0	13	4	0
0	22		1
0	31		2
1	03		3
1	12		4
⋮	⋮	⋮	⋮

Example 4.7: *Shell Mapping Strategies for* $N = 6$

Figure 4.24 shows three strategies for shell mapping with frame size $N = 6$. In the first one, each iteration processes one dimension, i.e., $n_1 = 1$ is chosen. The second one uses $n_1 = 2$. Two steps are required to split the 6-dimensional problem into three two-dimensional problems. The two-dimensional mapping is done as explained above. Finally, the last strategy first splits the 6-dimensional problem into two three-dimensional problems, i.e., $n_1 = 3$. After two further steps employing $n_1 = 1$, the shell indices are obtained.

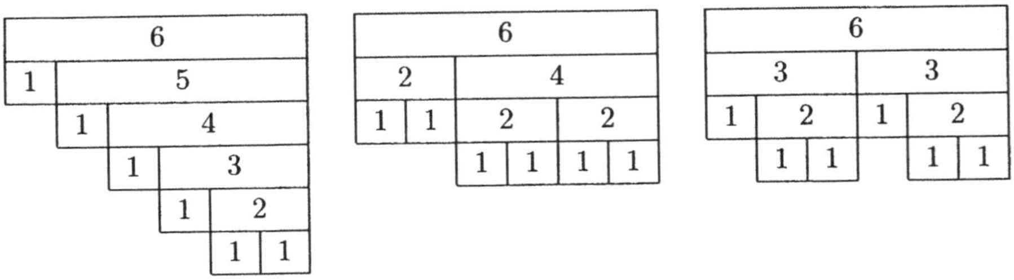

Fig. 4.24 Strategies for shell mapping with frame size $N = 6$. The numbers indicate the length of the respective frames.

Note that all strategies require five iteration steps. But for the first strategy, all generating functions from $G^{(1)}(x)$ to $G^{(6)}(x)$ have to be stored in memory. The second strategy requires generating functions $G^{(1)}(x)$, $G^{(2)}(x)$, $G^{(4)}(x)$, and $G^{(6)}(x)$, whereas the last method is based on $G^{(1)}(x)$, $G^{(2)}(x)$, $G^{(3)}(x)$, and $G^{(6)}(x)$.

Due to different sorting of vectors of the shell indices for the different strategies, the shaping gain obtained with these schemes may differ slightly.

4.3.5 General Cost Functions

In the above discussion we have focused on two-dimensional constituent constellations which are partitioned into rings of equal size. Only in this case, the simple cost function $C(s) = s$, $s = 0, 1, \ldots, M-1$, is applicable. We now turn to the problem of general cost functions, including the situation where we directly address signal points instead of rings.

When dealing with general cost functions or signal points themselves, we only have to modify a single item: The step of combining two one-dimensional problems into one two-dimensional problem—which is tightly connected with the special cost function—has to be replaced by the general algorithm (4.3.6) and (4.3.8). Instead of producing the shell index, the last encoding step then delivers the cost $C^{(1)}$ and an index $I^{(1)}$.

Again, shells (or signal points) shall be sorted according to their cost, i.e., $C(0) \leq C(1) \leq \cdots \leq C(M-1)$, which are assumed to be nonnegative integers. But, especially when treating signal points, more than one point can have a particular cost. Here, in addition to the cost, an index $I(s)$ is required for selecting the particular point or shell. Noteworthy, since for $C(s) = s$, i.e., shell index is identical to cost,

$I^{(1)} = 0$ holds, the index was of no interest in the above discussion. Table 4.5 shows an example with six shells and associated costs $C(s)$ and index $I(s)$.

Table 4.5 Example of association of costs $C(s)$ and index $I(s)$ to shells.

s	$C(s)$	$I(s)$
0	0	0
1	1	0
2	1	1
3	4	0
4	4	1
5	4	2

Based on such a table, the shell mapping encoder produces the shell number (or the signal point) from $C^{(1)}$ and $I^{(1)}$. Conversely, the shell mapping decoder translates shell index s (or signal point) into cost and index for further processing.

Example 4.8: *Shell Mapping for* 16*-ary QAM*

As a realistic example, consider 16-ary QAM. The shell mapper should directly produce a frame of N consecutive signal points. If real and imaginary parts of the signal points are taken from $\{\pm 1, \pm 3\}$, the following association (Table 4.6) of signal points to costs and indices can be set up. Note that any scaling and shifting of the cost is allowed for.

Table 4.6 Association of signal point to cost $C(s)$ and index $I(s)$ for 16-ary QAM.

Signal Point	$C(s)$	$I(s)$
$1 + \mathrm{j}$	2	0
$1 - \mathrm{j}$	2	1
$-1 - \mathrm{j}$	2	2
$-1 + \mathrm{j}$	2	3
$3 + \mathrm{j}$	10	0
$3 - \mathrm{j}$	10	1
$-1 + 3\mathrm{j}$	10	2
$-1 - 3\mathrm{j}$	10	3
$-3 - \mathrm{j}$	10	4
$-3 + \mathrm{j}$	10	5
$1 - 3\mathrm{j}$	10	6
$1 + 3\mathrm{j}$	10	7
$3 + 3\mathrm{j}$	18	0
$3 - 3\mathrm{j}$	18	1
$-3 - 3\mathrm{j}$	18	2
$-3 + 3\mathrm{j}$	18	3

Example 4.9: *Shell Mapping for One-Dimensional Constellations*

As a second example for general cost functions, we regard shell mapping based on one-dimensional constituent constellations. Although real and imaginary part of a two-dimensional signal point can be transmitted consecutively, i.e., in time multiplex, the result may be different from shell mapping directly based on one-dimensional signal constellations. Here, the total energy constraint only applies to the whole frame, whereas in QAM signaling the energy of pairs of coordinates is also limited.

We assume a uniformly spaced ASK constellation and an equal number of signal points within each shell. Then the points are grouped according to their magnitude m, whereby shell s, $s = 0, 1, \ldots, M-1$, comprises all points within the interval $A \cdot [s,\ s+1)$. If the average energy of shell s is approximated by the energy of the lower boundary, cost $C(s)$ is proportional to the squared shell index s

$$C(s) \sim s^2\,, \qquad s = 0, 1, \ldots, M-1\,. \tag{4.3.14}$$

This dependency is shown on the left-hand side of Figure 4.25.

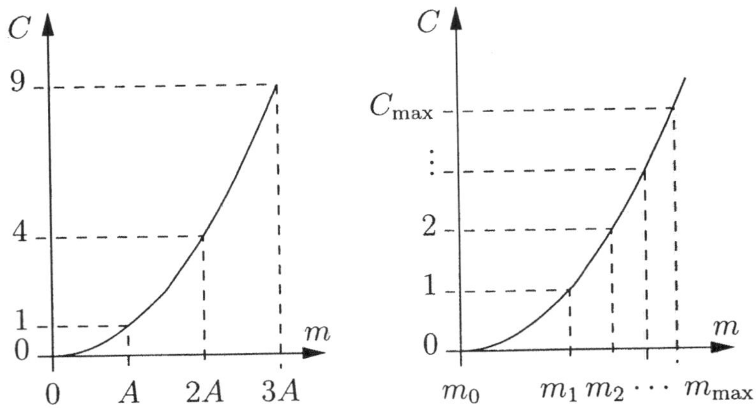

Fig. 4.25 Cost functions for shell mapping based on one-dimensional constituent constellations.

Cost function (4.3.14) has a disadvantage: because costs are nonuniformly spaced and large values can occur, sparse generating functions $G^{(N)}(x)$ of large order result. If $C_{\max} = C(s = M-1)$ denotes the maximum possible cost, $G^{(N)}(x)$ has $N \cdot C_{\max} + 1$ coefficients $g^{(N)}(c)$, from $c = 0$ to $c = N \cdot C_{\max}$. As an example, for $N = 8$ and $M = 10$ shells, $G^{(8)}(x)$ requires a table with $8 \cdot (10-1)^2 - 1 = 647$ elements, which is undesirably large.

A possible solution for requiring only small tables—but perhaps sacrificing some shaping gain—is as follows: First, M shells with an equal number of points (preferably a power of two) are formed. Then, to each shell an integer cost $C = 0, 1, \ldots, C_{\max} - 1$ is assigned. We will usually choose $C_{\max} < M$, and hence each cost C occurs more than once. Let $m_{\max}$ be the upper boundary of the outermost shell. Then an optimal assignment is made in the following way: Taking the parabola on the right of Figure 4.25 into account, the corresponding magnitude is given by $m = m_{\max} \cdot \sqrt{C/C_{\max}}$ for a given cost C. Hence, we assign cost C

to all shells whose lower boundary falls into the interval

$$\mathcal{I} = \left[\sqrt{C},\, \sqrt{C+1}\right) \cdot \frac{m_{\max}}{\sqrt{C_{\max}}}, \qquad C = 0, 1, \ldots, C_{\max} - 1\,. \tag{4.3.15}$$

This situation is shown on the right-hand side of Figure 4.25.

The advantage of this construction is that it is suited for all types of constellations, even nonuniformly spaced ones. Because $C_{\max}$ can be chosen small and the costs are again integers ranging from 0 to $C_{\max} - 1$, as in QAM signaling, the tables required for the generating functions are very small. For example, for $C_{\max} = 4$ and $N = 8$, $G^{(8)}(x)$ has only $3 \cdot 8 + 1 = 25$ coefficients.

Finally, Table 4.7 shows cost $C(s)$ and associated index $I(s)$ for $M = 8$ shells and various values for the maximum cost, and hence for the required memory size.

Table 4.7 Cost $C(s)$ and associated index $I(s)$ for shell mapping based on one-dimensional constituent constellations: $M = 8$, variation of $C_{\max}$.

$C(s)$, $I(s)$	Shell s							
	0	1	2	3	4	5	6	7
$C_{\max} = 2$	0, 0	1, 0	1, 1	1, 2	1, 3	2, 0	2, 1	2, 2
4	0, 0	1, 0	1, 1	1, 2	2, 0	3, 0	3, 1	4, 0
8	0, 0	1, 0	1, 1	2, 0	3, 0	5, 0	6, 0	8, 0
16	0, 0	1, 0	2, 0	3, 0	6, 0	9, 0	12, 0	16, 0

4.3.6 Shell Frequency Distribution

In order to assess shell mapping, in particular to give the shaping gain, the frequencies of the shells have to be known. We concentrate on two-dimensional constituent signal constellations partitioned into M shells, each of which contains 2^q points. The mapping frame size is N symbols, and K bits are mapped to the shell indices.

Let $H(s,i)$, $s = 0, 1, \ldots, M-1$, $i = 1, \ldots, N$, denote the number of occurrences of shell s in position i within all 2^K combinations. Then, a signal point $a_{s,l}$, $l = 1, 2, \ldots, 2^q$, in shell s is transmitted with probability

$$\Pr\{a_{s,l}\} = 2^{-q} \cdot 2^{-K} \cdot H(s,i) \tag{4.3.16}$$

in position i within the frame of N symbols. The transmit power is then proportional to the average energy σ_a^2 of the signal points, which is

$$\sigma_a^2 \triangleq \mathsf{E}\{|a_{s,l}|^2\} = 2^{-q} \cdot 2^{-K} \sum_{l=1}^{2^q} \sum_{s=0}^{M-1} |a_{s,l}|^2 \, \frac{1}{N} \sum_{i=1}^{N} H(s,i) \,. \tag{4.3.17}$$

Hence, for the assessment of shell mapping, the key point is the calculation of the histograms $H(s,i)$. In principle, this can be done by tabulating all possible shell combinations and counting the occurrences of the shells. But for large K it is impractical or even impossible to count over all 2^K combinations.

In Appendix D a simple but general method for the calculation of these histograms is given, cf. [Fis99]. Using *partial histograms* which give the number of occurrences of shells within all possible combinations of n-tuples that have some fixed total cost, the histogram $H(s,i)$ can be computed very easily. On the one hand, the partial histograms are easily calculated from the generating functions, and, on the other hand, only a small number of these histograms have to be combined to obtain the final result. The method has approximately the same complexity as the shell-mapping encoder, and for arbitrary parameters N, K, M, the shell frequency distribution can readily be given.

As an example, we discuss the calculation of shell frequencies in detail for the shell-mapping scheme specified for the international telephoneline V.34 modem standard [ITU94].

Example 4.10: *Histogram* $H(s,i)$

The meaning of the histogram $H(s,i)$ is shown for the V.34 shell mapper [ITU94] ($N = 8$) using $M = 3$ and $K = 5$ in order to get clear results.

Table 4.8 shows the mapping of all 2^5 possible inputs to the shell mapper (index) to the corresponding shell 8-tuples. The corresponding histogram $H(s,i)$ is displayed in Table 4.9. It results from simply counting the number of "0," "1," and "2" in each of the columns. Note that the columns correspond to the positions within one mapping frame of length 8, and the number of occurrences of the shells varies over the positions.

Table 4.8 Mapping of the input to the shell mapper to the shell 8-tuple. $M = 3$, $K = 5$.

Index	Shell 8-tuple							
00000	0	0	0	0	0	0	0	0
00001	0	0	0	0	0	0	0	1
00010	0	0	0	0	0	0	1	0
00011	0	0	0	0	0	1	0	0
00100	0	0	0	0	1	0	0	0
00101	0	0	0	1	0	0	0	0
00110	0	0	1	0	0	0	0	0
00111	0	1	0	0	0	0	0	0
01000	1	0	0	0	0	0	0	0
01001	0	0	0	0	0	0	0	2
01010	0	0	0	0	0	0	1	1
01011	0	0	0	0	0	0	2	0
01100	0	0	0	0	0	1	0	1
01101	0	0	0	0	1	0	0	1
01110	0	0	0	0	0	1	1	0
01111	0	0	0	0	1	0	1	0
10000	0	0	0	0	0	2	0	0
10001	0	0	0	0	1	1	0	0
10010	0	0	0	0	2	0	0	0
10011	0	0	0	1	0	0	0	1
10100	0	0	1	0	0	0	0	1
10101	0	1	0	0	0	0	0	1
10110	1	0	0	0	0	0	0	1
10111	0	0	0	1	0	0	1	0
11000	0	0	1	0	0	0	1	0
11001	0	1	0	0	0	0	1	0
11010	1	0	0	0	0	0	1	0
11011	0	0	0	1	0	1	0	0
11100	0	0	1	0	0	1	0	0
11101	0	1	0	0	0	1	0	0
11110	1	0	0	0	0	1	0	0
11111	0	0	0	1	1	0	0	0

Table 4.9 Histogram $H(s,i)$ corresponding to Table 4.8.

$H(s,i)$	$i =$							
	1	2	3	4	5	6	7	8
$s = 0$	28	28	28	27	26	23	23	23
1	4	4	4	5	5	8	8	8
2	0	0	0	0	1	1	1	1

Approximation In some applications an approximation to the shell frequencies is sufficient. In particular, the dependency on the position can often be ignored, and an average frequency distribution $H_{\text{avg}}(s) \triangleq \frac{1}{N}\sum_{i=1}^{N} H(s,i)$ is adequate. Note that in (4.3.17), only this average distribution comes into operation.

Using the above derivations, an approximation can be calculated very easily. In shell mapping usually $K \gg 1$ holds, e.g., K can be as large as 31 in V.34 [ITU94]. As a consequence, the total number, 2^K, of combinations of shells is well approximated by $z^{(N)}(C^{(N)})$, where the integer $C^{(N)}$, $C^{(N)} \geq 1$, is chosen such that $|z^{(N)}(C^{(N)} - 2^K|$ is minimized (here, $z^{(N)}(C^{(N)}) \geq 2^K$ is admitted). The remaining or surplus combinations are neglected. Hence, the frequencies are simply proportional to the histogram $H_s^{(N)}(C^{(N)})$ of all N-tuples with total cost less than $C^{(N)}$. From (4.3.16) and (D.1.8), the approximate frequency of signal points $a_{s,l}$, $l = 1, \ldots, 2^q$, in shell s, is thus given by

$$\begin{aligned} H_{\text{app}}(s) &\triangleq 2^q \cdot \Pr\{a_{s,l}\} \\ &\approx \frac{1}{z^{(N)}(C^{(N)})} \sum_{m=0}^{\infty} g^{(N)}(C^{(N)} - s - 1 - mM) \,, \end{aligned} \tag{4.3.18a}$$

independently of the position, where

$$C^{(N)} = \underset{c=1,2,3,\ldots}{\operatorname{argmin}} \left| z^{(N)}(c) - 2^K \right| . \tag{4.3.18b}$$

If $z^{(N)}(C^{(N)}) = 2^K$, then approximation (4.3.18) becomes exact.

Another approximation can be given from a different point of view. For large M and K, we expect the shell frequency distribution $H(s,i)$ to approach the shell frequency distribution that minimizes the average cost (energy) for a given entropy (rate). This *Maxwell-Boltzmann distribution*, derived in Section 4.1.2, is

$$H_{\text{M-B}}(s) \triangleq \Pr\{s\} = K(\lambda) \cdot e^{-\lambda C(s)} \,, \quad \lambda \geq 0 \,, \tag{4.3.19}$$

where $K(\lambda) = \left(\sum_{s=0}^{M-1} e^{-\lambda C(s)}\right)^{-1}$ again normalizes the distribution. The parameter λ is chosen so that the entropy $-\sum_{s=0}^{M-1} H_{\text{M-B}}(s)\log_2(H_{\text{M-B}}(s))$ of the distribution is equal to the desired rate.

Example 4.11: ***Shell Frequencies and Transmit Power in V.34***

In order to illustrate the validity of the approximations, examples valid for the V.34 shell-mapping scheme are given.

First, in Figure 4.26 $H(s,i)$ is plotted for $M = 9$ and $K = 18$. For comparison, the approximation $H_{\text{app}}(s)$ is given, too.

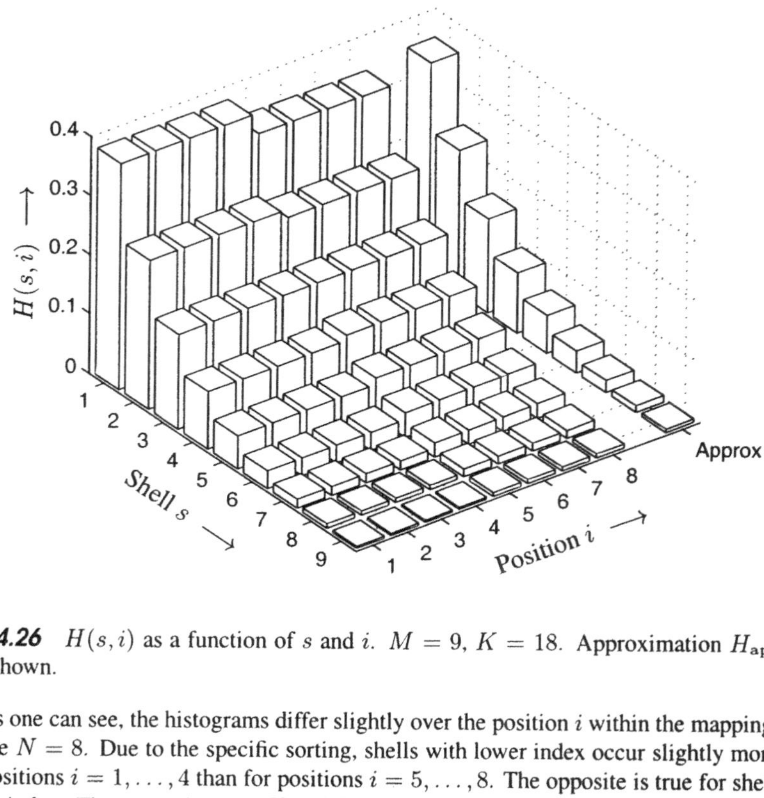

Fig. 4.26 $H(s,i)$ as a function of s and i. $M = 9$, $K = 18$. Approximation $H_{\text{app}}(s)$ is also shown.

As one can see, the histograms differ slightly over the position i within the mapping frame of size $N = 8$. Due to the specific sorting, shells with lower index occur slightly more often for positions $i = 1, \ldots, 4$ than for positions $i = 5, \ldots, 8$. The opposite is true for shells with larger index. The approximation $H_{\text{app}}(s)$ is very close to the average frequency distribution $H_{\text{avg}}(s)$.

The behavior of the distributions for different values of K is discussed in Figure 4.27. For $M = 9$ and $K = 12, 16, 20, 24$ the average frequency distribution $H_{\text{avg}}(s)$ and the approximation $H_{\text{app}}(s)$ are compared to the Maxwell-Boltzmann distribution $H_{\text{M-B}}(s)$. Here, the parameter λ is chosen so that the entropy is equal to $K/8$.

Even for low K the approximation $H_{\text{app}}(s)$ is very close to the true average frequency distribution $H_{\text{avg}}(s)$. The approximation improves as K increases. Unfortunately, the Maxwell-Boltzmann distribution $H_{\text{M-B}}(s)$ does not provide a good estimate of $H_{\text{app}}(s)$. Shells with low index occur less often than expected from the optimal entropy-power trade-off.

Finally, Table 4.10 summarizes the average energies σ_a^2 of the signal points in V.34. For a symbol rate of 3200 Hz the true average energy σ_a^2 (cf. (4.3.17)), the approximate energy $\sigma_{a,\text{app}}^2$ based on $H_{\text{app}}(s)$, and the energy $\sigma_{a,\text{M-B}}^2$ derived from the Maxwell-Boltzmann distribution are given for all possible data rates and associated mapping parameters K, M, and q [ITU94, Table 10, expanded]. The underlying signal constellation is specified in [ITU94, Figure 5].

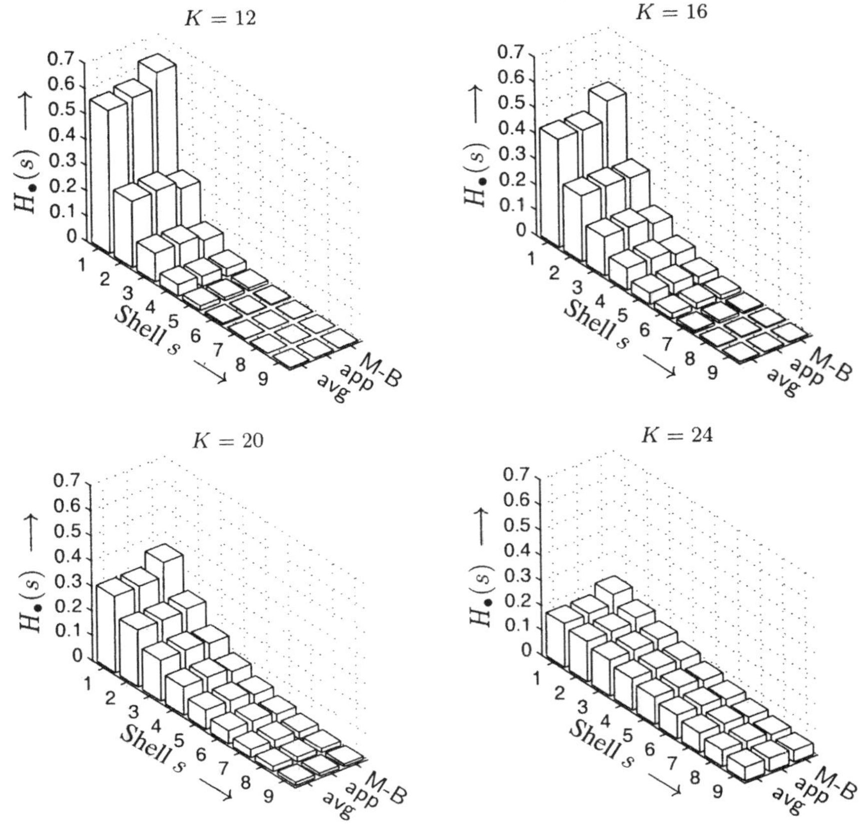

Fig. 4.27 Average frequency distribution $H_{\text{avg}}(s)$, approximation $H_{\text{app}}(s)$, and Maxwell-Boltzmann distribution $H_{\text{M-B}}(s)$ as a function of s. $M = 9$, $K = 12, 16, 20, 24$.

Again, the exact calculation and the approximation are very close. Obviously, the energies derived from the Maxwell-Boltzmann distribution underestimate the actual energies, as they are lower bounds. The approximation (4.3.18) provides much better results.

Table 4.10 Average energy σ_a^2 of the signal points in V.34. σ_a^2: true average energy, $\sigma_{a,\text{app}}^2$: approximate energy, $\sigma_{a,\text{M-B}}^2$ energy applying the Maxwell-Boltzmann distribution. Mapping parameters K, M, and q according to [ITU94, Table 10, Symbol rate 3200, expanded].

Rate [bit/s]	K	M	$4M2^q$	q	σ_a^2	$\sigma_{a,\text{app}}^2$	$\sigma_{a,\text{M-B}}^2$
4800	0	1	4	0	2.00	2.00	2.00
5000	1	2	8	0	2.50	2.00	2.14
7200	6	2	8	0	4.27	3.73	3.72
7400	7	2	8	0	4.91	4.49	4.36
9600	12	4	16	0	6.31	6.45	5.62
9800	13	4	16	0	6.89	6.95	6.19
12000	18	6	24	0	10.67	10.86	9.58
12200	19	6	24	0	11.86	11.81	10.74
14400	24	10	40	0	17.96	18.16	16.12
14600	25	11	44	0	19.44	19.25	17.44
16800	30	17	68	0	29.82	29.71	26.74
17000	31	18	72	0	32.73	32.87	29.42
19200	28	14	112	1	50.22	50.22	45.13
19400	29	15	120	1	54.96	54.48	49.43
21600	26	12	192	2	84.20	83.14	75.50
21800	27	13	208	2	91.77	91.70	82.37
24000	24	10	320	3	141.95	143.51	127.41
24200	25	11	352	3	154.51	152.95	138.60
26400	30	17	544	3	237.65	236.84	213.10
26600	31	18	576	3	260.97	262.10	234.46
28800	28	14	896	4	401.57	401.56	360.72
29000	29	15	960	4	439.92	436.09	395.52

4.4 TRELLIS SHAPING

Trellis shaping, proposed by Forney [For92], is an efficient scheme for signal shaping. The main idea is to employ tools well known from channel coding. In particular, mapping of information to the transmit sequence is done via a search through the trellis of a *shaping code*.

First, we give a motivation for such an approach. Starting from shell mapping and shaping on regions, shaping codes and trellis-based shaping are introduced. Then, shaping using convolutional codes is explained. Thereafter, a generalization to shaping regions other than a shell construction is given. In particular, lattice-based shaping is considered. Finally, numerical results showing the shaping gain are given, and some practical considerations, such as peak-to-average energy ratio, are addressed.

To conclude this section, other shaping aims than reducing average transmit power, especially spectral shaping, are briefly discussed.

4.4.1 Motivation

The following discussion relates to *shaping on regions*, introduced in Section 4.2.5. The signal constellation is partitioned into M regions (shells), and, taking a frame of N consecutive symbols into consideration, out of the M^N possible shell N-tuples, only the 2^K tuples with least total cost (energy) are selected.

Shaping Code It is natural to view the set of permitted shell N-tuples as some kind of *shaping code*. The codewords are of length N and the code symbols stem from an M-ary alphabet; the cardinality of the code is 2^K. The shaping algorithm maps the information to a codeword and the shaping decoder again extracts this information carried in the shell indices. Since the number of codewords is typically very large, encoding and decoding by resorting to code tables is not feasible. In this respect, the *shell-mapping* algorithm (cf. Section 4.3) is a very efficient way to assign shell N-tuples to information (and vice versa) without the need of excessive storage capacity.

As a very simple example for a shaping code, we consider $M = 2$ rings, a frame size of $N = 3$, and $K = 2$ information bits to be represented in the shell selection. The shaping code is then a binary code of length 3, and consists of the following $2^2 = 4$ words:

$$[0\ 0\ 0], \quad [0\ 0\ 1], \quad [0\ 1\ 0], \quad [1\ 0\ 0]\,; \tag{4.4.1}$$

i.e., only shell combinations with at most one "outer shell" are used.[9]

In the introduction to signal shaping (see Section 4.1) we have stated that shaping and channel coding are dual to each other. This fact is now exploited for designing shaping algorithms. In order to see the connection between these two fields, we

[9] For d-dimensional constituent constellations, this example results in a $3d$-dimensional constellation which is very similar to the "generalized cross constellations" discussed in [FW89].

first recall that the traditional design parameter for channel codes is their *minimum Hamming distance*.

Restricting ourselves to *linear codes*, the minimum Hamming distance is given by the *minimum Hamming weight* of the codewords (excluding the all-zero word, which is always present in linear codes). Hence, the minimum Hamming weight should be large. To the contrary, the binary shaping code (4.4.1) is designed such that its Hamming weight is at most 1.

The connection between channel and shaping codes becomes clearer, if we look at the *standard array* [Wic95]. The standard array lists the code (first row) and all its cosets. Table 4.11 displays the situation for the binary $(3, 1)$ repetition code. The

Table 4.11 Standard array of the binary repetition code of length 3.

[0 0 0]	[1 1 1]
[0 0 1]	[1 1 0]
[0 1 0]	[1 0 1]
[1 0 0]	[0 1 1]

code has two codewords ([0 0 0] and [1 1 1]) and—including the code itself—four cosets. Interestingly, the coset leaders are the code words of the above mentioned shaping code.

This principle can be generalized: Starting from a channel code, we can derive a shaping code as the set of minimum weight coset leaders. That is, a shaping code can be defined as the set of coset members which lie in the *Voronoi region* of the all-zero word. Information is then carried in the number of the coset. Figure 4.28 shows parts of a code (either in Hamming or Euclidean space) and the coset leaders constituting a shaping code.

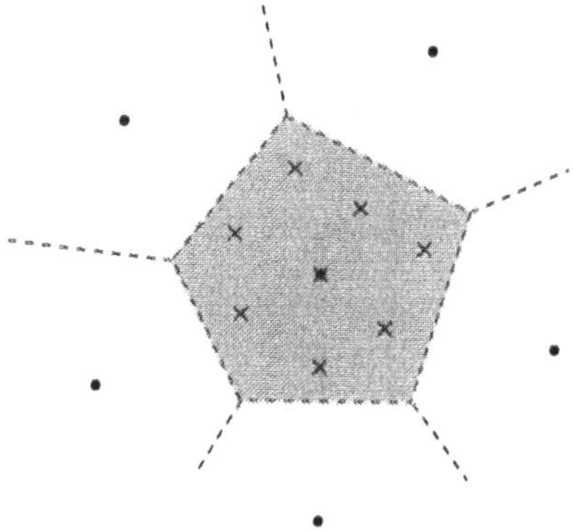

Fig. 4.28 Channel code and its cosets. Codewords are marked with •; the minimum weight coset leaders, i.e., the coset members lying in the Voronoi region of the all-zero codeword (shaded area), which give the shaping code are marked by ×.

When selecting shells, codes in Hamming space are a natural choice. But if signal points should be addressed directly, we may regard Euclidean space and its associated norm. Here, *lattice codes* are of special interest. Given an N-dimensional lattice $\boldsymbol{\Lambda}$ and a sublattice $\boldsymbol{\Lambda}'$ thereof, an N-dimensional constellation can be defined as all points of $\boldsymbol{\Lambda}$ lying in the Voronoi region of $\boldsymbol{\Lambda}'$ (which are the minimum weight coset leaders). Such so-called *Voronoi constellations* are investigated in detail in [CS83, For89]. *Voronoi shaping* would be feasible if low-complexity decoders for the underlying lattice were known. Then, shaping is performed *algorithmically* rather than by table lookup. The minimum weight point in an equivalence class (the coset) is "decoded."

Shaping by Channel Decoding We now turn to the problem of *addressing* the minimum weight coset member, which can be done in two steps.

A: Let G be the $(N-K) \times N$ *generator matrix* of the channel code, and let $\boldsymbol{H}$ be the respective $K \times N$ *parity-check matrix*, such that $\boldsymbol{G}\boldsymbol{H}^{\mathsf{T}} = \boldsymbol{0}_{(N-K)\times K}$, where $\boldsymbol{0}$ is the null matrix of the given dimension. The K-digit *syndrome*, $\boldsymbol{s}$, corresponding to a word $\boldsymbol{z}$ of length N is defined as

$$\boldsymbol{s} = \boldsymbol{z}\boldsymbol{H}^{\mathsf{T}} \, . \tag{4.4.2}$$

Sometimes, because of (4.4.2) the transpose $\boldsymbol{H}^{\mathsf{T}}$ of the parity-check matrix is called *syndrome former* [For73a, For92].

Note, arithmetic is here assumed to be done over the Galois field (finite field) $\mathbb{F}_2$. In order to distinguish between operations with real/complex number and that over the Galois fields, addition is indicated as $\oplus$, and multiplication either as juxtaposition or explicitly as $\odot$. The generalization to arbitrary Galois fields $\mathbb{F}_M$ (for the case of M shells) is straightforward.

A particular property of a coset is that all its members result in the same *syndrome*, and that the syndromes for different cosets differ. Hence, it is the syndrome which carries the information in the context of signal shaping.

Moreover, for any codeword $\boldsymbol{c} \triangleq \boldsymbol{r}\boldsymbol{G}$, the syndrome equals the all-zero word, since $\boldsymbol{s} = \boldsymbol{c}\boldsymbol{H}^{\mathsf{T}} = \boldsymbol{r}\boldsymbol{G}\boldsymbol{H}^{\mathsf{T}} = \boldsymbol{r}\boldsymbol{0} = \boldsymbol{0}$. Conversely, if we want the syndrome to carry the information, a left inverse $\left(\boldsymbol{H}^{-1}\right)^{\mathsf{T}}$ for $\boldsymbol{H}^{\mathsf{T}}$, i.e., $\left(\boldsymbol{H}^{-1}\right)^{\mathsf{T}} \boldsymbol{H}^{\mathsf{T}} = \boldsymbol{I}_{K\times K}$, can be used as a *coset representative generator*. Given the syndrome $\boldsymbol{s}$, a coset representative $\boldsymbol{z}$ can be obtained as

$$\boldsymbol{z} = \boldsymbol{s}\left(\boldsymbol{H}^{-1}\right)^{\mathsf{T}} \, , \tag{4.4.3}$$

since $\boldsymbol{z}\boldsymbol{H}^{\mathsf{T}} = \boldsymbol{s}\left(\boldsymbol{H}^{-1}\right)^{\mathsf{T}} \boldsymbol{H}^{\mathsf{T}} = \boldsymbol{s}$. Hence, there is an easy way to address a particular, but not necessarily minimum weight, coset member.

B: From its definition, each coset is given by the sum of any coset representative and all possible code words. With the coset representative $\boldsymbol{z}$, any codeword $\boldsymbol{c}$ can be added without changing the coset or the syndrome, and thus preserve information, since

$$(\boldsymbol{z} \oplus \boldsymbol{c})\,\boldsymbol{H}^{\mathsf{T}} = \boldsymbol{z}\boldsymbol{H}^{\mathsf{T}} \oplus \boldsymbol{c}\boldsymbol{H}^{\mathsf{T}} = \boldsymbol{s} \oplus \boldsymbol{0} = \boldsymbol{s} \, . \tag{4.4.4}$$

Consequently, in signal shaping the task of a "channel decoder" matched to G is to find a valid codeword which, added to the coset representative, results in the desired minimum weight coset member. But this is exactly the codeword closest to z, i.e., the codeword that will be found by a minimum-distance decoder. With regard to channel decoding, the requested minimum weight coset member corresponds to the noise added by the channel. In summary, mapping of information to a block of shell indices is thus transformed into the problem of decoding a (linear) channel code.

Finally, we recall that favorable implementations of channel decoders are based on the *trellis representation* of the code. Block codes can be described by a trellis as well, see, e.g., [Wol78, LKFF98]. Each codeword corresponds to a path through the trellis, starting and ending in the zero state. The trellis can be searched efficiently with respect to any additive cost. A particular cost is assigned to each branch (symbol); the total cost is the sum of all costs along the path. For the present situation of signal shaping based on two rings, the Hamming weight is used as cost. The "best" (minimum-cost) path is preferably found, i.e., "decoded," by applying the *Viterbi algorithm* [For73b].

Figure 4.29 sketches the basic concept of shaping by channel decoding. K bits

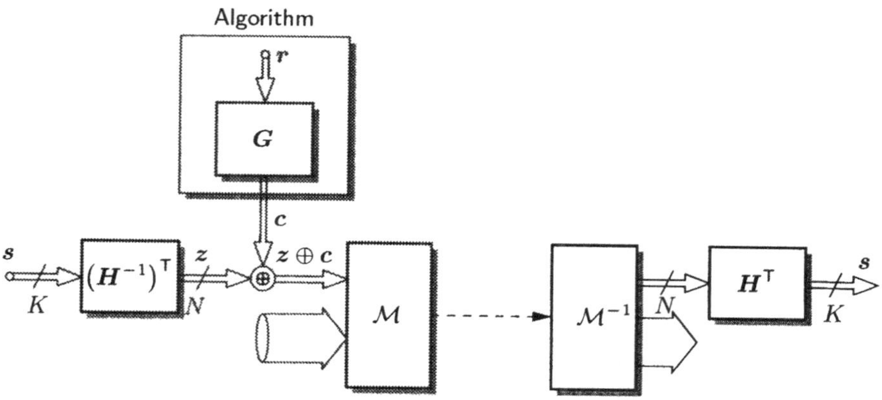

Fig. 4.29 Shaping based on channel decoding.

of information are contained in the syndrome s. Using the matrix $(H^{-1})^{\mathsf{T}}$, a coset representative z is generated. A channel decoder searches for a suitable codeword c, or equivalently for a vector r, such that $c = rG$. The sum[10] $z \oplus c$ determines a block of N shell indices, and the actual signal points are generated via mapping. At the receiver side, the inverse mapping extracts the binary labels. From the block of N shell indices, the syndrome is recovered by the parity-check matrix H^{T}.

[10]Since we operate on the Galois field $\mathbb{F}_2$, addition is the same as subtraction.

Discussion and Example The first point to be mentioned is that a generalization to an arbitrary number M of shells is straightforward. The binary channel codes are simply replaced by M-ary codes. Using the shell arrangement of Figure 4.13, as in shell mapping, the shell index may directly serve as cost in the decoding algorithm.

The addition of the suitably chosen codeword to the coset representative to obtain the final shell indices can be expressed as

$$\begin{aligned} \boldsymbol{z} \oplus \boldsymbol{c} &= \boldsymbol{s} \left(\boldsymbol{H}^{-1}\right)^{\mathsf{T}} \oplus \boldsymbol{r}\boldsymbol{G} \\ &= [\boldsymbol{s}\ \boldsymbol{r}] \begin{bmatrix} \left(\boldsymbol{H}^{-1}\right)^{\mathsf{T}} \\ \boldsymbol{G} \end{bmatrix} \\ &\triangleq [\boldsymbol{s}\ \boldsymbol{r}]\, \boldsymbol{S}\,. \end{aligned} \tag{4.4.5}$$

That is, K bits information carried in $\boldsymbol{s}$ are supplemented by $N - K$ bits *shaping redundancy* contained in the vector $\boldsymbol{r}$. Information and redundancy are then mixed, or *scrambled*, by the *scrambler matrix* $\boldsymbol{S}$ composed of coset representative generator matrix $\left(\boldsymbol{H}^{-1}\right)^{\mathsf{T}}$ and code generator matrix $\boldsymbol{G}$. For signal shaping, scrambling of (parts of) the user data and the shaping redundancy is indispensable.

Note that to recover data the scrambling matrix $\boldsymbol{S}$ does not need to have an inverse. Only the portion associated with the information $\boldsymbol{s}$ has to be invertible, which holds by construction.

Finally, due to the syndrome generation by $\boldsymbol{H}^{\mathsf{T}}$, transmission errors may be multiplied. This effect is limited to the current block, however, and also holds for shell mapping.

We illustrate this shaping technique in the following, more detailed example.

Example 4.12: *Shaping Using Hamming Codes*

As an example, we study the binary $(7, 4)$ Hamming code. Generator matrix $\boldsymbol{G}$ and the respective parity-check matrix $\boldsymbol{H}$ may be given as [Wic95]

$$\boldsymbol{G} = \begin{bmatrix} 1&1&0&1&0&0&0 \\ 0&1&1&0&1&0&0 \\ 0&0&1&1&0&1&0 \\ 0&0&0&1&1&0&1 \end{bmatrix}, \qquad \boldsymbol{H} = \begin{bmatrix} 1&0&1&1&1&0&0 \\ 0&1&0&1&1&1&0 \\ 0&0&1&0&1&1&1 \end{bmatrix}. \tag{4.4.6}$$

The $(7, 4)$ Hamming code has $2^{(7-4)} = 8$ cosets—the coset leaders are all binary 7-tuples with Hamming weight at most 1. Hence, using this Hamming code, a shaping code for frame size $N = 7$, carrying $K = 3$ information bits, can be defined. The corresponding trellis representation is shown in Figure 4.30.
It is easy to verify that a possible choice for the left inverse for $\boldsymbol{H}^{\mathsf{T}}$ is

$$\left(\boldsymbol{H}^{-1}\right)^{\mathsf{T}} = \begin{bmatrix} 1&0&0&0&0&0&0 \\ 0&1&0&0&0&0&0 \\ 0&0&0&0&0&0&1 \end{bmatrix}. \tag{4.4.7}$$

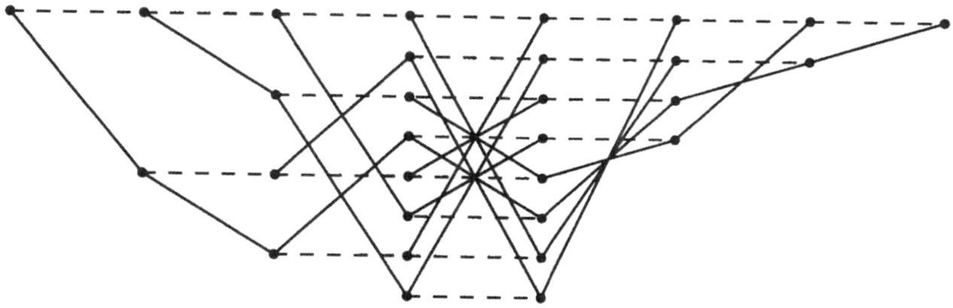

Fig. 4.30 Trellis diagram of the $(7,4)$ binary Hamming code. Solid branches represent code symbol 1, and dashed branches represent code symbol 0.

If, for example, information $\boldsymbol{s} = [0\ 1\ 1]$ should be encoded in the shell selection, first the coset representative is generated, and reads

$$\boldsymbol{z} = \boldsymbol{s}\left(\boldsymbol{H}^{-1}\right)^{\mathsf{T}} = [0\,1\,1] \begin{bmatrix} 1 & 0 & 0 & 0 & 0 & 0 & 0 \\ 0 & 1 & 0 & 0 & 0 & 0 & 0 \\ 0 & 0 & 0 & 0 & 0 & 0 & 1 \end{bmatrix} = [0\,1\,0\,0\,0\,0\,1]\,. \tag{4.4.8}$$

This vector has to be added to all possible codewords, and then the minimum weight word of this coset has to be determined. Using the code trellis, the trellis representation of the coset is simply obtained by adding the symbols of $\boldsymbol{z}$ to the corresponding code symbols, i.e., branches, of the trellis. Since we deal with binary codes, adding a 1 is the same as complementing the binary symbol. Thus, the trellis of the current coset is given by interchanging the 0's and 1's of trellis sections 2 and 7. This trellis, together with the minimum weight path (word $\boldsymbol{z} \oplus \boldsymbol{c}$) is plotted in Figure 4.31.

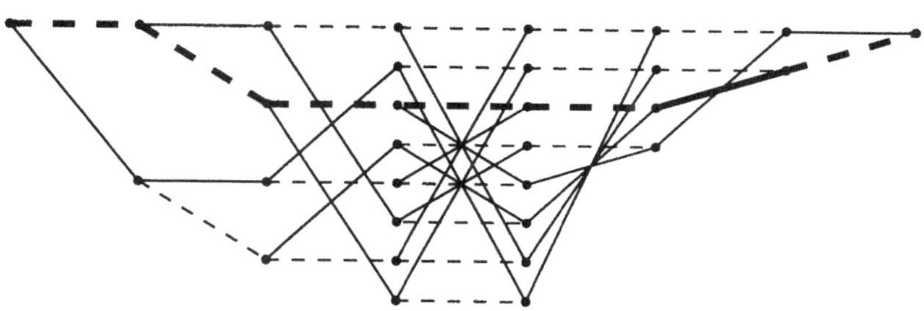

Fig. 4.31 Trellis diagram of the coset with syndrome $\boldsymbol{s} = [0\ 1\ 1]$ of the $(7,4)$ binary Hamming code. Solid branches represent code symbol '1, and dashed branches represent code symbol 0. The minimum weight word $[0\,0\,0\,0\,0\,1\,0]$ is marked.

Searching the above trellis for the minimum weight codeword, i.e., using the Hamming weight as cost, e.g., by applying the Viterbi algorithm, results in the decoded word

$$[0\,0\,0\,0\,0\,1\,0]\,,$$

which is the desired 7-tuple of shell indices. The word can be expressed as

$$[0\,0\,0\,0\,0\,1\,0] = \underbrace{[0\,1\,0\,0\,0\,0\,1]}_{z} \oplus \underbrace{[0\,1\,0\,0\,0\,1\,1]}_{c} \tag{4.4.9a}$$

$$= [0\,1\,0\,0\,0\,0\,1] \oplus \underbrace{[0\,1\,1\,1]}_{r} G\,. \tag{4.4.9b}$$

At the receiver, the sequence of shells (0 or 1) is observed, and from this 7-tuple data are recovered correctly to be

$$s = [0\,0\,0\,0\,0\,1\,0] \begin{bmatrix} 1 & 0 & 0 \\ 0 & 1 & 0 \\ 1 & 0 & 1 \\ 1 & 1 & 0 \\ 1 & 1 & 1 \\ 0 & 1 & 1 \\ 0 & 0 & 1 \end{bmatrix} = [0\,1\,1]\,. \tag{4.4.10}$$

Performing this procedure for all information triples s, the following mapping (Table 4.12) of information of shell 7-tuples is achieved by the shaping algorithm. Of course, this table is still tractable, but going to longer and nonbinary codes, the shaping algorithm proves its usefulness.

Table 4.12 Mapping of binary triples to 7-tuples of shell indices ($M = 2$ possible shells) by the shaping algorithm based on the binary $(7, 4)$ Hamming code.

s	z	$z \oplus c$
[0 0 0]	[0 0 0 0 0 0 0]	[0 0 0 0 0 0 0]
[0 0 1]	[0 0 0 0 0 0 1]	[0 0 0 0 0 0 1]
[0 1 0]	[0 1 0 0 0 0 0]	[0 1 0 0 0 0 0]
[0 1 1]	[0 1 0 0 0 0 1]	[0 0 0 0 0 1 0]
[1 0 0]	[1 0 0 0 0 0 0]	[1 0 0 0 0 0 0]
[1 0 1]	[1 0 0 0 0 0 1]	[0 0 1 0 0 0 0]
[1 1 0]	[1 1 0 0 0 0 0]	[0 0 0 1 0 0 0]
[1 1 1]	[1 1 0 0 0 0 1]	[0 0 0 0 1 0 0]

4.4.2 Trellis Shaping on Regions

All shaping techniques up to now work on blocks of symbols—mapping is done in a signal space of finite dimension. Since trellises are usually more tightly connected to *convolutional codes*, the method based on trellis decoding discussed above easily extends to this class of channel codes. This leads to a shaping method proposed by Forney [For92], which he called *trellis shaping*. As the scheme is sequence-oriented, the resulting signal constellation cannot be described in a finite-dimensional space, but has to be characterized in a infinite-dimensional sequence space. In this context, trellis shaping is to block-oriented shaping what convolutional codes are compared to block codes.

For trellis shaping, we need the definition of cosets of convolutional codes and their associated syndromes. If the entire sequence has to be known, the delay would go to infinity, which, of course, is unacceptable in a practical system. Fortunately, cosets and their respective syndromes can be specified on a symbol-by-symbol basis by applying some kind of filtering operation over a finite field. The generator matrix and syndrome former are replaced by matrices of transfer functions when dealing with convolutional codes.

First we review some basics of convolutional codes and then describe trellis shaping in its form as shaping on regions.

Preliminaries on Convolutional Codes For the characterization of convolutional codes, we have to deal with sequences whose elements are taken from a finite field. Here, we restrict ourselves to binary convolutional codes, hence we consider sequences of binary symbols or sequences of binary vectors. In order to distinguish such sequences from sequences of real/complex numbers, we use D-transform notations instead of the z-transform. A sequence $\langle s[k] \rangle$ of Galois field elements, i.e., $s[k] \in \mathbb{F}_2$, is given in terms of its D-transform as the formal power series[11]

$$s(D) \triangleq \bigoplus_{k} s[k]\, D^k \;. \tag{4.4.11}$$

The same definition applies to sequences of tuples (vectors), and we occasionally write the correspondence as $\boldsymbol{s}[k] \circ\!\!-\!\!\bullet\, \boldsymbol{s}(D)$. Structurally, the D-transform is obtained from the z-transform by substituting $D = z^{-1}$. In particular, all properties[12] of the z-transform directly translate to the D-transform.

A linear time-invariant binary rate-κ/η convolutional code $\mathcal{C}$ can be defined by a *generator matrix* $\boldsymbol{G}(D)$. This matrix has dimensions $\kappa \times \eta$, and its elements are polynomials or rational functions of polynomials in D. The code is the set of all legitimate *code sequences* which are generated as

$$\boldsymbol{c}(D) = \boldsymbol{r}(D)\boldsymbol{G}(D) \;, \tag{4.4.12}$$

[11]For clarity, operations with Galois field elements are again distinguished from operations with real/complex numbers.

[12]Except for the inverse transform, which needs a more in-depth study.

when $\boldsymbol{r}(D)$ ranges through all possible sequences of binary κ-tuples. The convolutional encoder can be realized by a κ-input η-output linear system with a finite number of encoder states. The elements of $\boldsymbol{G}(D)$ give the transfer characteristics from the inputs to the outputs.

A *syndrome former* associated with this rate-κ/η convolutional code is a linear system specified by an $\eta \times \rho$, with $\rho \triangleq \eta - \kappa$, transfer matrix $\boldsymbol{H}^{\mathsf{T}}(D)$ with maximum rank, such that

$$\boldsymbol{G}(D)\boldsymbol{H}^{\mathsf{T}}(D) = \boldsymbol{0}_{\kappa \times \rho} \,. \tag{4.4.13}$$

The *syndrome sequence* $\boldsymbol{s}(D)$ corresponding to $\boldsymbol{z}(D)$ is then

$$\boldsymbol{s}(D) = \boldsymbol{z}(D)\boldsymbol{H}^{\mathsf{T}}(D) \,. \tag{4.4.14}$$

In [For70, For73a] it is proved that for each linear time-invariant binary convolutional code there exists a feedback-free syndrome former which can be realized with the same number of states as the convolutional encoder.

Finally, a *coset representative generator* can be used to specifying a *coset representative sequence* $\boldsymbol{z}(D)$. Its $\rho \times \eta$ transfer matrix $\left(\boldsymbol{H}^{-1}(D)\right)^{\mathsf{T}}$ has to satisfy

$$\left(\boldsymbol{H}^{-1}(D)\right)^{\mathsf{T}} \boldsymbol{H}^{\mathsf{T}}(D) = \boldsymbol{I}_{\rho \times \rho} \,. \tag{4.4.15}$$

An elaboration on convolutional codes can be found in [Dho94, JZ99] or in [For70].

The Shaping Algorithm For trellis shaping we assume that the constituent constellation is partitioned into $M = 2^{\eta}$ *regions*, e.g., shells as used up to now. Moreover, $\rho < \eta$, $\rho \in \mathbb{N}$, bits should be transmitted in the shell selection.

Following Section 4.4.1, the choice of a linear time-invariant binary rate-κ/η convolutional code $\mathcal{C}_{\mathrm{S}}$ with generator matrix $\boldsymbol{G}_{\mathrm{S}}(D)$ meets our requirements. In each time instant, $\rho = \eta - \kappa$ bits can be carried by specifying the coset.

The resulting transmission scheme with trellis shaping is given in Figure 4.32. Binary information to be transmitted is split into two parts: the less significant bits are processed as usual. If trellis shaping should be combined with coded modulation, e.g., trellis-coded modulation, the data are encoded by the channel encoder. As explained in the context of shaping on regions (cf. Section 4.2.5), shaping and channel coding combine seamlessly if care is taken that each region contains an equal number of cosets of the coding lattice.

The most significant bits, combined into the sequence $\boldsymbol{s}(D)$ of binary ρ-tuples is transformed into the sequence of coset representatives $\boldsymbol{z}(D)$ by filtering with the inverse of the syndrome former

$$\boldsymbol{z}(D) = \boldsymbol{s}(D)\left(\boldsymbol{H}_{\mathrm{S}}^{-1}(D)\right)^{\mathsf{T}} \,. \tag{4.4.16}$$

Given this sequence, a trellis decoder for $\mathcal{C}_{\mathrm{S}}$—preferably a Viterbi algorithm—searches through the code trellis and determines a valid code sequence $\boldsymbol{c}(D)$. Note that the delay inherent in this decoding process is not shown in the figures. In

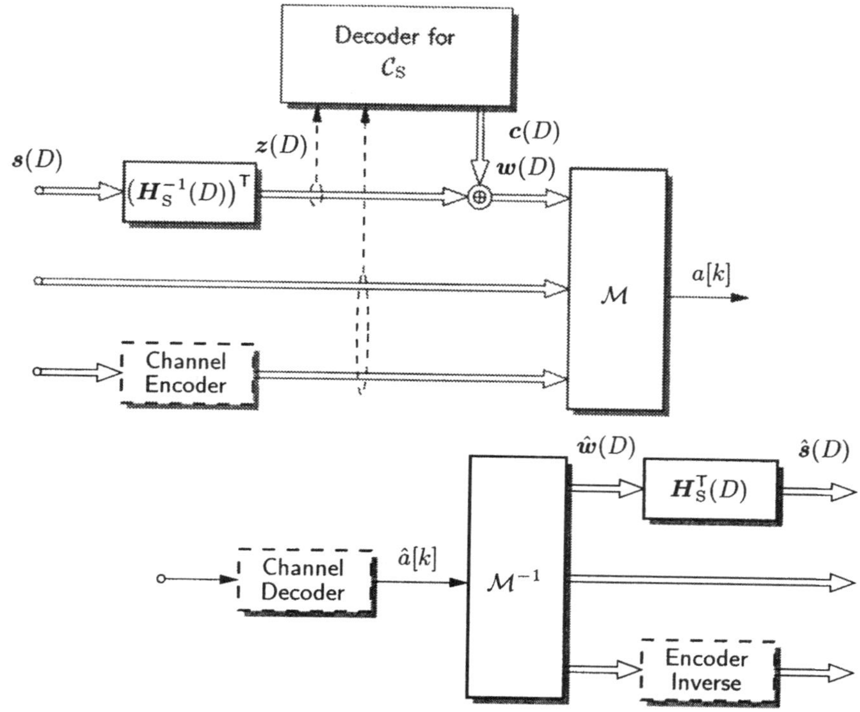

Fig. 4.32 Trellis shaping. Top: transmitter; bottom: receiver structure. Decoding delay not shown.

implementation, all signals have to be delayed by the same amount. The sum $w(D) \triangleq z(D) \oplus c(D)$ finally constitutes the sequence of shell indices or regions (binary η-tuples).

Assuming the "classical" shell arrangement (cf. Figure 4.13) and expressing the shell index as an integer number, the Viterbi decoder determines a code sequence, such that the sum over the shell indices is minimum. Thus, the branch metrics applied in the trellis decoder are simply the elements of $w(D)$ expressed as integer numbers, and the decoder looks for the minimum accumulated metrics. If only two shells are present, metric calculation is even more simple. In this case, the Hamming weight of the sequence $w(D)$ has to be minimum; cf. the discussion in Section 4.4.1.

A considerable advantage of using a decoder in the shaping algorithm is that the lower levels, i.e., the less significant bits entering the mapping, can be easily taken into account. Hence, shaping does not have to rely on approximating the signal energy by the shell number, but may consider the actual signal-point energy. This extension was already shown in Figure 4.32.

The branch metric $\lambda[k]$ in the trellis decoder is then simply chosen equal to the Euclidean norm $|a[k]|^2$ of the corresponding constellation point $a[k]$, i.e.,

$$\lambda[k] = |a[k]|^2 \, . \tag{4.4.17}$$

Using this branch metric, there is no longer any need to arrange the regions as concentric rings. Any (regular) partitioning of the signal constellation into shaping regions is possible.

At the receiver, an estimate of the PAM data symbol $a[k]$ is generated by a channel decoder or, in the case of uncoded transmission, by a threshold device. Since shaping only modifies the most significant bits entering the mapping, which are decoupled from the channel coding process, channel decoding can be done as usual. In particular, trellis shaping does not affect the complexity or the potential gain of channel coding.

The binary symbols are recovered via the inverse mapping, $\mathcal{M}^{-1}$. The less significant bits immediately determine the corresponding data bits, as they do without signal shaping. The estimates of the bits addressing the region are cotained in the sequence $\hat{\boldsymbol{w}}(D)$. From it, information is retrieved by the syndrome former. Assuming error-free transmission, i.e., $\hat{\boldsymbol{w}}(D) = \boldsymbol{w}(D)$, we have

$$\begin{aligned} \boldsymbol{w}(D)\boldsymbol{H}_{\mathrm{S}}^{\mathsf{T}}(D) &= \left(\boldsymbol{s}(D)\left(\boldsymbol{H}_{\mathrm{S}}^{-1}(D)\right)^{\mathsf{T}} \oplus \boldsymbol{r}(D)\boldsymbol{G}_{\mathrm{S}}(D)\right)\boldsymbol{H}_{\mathrm{S}}^{\mathsf{T}}(D) \\ &= \boldsymbol{s}(D)\underbrace{\left(\boldsymbol{H}_{\mathrm{S}}^{-1}(D)\right)^{\mathsf{T}}\boldsymbol{H}_{\mathrm{S}}^{\mathsf{T}}(D)}_{\boldsymbol{I}} \oplus \boldsymbol{r}(D)\underbrace{\boldsymbol{G}_{\mathrm{S}}(D)\boldsymbol{H}_{\mathrm{S}}^{\mathsf{T}}(D)}_{\boldsymbol{0}} \\ &= \boldsymbol{s}(D)\,, \end{aligned} \tag{4.4.18}$$

which recovers the initial syndrome sequence.

Since the syndrome former operates on a possibly disturbed sequence, it is essential that its implementation is feedback free. Otherwise catastrophic error propagation may occur. Fortunately, a feedback-free realization is always possible [For70, For73a].

Nevertheless, some error multiplication takes place when recovering data. Since error multiplication grows with the number of active taps in the syndrome former $\boldsymbol{H}_{\mathrm{S}}^{\mathsf{T}}(D)$, realizations with the least number of taps are preferable.

Finally we note that care should be taken that the shaping decoder produces a legitimate code sequence $\boldsymbol{c}(D)$. Under any circumstances, the "decoded" sequence has to correspond to a continuous path through the trellis. If no extra effort is made, a noncontinuous survivor path may occur due to finite path register length of the Viterbi algorithm. This in turn leads to errors in data recovery which are not caused by channel errors.

We describe a particularly simple form of trellis shaping in the subsequent example.

Example 4.13: *Sign-Bit Shaping*

Following [For92], we now give a detailed example for trellis shaping. Since "large" signal sets are preferable for shaping, we apply a 16×16 two-dimensional (256QAM) signal constellation. Since we are only interested in the performance of signal shaping, uncoded transmission is considered. Using one redundant bit for shaping, the transmission rate is 7 bits per two-dimensional symbol.

The two most significant bits involved in shaping should simply give the sign of the in-phase and quadrature component, respectively. In other words, the constellation is divided into four shaping regions which are chosen equal to the quadrants. Since only the sign of the signal is affected by shaping, this special case is called *sign-bit shaping*.

As shaping code we employ Ungerböck's 4-state, rate-1/2 convolutional code [Ung82], whose generator matrix is given by

$$\boldsymbol{G}_{\mathrm{S}}(D) = [D \quad 1 \oplus D^2] \ . \tag{4.4.19a}$$

Straightforward examination shows that a feedback-free syndrome former can be specified by the transfer matrix

$$\boldsymbol{H}_{\mathrm{S}}^{\mathsf{T}}(D) = \begin{bmatrix} 1 \oplus D^2 \\ D \end{bmatrix} , \tag{4.4.19b}$$

and an adequate inverse of the syndrome former is

$$\left(\boldsymbol{H}_{\mathrm{S}}^{-1}(D)\right)^{\mathsf{T}} = [1 \quad D] \ . \tag{4.4.19c}$$

The block diagram of the sign-bit shaping scheme is shown in Figure 4.33. The information bits q^i are divided into two parts: 6 bits enter the mapping directly, and one bit is fed to the coset representative generator. In each symbol interval, the shaping algorithm determines a redundant bit r which is input to the shaping code encoder to produce a valid code sequence. The sum of code sequence and coset representative sequence gives the "sign bits" w^1 and w^2.

Finally, the 8 bits w^1, w^2, and $q^1, \ldots, q^6$ are mapped to the complex-valued PAM signal point. Here, a standard Ungerböck set partitioning is used for labeling the points. The least significant bits select one of 64 points in each quadrant and the most significant bits address the quadrant as shown in Figure 4.34. This labeling is consistent with the set partitioning principle [Ung82].

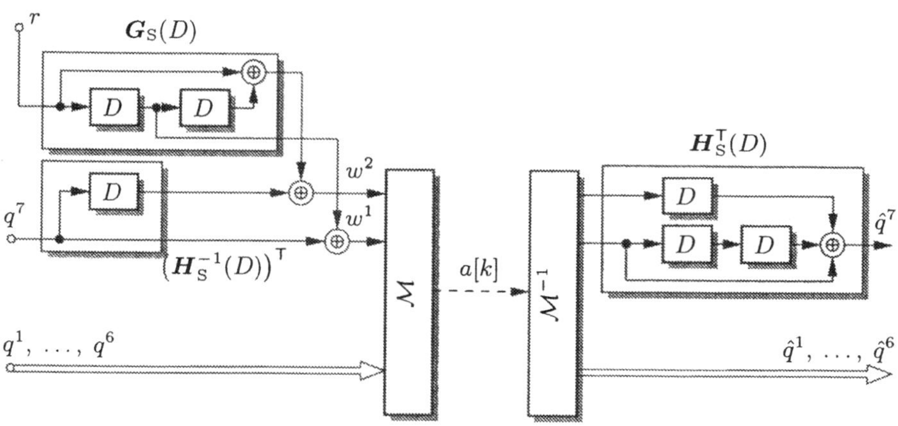

Fig. 4.33 Block diagram of sign-bit shaping scheme using a 4-state rate-1/2 convolutional shaping code.

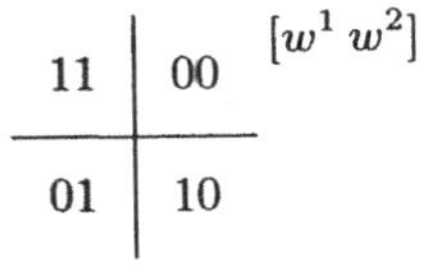

Fig. 4.34 Labeling of the quadrants in the sign-bit shaping scheme.

Shaping code encoder and coset representative generator can be combined into the *scrambler matrix* $S(D)$, which in the present example reads

$$S(D) \triangleq \begin{bmatrix} (\boldsymbol{H}_{\mathrm{S}}^{-1}(D))^{\mathsf{T}} \\ \boldsymbol{G}_{\mathrm{S}}(D) \end{bmatrix} = \begin{bmatrix} 1 & D \\ D & 1 \oplus D^2 \end{bmatrix} . \tag{4.4.20}$$

At the receiver, the six least significant bits are obtained immediately. The estimated sign bits are fed to the syndrome former which recovers the seventh information bit.

The distribution of the PAM symbols achieved by this shaping technique is shown in Figure 4.35. For each signal point a square bar is drawn, whose height corresponds to the probability. The path register length of the Viterbi algorithm, and hence the decoding delay, is adjusted to 256 symbols. The average energy of the signal points in the present example is $\mathsf{E}\{|a[k]|^2\} = 68.20$, which translates to a shaping gain of $G_s = 0.94$ dB compared to the baseline energy $2 \cdot \frac{2^7-1}{3} = 84.67$ of uniform signaling with the same rate. Please note that we use a slightly more conservative calculation of the shaping gain than in [For92].

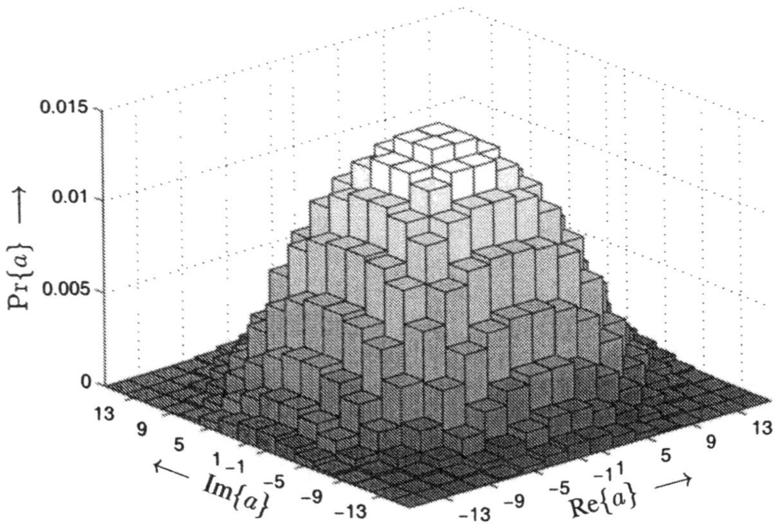

Fig. 4.35 Two-dimensional distribution of the PAM symbols $a[k]$.

As expected, the distribution resembles a Gaussian one. Compared to a 128-ary constellation with unifrom distribution (probability of the signal points equal to $1/128 \approx 0.0078$), the inner signal points are used more frequently, whereas the points at the perimeter of the constellation are rarely selected.

The shaping gain G_s over the decoding delay for the 4-state sign-bit shaping scheme is depicted in Figure 4.36. For reference, the shaping gain of a hypersphere with dimension equal to twice the decoding delay (because of the two-dimensional constituent constellation, the dimensionality is twice the length of the path register) is shown. The decisions are made over a (sliding) window whose length equals the delay. Hence, we can judge the shaping scheme to have a dimensionality given by the delay times the dimensionality of the constituent constellation, which in our case is two.

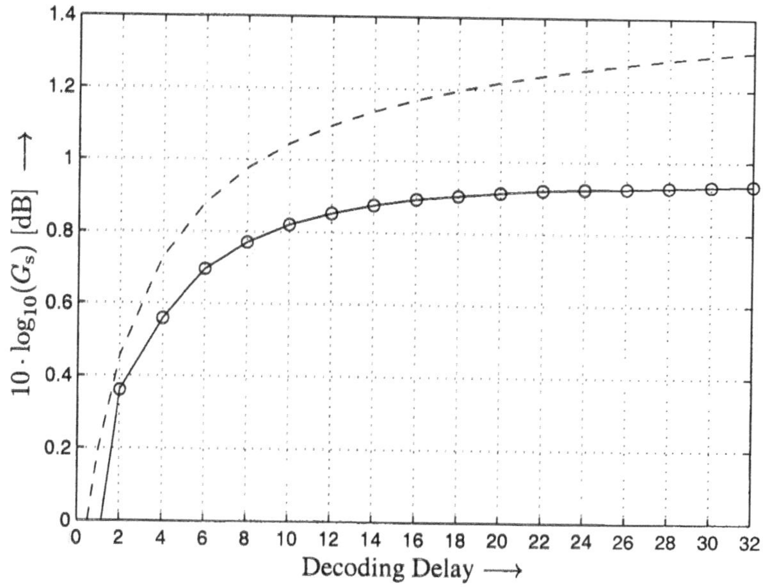

Fig. 4.36 Shaping gain G_s over the decoding delay (in two-dimensional symbols) for the 4-state sign-bit shaping scheme (solid). Dashed: shaping gain of a hypersphere with dimension equal to twice the decoding delay.

Remarkably, for small decoding window sizes large shaping gains are already achievable. For delays larger than approximately 20, the shaping gain saturates and approaches 0.94 dB. This gain is realized by a simple 4-state Viterbi algorithm.

Shaping Regions Derived from Lattice Partitions We have seen that in contrast to shell mapping, trellis shaping does not rely on a specific arrangement of regions. In principle, trellis shaping works with any partitioning of the signal set into regions. A systematic and neat approach is to base the selection of the regions on lattices and lattice partitions. Again, for an introduction to lattices, see Appendix C.

Often, the signal constellation can be described as the points of a *signal lattice* Λ_a (or a translate thereof) lying within the *Voronoi region* of some lattice Λ_b, which we here call *boundary lattice*. For example, the usual odd-coordinate square QAM constellations with 2^{2m} points emerge as the intersection of a translate (by $\boldsymbol{t} = \left[\begin{smallmatrix}1\\1\end{smallmatrix}\right]$) of the signal lattice $2\mathbb{Z}^2$ with the Voronoi region of the lattice $2 \cdot 2^m\mathbb{Z}^2$.

The shaping regions may then be specified by the *coset representatives* of a *shaping lattice* $\boldsymbol{\Lambda}_\mathrm{s}$. This, of course, requires the boundary lattice to be a sublattice of the shaping lattice. The shaping regions are translates (by the coset representative) of the fundamental region—not necessarily the Voronoi region—of $\boldsymbol{\Lambda}_\mathrm{s}$, reduced modulo $\boldsymbol{\Lambda}_\mathrm{b}$ to its Voronoi region $\mathcal{R}_\mathrm{V}(\boldsymbol{\Lambda}_\mathrm{b})$.

Moreover, if channel coding is applied, the *coding lattice* $\boldsymbol{\Lambda}_\mathrm{c}$ has to be a sublattice of the signal lattice $\boldsymbol{\Lambda}_\mathrm{a}$. Since signal shaping should not interfere with channel coding, i.e., the choice of the shaping regions may not change the coding coset, the shaping lattice has to be a sublattice of the coding lattice. Then adding an element of $\boldsymbol{\Lambda}_\mathrm{s}$ does not change the current coding coset. In summary

$$\boldsymbol{\Lambda}_\mathrm{a}/\boldsymbol{\Lambda}_\mathrm{c}/\boldsymbol{\Lambda}_\mathrm{s}/\boldsymbol{\Lambda}_\mathrm{b} \tag{4.4.21}$$

should constitute a lattice partition chain.

This property is also reflected in the mapping of the binary data to the signal point to be transmitted. The mapping may be split into three independent parts. First, a mapping $\mathcal{M}_\mathrm{c}$ comprising the portion involved in channel coding. Here a coset representative of the partition $\boldsymbol{\Lambda}_\mathrm{a}/\boldsymbol{\Lambda}_\mathrm{c}$ is selected by the binary tuple $\boldsymbol{q}^{(c)}$. Next, the uncoded levels (binary tuple $\boldsymbol{q}^{(u)}$) determine a coset representative of $\boldsymbol{\Lambda}_\mathrm{c}/\boldsymbol{\Lambda}_\mathrm{s}$ via a mapping $\mathcal{M}_\mathrm{u}$. Finally, shaping is done by choosing a coset of $\boldsymbol{\Lambda}_\mathrm{s}$ in $\boldsymbol{\Lambda}_\mathrm{b}$—mapping $\mathcal{M}_\mathrm{s}$ of binary label $\boldsymbol{q}^{(s)}$. In summary, when using regions derived from lattice partitions, mapping can be written as

$$\mathcal{M}([\boldsymbol{q}^{(c)}\ \boldsymbol{q}^{(u)}\ \boldsymbol{q}^{(s)}]) = \left(\boldsymbol{t} + \mathcal{M}_\mathrm{c}(\boldsymbol{q}^{(c)}) + \mathcal{M}_\mathrm{u}(\boldsymbol{q}^{(u)}) + \mathcal{M}_\mathrm{s}(\boldsymbol{q}^{(s)})\right) \bmod \boldsymbol{\Lambda}_\mathrm{b}\ . \tag{4.4.22}$$

The separation of the mapping can be utilized further. In [For92] it is shown that restricting ourselves to linear shaping codes and so-called linear labelings $\mathcal{M}_\mathrm{s}(\cdot)$ [For88], for which $\mathcal{M}_\mathrm{s}(\boldsymbol{q} \oplus \boldsymbol{q}') + \boldsymbol{\Lambda}_\mathrm{b} = \mathcal{M}_\mathrm{s}(\boldsymbol{q}) + \mathcal{M}_\mathrm{s}(\boldsymbol{q}') + \boldsymbol{\Lambda}_\mathrm{b}$ has to hold for every $\boldsymbol{q}$ and $\boldsymbol{q}'$. Then, the decoder for the shaping convolutional code $\mathcal{C}_\mathrm{S}$ can be replaced by a minimum-distance decoder for the trellis code based on $\mathcal{C}_\mathrm{S}$ and the lattice partition $\boldsymbol{\Lambda}_\mathrm{s}/\boldsymbol{\Lambda}_\mathrm{b}$. Information is mapped onto an initial sequence of signal points. Given this sequence, a trellis decoder searches for the nearest—with respect to Euclidean distance—trellis-code sequence. The difference between both sequences, i.e., the error sequence, is actually transmitted. By construction, this sequence lies in the infinite-dimensional Voronoi region of the trellis code. For details on lattice-theoretic shaping, the reader is referred to [For92].

Example 4.14: *Shaping and Lattices*

We continue Example 4.13, which can be interpreted as based on the lattice partition chain [For92]

$$\boldsymbol{\Lambda}_\mathrm{a}/\boldsymbol{\Lambda}_\mathrm{s}/\boldsymbol{\Lambda}_\mathrm{b} = 2\mathbb{Z}^2/16\mathbb{Z}^2/32\mathbb{Z}^2\ . \tag{4.4.23}$$

The signal points are drawn from a translate of the signal lattice $\Lambda_a = 2\mathbb{Z}^2$. The 256-point signal set is bounded by the Voronoi region of the lattice $\Lambda_b = 32\mathbb{Z}^2$, which is a square of length 32, centered at the origin.

If four regions are desired, the lattice $16\mathbb{Z}^2$ may serve as shaping lattice Λ_s, since $16\mathbb{Z}^2/32\mathbb{Z}^2$ is a 4-way partition. The quadrants of the signal constellation correspond to the fundamental region $\mathcal{R}(16\mathbb{Z}^2) = [0,\ 16)^2$, translated by the four coset representatives of $16\mathbb{Z}^2/32\mathbb{Z}^2$, namely

$$[16\mathbb{Z}^2/32\mathbb{Z}^2] = \left\{ \begin{bmatrix} 0 \\ 0 \end{bmatrix}, \begin{bmatrix} 16 \\ 0 \end{bmatrix}, \begin{bmatrix} 0 \\ 16 \end{bmatrix}, \begin{bmatrix} 16 \\ 16 \end{bmatrix} \right\} . \qquad (4.4.24)$$

Alternatively, the fundamental region $\mathcal{R}(16\mathbb{Z}^2) = [-8,\ 8)^2$ is applicable. The resulting scheme can no longer be interpreted as "sign bit" shaping, but has effectively the same performance. In particular, some regions here are noncontiguous. The difference between the shaping regions is depicted in Figure 4.37, cf. [For92].

Coding, if any, may be either based on the four-way partition $\Lambda_a/\Lambda_c = 2\mathbb{Z}^2/4\mathbb{Z}^2$ or, preferably, on the eight-way partition $\Lambda_a/\Lambda_c = 2\mathbb{Z}^2/4\boldsymbol{R}\mathbb{Z}^2$. In each case the coding lattice is a sublattice of Λ_a, but a superlattice of Λ_s, and hence of Λ_b.

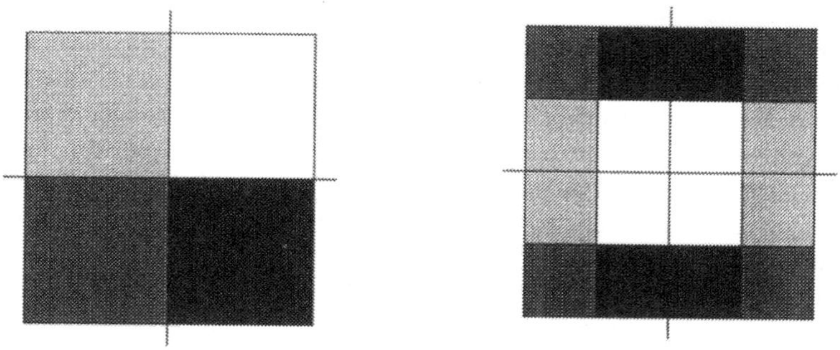

Fig. 4.37 Shaping regions for the four-way partitioning $c \cdot \mathbb{Z}^2/c \cdot 2\mathbb{Z}^2$ as translates of a fundamental region. Left: fundamental region $[0,\ c)^2$; right: fundamental region $[-c/2,\ c/2)^2$. $c \in \mathbb{N}$.

4.4.3 Practical Considerations and Performance

After having explained the basic principles and operations of trellis shaping, we now address practical consideration and give some performance results.

Choice of the Shaping Code The first question to be discussed is what shaping codes are suited for trellis shaping; this is equivalent the choice of the scrambler $S(D)$.

In Section 4.2 on bounds on shaping, we have shown (cf. Figure 4.10) that 0.5 bit redundancy per dimension is sufficient in order to achieve almost the entire shaping gain. Hence, using two-dimensional signaling, one redundant shaping bit suffices. In turn, shaping is preferably based on rate-$1/\eta$ convolutional codes. Coset representative generator and syndrome former are then $(\eta - 1)$-input η-output and η-input $(\eta - 1)$-output systems, respectively.

Using such shaping codes for one-dimensional transmission gives one redundant bit per dimension and thus leads to a larger constellation expansion. To avoid this, two consecutive symbols can be combined and shaping be applied as for two-dimensional constellations. This construction, of course, can be generalized. Combining N symbols drawn from a D-dimensional constellation into an ND-dimensional one, and selecting a rate-κ/η convolutional shaping code, the redundancy per dimension calculates to $\kappa/(ND)$. This approach allows a fine-grained adjustment of the redundancy, and hence of the constellation expansion when it is a major concern. In [For92], for example, in addition to one and two-dimensional codes, four-dimensional and eight-dimensional shaping codes are studied which allow lower CER at the price of some shaping gain.

The next point to be addressed is how large η, i.e., the size of the scrambler, should be. This question translates to that for the number 2^η of shaping regions. Again we can resort to the results of Section 4.2, although the situation is slightly different here. From Figure 4.14 we have concluded that for practice it is sufficient to use 8 or 16 shells. Since in trellis shaping the lower (unshaped) levels are taken into account in the decoding algorithm, for choice of the cost function we do not have to rely on the validity of approximating the energy by the shell number. In turn, the number of regions (shells) can be somewhat lower than in shell mapping, and four to eight regions seem to be a reasonable choice. Hence, 2×2 or 3×3 scramblers are preferable.

Finally, the number of states can be adjusted. As in channel coding, we expect the shaping gain to grow monotonously with the number of state.

Having designed all relevant parameters of the shaping code, a particular code has to be selected. With respect to shaping gain, no significant differences among the various codes are reported in [For92]. Ungerböck codes [Ung82], originally designed for channel coding, also work well for signal shaping. This, however, requires a labeling of the regions based on the set-partitioning principle. Since signal shaping and channel coding are dual operations, in [For92] it is conjectured that *dual* Ungerböck codes and a dual lattice partition—which is usually simply a scaled version of the initial lattice partition—are appropriate. The gains shown in [For92], however, are basically due to a larger scrambler and a larger number of shaping regions rather than to a better shaping code design.

The following example shows numerical results for some shaping codes.

Example 4.15: *Variation of the Shaping Code*

This example shows the shaping gain of various shaping codes. In view of the above discussions, we consider Ungerböck codes. In particular, one-dimensional (rate-$1/2$) Ungerböck codes [Ung82, Table I] are considered. Scrambler matrices for 2-state, 4-state, 8-state, and 16-state shaping codes are listed in Table 4.13. As in Example 4.13 on sign-bit shaping, the four regions are the quadrants and are labeled according to the set-partitioning principle.

Figure 4.38 shows the shaping gain (in dB) over the decoding delay of the shaping convolutional decoder (Viterbi algorithm). Except for the 2-state code, shaping gains of 0.9 to more than 1 dB can be achieved at moderate decoding delays ($\geq$ 16 symbols). Clearly, as the number of states increases, the gain improves. Even for a delay of 32 symbols, the gain has not yet saturated and slight improvements by further enlarging of the decoding window size are possible. Hence it seems that even for such large delays not all processed paths have merged into a definite survivor path.

Table 4.13 Scrambler matrix and respective syndrome former used in simulations.

Number of States	Scrambler $\boldsymbol{S}(D)$	Syndrome Former $\boldsymbol{H}_{\mathrm{S}}^{\mathsf{T}}(D)$
2	$\begin{bmatrix} 0 & 1 \\ 1 & D \end{bmatrix}$	$\begin{bmatrix} D \\ 1 \end{bmatrix}$
4	$\begin{bmatrix} 1 & D \\ D & 1 \oplus D^2 \end{bmatrix}$	$\begin{bmatrix} 1 \oplus D^2 \\ D \end{bmatrix}$
8	$\begin{bmatrix} 1 \oplus D & 1 \oplus D \oplus D^2 \\ D^2 & 1 \oplus D \oplus D^3 \end{bmatrix}$	$\begin{bmatrix} 1 \oplus D \oplus D^3 \\ D^2 \end{bmatrix}$
16	$\begin{bmatrix} 1 \oplus D & 1 \oplus D^2 \oplus D^3 \\ D^2 & 1 \oplus D \oplus D^4 \end{bmatrix}$	$\begin{bmatrix} 1 \oplus D \oplus D^4 \\ D^2 \end{bmatrix}$

Table 4.14 Scrambler matrix and respective syndrome former used in simulations.

Size of Scrambler	Scrambler $\boldsymbol{S}(D)$	Syndrome Former $\boldsymbol{H}_{\mathrm{S}}^{\mathsf{T}}(D)$
2×2	$\begin{bmatrix} 1 \oplus D & 1 \oplus D \oplus D^2 \\ D^2 & 1 \oplus D \oplus D^3 \end{bmatrix}$	$\begin{bmatrix} 1 \oplus D \oplus D^3 \\ D^2 \end{bmatrix}$
3×3	$\begin{bmatrix} D & 1 & D^2 \\ 1 & 0 & D \\ D^2 & D & 1 \oplus D^3 \end{bmatrix}$	$\begin{bmatrix} D^2 & 1 \\ 1 & D \\ D & 0 \end{bmatrix}$

The gain of using eight instead of four shaping regions, i.e., using a 3×3 scrambler instead of a 2×2 scrambler, is now assessed. The shaping code is selected to be dual to the 8-state two-dimensional Ungerböck code [Ung82, Table III]. The generator matrix $\boldsymbol{G}_{\mathrm{S}}(D)$ is thus given directly by the parity-check polynomial tabulated in the literature. Table 4.14 compares the scrambler matrix and respective syndrome former for 8-states codes. Note that the shaping regions are again obtained by standard set partitioning.

For these 8-state codes, the shaping gain (in dB) over the decoding delay is plotted in Figure 4.39. Increasing the number of information bits involved in the shaping operation provides

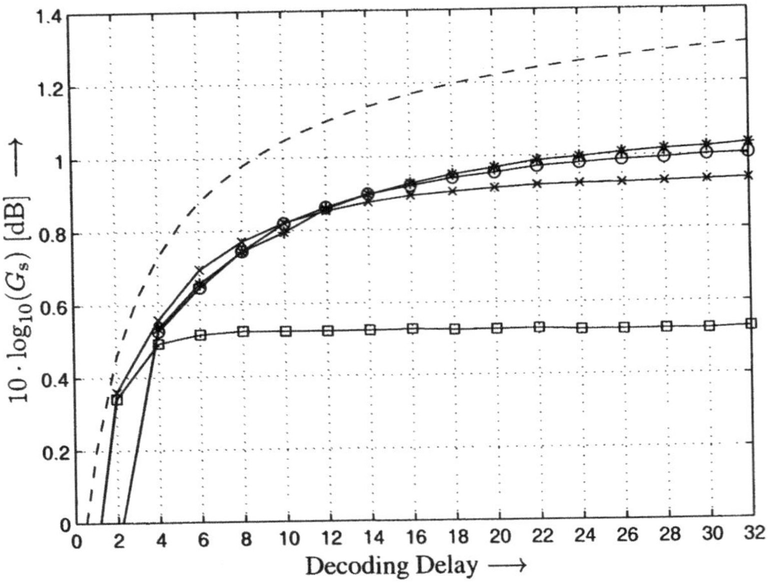

Fig. 4.38 Shaping gain G_s over the decoding delay (in two-dimensional symbols) for various shaping codes. □: 2-state code; ×: 4-state code; o: 8-state code; ∗: 16-state code. Dashed: shaping gain of a hypersphere with dimension equal to twice the decoding delay.

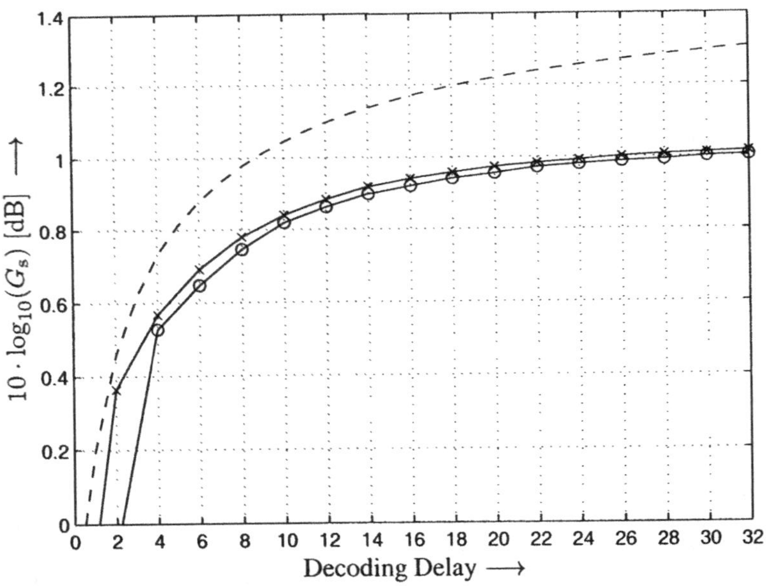

Fig. 4.39 Shaping gain G_s over the decoding delay (in two-dimensional symbols) for 8-state codes. o: 2 × 2 scrambler; ×: 3 × 3 scrambler. Dashed: shaping gain of a hypersphere with dimension equal to twice the decoding delay.

only a marginal additional gain of about 0.01 dB. This effect can again be attributed to the fact that in trellis shaping the lower levels are taken into account anyway.

To summarize, one can state that trellis shaping based on rate-$1/2$ codes with a relatively low—when compared to trellis-coded modulation—number of states (at most 16) can achieve a significant portion of the maximum possible shaping gain of 1.53 dB.

Choice of the Shaping Regions Besides the shaping code, the choice of the shaping regions is a major design parameter in trellis shaping. As shown above, a possible approach is to select the regions based on pure lattice theoretic considerations. This concept, moreover, emphasizes the duality of signal shaping and channel coding (coset codes).

But trellis shaping allows a very flexible choice of the regions. In [For92] some variants of four-way partitions are discussed. Again, no significant differences in performance are reported. But it has to be noted, that labeling of the regions and shaping code have to match. If Ungerböck codes are used, mapping by set partitioning is appropriate. For other types of labeling, corresponding codes may have to be used. For example, conventional convolutional codes designed for binary signaling fit to Gray labeling of the regions (cf. the differences between the sign-bit shaping given in Example 4.13 and Forney's sign-bit shaping [For92]).

As in shell mapping, an obvious approach is again to use concentric rings as shaping regions in trellis shaping. Spherical regions offer a substantially better peak-to-average energy ratio compared to that of a rectangular one. Using spherical regions, the labeling of the points within the rings is important. A method is given in [For92], which seems to be the only one which really works: the points of the innermost shell are labeled from the lowest-energy point to the highest-energy point. Within the next shell the points are labeled in reverse order (highest energy to lowest energy). This labeling is alternatingly repeated up to the outermost shell.

Going back to the discussion at the beginning of this section, when using a shell construction, we can do trellis shaping solely on the shell indices and ignore the actual point within the shell. The gain can be evaluated via numerical simulations by taking the actual signal energy into account in preference to using an approximate shell energy. This is done in the next example.

Example 4.16: *Trellis Shaping Using Spherical Regions*

This example assesses the shaping gain when using spherical regions. For that, the signal constellation specified in the ITU V.34 recommendation [ITU94] with 512 signal points is used, because here the points are already arranged in shells. The labeling within shells 0, 2, ... is kept, whereas the labeling of the point within shells 1, 3, ... is complemented. The resulting labeling complies with the above-mentioned requirement.

In Figure 4.40 the shaping gain (relative to uniform signaling using a square 256QAM constellation) over the decoding delay is shown for different shaping strategies, all using the same rate-$1/2$ 4-state shaping code. The curve marked by crosses is valid for trellis shaping based on spherical regions. For reference, the results for sign-bit shaping (boxes) are repeated.

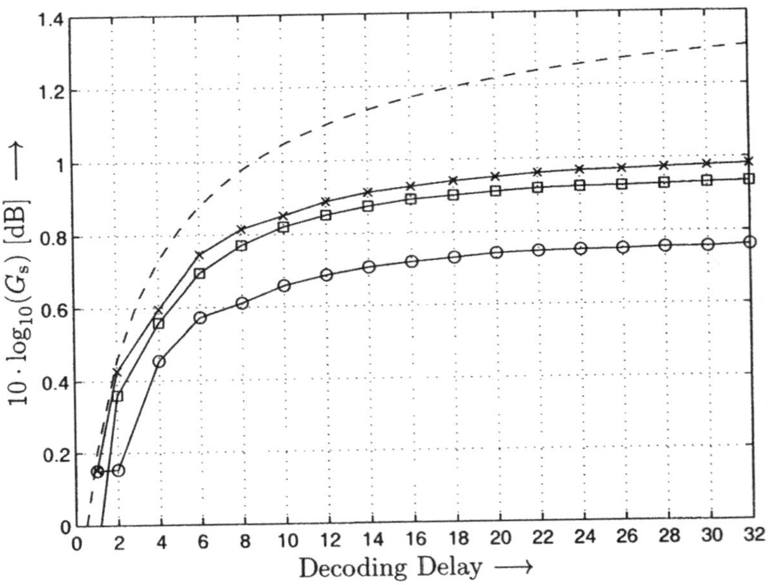

Fig. 4.40 Shaping gain G_s over the decoding delay (in two-dimensional symbols) for different shaping strategies and 4-state code. □: rectangular shaping regions; ×: spherical regions; o: spherical regions with shell index as cost. Dashed: shaping gain of a hypersphere with dimension equal to twice the decoding delay.

Remarkably, the performance differs only slightly, and only about 0.02 dB additional shaping gain is achieved when using spherical regions. But with respect to peak power, both schemes differ significantly.

In addition, the situation when disregarding the lower levels is plotted, too (circles). Here, the metric for decoding is simply given by the shell index, which ranges from 0 (innermost shell) to 3 (outermost shell). Obviously, relying on an approximation of the signal energy leads to a loss—in the present example of about 0.2 dB. Since taking the lower levels into consideration does (almost) not cost any additional complexity, trellis shaping always should use the actual signal energy. Besides this, a possible way to bridge the 0.2-dB gap is to partition the signal set into a larger number of shells, e.g., using a 3×3 scrambler together with eight shells.

Finally, Figure 4.41 shows the probability distribution obtained with trellis shaping based on spherical regions. The upper left part shows the situation when using the shell indices as costs, whereas in the lower right part the actual signal energy is used in the shaping algorithm. As in Figure 4.35, for each signal point a square bar whose height corresponds to the probability is drawn, which reflects the situation of discrete signal points.

Since the signal points within each shell are used with the same frequency, a stairstep function is visible when only the regions are considered in shaping. The different shells—each shell containing 64 signal points—can be easily distinguished. Conversely, even though the lower levels are not modified by the shaping algorithm, an almost Gaussian distribution results when the signal-point energy is taken as the branch metric. Here, the partitioning into four shells is not visible.

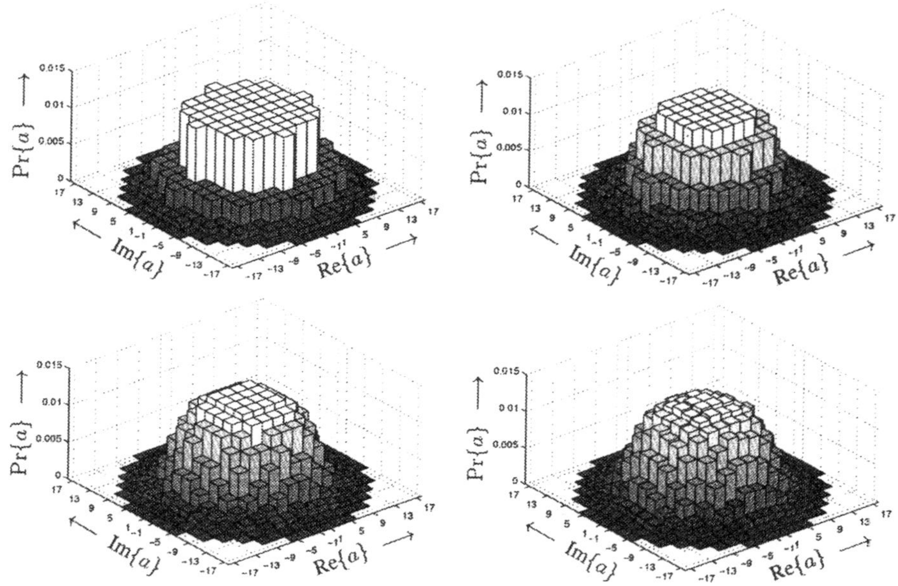

Fig. 4.41 Two-dimensional distribution of the PAM symbols $a[k]$. Top left: disregarding the lower levels; top right: taking the most significant bit of the lower levels into consideration; bottom left: considering the two most significant bits of the lower levels; bottom right: optimal shaping using signal energy for metric.

Since by simply taking the lower levels into account, but without modifying them, a higher shaping gain can be achieved, it is natural to combine both the aforementioned strategies into one design. Now, the most significant bit of the bits addressing the points within the shells is also included in the metric generation. Instead of four shells which are actually present, for metric generation eight virtual shells are used. The metric is still an integer number, now in the range of 0 to 7. The respective density of the signal points is shown in the upper right part of Figure 4.41. This construction can be continued and, e.g., the two most significant bits addressing the points within the shells are included in metric generation, too. Now, the metric increments in the Viterbi decoder are integers, ranging from 0 to 15. The lower left part of Figure 4.41 shows the corresponding density.

Surprisingly, almost the entire shaping gain can be achieved by this simple approach. The shaping gains of the four different metric generation approaches are 0.76 dB, 0.92 dB, 0.97 dB, and 0.98 dB, respectively (left to right and top to bottom in Figure 4.41). Visually, the densities become much smoother, reflecting the number of 8 and 16 shells assumed for metric generation. As already predicted in Section 4.2, using 16 shells the differences compared to optimal shaping is negligible.

Hence, assuming suitable labeling of the signal points, an actual mapping of the address label with regard to the signal energy is dispensable. A suitable part of the address bits—interpreted as a natural number—can serve as metric for trellis shaping. This saves storage and computational complexity, and the metrics in the Viterbi decoder exhibit a much smaller dynamic range.

Peak Constraints The price to be paid for achieving shaping gain is an increase in peak-to-average energy ratio and some constellation expansion. Especially when using rectangular shaping regions, the peak-to-average energy ratio can be unacceptably large. Since shaping produces a Gaussian-like distribution of the signal points, it is self-evident that high-energy signal points, which occur very rarely, can be eliminated without sacrificing much shaping gain.

Peak constraints can be easily incorporated into the trellis-shaping algorithm. The decoder has to block branches in the trellis which correspond to signal points $a[k]$ whose energy $|a[k]|^2$ exceeds the given threshold $E_{\max}$. Note that it has to be guaranteed that at least one branch leaving each state corresponds to an allowed signal point. Otherwise, the decoder may run into a dead end.

The performance of peak constraints is shown in the following example.

Example 4.17: *Shaping Under Peak Constraints*

Continuing Example 4.12 on sign-bit shaping using a rate-$1/2$ 4-state code, peak constraints are now incorporated. For a 256-ary QAM signal set (the coordinates are odd integers), Table 4.15 summarizes the number of signal points with energy at most equal to some threshold $E_{\max}$. Note that choosing $E_{\max}$ lower than 226 results in dead ends in the Viterbi algorithm.

Table 4.15 Number of signal points and constellation expansion ratio for various peak constraints.

$E_{\max}$	226	234	242	250	274	290	306	338	346	394	450
Nr. of points	180	188	192	208	216	224	232	236	244	252	256
$\mathrm{CER}^{(2)}$	1.41	1.47	1.50	1.63	1.69	1.75	1.82	1.84	1.91	1.97	2.00

Figure 4.42 shows the shaping gain over the constellation expansion ratio (left-hand side) and over the peak-to-average energy ratio (right-hand side). The decoding delay is adjusted to 32 symbols. For comparison, the trade-off for a hypersphere and the optimum trade-off under a peak constraint in 64 dimensions (32 two-dimensional symbols) are depicted.

We note that constellation expansion and peak-to-average energy ratio can be lowered significantly by sacrificing almost no shaping gain. For $\mathrm{CER}^{(2)} = 1.41$ and corresponding $\mathrm{PAR}^{(2)} = 3.27$ or 5.15 dB, about 0.9 dB shaping gain is still possible. This compares to $\mathrm{CER}^{(2)} = 2.0$ and $\mathrm{PAR}^{(2)} = 6.59$ or 8.19 dB if no peak limitation is present.

Again, the probability distribution of the signal points are shown; see Figure 4.43. Here, only points which are actually used are shown. The peak constraint is chosen to $E_{\max} = 226$, i.e., the lowest possible value. Except for the vertices of the constellation which are suppressed by the peak constraint, the distributions look almost identical.

In summary, applying peak constraints, trellis shaping is in principle able to offer a trade-off between shaping gain and constellation expansion and between shaping gain and peak-to-average energy ratio, respectively, close to theoretical limits.

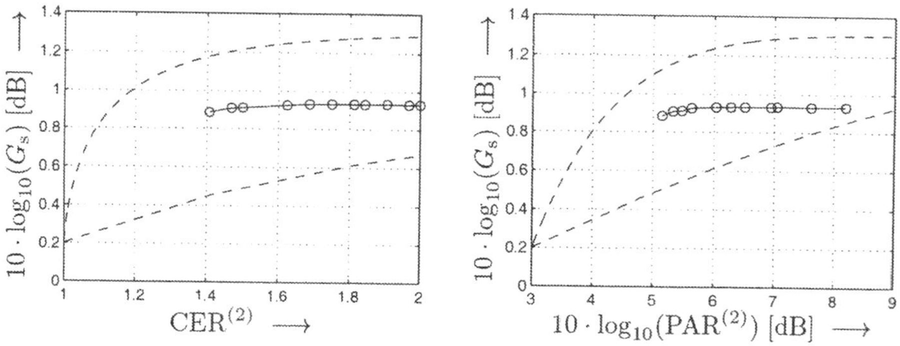

Fig. 4.42 Trade-off between G_s and $\mathrm{CER}^{(2)}$ (left) and G_s and $\mathrm{PAR}^{(2)}$ (right) for trellis shaping applying peak constraints. Dashed: optimum trade-off under peak constraint (64 dimensions) and for spherical shaping (cf. Figures 4.10, 4.11).

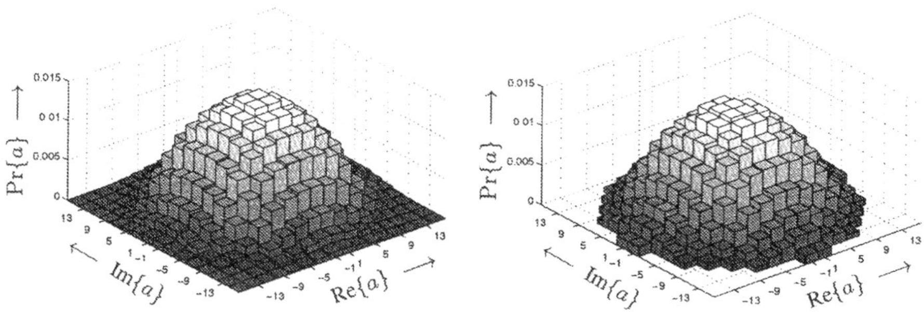

Fig. 4.43 Two-dimensional distribution of the PAM symbols $a[k]$. Left: without any restriction; right: applying peak constraint $E_{\max} = 226$.

Error Multiplication To conclude this section on trellis shaping, we remark that for data recovery a syndrome former $\boldsymbol{H}_{\mathrm{S}}^{\mathsf{T}}(D)$, i.e., a linear dispersive system, is required. Fortunately, a feedback-free realization of $\boldsymbol{H}_{\mathrm{S}}^{\mathsf{T}}(D)$ is guaranteed to exist [For70]. Hence, no catastrophic error propagation can occur, but transmission errors propagate through this filter and become effective multiple times.

The following example shows the symbol error rate for trellis shaping using different constellation sizes and the actual net gain of signal shaping.

Example 4.18: *Error Rate of Trellis-Shaping Schemes*

The symbol error rate of trellis-shaped transmission over the AWGN channel is shown in Figure 4.44. Again, the sign-bit shaping scheme using a 4-state convolutional code of Example 4.15 is employed. The constellation size is chosen to 32, 64, 128, and 256 points, corresponding to a rate of 4, 5, 6, and 7 bits, respectively.

For reference, the symbol error rates for uniform signaling over the AWGN channel are given (dashed lines). Shifting these curves by the (measured) shaping gain gives predictions for the performance of the shaped system (dashed-dotted line).

It is clearly visible that the net gain due to signal shaping is lower than the pure shaping gain, i.e., considering solely average transmit power. The whole shaping gain can be utilized only asymptotically for very low error rates. As the constellation (and hence the data rate) becomes larger, the levels involved in trellis shaping become more and more reliable. In turn, for large constellations, error multiplication is of minor impact. In contrast to this, for "small" constellations the achievable shaping is typically lower and, in addition, the error rate is increased by error propagation. Even a loss can occur for high error rates. This once again emphasizes that shaping is most suitable for "large" constellations.

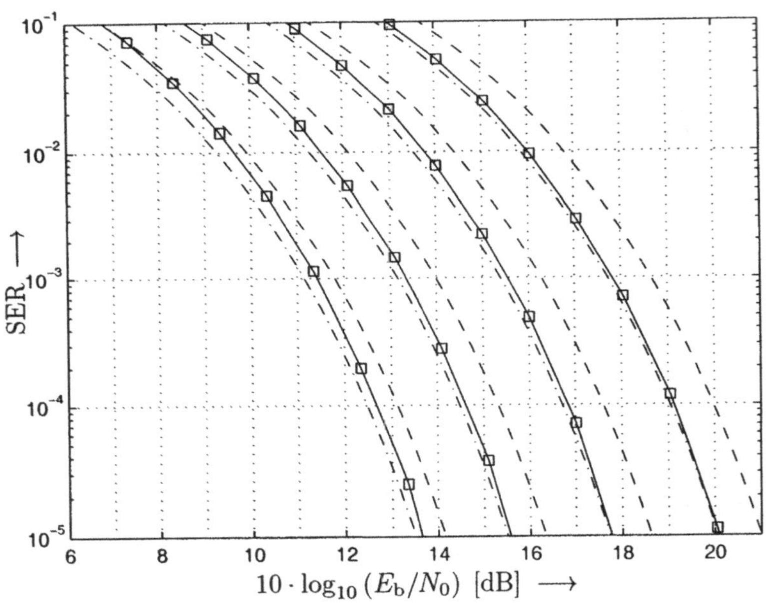

Fig. 4.44 Symbol error rates versus signal-to-noise ratio for trellis-shaped transmission over the AWGN channel. QAM constellations with 32, 64, 128, and 256 points, respectively. Dashed: symbol error rate of uniform transmission; dash-dotted: symbol error rate predicted by shaping gain.

4.4.4 Shaping, Channel Coding, and Source Coding

In the introduction to signal shaping we have argued that channel coding, source coding, and signal shaping are all dual operations. Converting a scheme, which is successfull in one of these areas, presumably leads to a good scheme in the other fields.

Trellis shaping (TS), presented by Forney [For92], is a good example of this procedure. As explained earlier, the aim of trellis shaping is to find the minimum-energy sequence out of a class of sequences representing the same message. A favorable implementation of trellis shaping is to select *regions* of a signal constellation by a shaping "decoder," and memorylessly address the points within the regions. The branch metrics in the trellis decoder are the signal-point energies.

Conversely, the aim of *trellis coded modulation (TCM)*, introduced by Ungerböck [Ung82], is to generate sequences of signal points with Euclidean distance as large as possible. The signal set is partitioned into subsets, sequences of which are selected by a convolutional code. The actual point of the subset is chosen by "uncoded" bits. At the receiver, a trellis decoder has to find the allowed code sequence which produces the observed channel output sequence, corrupted by noise, with highest probability. In the case of additive Gaussian noise, the decoding problem translates to minimizing the (squared) Euclidean distance between the received sequence and the estimated sequence in signal space.

Starting from trellis-coded modulation, and in the same spirit, Marcellin and Fischer [MF90] developed *trellis-coded quantization (TCQ)*, an efficient source-coding scheme. Here, codebooks are partitioned, and sequences of subcodebooks are selected. Again, a trellis characteristic of sequences is forced. The trellis branches are labeled with entire subcodebooks ("subsets") rather than with individual reproduction levels. The source encoder searches through the trellis to find the code sequence closest to the source sequence with respect to some distortion measure—e.g., squared Euclidean distance or Hamming distance. The encoder output consists of (a) the bit characterizing the trellis branch, and (b) the bits selecting the actual reconstruction level out of the subset. Given these two parts, data reconstruction is easily possible using a convolutional encoder which is fed with the bit sequence specifying the trellis path. The output of this encoder is a (partial) codebook, from which the representative is selected memoryless.

Finally, it should be noted that the duality between TCQ and TS can be exploited to convert quantization mean-squared errors into shaping gains. In particular, the potential gain of both schemes is limited to 1.53 dB compared to an appropriately defined baseline system [For92, ZF96, GN98].

Without further discussion, Table 4.16 compares the main features of TCM, TCQ, and TS. For a fuller treatment on these three related schemes, the reader is referred to the original literature.

Table 4.16 Comparison of trellis-coded modulation (TCM) [Ung82], trellis-coded quantization (TCQ) [MF90], and trellis shaping (TS) [For92].

	Trellis-Coded Modulation	Trellis-Coded Quantization	Trellis Shaping
Field	Channel Coding	Source Coding	Signal Shaping
Design Criterion	*Maximum Euclidean distance* between sequences of signal points	*Minimum average distortion* between original and quantized sequence	*Minimum average energy* of sequence of signal points
Lattice-Theoretic View	*Channel coding problem:* lattice with minimum probability of error *Packing problem:* consider only minimum Euclidean distance	*Quantization problem:* lattice whose Voronoi region has min. normalized second moment *Covering problem:* consider only peak distortion	*Quantization problem:* lattice whose Voronoi region has min. normalized second moment
Information Represented in	Sequence of *subsets* and point from subset	Sequence of *subsets* and point from subset	Sequence of *regions* and point from region
Encoder	Linear shift register (convolutional encoder)	Viterbi algorithm Metric: distortion	Viterbi algorithm Metric: signal-point energy
Decoder	Viterbi algorithm Metric: Euclidean distance	Linear shift register (convolutional encoder)	Linear shift register (syndrome former)
Main Complexity	Channel decoder	Source encoder	Shaping encoder
Gains	Typically 3–6 dB over uncoded transmission	Maximum 1.53 dB over scalar quantization of uniform distribution	Maximum 1.53 dB over uniform signaling

4.4.5 Spectral Shaping

Up to now, the term signal shaping has been used synonymously with "power shaping," i.e., with the aim of reducing average transmit power. This shaping property, of course, is the most important. However, sometimes other parameters than power should be influenced. Now, the signal is designed or "shaped" to comply with any specific demand.

Besides the average transmit power, in some applications the location of the power over frequency, i.e., the power spectral density (PSD) is of interest. Specifically, often low-frequency components are unwanted—the spectral components around DC should be highly attenuated. When using transformer coupling to connect a line to the transmitter, low frequency components can cause the magnetic field of the transformer to saturate, which in turn leads to nonlinear distortion, and hence a degradation of the performance. The generation of a DC-free transmit signal is a means to prevent such effects.

In the literature, the generation of DC-free signals is a large field, and is a special case of *line coding* or *data translation* [Bla90, Imm91, Ber96, GG98, And99]. Usually, some kind of *coding* is applied—either block codes or trellis/tree codes [Bla90]—which introduces redundancy. Examples for DC-free codes are the alternate mark inversion (AMI) code and HDB3 or MMS43 codes; e.g., [GG98].

Here, we show that trellis shaping is very well suited for generating spectral zeros; cf. [HSH93, FM93]. Since a trellis decoder is used, by appropriately adjusting the branch metrics, (almost) any desired property can be generated. We now derive a suitabled branch metric for spectral shaping and assess the performance of the scheme.

Basic Properties of DC-free Sequences Figure 4.45 shows the basic model for the analysis of DC-free sequences. Here, we restrict the discussion to so-called *first-order* spectral nulls, and follow the exposition in [Jus82]. We can model the DC-free sequence $\langle x[k] \rangle$ to be generated by filtering an *innovations sequence* $\langle u[k] \rangle$, which is i.i.d., with variance σ_u^2. The spectral shaping filter has a first-order rational transfer function

$$S(z) = \frac{z-1}{z-r} , \qquad |r| \leq 1 . \tag{4.4.25}$$

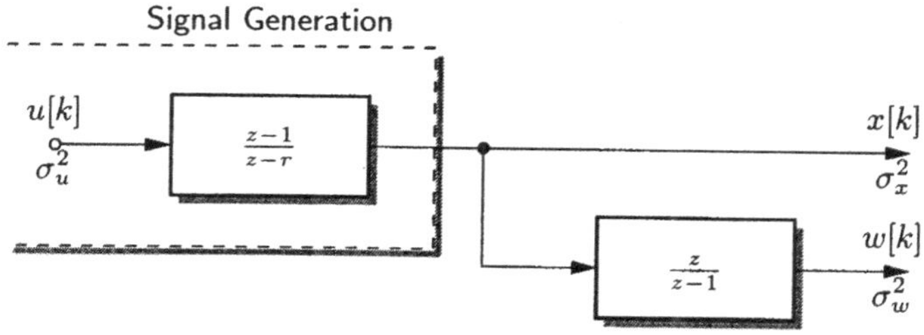

Fig. 4.45 Block diagram for the characterization of DC-free sequences.

The parameter r, the magnitude of which has to be smaller than one for stable filters, controls the width of the spectral null. By construction, the power spectral density and its first derivative vanish at $f = 0$. In general, for a *spectral null of order* Ω at frequency f, the spectral density and its first $2\Omega - 1$ derivatives have to vanish at f [EC91].

Now, the PSD of $\langle x[k] \rangle$ calculates to

$$\begin{aligned} \Phi_{xx}(\mathrm{e}^{\mathrm{j}2\pi fT}) &= \sigma_u^2 \cdot \left|S(\mathrm{e}^{\mathrm{j}2\pi fT})\right|^2 = \sigma_u^2 \cdot \left|\frac{\mathrm{e}^{\mathrm{j}2\pi fT} - 1}{\mathrm{e}^{\mathrm{j}2\pi fT} - r}\right|^2 \\ &= \sigma_u^2 \cdot 2 \frac{1 - \cos(2\pi fT)}{1 + r^2 - 2r\cos(2\pi fT)} \\ &= \sigma_x^2 (1 + r) \frac{1 - \cos(2\pi fT)}{1 + r^2 - 2r\cos(2\pi fT)} \,. \end{aligned} \tag{4.4.26}$$

In the last step, we introduced $\sigma_x^2 = T \int_{-1/(2T)}^{1/(2T)} \Phi_{xx}(\mathrm{e}^{\mathrm{j}2\pi fT})\, \mathrm{d}f$, which by using the general expression [Sch94, p. 102]

$$\sigma^2 = T \int\limits_{-1/(2T)}^{1/(2T)} \left|\frac{b_1 \mathrm{e}^{\mathrm{j}2\pi fT} + b_0}{\mathrm{e}^{\mathrm{j}2\pi fT} + c_0}\right|^2 \mathrm{d}f = \frac{b_1^2 + b_0^2 - 2b_1 b_0 c_0}{1 - c_0^2} \,, \tag{4.4.27}$$

calculates to $\sigma_x^2 = 2\sigma_u^2/(1 + r)$.

It is common to define the cutoff frequency f_0 for discrete-time processes as the "half-power frequency," i.e., as [Jus82, FC89]

$$\Phi_{xx}(\mathrm{e}^{\mathrm{j}2\pi f_0 T}) \overset{!}{=} \frac{\sigma_x^2}{2} \,. \tag{4.4.28}$$

Considering (4.4.26), this leads to the condition

$$1 - \cos(2\pi f_0 T) = \frac{(1 - r)^2}{2} \,, \tag{4.4.29}$$

which, by using the Taylor series expansion $1 - \cos(x) \approx x^2/2$, can be approximated to be

$$2\pi f_0 T \approx 1 - r \,. \tag{4.4.30}$$

Next, we study the *running digital sum (RDS)* of the DC-free sequence. The RDS $\langle w[k] \rangle$ is obtained by accumulating $\langle x[k] \rangle$,

$$w[k] = \mathrm{RDS}\left\{x[k]\right\} \triangleq \sum_{\kappa \le k} x[\kappa] \,, \tag{4.4.31}$$

i.e., filtering the sequence with a filter with transfer function $z/(z-1)$.

Since the accumulator exhibits a spectral pole at DC, its output can only be bounded if its input has no DC components. Conversely, if the RDS of a sequence assumes only values within a finite range, the power spectral density of this sequence vanishes as DC. It is proved (see, e.g., [Jus82, Pie84]) that the finite running digital sum condition is necessary and sufficient for spectral nulls at $f = 0$.

The PSD of the RDS sequence is given by

$$\begin{aligned}\Phi_{ww}(e^{j2\pi fT}) &= \sigma_u^2 \cdot \left|S(e^{j2\pi fT})\right|^2 \cdot \left|\frac{e^{j2\pi fT}}{e^{j2\pi fT}-1}\right|^2 = \sigma_u^2 \cdot \left|\frac{e^{j2\pi fT}}{e^{j2\pi fT}-r}\right|^2 \\ &= \sigma_u^2 \cdot \frac{1}{1+r^2-2r\cos(2\pi fT)} \\ &= \sigma_w^2(1-r^2)\frac{1}{1+r^2-2r\cos(2\pi fT)} \ . \end{aligned} \tag{4.4.32}$$

Again (4.4.27) has been used to substitute σ_u^2 by the variance $\sigma_w^2 = \sigma_u^2/(1-r^2)$ of the RDS sequence.

Using $\sigma_x^2 = 2\sigma_u^2/(1+r)$ and $\sigma_w^2 = \sigma_u^2/(1-r^2)$, we can eliminate the innovations power σ_u^2 and obtain a relationship between the variance of the sequence and the variance of its RDS, which reads

$$\frac{\sigma_x^2}{\sigma_w^2} = 2(1-r) \ . \tag{4.4.33}$$

Using this relation, (4.4.30) can finally be rewritten as

$$f_0 \approx \frac{1}{2\pi T}\frac{\sigma_x^2}{2\sigma_w^2} \ . \tag{4.4.34}$$

Hence, for fixed variance σ_x^2, we can state that the cutoff frequency f_0 becomes larger, and the corresponding notch in the PSD becomes larger as the variance σ_w^2 of the RDS decreases. Additionally, by measuring σ_x^2 and σ_w^2, the width of the null can be estimated very accurately.

The above statements hold as long as the PSD is not too dissimilar from a first-order power spectrum [FC89]. In particular, it is irrelevant how the sequence is generated—it is not required that the DC-free sequence is obtained through filtering. Figure 4.45 was just a model for the derivation. A disadvantage of using a linear filter is that the output samples, in general, are real numbers, even if the input assumes only a finite number of levels.

Returning to signal shaping, we may control the variance of the RDS, and hence create a spectral null at DC. Trellis shaping can be applied as usual, but the "power metric" $|a[k]|^2$ is replaced by the *instantaneous energy* of the RDS, i.e., $|\sum_{\kappa\leq k} a[\kappa]|^2$. In other words, power shaping is done with respect to the running digital sum. Doing this, we can expect the power spectral density to tend to a first-order spectrum.

The above discussion can easily be extended to spectral nulls at other frequencies than $f = 0$. If a spectral null at frequency f_0 is desired for complex sequences, we replace the spectral shaping filter by $S(z) = (ze^{-j2\pi f_0 T}-1)/(ze^{-j2\pi f_0 T}-r)$, which is the initial filter shifted in frequency by f_0, i.e., modulated. The integrator is consequently replaced by the system $ze^{-j2\pi f_0 T}/(ze^{-j2\pi f_0 T}-1)$, which performs an accumulation of modulated samples: $\sum_{\kappa \leq k} x[\kappa] e^{-j2\pi f_0 T \kappa}$. In particular, a null at Nyquist frequency $1/(2T)$—so-called *"Nyquist-free sequence"* [ISW98]—may be generated by looking at the alternating RDS $\sum_{\kappa \leq k} (-1)^\kappa x[\kappa]$.

Finally we note that higher-order spectral nulls can be obtained by taking the Ωth-*order running digital sum* into account, which is defined as [EC91]

$$\mathrm{RDS}^{(\Omega)}\left\{x[k]\right\} \triangleq \sum_{\kappa_1 \leq k} \sum_{\kappa_2 \leq \kappa_1} \cdots \sum_{\kappa_\Omega \leq \kappa_{\Omega-1}} x[\kappa_\Omega] \,. \tag{4.4.35}$$

Second-order spectra (DC^2 constraint) are, e.g., obtained by controlling the *running digital sum sum* [Imm85, MP89].

Trellis Shaping for Spectral Nulls We now show the performance of spectral shaping using trellis shaping with an example. Again, sign-bit shaping employing the simple 4-state rate-1/2 convolutional code is considered. Following the above discussions, the instantaneous energy $|\sum_{\kappa \leq k} a[\kappa]|^2$ of the running digital sum gives the branch metrics.

Example 4.19: *Trellis Shaping for Spectral Nulls*

Continuing Example 4.13 on sign-bit shaping, the energy of the PAM symbols is replaced by the energy of the running digital sum as branch metrics. All other parameters, e.g., constellation and/or shaping convolutional code, are unchanged; the decoding delay is chosen to be 16 2D symbols.

Figure 4.46 shows the numerical results. In the upper part, the situation for spectral shaping is plotted. On the right-hand side, the estimated (using the Welch-Bartlett method [OS75, Kay88] employing a Hann window) power spectral density is shown, and on the left-

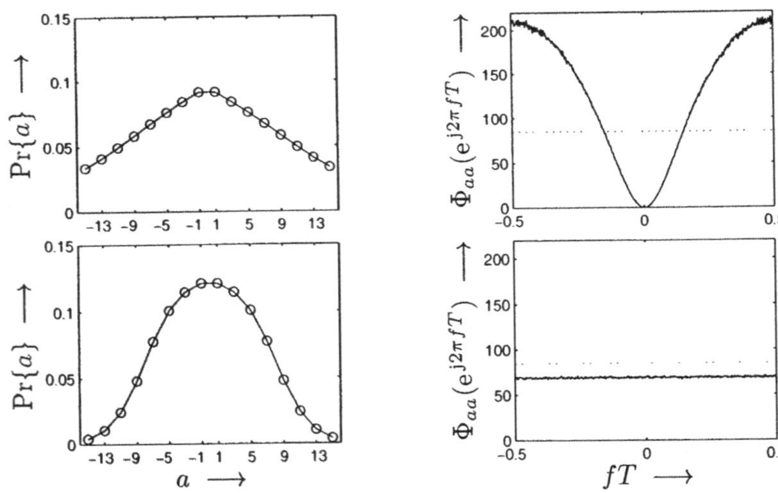

Fig. 4.46 Marginal one-dimensional distribution (left) and estimated power spectral density (right) of the PAM symbols $a[k]$. Top: spectral shaping; Bottom: power shaping.

hand side, the one-dimensional marginal distribution of the PAM symbols $a[k]$ is given. For reference, in the lower part, the respective graphs for power shaping are displayed. It can be clearly seen that spectral shaping is able to generate the desired spectral null at DC. But the spectral null is bought with an increase in average transmit power. Compared to uniform signaling with the same rate (7 bits per QAM symbol), a loss of of 1.79 dB occurs; the shaping "gain" (in dB) is negative. The distribution of the channel symbols is neither uniform nor Gaussian.

In contrast, for power shaping the distribution is close to Gaussian. In addition, the power spectral density is flat, i.e., trellis shaping produces a white transmit sequence. In spite of the shaping algorithm, the sequence of regions appears to be uncorrelated.

The dotted line shown in the power spectral densities holds for uniform, uncoded, and unshaped signaling; the variance of the PAM symbols is here given by $\sigma_a^2 = 2\frac{2^7-1}{3} = 84.67$. For white discrete-time processes, the PSD is constant with value σ_a^2. The shaping gain of 0.89 dB for trellis shaping is visible as a lowering of the power spectrum.

The preceding example shows that the straightforward generation of a spectral null is accompanied by an undesired increase in average transmit power. This is evident, since average transmit power is no longer regarded in the shaping algorithm.

To mitigate this effect, the metrics for spectral shaping with that for power shaping may be combined [FM93]. The trellis decoder then uses

$$\lambda[k] = \mu \cdot \left|\sum\nolimits_{\kappa \leq k} a[\kappa]\right|^2 + (1-\mu) \cdot |a[k]|^2 \tag{4.4.36}$$

as branch metrics. The parameter μ, $0 \leq \mu \leq 1$, governs the trade-off between spectral shaping ($\mu = 1$) and pure power shaping ($\mu = 0$).

The following example shows the exchange of shaping gain (average transmit power) and width of the spectral null.

Example 4.20: *Trellis Shaping for Spectral Nulls II*

Once again, Examples 4.13 and 4.19 on sign-bit shaping are continued. The metrics in the trellis decoder is now the linear combination (4.4.36) of power metrics and spectral metrics. In Figure 4.47 the variation of the parameter μ is assessed.

The parameter μ is selected to be 0.0, 0.2, 0.4, 0.6, 0.8, and 1.0. As μ increases, spectral shaping becomes more dominant, and as a result the width of the notch around DC increases. At the same time, the average transmit power also increases, and the initial 0.89-dB shaping gain turns into a loss of 1.79 dB. The distribution of the PAM symbols changes from Gaussian to a more uniform one.

The right column shows the power spectral density predicted by theory. From the measured variance σ_a^2 of the PAM symbols and the variance σ_w^2 of the RDS sequence cutoff frequency, f_0 and position r of the pole of the spectral shaping filter $S(z)$ are calculated using (4.4.33) and (4.4.34). Then, (4.4.26) gives the expected power spectral density. Cutoff frequency $\pm f_0$, where $\Phi_{aa}(e^{j2\pi f_0 T}) = \sigma_a^2/2$, is marked with circles. An almost perfect match of theory and numerical results is visible.

Finally, Figure 4.48 quantifies the trade-off between cutoff frequency f_0 and shaping gain G_s. The parameter μ varies from 0.0, i.e., pure power shaping, in steps of 0.1 to 1.0, i.e., pure

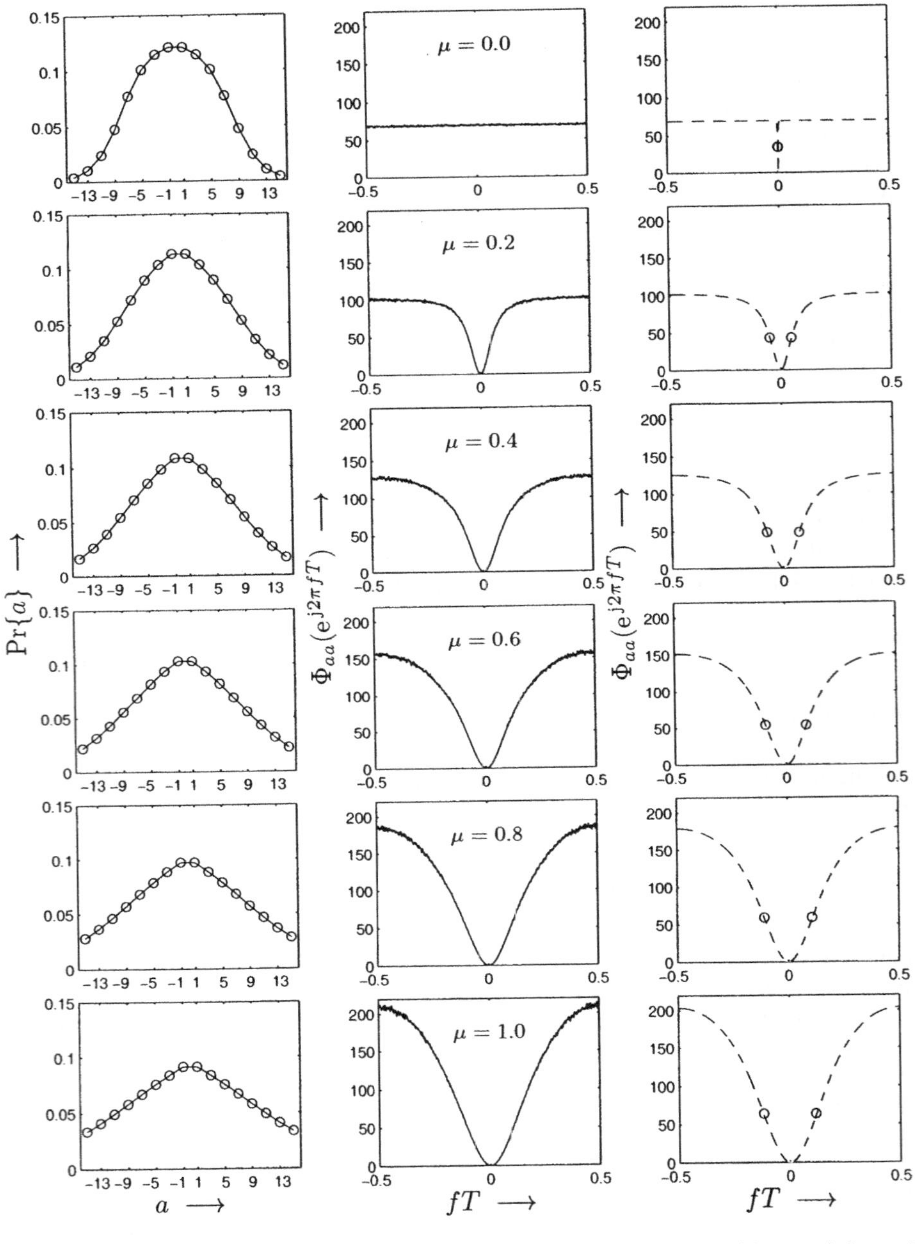

Fig. 4.47 Marginal one-dimensional distribution (left), estimated (middle), and theoretic (right) power spectral density of the PAM symbols $a[k]$. Top to bottom: variation of the parameter μ.

spectral shaping. The broader the spectral null desired, the higher the average transmit power. For cutoff frequencies above approximately $0.04/T$ the shaping gain turns into a loss.

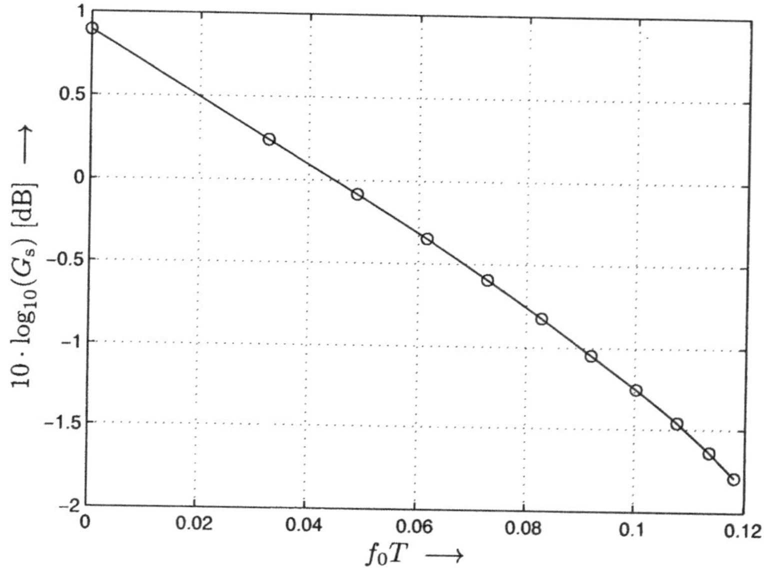

Fig. 4.48 Trade-off between cutoff frequency f_0 and shaping gain G_s (in dB) for combined spectral and power shaping.

Convolutional Spectral Shaping During the standardization activities for the ITU telephoneline modem standard V.90, a spectral shaping technique called *convolutional spectral shaping* was proposed [PCM97c, PCM97d, KE99]. This technique can be interpreted as trellis shaping with a particular choice for the scrambler and a particular mapping. Convolutional spectral shaping operates on blocks of N consecutive (one-dimensional) PAM symbols. As in sign-bit shaping, only the signs of the symbols are affected. Due to in-band signaling used in some telephone networks, which uses a frame size of 6 symbols, the frame size for spectral shaping is chosen to be $N = 6$ [ITU98] or an integer fraction thereof ($N = 3$ and $N = 2$). Here we concentrate on $N = 6$—the other cases are obtained by simply shortening the respective matrices. A simple 2-state rate-$1/6$ code with generator matrix[13]

$$G(D) = [1 \quad 1 \oplus D \quad 1 \quad 1 \oplus D \quad 1 \quad 1 \oplus D] \tag{4.4.37}$$

[13]Compared to the presentation here, in [ITU98] the order of the sign bits is reversed.

has been specified as the shaping convolutional code. The syndrome former can then be selected as

$$
\boldsymbol{H}^{\mathsf{T}}(D) = \begin{bmatrix} 1\oplus D & 0 & 0 & 0 & 0 \\ 0 & 1\oplus D & 0 & 0 & 0 \\ 0 & 0 & 1\oplus D & 0 & 0 \\ 0 & 0 & 0 & 1\oplus D & 0 \\ 0 & 0 & 0 & 0 & 1\oplus D \\ 1 & 1\oplus D & 1 & 1\oplus D & 1 \end{bmatrix}, \tag{4.4.38}
$$

and a suitable coset representative generator reads

$$
\left(\boldsymbol{H}^{-1}\right)^{\mathsf{T}} = \begin{bmatrix} 1/(1\oplus D) & 0 & 0 & 0 & 0 & 0 \\ 0 & 1/(1\oplus D) & 0 & 0 & 0 & 0 \\ 0 & 0 & 1/(1\oplus D) & 0 & 0 & 0 \\ 0 & 0 & 0 & 1/(1\oplus D) & 0 & 0 \\ 0 & 0 & 0 & 0 & 1/(1\oplus D) & 0 \end{bmatrix}. \tag{4.4.39}
$$

A close look at $\boldsymbol{H}^{\mathsf{T}}(D)$ and $\left(\boldsymbol{H}^{-1}\right)^{\mathsf{T}}$ reveals that the five initial sign bits are simply differentially encoded, i.e., accumulated, at the transmitter. The last sign bit is set to 0. Then, the block of six sign bits is modified according to the possible code symbols: they are either kept as they are or all are inverted if the encoder is in state 0, or the bits at even-numbered positions are inverted or the bits at odd-numbered positions are inverted if the encoder is in state 1. At the receiver the sign bits are basically differentiated and again modified as in the transmitter. This is easily possible, since the last sign bit directly gives the state of the encoder at the transmitter side.

The best transmit sequence can be decoded by a 2-state Viterbi algorithm. Because of the simple scheme, a brute-force search through the code tree is also possible; cf. [PCM97d]. Typically, the look-ahead depth (decoding delay) is chosen to be very small. In the V.90 standard, delays in the range from 0 (symbol-by-symbol decision) to 3 frames are specified.

Decoding is based on the *running filter sum (RFS)*, a generalization of the running digital sum. Instead of using an integrator, a rational, at most second-order transfer function, can be specified, the output of which serves as the shaping criterion. This allows more flexibility and, due to the second-order filter, two spectral zeros, e.g., at DC and Nyquist frequency, can be generated simultaneously.

4.4.6 Further Shaping Properties

To conclude this section on trellis shaping, we briefly discuss two further applications of trellis shaping other than reducing average transmit power or spectral shaping.

Since shaping is done via a search algorithm, any cost function can be used in trellis shaping. Hence, any shaping aim can be achieved as long as a suitable branch metric can be defined. The subsequent two examples for applications of trellis shaping give an impression of its flexibility and generality. In the next chapter, where combined precoding and shaping techniques are discussed, further design aims for signal shaping are introduced and corresponding metrics are presented.

Mitigation of Envelope Fluctuations A constant envelope of the transmit signal is desirable especially in mobile applications. This is because these signals can be amplified efficiently without suffering nonlinear distortions due to AM/AM or AM/PM conversion of the power amplifier. After pulse shaping, even conventional phase-shift keying (PSK) exhibits envelope fluctuations. The nonlinear behavior of the amplifier then leads to an increase of the out-of-band power, and hence causes adjacent channel interference.

In [Mor92, HSH93, LR94b, LR94a], trellis shaping is applied to reduce the envelope fluctuations of filtered 8-ary PSK. In particular, transitions crossing the origin or passing near, i.e., phase changes around 180°, have to be avoided. Trellis shaping then operates on phase differences rather than the PSK symbols themselves.

Since there is no analytical connection between phase changes and out-of-band power, the metric has to be chosen heuristically. In [HSH93] the branch metric of a $\pm x^\circ$ transition is selected to be proportional to the out-of-band power of a degenerated PSK signal with only two signal points spaced by x°. Conversely, in [LR94b, LR94a] the metric grows exponentially with the phase difference.

For this application of trellis shaping, two or more PSK symbols should be combined into one shaping step, leading to a redundancy lower than one bit per symbol. Given 8-ary PSK with one redundant bit, it is more efficient to apply simple $\pi/4$-shift (D)QPSK [Bak62, Rap96] rather than resorting to a much more complex trellis-shaping scheme. For details on trellis shaping for the mitigation of envelope fluctuations, the reader is referred to the above-cited papers.

PAR Reduction in OFDM It is well known that the transmit signal in multicarrier transmission, e.g., in *orthogonal frequency division multiplexing (OFDM)* or *discrete multitone (DMT)*, is near Gaussian distributed, and hence exhibits an extremely high peak-to-average power ratio. The signal is specified in frequency domain as a large number of individual carriers and then transformed to time domain. In the literature, a large number of so-called PAR reduction schemes are discussed (see, e.g., [MBFH97, Fri97, Tel98] and the references therein) which mitigate the PAR problem by avoiding OFDM blocks with high peaks.

Recently, trellis shaping was proposed for this application [HW96, HW97, HW00]. The main problem in doing so is that OFDM operates on blocks of symbols, and within a block each frequency-domain symbol influences each time-domain symbol; there is no causality as in single-carrier transmission. Hence, in principle, a decision on the actual transmit symbols has to be made taking the entire OFDM symbol into consideration.

Nevertheless, in [HW00] it is proposed that an OFDM symbol be subdivided into sub-symbols and that a metric for these blocks be derived. Basically, two different "metrics" are presented: one in the time domain (peak amplitude) and the other in the frequency domain, where block transitions are assessed. Note that the costs used here do not satisfy the criteria necessary for being a metric in the strict mathematical sense. In particular, additivity does not hold. Even so, simulation results show that trellis shaping is also in principle applicable to reduce the PAR of a multicarrier signal.

4.5 APPROACHING CAPACITY BY EQUIPROBABLE SIGNALING

It is well accepted that for approaching the capacity of the additive white Gaussian noise (AWGN) channel, the channel input has to have a (continuous) Gaussian distribution. The aim of constellation shaping, as discussed throughout this chapter, is to generate such a (discrete) Gaussian channel input signal. We will now briefly show that the capacity of the AWGN channel can also be approached by equiprobable signaling. Instead of generating nonuniform probabilities for uniformly spaced signal points, it can also be done the other way round. Equiprobable signal points are considered now, but the spacing of the signal points will be suitably adjusted.

In this section, the main statements concerning equiprobable signaling and channel capacity are sketched. A practical approach of generating such signals is given. Finally, the mapping of binary data to such constellations is addressed.

4.5.1 AWGN Channel and Equiprobable Signaling

We consider an AWGN channel with real input x and output y. It is well known that the capacity for an average energy constraint $\mathsf{E}\{|x|^2\} \le \sigma_x^2$ for the channel input symbols x is approached by a continuous Gaussian channel input [Gal68, CT91], i.e.,

$$\mathsf{f}_x(x) = \frac{1}{\sqrt{2\pi\sigma_x^2}} \cdot \mathrm{e}^{-x^2/(2\sigma_x^2)} \; . \tag{4.5.1}$$

The corresponding channel output is also Gaussian, with variance $\sigma_y^2 = \sigma_x^2 + \sigma_n^2$, where σ_n^2 denotes the variance of the additive white Gaussian noise n.

In [ST93, Sch96] it is proved that by using equiprobable signaling the capacity can also be approached asymptotically. We consider M discrete signal points, each used with probability $1/M$, but the coordinates x_i, $i = 1, 2, \ldots, M$, of the points are chosen appropriately.

Starting from the optimal Gaussian distribution (4.5.1), this density is partitioned into M intervals, such that the probability for each interval is the same. In other words, $M + 1$ boundaries ξ_i, $i = 0, 1, \ldots, M$, with

$$-\infty = \xi_0 < \xi_1 < \cdots < \xi_{M-1} < \xi_M = \infty \; , \tag{4.5.2}$$

are selected from the real line, such that

$$\int_{\xi_{i-1}}^{\xi_i} \frac{1}{\sqrt{2\pi\sigma_x^2}} \mathrm{e}^{-x^2/(2\sigma_x^2)} \, \mathrm{d}x \stackrel{!}{=} \frac{1}{M} \, , \qquad i = 1, 2, \ldots, M \tag{4.5.3}$$

holds for all intervals $i = 1, 2, \ldots, M$. The coordinates x_i of the signal points are then given as the *centroids* of the intervals $[\xi_{i-1}, \xi_i]$, namely,

$$x_i = \int_{\xi_{i-1}}^{\xi_i} x \cdot \frac{M}{\sqrt{2\pi\sigma_x^2}} \mathrm{e}^{-x^2/(2\sigma_x^2)} \, \mathrm{d}x \, , \qquad i = 1, 2, \ldots, M \; . \tag{4.5.4}$$

Constructing the signal set in such a way, the capacity of the underlying channel can be approached as the number M of signal points goes to infinity. The proof given in [ST93] first shows that the channel input still satisfies the average energy constraint. Using the Cauchy-Schwarz inequality (e.g., [BB91]), the average energy of the symbols x can be upper bounded by the variance σ_x^2, i.e.,

$$\frac{1}{M}\sum_{i=1}^{M} x_i^2 \leq \sigma_x^2 \tag{4.5.5}$$

holds. The probability density of the channel output is given by

$$\begin{aligned} \mathsf{f}_y(y) &= \mathsf{f}_x(y) * \mathsf{f}_n(y) = \frac{1}{M}\sum_{i=1}^{M} \mathsf{f}_n(y - x_i) \\ &= \frac{1}{M}\sum_{i=1}^{M} \frac{1}{\sqrt{2\pi\sigma_n^2}} \cdot \mathrm{e}^{-(y-x_i)^2/(2\sigma_n^2)} \; . \end{aligned} \tag{4.5.6}$$

A lengthy derivation in [ST93] reveals that the absolute deviation of this pdf relative to a Gaussian one with variance $\sigma_y^2 = \sigma_x^2 + \sigma_n^2$ is bounded by

$$\left| \mathsf{f}_y(y) - \frac{1}{\sqrt{2\pi\sigma_y^2}} \cdot \mathrm{e}^{-y^2/(2\sigma_y^2)} \right| \leq \frac{4}{\sqrt{2\pi\sigma_n^2 M}} \,, \quad \forall y \,. \tag{4.5.7}$$

Hence, as $M \to \infty$, the channel output pdf converges to that of an AWGN channel with Gaussian input, namely, a Gaussian one with variance σ_y^2. Then, as this convergence is uniform with respect to y, differential entropy $h(Y) \triangleq -\int \mathsf{f}_y(y) \log_2(\mathsf{f}_y(y))\, dy$ of the channel output tends to $\frac{1}{2}\log_2(2\pi\mathrm{e}\sigma_y^2)$, and mutual information $I(X;Y) = h(Y) - h(N)$ approaches that for Gaussian input.

The questions which remain are first, how to address such constellations. Since the number of signal points does not need to be a power of two, simple parsing of the binary data will not always be possible. Second, it is still unknown how to do efficient channel coding for such nonuniformly spaced signal sets. Since this nonuniform spacing can be viewed as the result of some kind of fading channel, it can be conjectured that at least very long, powerful codes can be used as on the conventional AWGN channel.

The following example shows some constellations according to (4.5.2) through (4.5.4).

Example 4.21: *Optimal Signal Set for Equiprobable Signaling*

In Figure 4.49 the evolution of the optimal signal set for equiprobable signaling and of the pdf of the channel output is displayed for various numbers M of signal points. For reference, the optimal Gaussian channel input is plotted as well. The variance of the channel input is limited to $\sigma_x^2 = 1.0$, and the variance of the additive Gaussian noise is set to $\sigma_n^2 = 0.1$.

Note that in each case the input energy constraint is met. The variance of the channel input is only $\sigma_x^2 = 0.6366$ for $M = 2$, but approaches one as M increases. In practice, after having

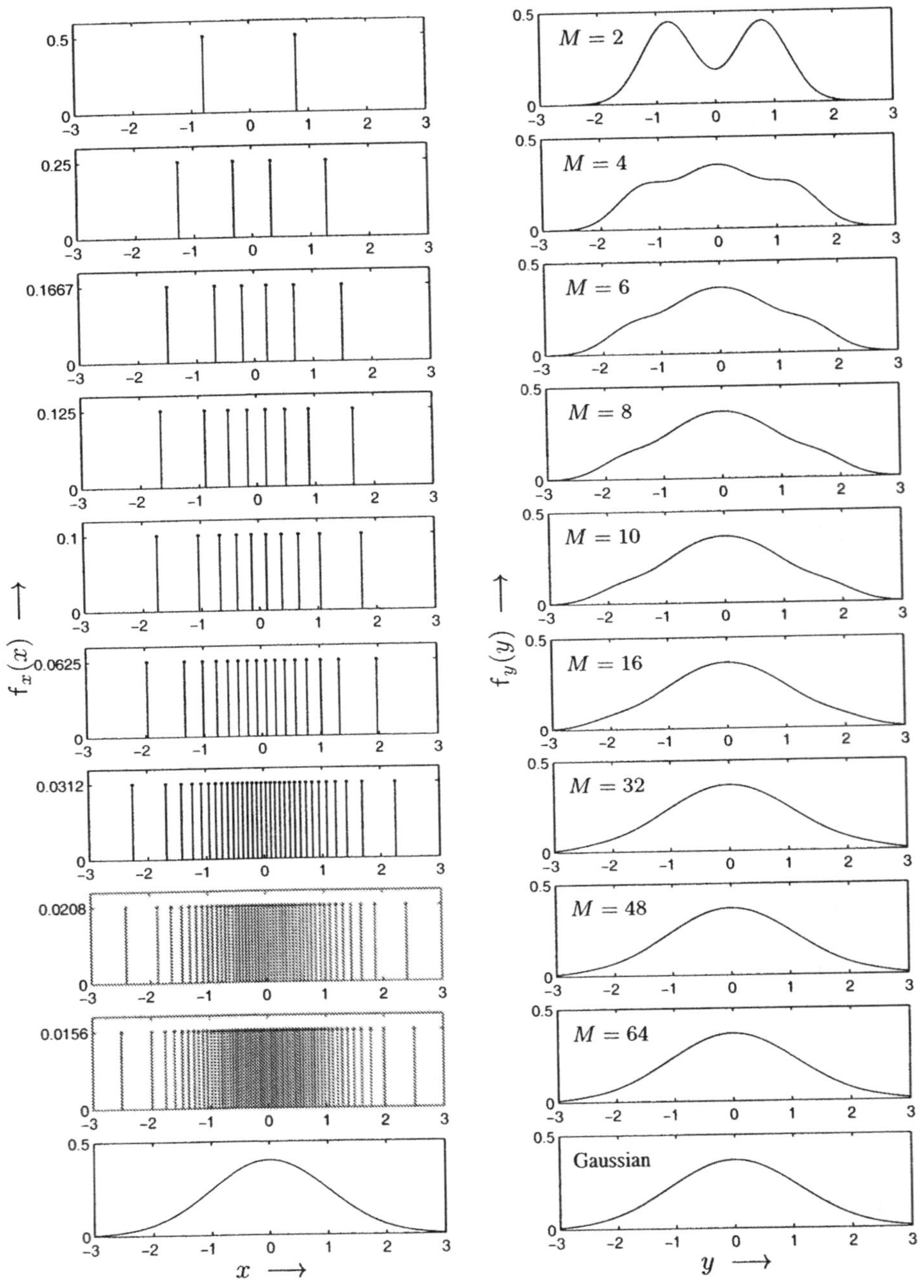

Fig. 4.49 Evolution of the optimal signal set for equiprobable signaling and corresponding pdf of the channel output. Reference: optimal Gaussian channel input. Parameters: $\sigma_x^2 \leq 1.0$, $\sigma_n^2 = 0.1$.

designed the constellation according to (4.5.2) through (4.5.4), the signal set can be scaled in order to adjust average transmit power to the desired value.

For $M = 16$ the channel output already looks quite Gaussian and is almost indistinguishable from the channel output for an optimal continuous Gaussian channel input. If the variance of the channel noise is lower, the convergence is somewhat slower (cf. (4.5.7)), but the basic statements still hold.

As in signal shaping, nonuniform signal spacing increases the peak-to-average energy ratio of the signal constellation. But in contrast to signal shaping, no constellation expansion with respect to the number of signal points is required. However, with regard to the support region of the constellation, some expansion compared to conventional signaling is present.

4.5.2 Nonuniform Constellations—Warping

The above construction of optimal signal sets results in nonuniformly spaced constellations. Such signal sets also emerge in other scenarios, in particular, in QAM data transmission over channels that include logarithmic quantization. This is of special interest for analog voice-band modems transmitting over the public switched telephone network.

When A-law or μ-law PCM codecs [ITU93, Hay94] are present, quantization noise grows with the distance of the signal point from the origin. Hence, points at the perimeter of the constellation are distorted more severely than points with small amplitude. To compensate for this effect, the spacing of the signal points should grow with their amplitude. Such a transformation from a uniform, i.e., regular, signal constellation to a nonuniform one is sometimes called *warping*. For a detailed discussion on QAM transmission over companding channels see, e.g., [KS94] and the references therein.

Later it was recognized that warping is also rewarding on the AWGN channel [BCL94]. We now briefly study the performance of warping, i.e., nonuniform signal constellations as discussed in [BCL94]. First, we restrict ourselves to one-dimensional signaling, after which we show the generalization to QAM and the specific implementation in the ITU voice-band modem standard V.34 [ITU94].

Optimization Criteria We start with a uniform constellation $\mathcal{A}_\mathrm{u}$ where all signal points are selected equiprobable. To clarify of the subsequent derivations, we normalize the constellation such that the support region is the interval $[-1,\ 1]$. The distance of adjacent signal points (minimum distance) is denoted by d_u. Since we are interested in large constellations, we resort to the *continuous approximation*, cf. Section 4.2.1. The initial signal x is thus characterized by a uniform probability density over the support region, i.e., its pdf reads

$$f_x(x) = \frac{1}{2}\,, \qquad x \in [-1,\ 1]\,, \tag{4.5.8}$$

and zero else.

This constellation is fed into a *warping function* ω^{-1} (we will see that it is more convenient to use ω^{-1}, the inverse function of ω, than to use the function itself). The warping function is assumed to perform a nonlinear transformation of the interval $[-1,\ 1]$ onto itself

$$\omega^{-1}:\ [-1,\ 1] \ni x \mapsto z \in [-1,\ 1]\ . \tag{4.5.9}$$

Moreover, it should be strictly monotonously increasing and, because of symmetry reasons, be an odd function: $\omega^{-1}(-x) = -\omega^{-1}(x)$, $x \in [-1,\ 1]$.

Figure 4.50 sketches the effect of warping, cf. [BCL94]. Depending on the

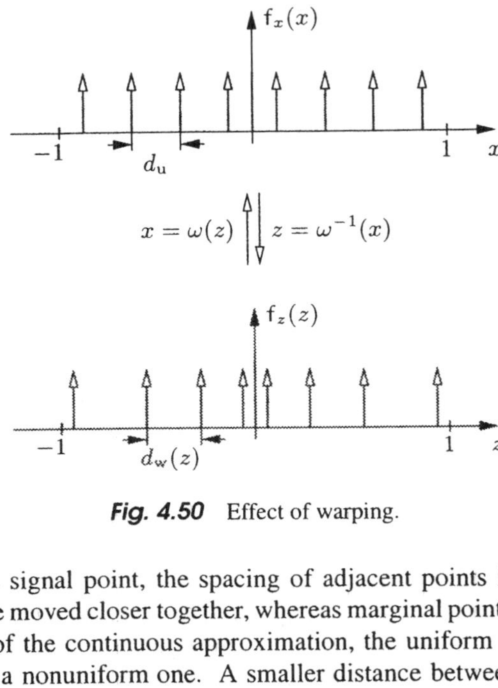

Fig. 4.50 Effect of warping.

magnitude of the signal point, the spacing of adjacent points is changed. Points near the origin are moved closer together, whereas marginal points are spaced further apart. In terms of the continuous approximation, the uniform density (shaded) is transformed into a nonuniform one. A smaller distance between the signal points corresponds to a higher density and vice versa.

Given the pdf $\mathsf{f}_x(x)$ of the initial signal x, the probability density function $\mathsf{f}_z(z)$ of the signal $z = \omega^{-1}(x)$ generated by warping can be determined. Following the general result in [Pap91] and considering that $\omega^{-1}(x)$ is strictly monotonously increasing, we have

$$\mathsf{f}_z(z) = \left.\frac{\mathsf{f}_x(x)}{\frac{\mathrm{d}\omega^{-1}(x)}{\mathrm{d}x}}\right|_{x=\omega(z)} = \frac{1}{2}\omega'(z)\ , \qquad z \in [-1,\ 1]\ . \tag{4.5.10}$$

The final manipulation is due to the fact that a function and its inverse have reciprocal slopes, $\omega'(z) \triangleq \frac{\mathrm{d}\omega(z)}{\mathrm{d}z} = \frac{1}{\left.\frac{\mathrm{d}\omega^{-1}(x)}{\mathrm{d}x}\right|_{x=\omega(z)}}$, and because of (4.5.8).

The distance of the points after warping depends on the position z. From $d_\mathrm{u} \cdot \mathsf{f}_x(x = \omega(z)) = d_\mathrm{w}(z) \cdot \mathsf{f}_z(z)$ (fixed probability when transforming an interval of width d_{u}), we have

$$d_{\mathrm{w}}(z) = d_{\mathrm{u}} \cdot \frac{\mathrm{f}_x(x = \omega(z))}{\mathrm{f}_z(z)} = d_{\mathrm{u}} \cdot \frac{1/2}{\omega'(z)/2} = \frac{d_{\mathrm{u}}}{\omega'(z)} \,. \tag{4.5.11}$$

Since we have fixed the support of the constellation, the signal energy is decreased by warping. Compared to the average energy of the initial signal x, $\sigma_x^2 = \int_{-1}^{1} \frac{1}{2} x^2 \, \mathrm{d}x = \frac{1}{3}$, the average energy of the warped signal z is given by

$$\sigma_z^2 = \int_{-1}^{1} z^2 \cdot \mathrm{f}_z(z) \, \mathrm{d}z = \frac{1}{2} \int_{-1}^{1} z^2 \, \omega'(z) \, \mathrm{d}z \,. \tag{4.5.12}$$

For preserving average energy, the signal z after the nonlinear function may be scaled by a factor ζ, with

$$\zeta^2 \triangleq \frac{\sigma_x^2}{\sigma_z^2} = \frac{2/3}{\int_{-1}^{1} z^2 \, \omega'(z) \, \mathrm{d}z} \,. \tag{4.5.13}$$

At the receiver this scaling is reversed by the factor $1/\zeta$. In turn, the underlying AWGN channel with noise variance $\bar{\sigma}_n^2$ is transformed into an AWGN channel with noise variance $\sigma_n^2 = \bar{\sigma}_n^2/\zeta^2$. Figure 4.51 shows the transmission scenario with warping.

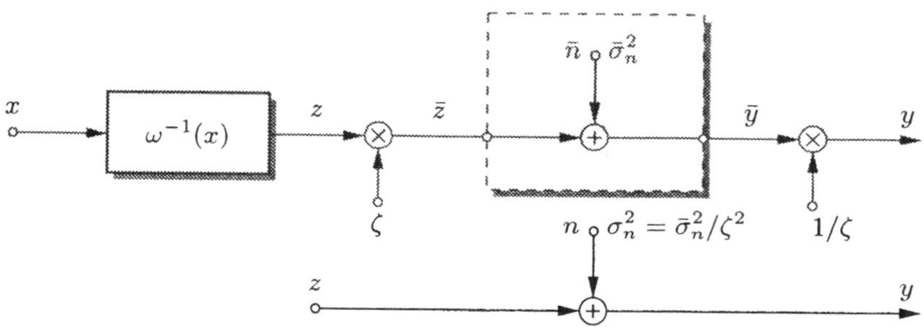

Fig. 4.51 Transmission over the AWGN channel with warping.

Using the uniform baseline constellation for transmission over the underlying AWGN channel, the symbol error rate is well approximated by

$$\mathrm{SER_u} \approx 2 \cdot \mathrm{Q}\left(\frac{d_{\mathrm{u}}/2}{\bar{\sigma}_n}\right) = 2 \cdot \mathrm{Q}\left(\frac{\lambda}{\zeta}\right) \,. \tag{4.5.14}$$

The expression on the right-hand side is written in terms of a normalized signal-to-noise ratio

$$\lambda^2 \triangleq \frac{d_{\mathrm{u}}^2}{4\sigma_n^2} = \frac{d_{\mathrm{u}}^2 \zeta^2}{4\bar{\sigma}_n^2} \,, \tag{4.5.15}$$

which proves to be a useful parameter. Conversely, using (4.5.10) and (4.5.11), the symbol error rate for transmission using warping reads

$$\mathrm{SER_w} \approx \int_{-1}^{1} \mathsf{f}_z(z) \cdot 2\mathrm{Q}\left(\frac{d_\mathrm{w}(z)/2}{\sigma_n}\right) \, \mathrm{d}z = \int_{-1}^{1} \omega'(z) \cdot \mathrm{Q}\left(\frac{\lambda}{\omega'(z)}\right) \, \mathrm{d}z \,. \quad (4.5.16)$$

We now compare transmission using uniform constellations over the underlying AWGN channel and transmission with warping over the scaled (by ζ) channel. The same average performance, i.e.,

$$\mathrm{SER_u} \stackrel{!}{=} \mathrm{SER_w} \quad (4.5.17a)$$

is achieved for

$$2 \cdot \mathrm{Q}\left(\frac{\lambda_\mathrm{u}}{\zeta}\right) \stackrel{!}{=} \int_{-1}^{1} \omega'(z) \cdot \mathrm{Q}\left(\frac{\lambda_\mathrm{w}}{\omega'(z)}\right) \, \mathrm{d}z \,. \quad (4.5.17b)$$

Here, the normalized signal-to-noise ratio for uniform signaling is denoted by λ_u, and the respective quantity for warping by λ_w. Solving (4.5.17b) for λ_u gives the signal-to-noise ratio where uniform signaling performs the same as using a warped constellation at signal-to-noise ratio λ_w.

The gain in noise immunity due to warping—which depends on the (inverse) warping function ω and the normalized signal-to-noise ratio λ_w—can then be defined as

$$G_\mathrm{w}(\omega, \lambda_\mathrm{w}) \triangleq \frac{\sigma_{n,\mathrm{w}}^2}{\sigma_{n,\mathrm{u}}^2} = \frac{\lambda_\mathrm{u}^2}{\lambda_\mathrm{w}^2} \,. \quad (4.5.18)$$

Using (4.5.13), (4.5.17b) and denoting the inverse function of $\mathrm{Q}(x)$ by $\mathrm{Q}^{-1}(x)$, the *warping gain* reads explicitly

$$G_\mathrm{w}(\omega, \lambda_\mathrm{w}) = \frac{\left[\mathrm{Q}^{-1}\left(\frac{1}{2}\int_{-1}^{1} \omega'(z) \cdot \mathrm{Q}\left(\frac{\lambda_\mathrm{w}}{\omega'(z)}\right) \, \mathrm{d}z\right)\right]^2}{\lambda_\mathrm{w}^2} \cdot \frac{2/3}{\int_{-1}^{1} z^2 \omega'(z) \, \mathrm{d}z} \,. \quad (4.5.19)$$

Discussion Obviously, the aim of an optimization is to find that warping function ω which maximizes the warping gain G_w, i.e., mathematically we demand

$$G_\mathrm{w}(\omega, \lambda_\mathrm{w}) \xrightarrow{\omega(z)} \max \,. \quad (4.5.20)$$

Note that the optimal warping function depends on the signal-to-noise ratio. Most important, it can be shown [BCL94] that by proper choice of ω the warping gain $G_\mathrm{w}(\omega)$ can always be made greater than one (positive in dB). Hence, using nonuniform signaling is rewarding on the AWGN channel.

Moreover, the gain for a particular, simple, piecewise-linear warping function is calculated in [BCL94]. The results reported there allow the study of the effects of warping. In particular, higher gains can be attained for smaller values of λ_w, whereas for very large signal-to-noise ratios warping becomes ineffective. The optimization focusing on error rate thus contrasts the results based on channel capacity (Section 4.5.1) where the potential shaping gain increases with the data rate, and hence, the required signal-to-noise ratio.

Using Taylor series expansions and disregarding all third- and higher-order terms, an approximation for the optimal warping function can be derived. The inverse of the optimal warping function $\omega^{-1}(x)$ is given by [BCL94]

$$\omega_{\text{opt}}(z) \approx (1+c) \cdot z - c \cdot z^3 , \qquad z \in [-1,\ 1] , \tag{4.5.21}$$

where $c = \sqrt{15/72}/\lambda_w^2$. The warping gain for this choice of ω then reads $10 \cdot \log_{10}(G_w) \approx 0.829/\lambda_w^2$ dB. Since this approximation is only valid for $\lambda_w \gg 1$, i.e., sufficiently large signal-to-noise ratios, the infinite gain for $\lambda_w \to 0$ should be ignored. A close look at (4.5.21) shows that the perturbation of the constellation is unexpectedly small; as λ_w increases, the case of no warping, $\omega(z) = z$, is approached.

Passband Transmission The concept of warping can of course be extended to passband transmission using two-dimensional constellations. This requires a two-dimensional or complex-valued warping function with two-dimensional or complex input. Depending on the pair of in-phase and quadrature component of the initial point, the warped signal point has to be generated. Here, an analytical optimization of the warping function is very difficult or even impossible.

A much simpler, but only slightly suboptimal, approach is to warp a QAM constellation only in the radial direction, i.e., only the absolute amplitude of the points is warped [KS94]. If the symbols x are drawn from a QAM constellation, the points of which have bounded amplitude, $X_{\max} = \max |x|$, warping is performed as

$$z = X_{\max} \cdot \omega^{-1}\left(\frac{|x|}{X_{\max}}\right) \cdot \mathrm{e}^{\mathrm{j}\arg(x)} \tag{4.5.22a}$$

$$= \frac{X_{\max}}{|x|} \cdot \omega^{-1}\left(\frac{|x|}{X_{\max}}\right) \cdot x . \tag{4.5.22b}$$

The amplitude warping function ω^{-1} is again real-valued and maps the interval $[0,\ 1]$ onto itself.

This procedure of considering only the amplitude of the signal can be justified by the fact that the optimal channel input distribution for a complex-valued AWGN channel—a complex Gaussian distribution—is circular symmetric, i.e., the distribution of the amplitude is relevant; phase is uniform.

In the context of channels that include logarithmic quantization, a further justification of the above warping function arises. Here, quantization of a modulated signal more severely affects the amplitude of the corresponding equivalent baseband signal than its phase.

Without warping, and taking points spaced by $\mathrm{d}|x|$ in the radial direction into account, performance on the AWGN channel is governed by the ratio $\frac{\mathrm{d}|x|}{\sigma_n}$, where σ_n denotes the standard deviation of the additive noise. However, in QAM transmission over channels including μ-law codecs, the noise variance in the radial direction increases with the amplitude $|z|$ and is well approximated by $\sigma_n^2 \cdot (\gamma^2|z|^2+1)$ [KS94]. The parameter γ depends on the shape of the pulse-shaping filter $H_{\mathrm{T}}(f)$ and receiver input filter $H_{\mathrm{R}}(f)$. Denoting the point spacing in the radial direction after warping by $\mathrm{d}|z|$, performance can be expressed in terms of the ratio $\frac{\mathrm{d}|z|}{\sigma_n\sqrt{\gamma^2|z|^2+1}}$.

Requiring the same proportions with and without warping, we have

$$\frac{\mathrm{d}|x|}{\sigma_n} \stackrel{!}{=} \text{const} \cdot \frac{\mathrm{d}|z|}{\sigma_n\sqrt{\gamma^2|z|^2+1}} . \tag{4.5.23}$$

Taking $\int \frac{\mathrm{d}x}{\sqrt{a^2x^2+1}}\,\mathrm{d}x = \frac{1}{a}\operatorname{arsinh}(ax) + \text{const}$ into consideration [BS98], (4.5.23) integrates to [KS94]

$$|x| = \text{const} \cdot \operatorname{arsinh}(\gamma|z|) , \tag{4.5.24}$$

where $\operatorname{arsinh}(x) = \log(x + \sqrt{x^2+1})$ is the inverse of the hyperbolic sine function $\sinh(x) = (\mathrm{e}^x - \mathrm{e}^{-x})/2$.

After appropriate normalization, the warping function for the amplitude may thus be given as

$$|z| = \omega^{-1}(|x|) = \frac{1}{\sinh(\sqrt{g})} \cdot \sinh(\sqrt{g}|x|) , \tag{4.5.25}$$

where g is a positive *warping factor*. The larger g is, the more pronounced warping takes place. This particular simple warping function is also proposed in [EFDL93], and an approximation thereof is adopted in the voice-band modem standard ITU recommendation V.34 [ITU94]. Based on the Taylor series expansion of the hyperbolic sine function

$$\sinh(x) = x + \frac{x^3}{3!} + \frac{x^5}{5!} + \cdots \tag{4.5.26}$$

warping in V.34 is done by (cf. (4.5.22b))

$$z = \left(1 + \frac{g|x|^2}{6X_{\max}^2} + \frac{g^2|x|^4}{120X_{\max}^4}\right) \cdot x , \tag{4.5.27}$$

where x and z are the complex-valued QAM symbols before and after warping, respectively. The constant $X_{\max}$ again denotes the constellation circumradius. According to [ITU94], the warping factor g can either be selected to be 0 (no warping) or $g = 0.625 = 10/16$. Note that after this nonlinear distortion, the average energy again has to be adjusted by appropriate scaling.

The following example shows the effect of warping as used in V.34.

Example 4.22: *Warping in V.34*

Figure 4.52 displays the 960-ary quasi-circular QAM constellation specified in [ITU94] before (left-hand side) and after warping (right-hand side) using the V.34 warping function (4.5.27). The warped signal constellation is normalized to the same average energy as without warping.

The perturbation caused by warping is clearly visible, even though it is not as pronounced as one would expect.

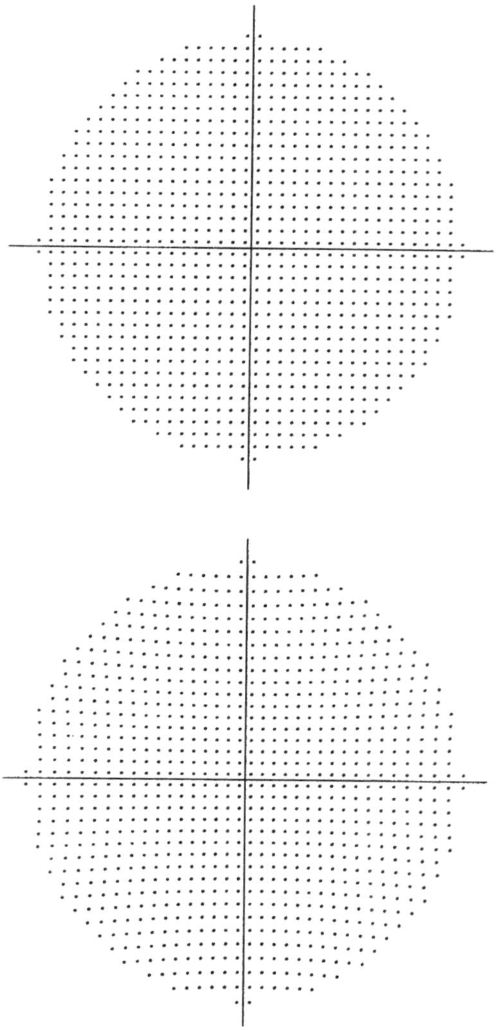

Fig. 4.52 Top: 960-ary quasi-circular QAM constellation of [ITU94]; bottom: signal constellation after warping, as specified in [ITU94].

4.5.3 Modulus Conversion

From the preceding discussion we have seen that by arranging the signal points appropriately, "shaping" gains can be achieved by equiprobable signaling, too. But besides achieving shaping gain, a second important property of shaping algorithms, in particular shell mapping, is that they are useful for accommodating noninteger numbers of bits per symbol. In other words, using equiprobable signaling, the sizes of the constellations should not be limited to powers of two.

We now show how binary information can be mapped to signal points drawn from signal sets with arbitrary numbers of points. The mapping scheme is known as *modulus conversion* or *radix mapping* [PCM97a, PCM97b, FU98]. This enumerative encoding technique orders blocks of signal points according to a lexicographical order.

As in shell mapping, modulus conversion is assumed to operate on a frame of N consecutive symbols, and K bits shall be mapped within one block. The constellations $\mathcal{A}_i$ used in the N phases may all differ, and their sizes—here also called *moduli*—are denoted by $M_i \triangleq |\mathcal{A}_i|$, $i = 1, 2, \ldots, N$. The signal points within one phase i are labeled by $s_i = 0, 1, \ldots, M_i - 1$. Obviously, for supporting the desired rate by the Cartesian product of the N D-dimensional constituent constellations, the number $\prod_{i=1}^{N} M_i$ of (ND)-dimensional signal point has to be greater than 2^K, i.e., we require

$$\prod_{i=1}^{N} M_i \geq 2^K . \tag{4.5.28}$$

As we will see later, the above product should be as close as possible to a power of two.

Mapping and Inverse Mapping The task of modulus conversion is to map the K-bit number

$$I = b_{K-1} \cdot 2^{K-1} + \cdots + b_2 \cdot 2^2 + b_1 \cdot 2 + b_0 , \quad b_j \in \{0, 1\} \tag{4.5.29}$$

to the vector (block)

$$\boldsymbol{s} = [s_1,\ s_2,\ \ldots,\ s_N] \tag{4.5.30}$$

of N signal-point labels s_i, $i = 1, 2, \ldots, N$. Of course, the inverse mapping aims to again extract the K bits b_j from the label vector $\boldsymbol{s}$.

The rationale behind this mapping technique is to represent an integer I_i as

$$I_i = Q_i \cdot M_i + R_i , \qquad Q_i \in \mathbb{N} , \quad R_i \in \{0, 1, \ldots, M_i - 1\} , \tag{4.5.31}$$

where M_i is the size of the constellation and Q_i and R_i are the quotient and remainder, respectively, of the division of I_i by M_i. The remainder is limited to the set $\{0, 1, \ldots, M_i - 1\}$ which coincides with the set of labels of an M_i-ary constellation. Hence, this quantity can be used directly as the signal-point label. If the frame size is $N = 2$, the quotient gives the second label and mapping is complete. Otherwise, the procedure is iterated until the labels of all phases are calculated. In all, this procedure requires $N - 1$ operations (4.5.31).

In summary, given the constellation sizes M_i, an integer I is mapped using the following modulus conversion algorithm:

1. Let $I_1 = I$, and $i = 1$.
2. Calculate
$$s_i = I_i \bmod M_i \qquad \text{and} \qquad I_{i+1} = (I_i - s_i)/M_i$$
3. Increment i. If $i < N$ go to Step 2.
4. Let $s_N = I_N$.

The inverse mapping, or demapping, can be done as easily as the mapping itself. It is performed by repeated multiplication and accumulation in contrast to repeated division and subtraction at the encoder. Given the sizes M_i of the constellations and the N signal-point labels s_i, the data, represented in the integer I, are recovered by

$$I = (\cdots((s_N \cdot M_{N-1}) + s_{N-1})\, M_{N-2} + \cdots)\, M_1 + s_1 \,. \tag{4.5.32}$$

The advantage of modulus conversion is its great flexibility. Any constellation sizes can be used and there are no restrictions to the constellations within the different phases. Moreover, frame size N and number of bits K can be adjusted arbitrarily. One disadvantage of this scheme is that implementation requires the division of long binary numbers.

If the constellation sizes are all powers of two, modulus conversion reduces to just dissecting the K-bit number I into parts of $\log_2(M_i)$ bits. The procedure is then the same as with the conventional, individual mapping of binary data to a constellation.

The following example shows mapping and demapping using modulus conversion.

Example 4.23: *Modulus Conversion*

In this example we consider mapping of $K = 10$ bits of binary data to frames of $N = 4$ signal points. The average transmission rate is thus $10/4 = 2.5$ bits per symbol. The four constellations are all of size $M_i = 6$, $i = 1, 2, 3, 4$, i.e., a 6-ary PAM transmission is assumed. Note that condition (4.5.28) is met since $6^4 = 1296 > 2^{10} = 1024$.

As an example, we study the mapping of the binary number $I = [0001111111]_2 = 127$ to the signal-point labels s_i. Performing the above algorithm gives

$$\begin{aligned} 127/6 &= 21 \text{ rem } 1 \rightarrow s_1 = 1 \\ 21/6 &= 3 \text{ rem } 3 \rightarrow s_2 = 3 \\ 3/6 &= 0 \text{ rem } 3 \rightarrow s_3 = 3 \\ & \qquad\qquad\quad \rightarrow s_4 = 0 \end{aligned} \,, \tag{4.5.33}$$

or $\boldsymbol{s} = [1,\ 3,\ 3,\ 0]$. Conversely, inverse mapping of $[1,\ 3,\ 3,\ 0]$ by repeated multiplication/accumulation results in

$$0 \cdot 6 = 0\,, \qquad (0+3)\cdot 6 = 18\,, \qquad (18+3)\cdot 6 = 126\,, \qquad 126 + 1 = 127\,, \tag{4.5.34}$$

which, of course, recovers the initial integer I.

In order to get an impression of the ordering of the label vectors $\boldsymbol{s}$, Table 4.17 contains some excerpts of data to be mapped and corresponding labels s_i. Reading the labels from right to left, the label vectors are sorted as words in a lexicon or dictionary (lexicographic order).

Table 4.17 Data I and corresponding label vector $\boldsymbol{s} = [s_1,\ s_2,\ s_3,\ s_4]$ for $N = 4$ and $M_i = 6, i = 1, 2, 3, 4$.

I	s_1	s_2	s_3	s_4
$0 = [0000000000]_2$	0	0	0	0
$1 = [0000000001]_2$	1	0	0	0
$2 = [0000000010]_2$	2	0	0	0
$3 = [0000000011]_2$	3	0	0	0
$4 = [0000000100]_2$	4	0	0	0
$5 = [0000000101]_2$	5	0	0	0
$6 = [0000000110]_2$	0	1	0	0
$7 = [0000000111]_2$	1	1	0	0
$\vdots$		$\vdots$		
$34 = [0000100010]_2$	4	5	0	0
$35 = [0000100011]_2$	5	5	0	0
$36 = [0000100100]_2$	0	0	1	0
$37 = [0000100101]_2$	1	0	1	0
$38 = [0000100110]_2$	2	0	1	0
$39 = [0000100111]_2$	3	0	1	0
$40 = [0000101000]_2$	4	0	1	0
$\vdots$		$\vdots$		
$1018 = [1111111010]_2$	4	1	4	4
$1019 = [1111111011]_2$	5	1	4	4
$1020 = [1111111100]_2$	0	2	4	4
$1021 = [1111111101]_2$	1	2	4	4
$1022 = [1111111110]_2$	2	2	4	4
$1023 = [1111111111]_2$	3	2	4	4

Probability of Signal Points As in the case of other mapping strategies, we are interested in the probabilities of the signal points induced by modulus conversion. If 2^K is large compared to the size M_i of the constellation, the remainder of a division of a K-bit number by M_i will assume all possible values approximately equally often. Hence, we expect an almost uniform distribution of the signal points.

For the exact calculation of the signal-point probabilities, we first derive a general result. Let I be a random variable assuming nonnegative integers within a given interval, i.e., $I \in \mathbb{N}$ and $0 \leq I \leq I_{\max}$. The distribution of I is given as

$$\Pr\{I\} = \begin{cases} p_1, & 0 \leq I < I_{\max} \\ p_2, & I = I_{\max} \\ 0, & \text{else} \end{cases}, \qquad (4.5.35)$$

where $p_1 \cdot I_{\max} + p_2 = 1$ has to hold in order for (4.5.35) to be a probability distribution.

Given an (positive) integer M, each I can be expressed uniquely using integers Q (quotient) and R (remainder) as

$$I = Q \cdot M + R\,, \qquad Q \geq 0\,, \quad 0 \leq R < M\,. \qquad (4.5.36)$$

We are now interested in the probability distributions $\Pr\{Q\}$ and $\Pr\{R\}$ of the random variables Q and R, respectively. These can be derived from geometrical considerations, see Figure 4.53.

If the distribution $\Pr\{I\}$ is rearranged as a two-dimensional probability distribution—folding a modulo M—the distributions $\Pr\{Q\}$ and $\Pr\{R\}$ are the marginal distributions, i.e., the projections to both of the dimensions. Writing $I_{\max}$ itself in the form

$$I_{\max} = Q_{\max} \cdot M + R_{\max}\,, \qquad Q_{\max} \geq 0\,, \quad 0 \leq R_{\max} < M\,, \qquad (4.5.37)$$

the desired pdfs are obtained as

$$\Pr\{Q\} = \begin{cases} M \cdot p_1, & 0 \leq Q < Q_{\max} \\ R_{\max} \cdot p_1 + p_2, & Q = Q_{\max} \\ 0, & \text{else} \end{cases}, \qquad (4.5.38a)$$

and

$$\Pr\{R\} = \begin{cases} (Q_{\max}+1) \cdot p_1, & 0 \leq R < R_{\max} \\ Q_{\max} \cdot p_1 + p_2, & R = R_{\max} \\ Q_{\max} \cdot p_1, & R_{\max} < R < M \\ 0, & \text{else} \end{cases}. \qquad (4.5.38b)$$

Since in modulus conversion the signal-point label is the remainder of a division by the number of signal points, (4.5.38b) gives the distribution of the signal points. In contrast, the distribution (4.5.38a) of the quotient Q is the starting point for the integer I in the next iteration. This also justifies the assumption in (4.5.35).

In summary, modulus conversion with respect to the moduli M_i, $i = 1, 2, \ldots, N$, of a K-bit number I, equiprobably drawn from the interval $0 \leq I < 2^K$, results in N

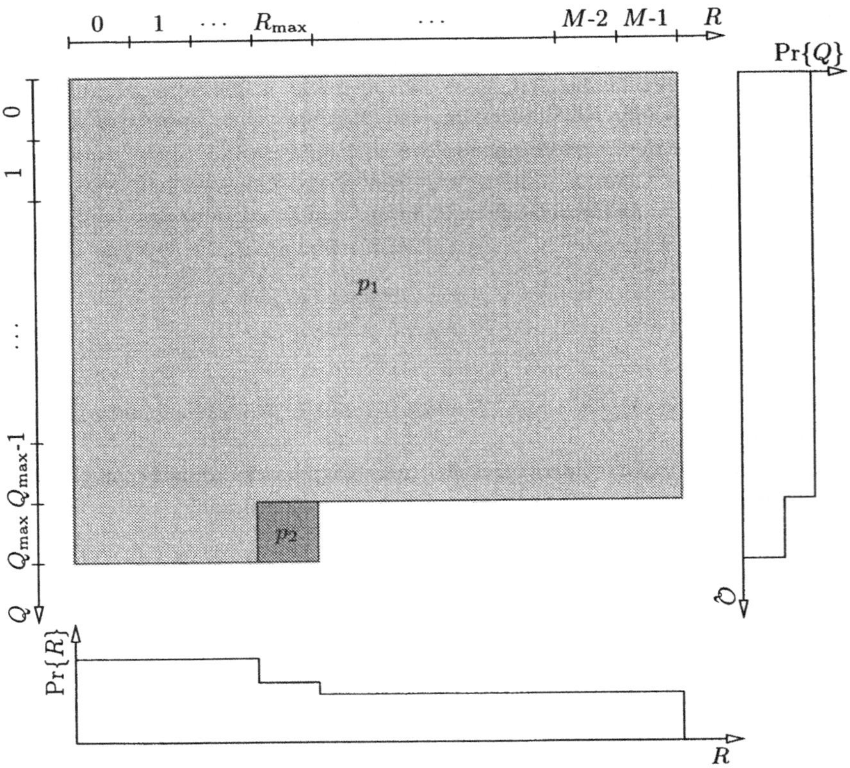

Fig. 4.53 Geometrical explanation of the pdf calculation.

distributions $\Pr\{s_i\}$, $i = 1, 2, \ldots, N$, of the signal points, which can be calculated by the following algorithm:

1. Let $i = 1$, $I_{\max}^{(1)} = 2^K - 1$, and $p_1^{(1)} = p_2^{(1)} = 2^{-K}$.

2. Calculate $R_{\max}^{(i)} = I_{\max}^{(i)} \bmod M_i$, and $Q_{\max}^{(i)} = (I_{\max}^{(i)} - R_{\max}^{(i)})/M_i$.

3. Let $\Pr\{s_i\} = Q_{\max}^{(i)} \cdot p_1^{(i)} + \begin{cases} p_1^{(i)}, & 0 \le s_i < R_{\max}^{(i)} \\ p_2^{(i)}, & s_i = R_{\max}^{(i)} \\ 0, & R_{\max}^{(i)} < s_i < M_i \end{cases}$.

4. Let $p_1^{(i+1)} = M_i \cdot p_1^{(i)}$, $p_2^{(i+1)} = R_{\max}^{(i)} \cdot p_1^{(i)} + p_2^{(i)}$, and $I_{\max}^{(i+1)} = Q_{\max}^{(i)}$,

5. Increment i. If $i \le N$ go to Step 2.

A close look at (4.5.38b) reveals that signal points with three different probabilities occur in each phase i—except for phase 1, where there are only two different probabilities. Moreover, since $p_1 > p_2$, the probability of the signal points decreases with the label of the point. Label 0 is thus (slightly) preferred over label $M-1$. The differences become stronger with the number of the phase. While in phase 1 the signal points are approximately uniformly distributed, the largest variations can be observed for phase N—it often happens that some points do not occur at all, in which case the sizes of the constellations may be reduced so $\prod_i M_i$ more closely matches a power of two.

Example 4.24 shows the distribution of signal points obtained with modulus conversion.

Example 4.24: Distribution Induced by Modulus Conversion

We continue Example 4.23 on modulus conversion. The parameters are still $N = 4$, $M_i = 6$, $i = 1, 2, 3, 4$, and $K = 10$. In Figure 4.54 the distributions of the signal points labeled by $s_i = 0, 1, \ldots, 5$, obtained by applying the above algorithm, are plotted.

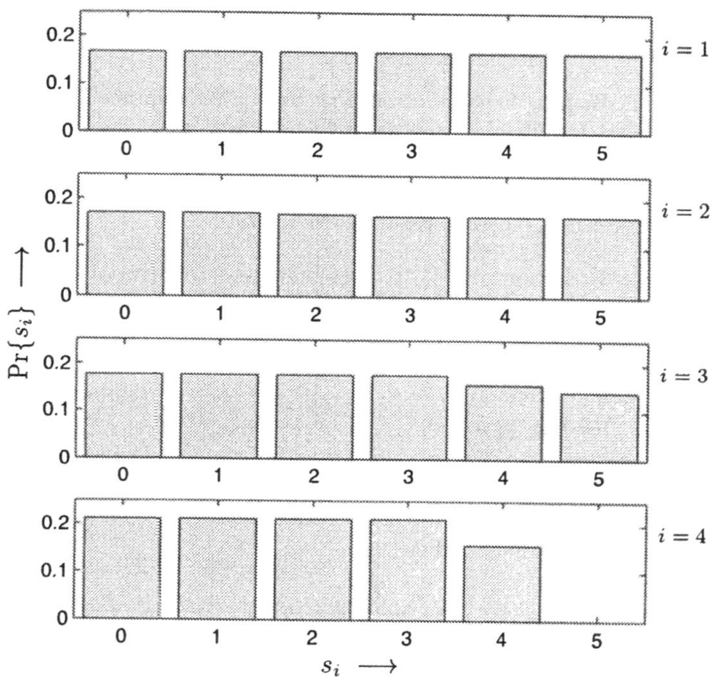

Fig. 4.54 Distribution of the signal points when using modulus conversion. Frame size $N = 4$; constellation sizes $M_i = 6$, $i = 1, 2, 3, 4$; $K = 10$ bits to be mapped.

In phases $i = 1$ and $i = 2$ the signal points are almost uniformly distributed. For phase $i = 3$, three different probabilities already can be clearly distinguished. Finally, in phase $i = 4$ the signal point labeled by 5 never occurs. This is because $6 \cdot 6 \cdot 6 \cdot 5 = 1080$ is still larger than $2^{10} = 1024$.

REFERENCES

[And99] J. B. Anderson. *Digital Transmission Engineering*. IEEE Press, Piscataway, NJ, 1999.

[Bak62] P. A. Baker. Phase modulation data sets for serial transmission at 2000 and 2400 bits per second, Part 1. *AIEE Transactions on Communications and Electronics*, pp. 166–171, July 1962.

[BB91] E. J. Borowski and J. M. Borwein. *The HarperCollins Dictionary of Mathematics*. HarperPerennial, New York, 1991.

[BCL94] W. Betts, A. R. Calderbank, and R. Laroia. Performance of Nonuniform Constellations on the Gaussian Channel. *IEEE Transactions on Information Theory*, IT-40, pp. 1633–1638, September 1994.

[Ber96] J. W. M. Bergmans. *Digital Baseband Transmission and Recording*. Kluwer Academic Publishers, Dordrecht, The Netherlands, 1996.

[Bla87] R. E. Blahut. *Principles and Practice of Information Theory*. Addison-Wesley Publishing Company, Reading, MA, 1987.

[Bla90] R. E. Blahut. *Digital Transmission of Information*. Addison-Wesley Publishing Company, Reading, MA, 1990.

[BS98] I. N. Bronstein and K. A. Semendjajew. *Handbook of Mathematics*. Springer Verlag, Berlin, Heidelberg, Reprint of the third edition, 1998.

[CO90] A. R. Calderbank and L. H. Ozarow. Nonequiprobable Signaling on the Gaussian Channel. *IEEE Transactions on Information Theory*, IT-36, pp. 726–740, 1990.

[CS83] J. H. Conway and N. J. A. Sloane. A Fast Encoding Method for Lattice Codes and Quantizers. *IEEE Transactions on Information Theory*, IT-29, pp. 820–824, November 1983.

[CS88] J. H. Conway and N. J. A. Sloane. *Sphere Packings, Lattices and Groups*. Springer Verlag, New York, Berlin, 1988.

[CT91] T. M. Cover and J. A. Thomas. *Elements of Information Theory*. John Wiley & Sons, Inc., New York, 1991.

[Dho94] A. Dholakia. *Introduction to Convolutional Codes with Applications*. Kluwer Academic Publishers, Norwell, MA, 1994.

[EC91] E. Eleftheriou and R. D. Cideciyan. On Codes Satisfying Mth-Order Running Digital Sum Constraints. *IEEE Transactions on Information Theory*, IT-37, pp. 1294–1313, September 1991.

[EFDL93] M. V. Eyuboğlu, G. D. Forney, P. Dong, and G. Long. Advanced Modulation Techniques for V.fast. *European Transactions on Telecommunications*, ETT-4, pp. 243–256, May/June 1993.

[FC89] G. D. Forney and A. R. Calderbank. Coset Codes for Partial Response Channels; or, Coset Codes with Spectral Nulls. *IEEE Transactions on Information Theory*, IT-35, pp. 925–943, September 1989.

[FGL+84] G. D. Forney, R. G. Gallager, G. R. Lang, F. M. Longstaff, and S. U. H. Qureshi. Efficient Modulation for Band-Limited Channels. *IEEE Journal on Selected Areas in Communications*, JSAC-2, pp. 632–647, September 1984.

[Fis99] R. Fischer. Calculation of Shell Frequency Distributions Obtained with Shell-Mapping Schemes. *IEEE Transactions on Information Theory*, IT-45, pp. 1631–1639, July 1999.

[FM93] J. Forster and R. Matzner. Trellis Shaping als ein Verfahren zur Leitungscodierung — Theorie, Ergebnisse und Vergleich mit anderen Verfahren. In *Kleinheubacher Tagung*, Vol. 37, pp. 157–166, Kleinheubach, Germany, October 1993. (In German.)

[For70] G. D. Forney. Convolutional Codes I: Algebraic Structure. *IEEE Transactions on Information Theory*, IT-16, pp. 720–738, November 1970.

[For73a] G. D. Forney. Structural Analysis of Convolutional Codes via Dual Codes. *IEEE Transactions on Information Theory*, IT-19, pp. 512–518, July 1973.

[For73b] G. D. Forney. The Viterbi Algorithm. *Proceedings of the IEEE*, 61, pp. 268–278, March 1973.

[For88] G. D. Forney. Coset Codes — Part I: Introduction and Geometrical Classification. *IEEE Transactions on Information Theory*, IT-34, pp. 1123–1151, September 1988.

[For89] G. D. Forney. Multidimensional Constellations — Part II: Vornonoi Constellations. *IEEE Journal on Selected Areas in Communications*, JSAC-7, pp. 941–958, August 1989.

[For92] G. D. Forney. Trellis Shaping. *IEEE Transactions on Information Theory*, IT-38, pp. 281–300, March 1992.

[Fri97] M. Friese. Multitone Signals with Low Crest Factor. *IEEE Transactions on Communications*, COM-45, pp. 1338–1344, October 1997.

[FTC00] G. D. Forney, M. D. Trott, and S.-Y. Chung. Sphere-Bound-Achieving Coset Codes and Multilevel Coset Codes. *IEEE Transactions on Information Theory*, IT-46, pp. 820–850, May 2000.

[FU98] G. D. Forney and G. Ungerböck. Modulation and Coding for Linear Gaussian Channels. *IEEE Transactions on Information Theory*, IT-44, pp. 2384–2415, October 1998.

[FW89] G. D. Forney and L.-F. Wei. Multidimensional Constellations — Part I: Introduction, Figures of Merit, and Generalized Cross Constellations. *IEEE Journal on Selected Areas in Communications*, JSAC-7, pp. 877–892, August 1989.

[Gal68] R. G. Gallager. *Information Theory and Reliable Communication*. John Wiley & Sons, Inc., New York, London, 1968.

[GG98] I. A. Glover and P. M. Grant. *Digital Communications*. Prentice-Hall, Inc., Englewood Cliffs, NJ, 1998.

[GN98] R. M. Gray and D. L. Neuhoff. Quantization. *IEEE Transactions on Information Theory*, IT-44, pp. 2325–2383, October 1998.

[GR80] I. S. Gradshteyn and I. M. Ryzhik. *Table of Integrals, Series, and Products*. Academic Press, Orlando, FL, 1980.

[Hay94] S. Haykin. *Communication Systems*. John Wiley & Sons, Inc., New York, 3ed edition, 1994.

[HSH93] W. Henkel, R. Schramm, and J. Hofmann. A Modified Trellis-Shaping without Doubling of the Symbol Alphabet. In *Proceedings of the 6th Joint Swedish-Russian International Workshop on Information Theory*, Mölle, Sweden, August 1993.

[HW96] W. Henkel and B. Wagner. Trellis–Shaping zur Reduktion des Spitzen–Mittelwert–Verhältnisses bei DMT/OFDM. In *OFDM–Fachgespräch*, Braunschweig, Germany, September 1996. (In German.)

[HW97] W. Henkel and B. Wagner. Trellis Shaping for Reducing the Peak-to-Average Ratio of Multitone signals. In *Proceedings of the IEEE International Symposium on Information Theory*, pp. 519, Ulm, Germany, June/July 1997.

[HW00] W. Henkel and B. Wagner. Another Application of Trellis Shaping: PAR Reduction for DMT (OFDM). *IEEE Transactions on Communications*, COM-48, pp. 1471–1476, September 2000.

[Imm85] K. A. S. Immink. Spectrum Shaping with Binary DC^2-Constraint Channel Codes. *Philips Journal of Research*, 40, pp. 40–53, 1985.

[Imm91] K. A. S. Immink. *Coding Techniques for Digital Recorders*. Prentice-Hall, Inc., Hertfordshire, UK, 1991.

[ISW98] K. A. S. Immink, P. H. Siegel, and J. K. Wolf. Codes for Digital Recoders. *IEEE Transactions on Information Theory*, IT-44, pp. 2260–2299, October 1998.

[ITU93] ITU-T Recommendation G.711. *Pulse Code Modulation (PCM) of Voice Frequencies*. International Telecommunication Union (ITU), Geneva, Switzerland, 1994.

[ITU94] ITU-T Recommendation V.34. *A Modem Operating at Data Signalling Rates of up to 28800 bit/s for Use on the General Switched Telephone Network and on Leased Point-to-Point 2-Wire Telephone-Type Circuits*. International Telecommunication Union (ITU), Geneva, Switzerland, September 1994.

[ITU98] ITU-T Recommendation V.90. *A Digital Modem and Analog Modem Pair for Use on the Public Switched Telephone Network at Data Signalling Rates of up to 56000 bit/s Downstream and up to 33600 bit/s Upstream*. International Telecommunication Union (ITU), Geneva, Switzerland, September 1998.

[Jus82] J. Justesen. Information Rates and Power Spectra of Digital Codes. *IEEE Transactions on Information Theory*, IT-28, pp. 457–472, May 1982.

[JZ99] R. Johannesson and K. Sh. Zigangirov. *Fundamentals of Convolutional Coding*. IEEE Press, Piscataway, NJ, 1999.

[Kay88] S. M. Kay. *Modern Spectral Estimation: Theory and Application*. Prentice-Hall, Inc., Englewood Clifs, NJ, 1988.

[KE99] D. Kim and M. V. Eyuboğlu. Convolutional Spectral Shaping. *IEEE Communications Letters*, COMML-3, pp. 9–11, January 1999.

[KK93] A. K. Khandani and P. Kabal. Shaping Multidimensional Signal Spaces — Part I: Optimum Shaping, Shell Mapping, Part II: Shell-Addressed Constellations. *IEEE Transactions on Information Theory*, IT-39, pp. 1799–1819, November 1993.

[KP93] F. R. Kschischang and S. Pasupathy. Optimal Nonuniform Signaling for Gaussian Channels. *IEEE Transactions on Information Theory*, IT-39, pp. 913–929, May 1993.

[KP94] F. R. Kschischang and S. Pasupathy. Optimal Shaping Properties of the Truncated Polydisc. *IEEE Transactions on Information Theory*, IT-40, pp. 892–903, May 1994.

[KS94] I. Kalet and B. R. Saltzberg. QAM Transmission Through a Companding Channel — Signal Constellation and Detection. *IEEE Transactions on Communications*, COM-42, pp. 417–429, February/March/April 1994.

[LF93a] R. Laroia and N. Farvardin. A Structured Fixed-Rate Vector Quantizer Derived from a Variable-Length Scalar Quantizer: Part I—Memoryless Sources. *IEEE Transactions on Information Theory*, IT-39, pp. 851–867, May 1993.

[LF93b] R. Laroia and N. Farvardin. A Structured Fixed-Rate Vector Quantizer Derived from a Variable-Length Scalar Quantizer: Part II—Vector Sources. *IEEE Transactions on Information Theory*, IT-39, pp. 868–876, May 1993.

[LFT94] R. Laroia, N. Farvardin, and S. A. Tretter. On Optimal Shaping of Multidimensional Constellations. *IEEE Transactions on Information Theory*, IT-40, pp. 1044–1056, July 1994.

[Liv92] J. N. Livingston. Shaping Using Variable-Size Regions. *IEEE Transactions on Information Theory*, IT-38, pp. 1347–1353, July 1992.

[LKFF98] S. Lin, T. Kasami, T. Fujiwara, and M. Fossorier. *Trellises and Trellis-Based Decoding Algorithms for Linear Block Codes*. Kluwer Academic Publishers, Norwell, MA, 1998.

[LL89] G. R. Lang and F. M. Longstaff. A Leech Lattice Modem. *IEEE Journal on Selected Areas in Communications*, JSAC-7, pp. 986–973, August 1989.

[LR94a] M. Litzenburger and W. Rupprecht. A Comparison of Trellis Shaping Schemes for Controlling the Envelope of a Bandlimited PSK-Signal. In *Proceedings of the IEEE Vehicular Technology Conference*, pp. 982–986, Stockholm, Sweden, September 1994.

[LR94b] M. Litzenburger and W. Rupprecht. Combined Trellis Shaping and Coding to Control the Envelope of a Bandlinited PSK-Signal. In *Proceedings of the IEEE International Conference on Communications (ICC'94)*, pp. 630–634, New Orleans, LA, June 1994.

[MBFH97] S. H. Müller, R. W. Bäuml, R. F. H. Fischer, and J. B. Huber. OFDM with Reduced Peak-to-Average Power Ratio by Multiple Signal Representation. *Annals of Telecommunications*, Vol. 52, pp. 58–67, February 1997.

[MF90] M. W. Marcelin and T. R. Fischer. Trellis Coded Quantization of Memoryless and Gauss-Markov Sources. *IEEE Transactions on Communications*, COM-38, pp. 82–93, January 1990.

[Mor92] I. S. Morrison. Trellis Shaping Applied to Reducing the Envelope Fluctuations of MQAM and bandlimited MPSK. In *Proceedings of the International Conference on Digital Satellite Communications (ICDSC)*, pp. 143–149, Copenhagen, Denmark, May 1992.

[MP89] C. M. Monti and G. L. Pierobon. Codes with a Multiple Spectral Null at Zero Frequency. *IEEE Transactions on Information Theory*, IT-35, pp. 463–472, March 1989.

[OS75] A. V. Oppenheim and R. W. Schafer. *Digital Signal Processing*. Prentice-Hall, Inc., Englewood Cliffs, NJ, 1975.

[OW90] L. H. Ozarow and A. D. Wyner. On the Capacity of the Gaussian Channel with a Finite Number of Input Levels. *IEEE Transactions on Information Theory*, IT-36, pp. 1426–1428, November 1990.

[Pap91] A. Papoulis. *Probability, Random Variables, and Stochastic Processes*. McGraw-Hill, New York, 3rd edition, 1991.

[PCM97a] D. Walsh. Multiple Modulus Conversion for Robbed Bit Signaling Channels. TIA TR30.1 Ad-hoc Meeting, Orange County, CA, March 1997.

[PCM97b] R. G. C. Williams. Mixed Base Mapping. TIA TR30.1 Ad-hoc Meeting, Orange County, CA, March 1997.

[PCM97c] N. Dagdeviren, V. Eyuboğlu, and S. Olafsson. Draft Text for Downstream Signal Encoding. V.PCM Rapporteur Meeting, La Jolla, CA, May 1997.

[PCM97d] V. Eyuboğlu. More on Convolutional Spectral Shaping. V.PCM Rapporteur Meeting, La Jolla, CA, May 1997.

[Pie84] G. L. Pierobon. Codes for Zero Spectral Density at Zero Frequency. *IEEE Transactions on Information Theory*, IT-30, pp. 435–439, March 1984.

[Pro01] J. G. Proakis. *Digital Communications*. McGraw-Hill, New York, 4th edition, 2001.

[Rap96] T. S. Rappaport. *Wireless Communications — Principles & Practice*. Prentice-Hall, Inc., Upper Saddle River, NJ, 1996.

[Sch94] H. W. Schüßler. *Digitale Signalverarbeitung, Band I*. Springer Verlag, Berlin, Heidelberg, 4th edition, 1994. (In German.)

[Sch96] H. Schwarte. Approaching Capacity of a Continuous Channel by Discrete Input Distributions. *IEEE Transactions on Information Theory*, IT-42, pp. 671–675, March 1996.

[Sha49] C. E. Shannon. Communications in the Presence of Noise. *Proceedings of the Institute of Radio Engineers*, 37, pp. 10–21, 1949.

[ST93] F.-W. Sun and H. C. A. van Tilborg. Approaching Capacity by Equiprobable Signaling on the Gaussian Channel. *IEEE Transactions on Information Theory*, IT-39, pp. 1714–1716, September 1993.

[Tel98] C. Tellambura. Phase Optimization Criterion for Reducing Peak-to-Average Power Ratio in OFDM. *Electronics Letters*, Vol. 34, pp. 169–170, January 1998.

[Ung82] G. Ungerböck. Channel Coding with Multilevel/Phase Signals. *IEEE Transactions on Information Theory*, IT-28, pp. 55–67, January 1982.

[WFH99] U. Wachsmann, R. F. H. Fischer, and J. B. Huber. Multilevel Codes: Theoretical Concepts and Practical Design Rules. *IEEE Transactions on Information Theory*, IT-45, pp. 1361–1391, July 1999.

[Wic95] S. B. Wicker. *Error Control Systems for Digital Communications and Storage*. Prentice-Hall, Inc., Englewood Cliffs, NJ, 1995.

[Wol78] J. K. Wolf. Efficient Maximum-Likelihood Decoding of Linear Block Codes Using a Trellis. *IEEE Transactions on Information Theory*, IT-24, pp. 76–80, January 1978.

[ZF96] R. Zamir and M. Feder. On Lattice Quantization Noise. *IEEE Transactions on Information Theory*, IT-42, pp. 1152–1159, July 1996.

5

Combined Precoding and Signal Shaping

Recall the precoding schemes presented in Chapter 3, namely *Tomlinson-Harashima precoding* and *flexible precoding*. The schemes have in common that they produce discrete-time transmit symbols, uniformly distributed over some support region. However, replacing this uniform distribution by a Gaussian one allows transmission at the same rate and the same reliability but with average energy reduced by up to 1.53 dB, cf. Section 4.1.3. Hence, a natural question is how to combine *precoding schemes* with *signal shaping algorithms*.

In its broadest definition, the aim of signal shaping is to generate signals with some desired properties. Mostly, we concentrated on the reduction of average transmit energy and aimed to achieve *shaping gain*. But precoding also can be seen as a particular form of signal shaping: here a low-power transmit signal is generated, such that the intersymbol-interference channel is preequalized, i.e., the channel output symbols should be free of intersymbol interference.

Combined precoding/shaping techniques therefore extend the properties which should be controlled. Besides a preequalization of the channel, the transmit signal should exhibit desirable characteristics. Again, we mainly focus on the reduction of average energy, and try to achieve shaping gain. Later, additional signal parameters are included in the shaping algorithm. Furthermore, a geometrical interpretation of combined precoding and shaping schemes is given. Finally, we briefly discuss the duality of precoding/shaping to source coding of sources with memory.

As we anticipated in Chapter 3, flexible precoding can be combined with signal shaping with respect to average energy in a straightforward manner. Since the flexible precoder only adds a small "dither sequence," the basic characteristic of a nonuniform distribution of the data symbols is still visible for the channel symbols; cf., e.g., Figure 3.24. In fact, flexible precoding was developed to separate the operations of signal shaping and precoding, and to have "tools" in a "modulation toolbox" which work separately or in combination, and can be chosen adaptively [EFDL93].

A drawback of flexible precoding is that the channel input has a stairstep pdf. The density may come close to a continuous Gaussian one only for very large constellations corresponding to high spectral efficiencies. Conversely, a significant loss in maximum achievable shaping gain occurs for low rates (small constellations). Moreover, flexible precoding has the disadvantage of error multiplication at the receiver, and modifications are required in order for it to be applicable to channels with spectral nulls.

Because of these reasons, flexible precoding is not considered in this chapter. Here, we will study the combination of Tomlinson-Harashima precoding, i.e., precoding based on modulo-congruent signal points, and signal shaping. In particular, *trellis shaping* turns out to be best suited for the present situation.

Before we go into detail on combined precoding/shaping schemes, we should look at the classification of transmission systems in Figure 5.1. Three different

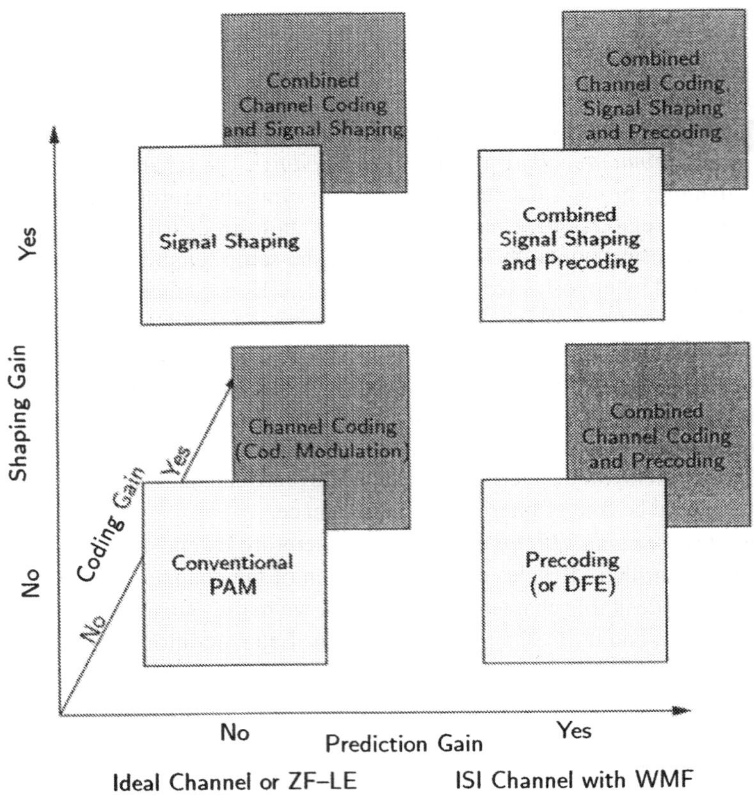

Fig. 5.1 Classification of transmission schemes.

items—shown as three dimensions—are considered, which aim to achieved three different and (almost) independent gains. The baseline for performance evaluation is conventional pulse amplitude modulation, i.e., transmission without channel coding and signal shaping (lower left entry). On intersymbol-interference channels (zero-forcing) *linear equalization (ZF-LE)* (cf. Section 2.2.1) is applied.

By introducing a noise whitening filter $H(z)$ and compensating for the introduced intersymbol interference either by *decision-feedback* at the receiver or *precoding* at the transmitter, *noise prediction gain* G_{p} is achieved. When increasing the order of the noise whitening filter, the prediction gain increases and approaches the ultimate prediction gain (2.3.28). Asymptotically, the whitened matched filter (WMF) constitutes the optimum receiver input filter. When applying precoding schemes, the noise prediction or noise whitening gain can be utilized. Such schemes (lower right part in Figure 5.1) were the topic of Chapters 2 and 3.

In Chapter 4, *signal shaping* for the AWGN channel was discussed, which can be found in the upper left part of Figure 5.1. Using shaping algorithms, a reduction of average transmit power is possible and *shaping gain* G_{s} is provided.

This chapter deals with a combination of precoding and signal shaping, which is located in the upper right part of Figure 5.1. Consequently, such schemes aim to simultaneously achieve both noise prediction gain over ZF-LE, and shaping gain over uniform signaling.

Finally, although not a major topic of this book, performance can always be enhanced by the application of channel coding. The third dimension in Figure 5.1 indicates whether the transmission system achieves *coding gain* G_{c} or not.

Since the combined precoding/shaping schemes to be developed in this chapter are based on Tomlinson-Harashima precoding, channel coding can be applied as for the AWGN channel, cf. Section 3.2.7. In summary, compared to a baseline system using linear equalization and uniform, uncoded transmission, the desired sophisticated schemes should provide *prediction gain* (up to 6 ... 10 dB, depending on the actual situation), *shaping gain* (up to 1.53 dB), and *coding gain* (typically 4 to 6 dB).

5.1 TRELLIS PRECODING

In Chapter 4 we saw that replacing the uniform probability density of the transmit signal by a Gaussian one results in a saving of average power of up to 1.53 dB. However, the techniques discussed in the previous chapter are primarily designed for transmission over the AWGN channel. When transmitting over an ISI channel, precoding schemes as discussed in Chapter 3 are an interesting approach to equalization. However, we have shown (cf. Section 3.2.2) that—except for some special cases—precoding schemes produce a pdf that is uniform over some region $\mathcal{R}$. Hence, an obvious aim is to look for precoding schemes which produce a Gaussian transmit pdf rather than a uniform one. In particular, here we focus on Tomlinson-Harashima precoding (Section 3.2) where precoding is done based on the modulo congruence of signal points. Contrary to flexible precoding, a simple modulo reduction is performed at the receiver to recover data, and no error multiplication of occasional transmission errors occurs.

Since, ignoring the modulo congruence, Tomlinson-Harashima precoding produces an end-to-end AWGN channel, an obvious approach to combine signal shaping and precoding would be to simply cascade both operations, as shown in the upper part of Figure 5.2. A shaping algorithm, e.g., trellis shaping, generates a signal which exhibits some nonuniform density over a boundary region $\mathcal{R}$. For precoding, this region $\mathcal{R}$ has to be suitably chosen and the modulo operation in Tomlinson-Harashima precoding has to match with $\mathcal{R}$.

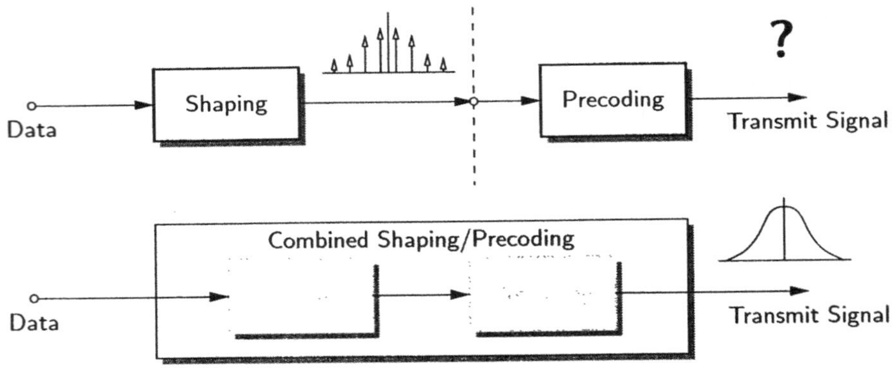

Fig. 5.2 Combination of signal shaping and precoding. Top: independently cascaded; bottom: combined shaping and precoding.

Unfortunately, this straightforward combination of Tomlinson-Harashima precoding and shaping techniques designed for the AWGN channel does not result in the desired Gaussian pdf at the output of the precoder. The nonlinear modulo device in the feedforward path of the precoder is a hindrance. Due to the modulo operation, the signal is randomized to some extent [Tom71], and the signal characteristics will be changed completely. Example 5.1 on page 347 will show the effect of simply cascading shaping and precoding.

The way to overcome this problem is to combine shaping and precoding into a single entity, cf. bottom of Figure 5.2. The shaping algorithm has to take the precoding operation into account. Enumerative shaping methods such as shell mapping (Section 4.3) do not have this ability since the mapping is fixed and cannot be changed. Conversely, as *trellis shaping* is based on a path search through a trellis, by properly selecting the branch metrics, combined shaping and precoding is possible.

The combination of trellis shaping and Tomlinson-Harashima precoding was proposed in [EF92, FE91] and named *trellis precoding*. In this section, we explain the operation of this combined precoding/shaping technique.

5.1.1 Operation of Trellis Precoding

The structural representation of trellis precoding is shown in Figure 5.3 [EF92]. A comparison with Figures 4.32 and 3.4 reveals that trellis precoding is basically the combination of trellis shaping and Tomlinson-Harashima precoding. Trellis shaping is based on a binary rate-κ/η convolutional code $\mathcal{C}_\mathrm{S}$. In addition, we require the signal constellation to be bounded by the region $\mathcal{R}$, which is a fundamental region of the precoding lattice $\boldsymbol{\Lambda}_\mathrm{p}$.

Note that trellis shaping and Tomlinson-Harashima precoding are included as special cases of trellis precoding: trellis shaping results for $H(z) = 1$, whereas choosing the shaping code $\mathcal{C}_\mathrm{S}$ as the trivial all-zero code leads to Tomlinson-Harashima precoding.

As in trellis shaping, the most significant bits of the binary information to be transmitted are combined into the sequence $\boldsymbol{s}(D)$ of binary $\rho = \eta - \kappa$-tuples, which in turn is transformed into the sequence of coset representatives $\boldsymbol{z}(D)$ by filtering with the inverse of the syndrome former, i.e., $\boldsymbol{z}(D) = \boldsymbol{s}(D)\left(\boldsymbol{H}_\mathrm{S}^{-1}(D)\right)^\mathsf{T}$. If coded modulation is active, the least significant bits are encoded by the channel encoder. With the knowledge of these sequences, a trellis decoder for $\mathcal{C}_\mathrm{S}$ determines a valid code sequence $\boldsymbol{c}(D)$, which modifies the sequence $\boldsymbol{z}(D)$ of coset representatives. Given this modified binary data, the PAM signal point $a[k] \in \mathcal{A}$ is obtained by ordinary mapping.

The PAM data symbols $a[k]$ are passed to the Tomlinson-Harashima precoder, which consists of a feedback part (transfer function $H(z) - 1$ for presubtraction of the ISI post cursors), and a modulo reduction with respect to the precoding lattice $\boldsymbol{\Lambda}_\mathrm{p}$ of the transmit symbols $x[k]$ into the boundary region $\mathcal{R}$. The aim of the trellis decoder is to minimize the average energy $\sum_{\nu \leq k} |x[\nu]|^2$ of the transmit symbols $x[k]$. Note that the delay inherent in the decoding process is not shown. As in trellis shaping, care has to be taken that the shaping decoder produces a legitimate code sequence $\boldsymbol{c}(D)$.

The signal $\langle x[k] \rangle$ is transmitted over the channel with transfer function $H(z) = 1 + \sum_{k=1}^{p} h[k]\, z^{-k}$ and disturbed by additive white Gaussian noise. Given the receive sequence $\langle y[k] \rangle$, an estimate of the PAM data symbol $a[k]$ is generated. As in Tomlinson-Harashima precoding, this can be done by first estimating the *effective data sequence* $\langle v[k] \rangle$ with symbols drawn from the expanded signal set $\mathcal{V} = \mathcal{A} + \boldsymbol{\Lambda}_\mathrm{p}$;

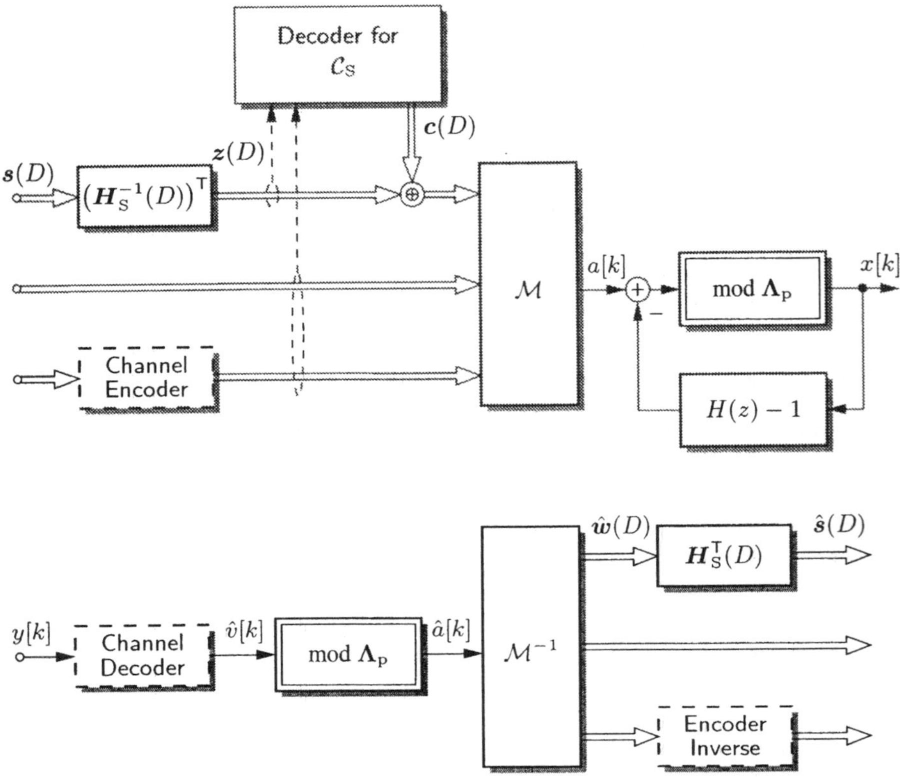

Fig. 5.3 Trellis precoding. Top: transmitter; bottom: receiver structure. Decoding delay not shown.

cf. Section 3.2. From $\hat{v}[k]$ the estimate $\hat{a}[k]$ of the data symbol $a[k]$ is obtained by a modulo-Λ_p reduction into the region $\mathcal{R}$.

Finally, binary data are recovered as in trellis shaping by applying the inverse mapping $\mathcal{M}^{-1}$. The least significant bits immediately determine the corresponding data bits. From the estimates of the most significant bits, contained in the sequence $\hat{w}(D)$, information is retrieved by the syndrome former as $\hat{s}(D) = \hat{w}(D)H_S^T(D)$. Since a feedback-free realization of the syndrome former always exists, the effect of transmission errors is limited to a finite period of time. Nevertheless, error multiplication takes place when recovering data.

5.1.2 Branch Metrics Calculation

Parallel Decision-Feedback Decoding For reducing the average energy of the transmit sequence $\langle x[k] \rangle$, the shaping decoder has to take the precoder part into account. Therefore, we remember that, as in trellis shaping, each path through the trellis of $\mathcal{C}_S$ corresponds to a valid sequence $\langle a[k] \rangle$ of data symbols. Each such possible sequence has then to be filtered separately by the Tomlinson-Harashima precoder. Hence, the precoder individually influences the survivors stored in the path memory of the trellis decoder. Since, in a conventional Viterbi decoder, each state possesses its own survivor, each state can be associated with a history of the precoding filter. For each state in the trellis, the Viterbi algorithm thus also stores the history of hypothetical transmit symbols $x[k]$. This decoding principle—which is known from the theory of maximum-likelihood sequence estimation (MLSE) as a special version of reduced-state sequence estimation (RSSE)—is called *parallel decision-feedback decoding (PDFD)* [Wes87, EQ88, EQ89, CE89, DH89].

In summary, the branch metric in trellis precoding is calculated as follows. For each state S, at time index k, the hypothetical past transmit symbols $x^{(S)}[k-1]$, $x^{(S)}[k-2]$, ..., $x^{(S)}[k-p]$ are stored in a register. The length of these parallel decision-feedback registers equals p, the order of $H(z)$. As in trellis shaping, for a branch emerging from state S and leading to state $\tilde{S}$, the data bits together with the current branch label determine the data symbol $a^{(S\to\tilde{S})}[k]$. Having this, the corresponding transmit symbol $x[k]$ is then calculated to be

$$x^{(S\to\tilde{S})}[k] = \left(a^{(S\to\tilde{S})}[k] - \sum\nolimits_{\nu=1}^{p} h[\nu]x^{(S)}[k-\nu]\right) \bmod \mathbf{\Lambda}_{\mathrm{p}} \,. \qquad (5.1.1)$$

Finally, for reducing average transmit energy, the branch metric is given as

$$\begin{aligned} \lambda^{(S\to\tilde{S})}[k] &= \left|x^{(S\to\tilde{S})}[k]\right|^2 \\ &= \left|\left(a^{(S\to\tilde{S})}[k] - \sum\nolimits_{\nu=1}^{p} h[\nu]x^{(S)}[k-\nu]\right) \bmod \mathbf{\Lambda}_{\mathrm{p}}\right|^2 . \qquad (5.1.2) \end{aligned}$$

As usual in Viterbi decoding, for each state $\tilde{S}$, among all merging paths the one with minimum total cost $\sum_{k\leq nu} \lambda[\nu]$ is determined, and only this survivor is retained in the path register. Additionally, the parallel decision-feedback register of each state is updated, i.e., shifted by one symbol, and the transmit symbol $x[k]$ corresponding to the last segment of the survivor path is stored.

After some finite decoding delay, a decision is made on the actual transmit symbol. Once again we remark that in trellis precoding it is essential to guarantee that the decoder only results in legitimate sequences of the shaping code $\mathcal{C}_{\mathrm{S}}$. Otherwise, error-free data recovery may be impossible at the receiver.

The following example shows the performance of trellis precoding using parallel decision-feedback decoding.

Example 5.1: *Trellis Precoding—Combined Shaping and Precoding*

In this example, the performance of trellis precoding is visualized. For clarity, we use one-dimensional signaling. As in Chapter 3, we use the simplified SDSL up-stream example (self-NEXT dominated environment). Here, three information bits per symbol are transmitted over 3-km-long twisted-pair lines. The end-to-end discrete-time channel model $H(z)$ is again of order $p = 10$. For details, see Chapters 2 and 3 and Appendix B.

Shaping is based on the 16-state code given in Table 4.13 and a partitioning of the signal set into regions according to Ungerböck's principle. The decoding delay is set to 32 symbols.

Figure 5.4 shows the probabilities of the PAM data symbols $a[k]$ (left column) and the probability densities of the transmit symbols $x[k]$ after precoding (right column) obtained from numerical simulations.

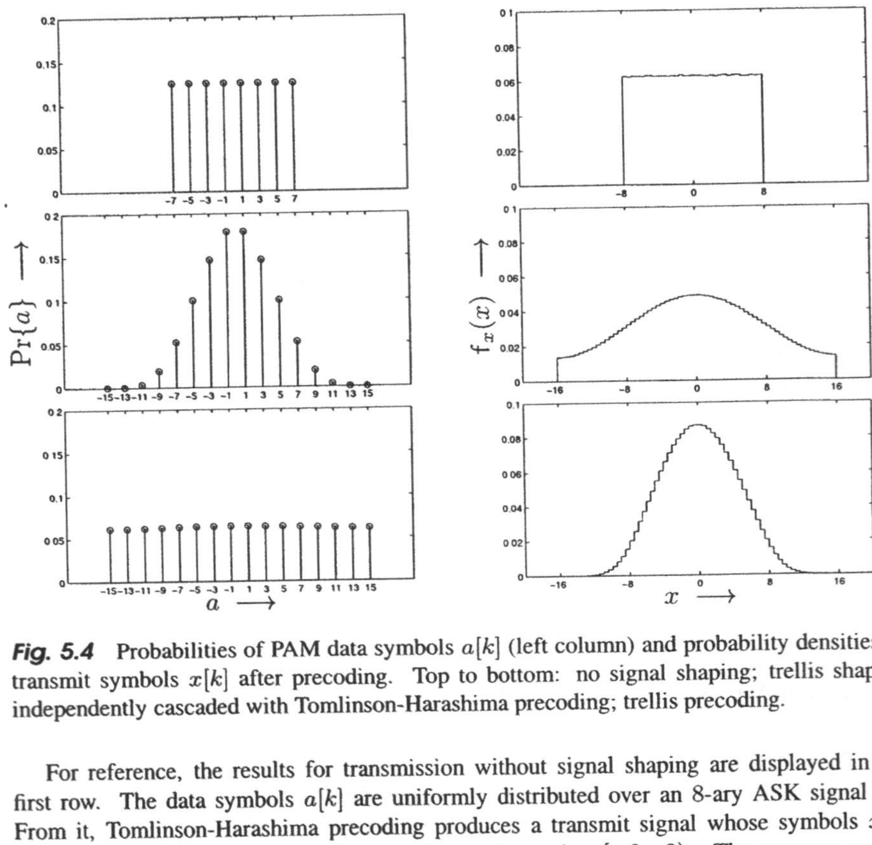

Fig. 5.4 Probabilities of PAM data symbols $a[k]$ (left column) and probability densities of transmit symbols $x[k]$ after precoding. Top to bottom: no signal shaping; trellis shaping independently cascaded with Tomlinson-Harashima precoding; trellis precoding.

For reference, the results for transmission without signal shaping are displayed in the first row. The data symbols $a[k]$ are uniformly distributed over an 8-ary ASK signal set. From it, Tomlinson-Harashima precoding produces a transmit signal whose symbols $x[k]$ are continuous[1] and uniformly distributed over the region $[-8,\ 8)$. The average energy $\sigma_x^2 = 8^2/3 = 21.33$ of these symbols is the baseline for shaping gain calculation.

Next, trellis shaping is simply cascaded with Tomlinson-Harashima precoding as in the top of Figure 5.2. To accommodate the shaping redundancy of one bit, the signal set is extended to a 16-ary ASK constellation. It can be clearly seen that shaping achieves almost a sampled Gaussian density over the data symbols $a[k]$. This signal is fed to the precoder, which is now adapted to the extended signal set. The boundary region is the interval $[-16,\ 16)$ and the precoding lattice is $\Lambda_p = 32\mathbb{Z}$. Due to the modulo operation in the precoder, the Gaussian density of the input is destroyed and, compared to the first row, the pdf of the transmit symbols is broadened. The average energy of the symbols $x[k]$ is increased, causing a loss of about 4.2 dB. This shows the necessity of combining shaping and precoding into one entity.

Finally, the results for trellis precoding are shown in the last row. Here, the parallel decision-feedback decoder explained above is used. As desired, the transmit symbols $x[k]$ exhibit a

[1]The stairstep character of the pdf $f_x(x)$ results from the unavoidable quantization when estimating the histogram of the symbols x in numerical simulations.

continuous, near Gaussian distribution restricted to the interval $[-16, 16)$. Compared to the baseline system, for this particular choice, trellis precoding achieves a shaping gain of 0.84 dB. Note, here the pdf of the data symbols $a[k]$—which in this example deviates only slightly from a uniform one—is only of minor interest.

It should be noted that because of the dispersive nature of $H(z)$ for each of the above transmission schemes, the noiseless channel output symbols $v[k]$ are approximately discrete Gaussian distributed.

Improved Shaping by Modified Trellis Decoding Simulation results (e.g., those given in [EF92] and Example 5.3) showed that employing the same shaping code, trellis precoding using parallel decision-feedback decoding achieves lower shaping gains than trellis shaping. Moreover, the potential gain depends on the impulse response $\langle h[k] \rangle$ of the end-to-end channel $H(z)$. We now study a method which improves the shaping gain at the cost of additional complexity.

In order to illustrate a possible approach to increase performance, the top part of Figure 5.5 shows trellis precoding in a slightly different way compared to Figure 5.3. Here, coset representative generator $\left(\boldsymbol{H}_{\mathrm{S}}^{-1}(D)\right)^{\mathsf{T}}$ and shaping code generator matrix $\boldsymbol{G}_{\mathrm{S}}(D)$ are combined into the *scrambler matrix*

$$\boldsymbol{S}(D) = \begin{bmatrix} \left(\boldsymbol{H}_{\mathrm{S}}^{-1}(D)\right)^{\mathsf{T}} \\ \boldsymbol{G}_{\mathrm{S}}(D) \end{bmatrix}, \tag{5.1.3}$$

as was done in Section 4.4.2 on trellis shaping. When using a rate-κ/η shaping convolutional code $\mathcal{C}_{\mathrm{S}}$, $\rho = \eta - \kappa$ bits of the user data together with κ bits of shaping redundancy are scrambled. The rest of the user data are fed to the mapping directly. Furthermore, the precoder part is drawn in its linearized version, cf. Figure 3.4. The modulo device is replaced by the addition of symbols $d[k]$, appropriately chosen from the precoding lattice Λ_{p}.

In the subsequent discussion, we assume that shaping is based on lattice partitions. The shaping regions are specified by the cosets of the shaping lattice Λ_{s} with respect to the precoding lattice Λ_{p}. The η bits involved in the shaping procedure therefore select one of the 2^{η} coset representatives of $\Lambda_{\mathrm{s}}/\Lambda_{\mathrm{p}}$. The rest of the user data address the point within the region—or a representative of $\Lambda_{\mathrm{a}}/\Lambda_{\mathrm{s}}$, where Λ_{a} is again the signal lattice. Finally, the selection of the precoding symbol $d[k] \in \Lambda_{\mathrm{p}}$ can be characterized as addressing one of the infinitely many coset representatives of $\Lambda_{\mathrm{p}}/\boldsymbol{0}$, where $\boldsymbol{0}$ is the trivial lattice consisting only of the origin. In summary, a lattice partition chain $\Lambda_{\mathrm{a}}/\Lambda_{\mathrm{s}}/\Lambda_{\mathrm{p}}$ is present. Trellis precoding can then be characterized as displayed in the bottom of Figure 5.5.

The aim of a shaping algorithm is then to determine the redundancy such that the average energy of the transmit symbols $x[k]$ is minimized. One part of the redundancy is the shaping bits which are scrambled with the user data. The other part is the redundancy involved in precoding which is directly mapped. This representation of trellis precoding has some advantages, which we now exploit in order to improve shaping gain.

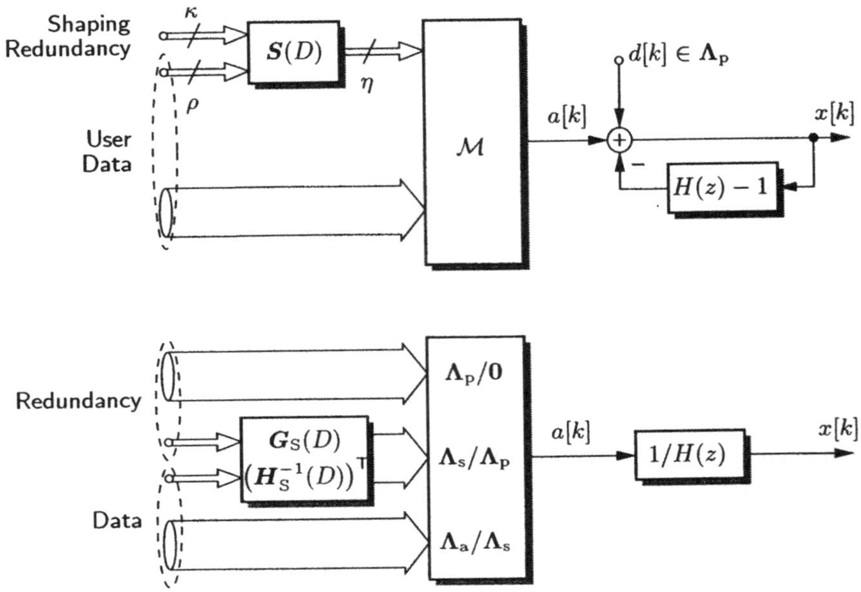

Fig. 5.5 Different representations of trellis precoding.

If we ignore for the moment the user data, i.e., interpret the contribution of the lower addressing levels as some kind of "noise," the process of finding the best sequence in trellis precoding is strikingly similar to maximum-likelihood sequence estimation of coded signaling over intersymbol interference (ISI) channels; see Figure 5.6. Interchanging the roles of redundancy and information, replacing $\boldsymbol{G}_{\mathrm{S}}(D)$ by the code generator matrix $\boldsymbol{G}_{\mathrm{C}}(D)$, and identifying $F(z)$ by $1/H(z)$, the analogy is perfect. In channel decoding, that code sequence is requested which produces the observed (noisy) channel output sequence with highest probability. Similarly, in shaping, a valid shaping code sequence is searched which—together with the lower addressing levels—produces a predistortion filter output sequence with minimum average energy, i.e., which is closest to the all-zero sequence.

We note that the optimal decoder for maximum-likelihood sequence estimation of coded transmission over intersymbol-interference channels operates on the *su-*

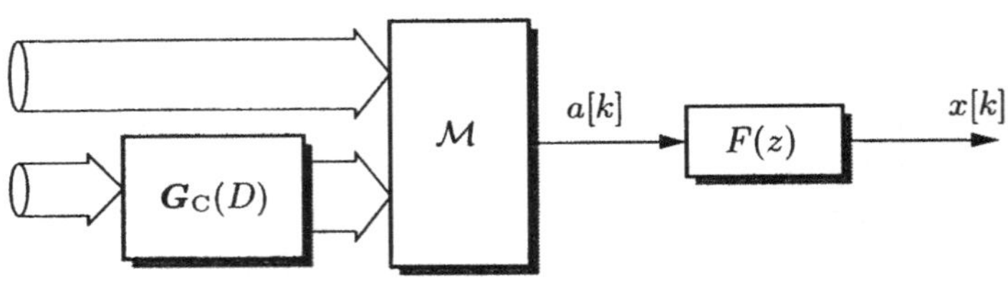

Fig. 5.6 Coded modulation over intersymbol-interference channel $F(z)$.

pertrellis composed of the code trellis and that of the ISI channel. The *superstates* $U[k] \triangleq [S[k];\, \boldsymbol{A}[k]]$ at time index k are the combination of the code states $S[k]$ and the states $\boldsymbol{A}[k] \triangleq [a[k-1],\ a[k-2],\ \ldots,\ a[k-q_f]]$ of the ISI channel. Here, the order of the ISI channel $F(z)$ is denoted by q_f. From the literature, e.g., [Wes87, CE89, EQ88, EQ89, Hub92], methods are known which usually dramatically reduce the complexity of sequence estimation without a significant loss in performance. We have already used one special strategy of reduced-state sequence estimation, namely, parallel decision-feedback decoding, for trellis precoding.

In reduced-state sequence estimation, a number of superstates are merged into *hyperstates* by ignoring parts of the portion $\boldsymbol{A}[k]$ associated with the ISI state. This state reduction is preferably done based on set partitioning or lattice partitioning principles [CE89, EQ88]. Instead of treating the actual symbol $a[k-\nu]$, only the subset to which it belongs is considered. Moreover, the number of subsets which are distinguished decreases (or stays the same) as ν increases, i.e., recent symbols are more precisely distinguished than past symbols.

As a consequence, the reduced-state trellis will exhibit parallel branches, i.e., branches leaving the same state and merging into a common state. These branches correspond to the points coming from the same subset. From these subset points, the actual representative is selected by a symbol-by-symbol decision.

Nevertheless, for metric calculation, the exact channel state has to be known. Since the channel state is not uniquely given by the hyperstates in the decoder, its path history $a[k-1],\ a[k-2],\ \ldots,\ a[k-q_f]$ has to be stored for each decoder state. By selecting the number of subsets which are processed, a flexible exchange between effort and performance can be achieved. For details on reduced-state sequence estimation the reader is referred to the literature, e.g., [Wes87, CE89, EQ88, EQ89, Hub92].

We now return to trellis precoding. As opposed to the direction taken in channel decoding/equalization, where a complexity reduction relative to optimum maximum-likelihood decoding is desired, we now admit a higher complexity in trellis precoding with the goal of achieving higher shaping gains. Here, the baseline is parallel decision-feedback decoding, where the entire part $\boldsymbol{A}$ associated with the ISI state is ignored in the state definition.

Starting from the shaping code trellis, and following the guidelines in [EQ89, CE89], expanded RSSE trellises can be designed. First, in the state $U[k]$ the four different subsets addressed by the rate-$1/2$ shaping convolutional code at time index $k-1$ are regarded. Since only two paths merge in each state—corresponding to two different subsets—the number of states is doubled compared to the number of code states. Next, the subsets of the last two symbols are used in defining the state. Here, four combinations are possible, and thus the number of states in the augmented trellis is four times that of the shaping code.

Increasing the number of states compared to parallel decision-feedback decoding by a factor of Υ can be viewed as processing this number Υ of survivor paths per code state. Instead of making a decision, both possible paths are kept in the path memory and none of them is discarded until the next time step.

Example 5.2: *Expanded Trellises*

For Ungerböck's 4-state rate-$1/2$ convolutional code, we now construct expanded 8- and 16-state trellises. Figure 5.7 shows the respective encoder. In parallel decision-feedback decoding, the original code trellis with states $S[k] = [s^2[k]\ s^1[k]]$ is used.

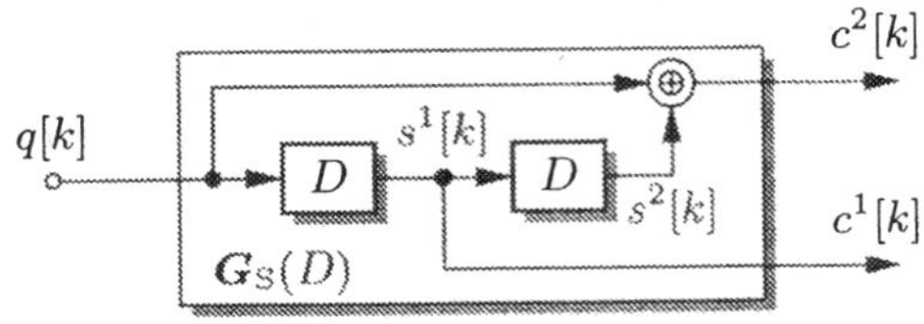

Fig. 5.7 Encoder for Ungerböck's 4-state rate-$1/2$ convolutional code.

The hyperstates of expanded 8- and 16-state trellises, regarding the subsets of the last and the last two symbols, respectively, are defined as

$$U^{(1)}[k] = [\,c^2[k-1]\ c^1[k-1]\ s^2[k]\ s^1[k]\,] \tag{5.1.4a}$$

$$U^{(2)}[k] = [\,c^2[k-2]\ c^1[k-2]\ c^2[k-1]\ c^1[k-1]\ s^2[k]\ s^1[k]\,]\,. \tag{5.1.4b}$$

The resulting trellises are drawn in Figure 5.8. The original 4-state trellis is shown on the left-hand side of the figure. The branches are labeled with the pair $[c^2[k]\ c^1[k]]$ of encoder output symbols. In the middle, an 8-state trellis is given. Besides the encoder state, the previous encoder output pair $[c^2[k-1]\ c^1[k-1]]$ is taken into account here as well. The hyperstates corresponding to the original encoder states are inclosed in the shaded area. Here, branches entering a state correspond to the same pair $[c^2[k]\ c^1[k]]$; this pair is only given once.

Finally, a 16-state trellis is depicted on the right hand side of the figure. In addition to the encoder state, the past encoder output pairs $[c^2[k-1]\ c^1[k-1]]$ and $[c^2[k-2]\ c^1[k-2]]$ are incorporated into the hyperstate definition. Again, branches entering a state correspond to the same pair $[c^2[k]\ c^1[k]]$.

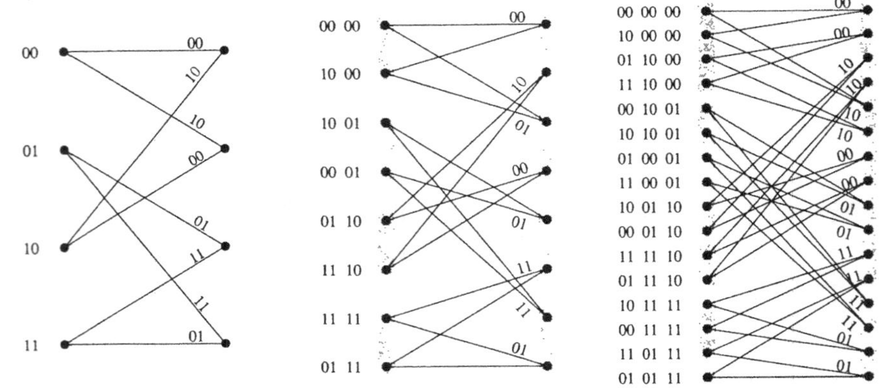

Fig. 5.8 Expanded 8- and 16-state trellises based on Ungerböck's 4-state rate-$1/2$ convolutional code.

The points from the subsets are selected on a symbol-by-symbol basis. From Figure 5.5 we see that this selection concerns the choice of the precoding symbol $d[k]$, drawn from the precoding lattice Λ_{p}. Hence, in each case the precoder part reduces the transmit symbols symbol-by-symbol into a fundamental region of Λ_{p}—this procedure is almost optimum. In Chapter 4 we saw that one shaping bit per two-dimensional signal point is sufficient to nearly achieve the maximum possible shaping gain. Hence, restricting the transmit symbols $x[k]$ to a region whose size is doubled compared to the case of uniform transmission cannot be the reason for reduced shaping gains.

As noted in [EF92], the shaping gain in trellis precoding depends on the interaction of the shaping code $\mathcal{C}_{\text{S}}$ and the impulse response of $H(z)$. This is the same as in channel decoding on ISI channels, where the minimum Euclidean distance of channel output sequences depends on the interaction of channel code and ISI channel.

We show the performance of improved decoding in the next example.

Example 5.3: *Performance of Trellis Precoding*

In this example, the performance of trellis precoding is studied. We look at the simplified ADSL down-stream transmission scenario with cable length of 5 km, already used in Chapter 3 (cf. also Appendix B). The discrete-time end-to-end channel model $H(z)$ is again of order $p = 10$. Since five information bits are transmitted per symbol, and one bit redundancy is added for shaping, the PAM data symbols are drawn from a 64-ary QAM alphabet. The precoding lattice therefore equals $\Lambda_{\text{p}} = 16\mathbb{Z}^2$.

The sign-bit shaping scheme introduced in Example 4.13 is adopted as the shaping part of trellis precoding. The only modification now is that four instead of six "lower" levels are present. Here, the rate-$1/2$ shaping convolutional code of Table 4.14 is used.

Figures 5.9 through 5.11 plot the shaping gain, obtained in numerical simulations, over the decoding delay (measured in two-dimensional symbols). In each case, the performance of parallel decision-feedback decoding (boxes) is shown. Additionally, the results of decoding based on expanded trellises—doubling (crosses) or quadruplicating (circles) the number of states—is given. For reference, the shaping gain of a hypersphere with dimension equal to twice the decoding delay is shown (dashed). Furthermore, the shaping gain of trellis shaping, i.e., shaping on the AWGN channel, using the same code is included (dash-dotted).

Compared to trellis shaping on the AWGN channel, trellis precoding employing parallel decision-feedback decoding achieves only a (significantly) lower shaping gain. Using expanded trellises derived by reversing RSSE principles, a higher shaping gain can be achieved at the cost of doubled or quadrupled complexity. Starting from a 4-state code, the additional gains are more significant than when shaping is already based on an 8- or 16-state code. Using improved decoding techniques, the shaping gain lies in the same region as the gain of trellis shaping using the same shaping code.

Finally, comparing decoding strategies with the same number of states, no significant difference in achievable shaping gain is seen. For example, for decoding delays of 20 to 22 symbols, the 4-state code decoded in an expanded 16-state trellis (Figure 5.9), the 8-state code decoded in an expanded 16-state trellis (Figure 5.10), and the 16-state code (Figure 5.11) provide about 0.9 dB shaping gain.

The probability density functions obtained by trellis precoding are plotted in Figure 5.12. The situation is the same as above and the 16-state code is used. The decision delay is fixed to

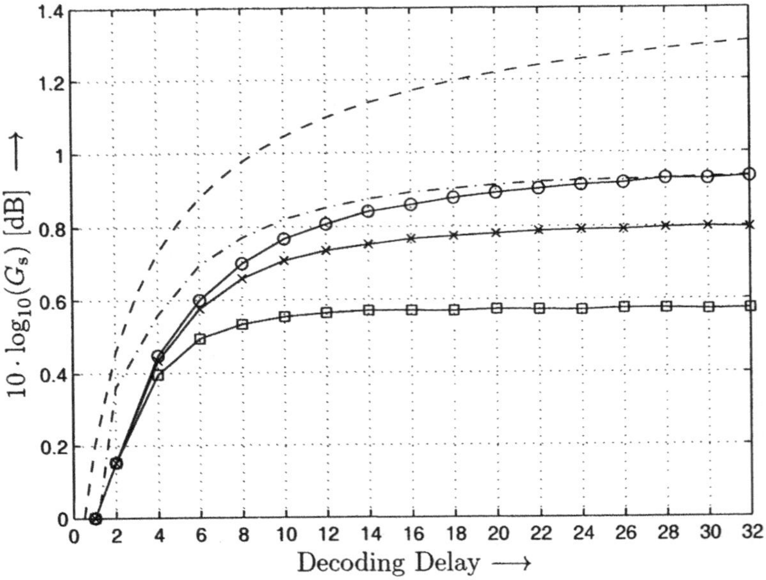

Fig. 5.9 Shaping gain G_s over the decoding delay (in two-dimensional symbols). 4-state shaping code; ADSL transmission scenario. □: Parallel decision-feedback decoding (4 states); ×/o: decoding using 8/16-state extended trellis; dash-dotted line: shaping gain of trellis shaping using the same code; dashed line: shaping gain of hypersphere.

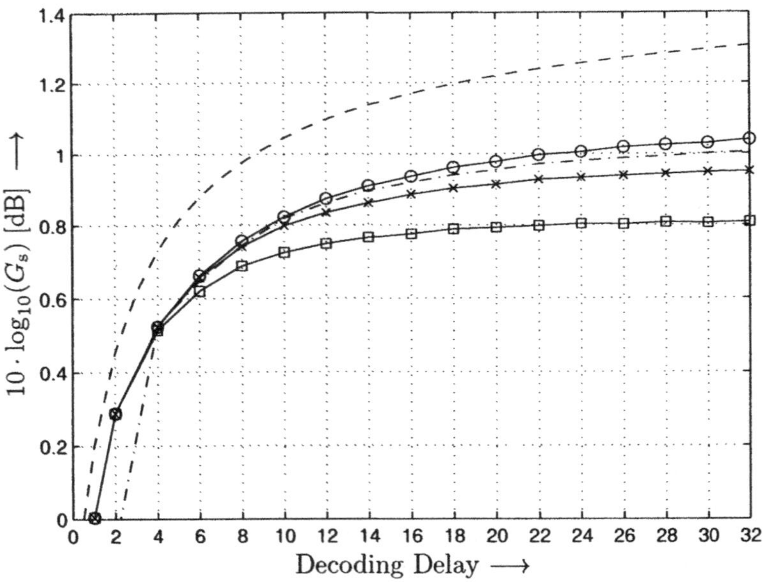

Fig. 5.10 Shaping gain G_s over the decoding delay (in two-dimensional symbols). 8-state shaping code; ADSL transmission scenario. □: Parallel decision-feedback decoding (8 states); ×/o: decoding using 16/32-state extended trellis; dash-dotted line: shaping gain of trellis shaping using the same code; dashed line: shaping gain of hypersphere.

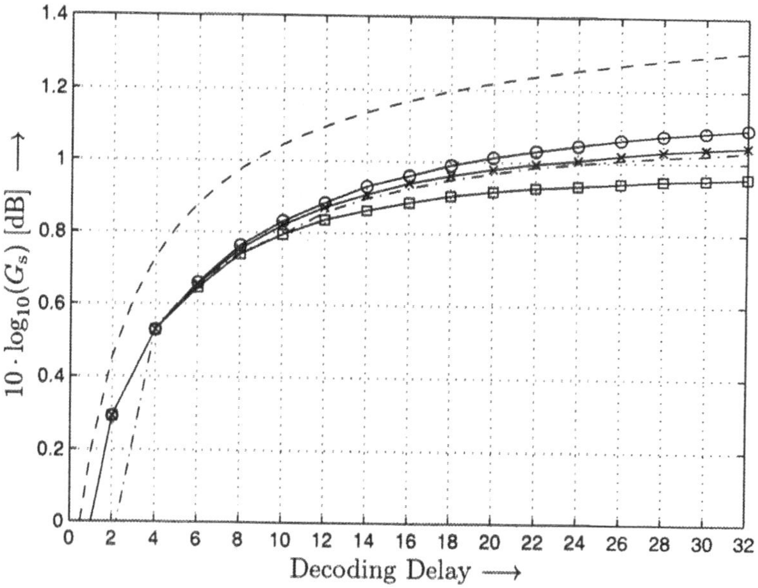

Fig. 5.11 Shaping gain G_s over the decoding delay (in two-dimensional symbols). 16-state shaping code; ADSL transmission scenario. □: Parallel decision-feedback decoding (16 states); ×/○: decoding using 32/64-state extended trellis; dash-dotted line: shaping gain of trellis shaping using the same code; dashed line: shaping gain of hypersphere.

20 symbols, which in turn leads to a shaping gain of 0.92 dB. Note that the symbols $a[k]$ are discrete, hence probabilities $\Pr\{a\}$ are given on the left-hand side. Conversely, the transmit symbols $x[k]$ are continuously distributed over the region $[-8,\ 8)$; here the pdf $f_x(x)$ is displayed. As expected, the channel symbols $x[k]$ are almost Gaussian distributed, whereas the data symbols $a[k]$ are neither uniform nor Gaussian.

To conclude we remark that, as in trellis shaping (see Section 4.4.3), applying a peak constraint, the peak-to-average energy ratio can be lowered in trellis precoding, too.

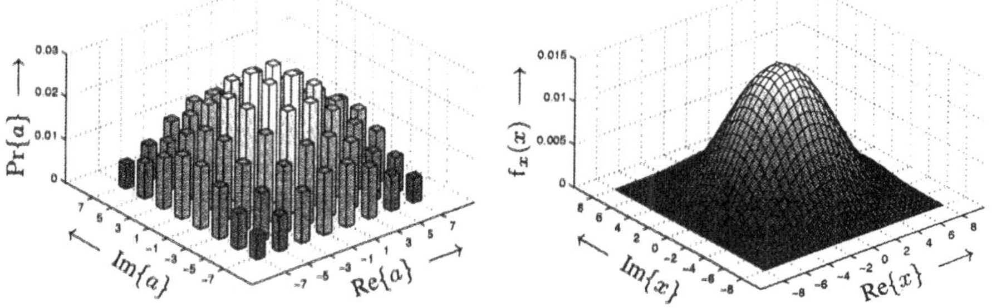

Fig. 5.12 Two-dimensional distribution of the PAM symbols $a[k]$ (left) and the transmit symbols $x[k]$ (right).

5.2 SHAPING WITHOUT SCRAMBLING

Trellis shaping and its extension to intersymbol-interference channels, trellis precoding, scramble user data with shaping redundancy. As a consequence, for data recovery a descrambler, in particular a syndrome former, is required at the receiver. Since it is a dispersive system, channel errors become effective several time in the decoded user data bits. Fortunately, a feedback-free realization of the syndrome former is always possible, hence error multiplication is finite. However, at least part of the shaping gain is lost by error propagation at the receiver.

In trellis shaping, i.e., when transmitting over the AWGN channel, explicit scrambling of data and redundancy is indispensable. However, when performing precoding, information and (precoding) redundancy are already intermixed at the predistortion filter $1/H(z)$. This suggests that shaping on intersymbol-interference channels may be done without a dedicated scrambler. We now present a technique for combined precoding and signal shaping which avoids an extra scrambling. This technique, developed in [Fis92, FGH95] and anticipated in [EF92], is here called *shaping without scrambling*.

5.2.1 Basic Principle

From Figure 5.3 we see that in trellis precoding—the straightforward combination of trellis shaping and Tomlinson-Harashima precoding—the shaping and the precoding parts are clearly separated. In order to avoid the scrambler at the transmitter and the respective descrambler at the receiver, the operations of shaping and precoding have to be combined into one entity.

This can be done by selecting the *precoding sequence* $\langle d[k]\rangle$, with $d[k] \in \mathbf{\Lambda}_{\mathrm{p}}$, suitably. Instead of performing a symbol-by-symbol decision, the *effective data sequence* $\langle v[k]\rangle$, $v[k] = a[k] + d[k]$, is selected in the long run so that the output of $1/H(z)$ (the formal inverse of $H(z)$) exhibits some desired properties. Thus, given the data sequence $\langle a[k]\rangle$, an algorithm has to determine or "decode" the best sequence $\langle d[k]\rangle$. In other words, among the symbols $v[k]$ modulo-congruent to the PAM data symbols $a[k]$, the representatives are chosen which form the best sequence $\langle v[k]\rangle$. The structure of this shaping without scrambling scheme is depicted in Figure 5.13.

The corresponding receiver structure is plotted in Figure 5.14. As in Tomlinson-Harashima precoding, a simple modulo device at the receiver is sufficient to eliminate redundancy and to recover the data. No dispersive system is necessary at the receiver, and hence no error multiplication occurs. The whole shaping gain directly translates to a gain in performance.

Besides the higher net shaping gain, the main advantage of shaping without scrambling is its compatibility with Tomlinson-Harashima precoding. Even in standardized systems, the precoder can be replaced by the shaping technique at hand—no modifications at the receiver are required. Whenever Tomlinson-Harashima precoding is used, it can be replaced by shaping without scrambling to achieve shaping gain, and hence improve performance.

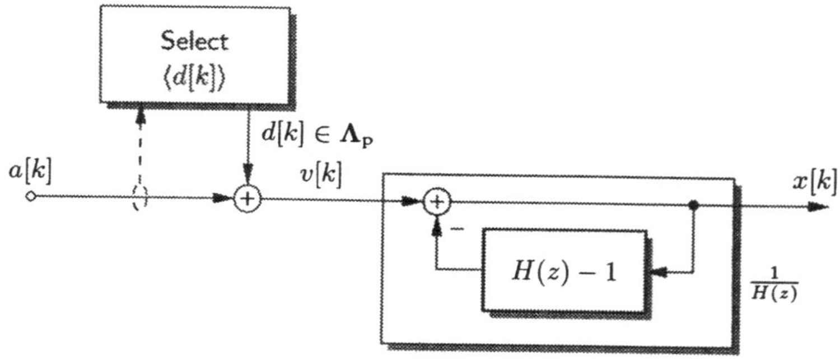

Fig. 5.13 Basic concept of shaping without scrambling.

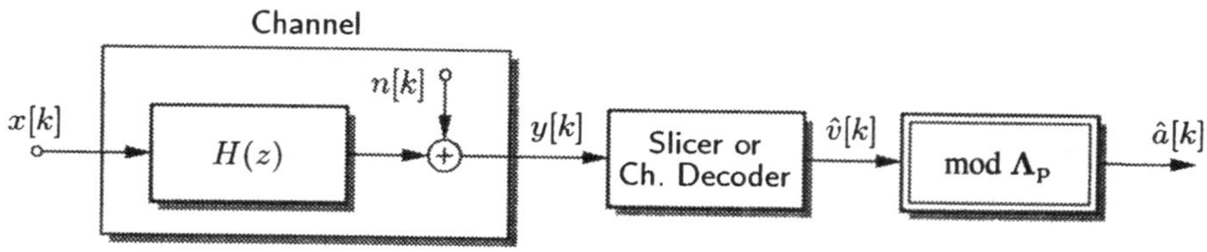

Fig. 5.14 Receiver structure for shaping without scrambling.

Unfortunately, as in Tomlinson-Harashima precoding and trellis precoding, circular constellations are not supported. This may result in a somewhat higher peak-to-average energy ratio or constellation expansion for the given shaping gain or decoding delay [LFT94]. But, applying peak constraints [For92, EF92] (cf. also Sections 4.4.3 and 5.3.4), this disadvantage of shaping without scrambling can be mitigated to a great extent.

Moreover, by choosing the right metric in the decoder, along with a reduction of average transmit power, almost any signal property can be created by shaping without scrambling. We return to this point in more detail in Section 5.3.

5.2.2 Decoding and Branch Metrics Calculation

Shaping without scrambling can be obtained from trellis precoding by employing the trivial all-zero convolutional code $\mathcal{C}_S$ for shaping [EF92]. Both, coset representative generator and syndrome former in trellis precoding then degenerate to identity matrices, i.e., are not present. Hence, in principle, decoding in shaping without scrambling could be done as described for trellis precoding. But here the use of expanded trellises (starting from a one-state trellis) is indispensable for achieving gains. Nevertheless, we now present a different, simpler approach for decoding the best sequence in shaping without scrambling.

Obviously, all possible precoding sequences $\langle d[k] \rangle$ can be described using a *tree*. In each time step, regardless of the preceding symbols, any of the precoding symbols $d[k] \in \mathbf{\Lambda}_\mathrm{P}$ may be selected. Hence, *sequential decoding algorithms*, e.g., Fano,

Stack, or M-algorithm [JZ99], may be applied to the present situation. The influence of the predistortion filter $1/H(z)$ has to be taken into account again by means of parallel decision-feedback, as shown above.

In practice, however, such decoding strategies cannot be implemented, because an infinite number of branches eminates from each state. Each symbol $d[k]$ can assume any point of the precoding lattice $\mathbf{\Lambda}_{\text{p}}$, which comprises an infinite number of points. Only if $d[k]$ would be restricted to a finite range, would such decoders be feasible in principle, but they are still much too complex.

In order to enable decoding in shaping without scrambling, we remember that for signal shaping one redundant bit is sufficient, see Section 4.2.4. Consequently, compared to uniform signaling, the support range of the constellation has at most to be doubled in order to achieve significant gains. But if we restrict the channel symbols $x[k]$ to a region whose size is doubled compared to that in Tomlinson-Harashima precoding—where the choice of $d[k]$ is unique—only two candidates for the precoding symbol $d[k]$ remain. In summary, a *binary* tree for decoding suffices.

The operation of shaping without scrambling is again preferably described in terms of lattices. An introduction to lattices is given in Appendix C. The PAM data symbols $a[k]$ are drawn from a signal constellation $\mathcal{A}$, which is the intersection of some (translated by $\boldsymbol{t}$) signal-point lattice $\mathbf{\Lambda}_{\text{a}}$ and a fundamental region $\mathcal{R}(\mathbf{\Lambda}_{\text{p}})$ of the precoding lattice $\mathbf{\Lambda}_{\text{p}}$, i.e., $\mathcal{A} = (\mathbf{\Lambda}_{\text{a}} + \boldsymbol{t}) \cap \mathcal{R}(\mathbf{\Lambda}_{\text{p}})$, cf. Section 3.2.3. Usually, in D-dimensional signaling ($D = 1, 2$), the signal lattice is $\mathbf{\Lambda}_{\text{a}} = 2\mathbb{Z}^D$, and the translation vector $\boldsymbol{t}$ equals the all-one vector of dimension D. The channel symbols $x[k]$ lie in a fundamental region—preferably the Voronoi region—of some *shaping lattice* $\mathbf{\Lambda}_{\text{s}}$. This lattice has to be a sublattice of the precoding lattice, and the order of the partition $\mathbf{\Lambda}_{\text{p}}/\mathbf{\Lambda}_{\text{s}}$ has to be two. Then, as required, $V(\mathbf{\Lambda}_{\text{s}}) = 2 \cdot V(\mathbf{\Lambda}_{\text{p}})$ holds, i.e., the size of the support region is doubled compared to Tomlinson-Harashima precoding.

Figure 5.15 depicts the resulting structure for shaping without scrambling. Here, the task of finding the best precoding sequence $\langle d[k] \rangle$ is split into two parts. First, in the long run, the shaping decoder only makes a binary decision $b[k]$. This binary

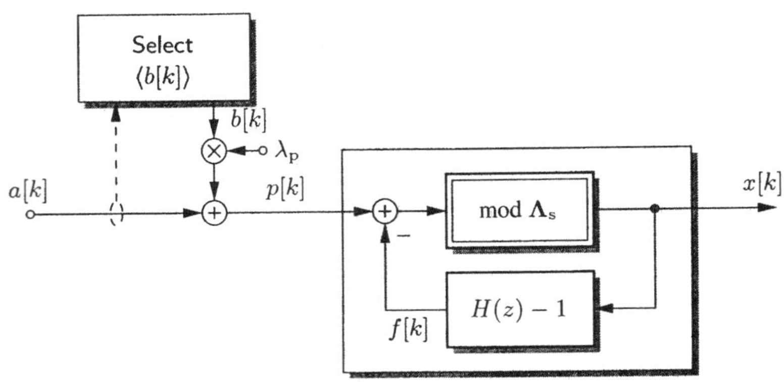

Fig. 5.15 Block diagram of shaping without scrambling. Decoding delay not shown.

symbol determines the coset of Λ_s with respect to Λ_p, from which the precoding symbol $d[k]$ is drawn. This coset is specified by the coset representative $b[k] \cdot \lambda_p$, where λ_p is a point in Λ_p, but not in Λ_s. Hence, $b[k]$ determines whether $d[k]$ is drawn from the set Λ_s ($b[k] = 0$) or its coset $\Lambda_s + \lambda_p$ ($b[k] = 1$). To shift the data symbols $a[k]$ to the appropriate coset, the coset representative, either $\mathbf{0}$ or λ_p, is simply added to $a[k]$.

Second, the actual precoding symbol is chosen symbol-by-symbol from the current coset, such that the channel symbols $x[k]$ lie in the fundamental region of Λ_s. This is done as in Tomlinson-Harashima precoding, but replacing the mod-Λ_p operation by mod-Λ_s. By construction, adding an arbitrary point from Λ_s—which is done in the modulo operation—does not change the selected coset. As in Tomlinson-Harashima precoding, the precoding symbols $d[k]$ are implicitly chosen, and are no longer visible in Figure 5.15.

Note that, due to the mod-Λ_s reduction in the precoder part of shaping without scrambling, stable operation is guaranteed, even if $H(z)$ has spectral zeros.

Example 5.4: *Shaping and Precoding Lattice*

This example gives the lattices involved in shaping without scrambling, for both one- and two-dimensional signaling.

A one-dimensional M-ary ASK signal constellation is employed in baseband transmission, which is given as

$$\mathcal{A} = \{\pm 1,\ \pm 3,\ \ldots,\ \pm(M-1)\} = (2\mathbb{Z}+1) \cap [-M,\ M)\ . \tag{5.2.1a}$$

Here, precoding lattice Λ_p, shaping lattice Λ_s, and coset representative λ_p read

$$\Lambda_p = 2M\mathbb{Z}\ , \qquad \Lambda_s = 4M\mathbb{Z}\ , \qquad \lambda_p = 2M\ . \tag{5.2.1b}$$

Figure 5.16 shows excerpts of the above one-dimensional lattices.

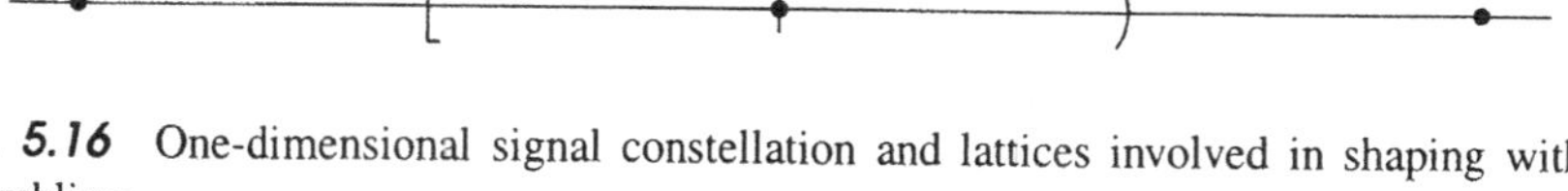

Fig. 5.16 One-dimensional signal constellation and lattices involved in shaping without scrambling.

For the present situation, the operation of shaping without scrambling is as follows. The shaping algorithm decides whether the precoding symbol $d[k]$ is drawn from the set $4M\mathbb{Z}$ ($b[k] = 0$) or from the set $4M\mathbb{Z} + 2M$ ($b[k] = 1$). The actual point from these sets is selected by the mod-$4M$ operation in the precoder part. By adding nothing or $2M$, the initial signal constellation is shifted to the appropriate coset of the shaping lattice Λ_s.

For passband transmission, a two-dimensional M-ary QAM signal constellation is given as

$$\mathcal{A} = (2\mathbb{Z}^2 + \left[\begin{smallmatrix}1\\1\end{smallmatrix}\right]) \cap \mathcal{R}(\boldsymbol{\Lambda}_{\mathrm{p}}) . \tag{5.2.2a}$$

If M is a square number, the signal constellation is also square, and precoding lattice $\boldsymbol{\Lambda}_{\mathrm{p}}$, shaping lattice $\boldsymbol{\Lambda}_{\mathrm{s}}$, and coset representative $\boldsymbol{\lambda}_{\mathrm{p}}$ read

$$\boldsymbol{\Lambda}_{\mathrm{p}} = 2\sqrt{M}\mathbb{Z}^2 , \qquad \boldsymbol{\Lambda}_{\mathrm{a}} = 2\sqrt{M}\boldsymbol{R}\mathbb{Z}^2 , \qquad \boldsymbol{\lambda}_{\mathrm{p}} = \left[\begin{smallmatrix}2\sqrt{M}\\0\end{smallmatrix}\right] . \tag{5.2.2b}$$

Otherwise, if $\log_2(M)$ is odd, we employ rotated square constellations (cf. Section 3.2.3), where the respective lattices are

$$\boldsymbol{\Lambda}_{\mathrm{p}} = \sqrt{2M}\boldsymbol{R}\mathbb{Z}^2 , \qquad \boldsymbol{\Lambda}_{\mathrm{a}} = 2\sqrt{2M}\mathbb{Z}^2 , \qquad \boldsymbol{\lambda}_{\mathrm{p}} = \left[\begin{smallmatrix}2\sqrt{M}\\2\sqrt{M}\end{smallmatrix}\right] . \tag{5.2.2c}$$

Here, $\boldsymbol{R} = \left[\begin{smallmatrix}1 & -1\\1 & 1\end{smallmatrix}\right]$ is the rotation and scaling operator in two dimensions.

Figure 5.17 shows excerpts of the above two-dimensional lattices and $M = 2^{2m}$, $m \in \mathbb{N}$, a square number. Rotating the lattices by 45°, the situation for $M = 2^{2m+1}$, $m \in \mathbb{N}$, i.e., rotated square constellations, is obtained.

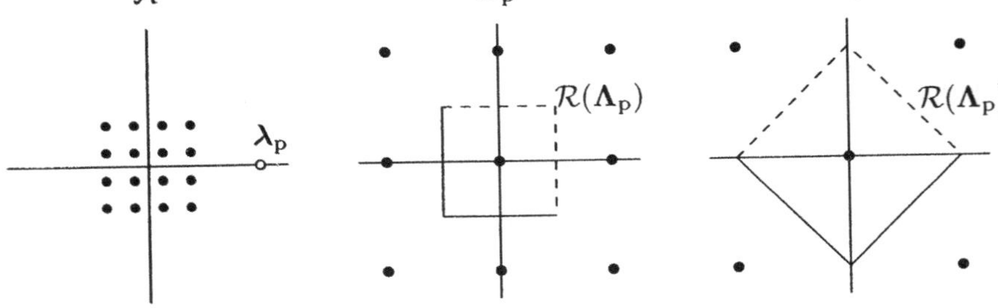

Fig. 5.17 Two-dimensional signal constellation and lattices involved in shaping without scrambling.

For reducing average transmit energy, the branch metric $\lambda[k]$ is again chosen as the squared magnitude of the channel symbols $x[k]$. Given some state S of the decoder with its associated hypothetical past transmit symbols $x^{(S)}[k-\kappa]$, $\kappa = 1, 2, \ldots, p$, and the binary symbols $b[k]$ determining the branch, we have, from Figure 5.15,

$$\lambda^{(S,b)}[k] = |x^{(S,b)}[k]|^2 = \left| \left[a[k] - b[k]\boldsymbol{\lambda}_{\mathrm{p}} - \sum\nolimits_{\kappa=1}^{p} h[\kappa] x^{(S)}[k-\kappa] \right] \bmod \boldsymbol{\Lambda}_{\mathrm{s}} \right|^2 . \tag{5.2.3}$$

In the present version, the shaping algorithm has to decode a binary tree. Unfortunately, because of their nonregular structure, sequential decoding algorithms are not very well suited for fast hardware implementation, as is required, e.g., for xDSL transmission. We now show how a Viterbi decoder can be used in shaping without scrambling.

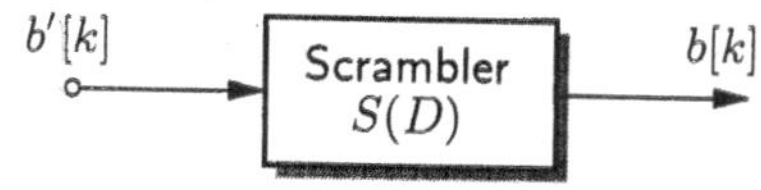

Fig. 5.18 Binary sequence $\langle b[k] \rangle$ as the outcome of an imaginary scrambler operating over $\mathbb{F}_2$.

Therefore, we *imagine* that the binary sequence $\langle b[k] \rangle$ is the output of an arbitrary *scrambler*, see Figure 5.18. This linear, dispersive, discrete-time system operates over the Galois field $\mathbb{F}_2$, where modulo-2 arithmetic is performed. The scrambler should have a finite number of states, but it does not matter whether the scrambler can be implemented feedback-free. Its transfer polynomial is denoted by $S(D)$.

Since here the scrambler is a rate-$1/1$ system, any output sequence $\langle b[k] \rangle$ can be generated by feeding an appropriate sequence $\langle b'[k] \rangle$ into the scrambler. Moreover, each sequence can uniquely be characterized by a path through the trellis, defined by the internal states and possible branches of the scrambler. Hence, instead of treating the sequence $\langle b[k] \rangle$ itself, we can resort to the states and branches of the scrambler. The sequential decoder may then be replaced by a *Viterbi algorithm* applied to the trellis of an *imaginary scrambler*. In other words, the tree for decoding is folded to a trellis [For73]. Different paths are arbitrarily forced to merge and a decision between them is made. The exponential growth of the tree over time is broken down and only a finite number of paths, equal to the number of states, is considered.

Note that no dispersive system is required to recover data at the receiver. The scrambler does not have to be inverted; its influence is eliminated at the modulo device. Hence, systematic errors as in trellis shaping and trellis precoding cannot occur. In shaping without scrambling, continuous-path integrity is thus of minor importance.

5.2.3 Performance of Shaping Without Scrambling

We now assess the performance of shaping without scrambling and compare this technique with Tomlinson-Harashima precoding and trellis precoding.

First, the choice of the scrambler polynomial $S(D)$ is studied. This imaginary scrambler defines the trellis on which decoding is based. Then, the signals generated by shaping without scrambling and Tomlinson-Harashima precoding, respectively, are visualized and compared. Finally, the error rates of transmission using shaping without scrambling, Tomlinson-Harashima precoding, and trellis precoding are contrasted.

Note that, as in trellis shaping and trellis precoding, applying peak constraints, the peak-to-average energy ratio of the transmit symbols can be lowered also in shaping without scrambling.

Example 5.5: Choice of the Imaginary Scrambler

In this example on shaping without scrambling, we again study the simplified ADSL downstream transmission scenario with a cable length of 5 km. The parameters are equal to Example 5.3 on trellis precoding.

Since five information bits are transmitted per symbol, the initial data symbols $a[k]$ are drawn from a 32-ary rotated square QAM constellation. Consequently, the precoding lattice equals $\mathbf{\Lambda}_\mathrm{p} = 8\boldsymbol{R}\mathbb{Z}^2$, and the shaping lattice is given by $\mathbf{\Lambda}_\mathrm{s} = 16\mathbb{Z}^2$. The point $\boldsymbol{\lambda}_\mathrm{p} = \left[\begin{smallmatrix}8\\8\end{smallmatrix}\right]$ can be used to specify the coset of $\mathbf{\Lambda}_\mathrm{s}$ with respect to $\mathbf{\Lambda}_\mathrm{p}$.

First, the shaping algorithm is based on an imaginary scrambler with polynomial

$$S(D) = 1 \oplus D \oplus D^s\ . \tag{5.2.4}$$

The number 2^s of states can be adjusted via the exponent $s \in \mathbb{N}$. Figure 5.19 plots the shaping gain, obtained in numerical simulations, over the decoding delay (measured in two-dimensional symbols). The number of states ranges from 4 to 64. For reference, the shaping gain of a hypersphere with dimension equal to twice the decoding delay is also shown (dashed).

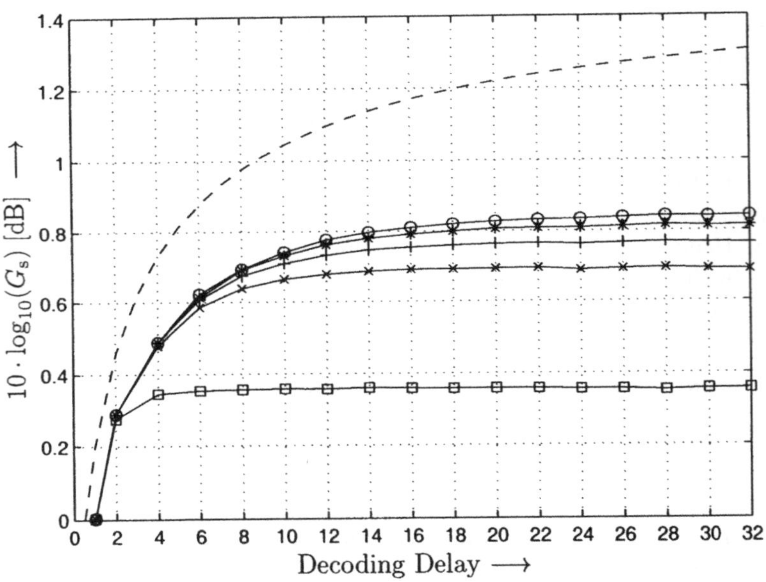

Fig. 5.19 Shaping gain G_s of shaping without scrambling over the decoding delay (in two-dimensional symbols). ADSL transmission scenario. Decoding is done in the trellis of an imaginary scrambler with □: 4-states; ×: 8-states; +: 16-states; ∗: 32-states; o: 64-states. Dashed line: shaping gain of hypersphere.

As can be seen, using a 16-state scrambler, a shaping gain of almost 0.8 dB is possible. Going to a larger number of states increases the shaping gain to about 0.85 dB.

The dependency of the shaping gain on the choice of the scrambler polynomial is depicted in Figure 5.20. Here, 16-state scramblers are compared. The scrambler polynomial $S(D)$ is selected to be

$$1 \oplus D \oplus D^4\,, \qquad 1 \oplus D^2 \oplus D^4\,, \qquad 1 \oplus D^4$$
$$1 \oplus D \oplus D^2 \oplus D^4\,, \qquad 1 \oplus D \oplus D^2 \oplus D^3 \oplus D^4 \tag{5.2.5}$$

The curves for the various scrambler polynomials are almost indistinguishable. Hence, there does not seem to be a strong dependency on the actual choice of the scrambler, i.e., the actual definition of the imaginary trellis for Viterbi decoding.

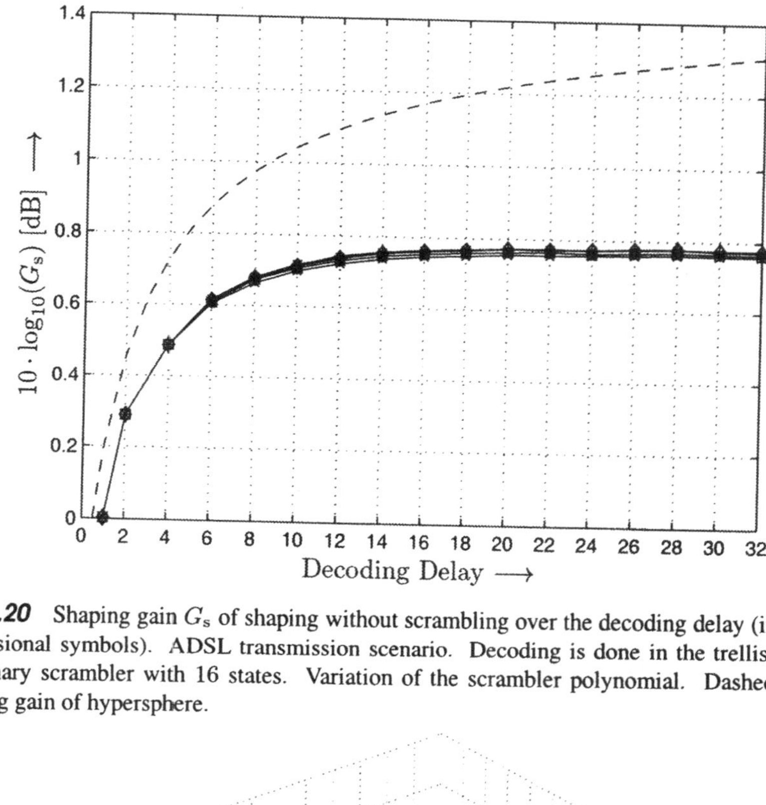

Fig. 5.20 Shaping gain G_s of shaping without scrambling over the decoding delay (in two-dimensional symbols). ADSL transmission scenario. Decoding is done in the trellis of an imaginary scrambler with 16 states. Variation of the scrambler polynomial. Dashed line: shaping gain of hypersphere.

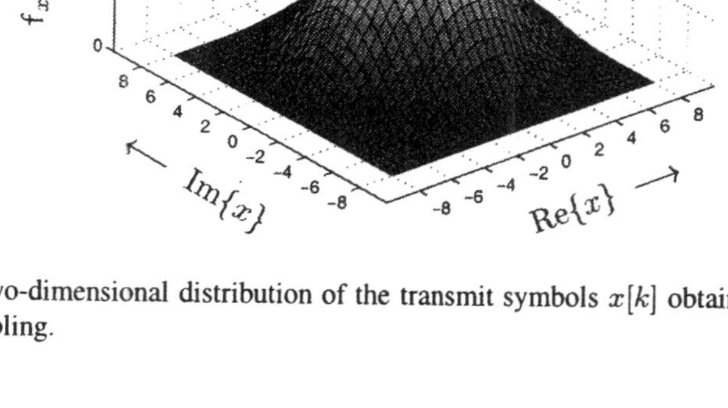

Fig. 5.21 Two-dimensional distribution of the transmit symbols $x[k]$ obtained by shaping without scrambling.

Finally, the probability density function of the channel symbols $x[k]$ obtained by shaping without scrambling is plotted in Figure 5.21. The situation is again the same as above, and a 16-state trellis is used. The decision delay is fixed at 20 symbols, which, in the present situation, leads to a shaping gain of 0.77 dB. As expected, the distribution of the channel symbols $x[k]$ looks quite Gaussian.

Example 5.6: *Comparison of Shaping and Precoding*

This example aims to compare Tomlinson-Harashima precoding and shaping without scrambling. For clarity, we restrict the discussion to one-dimensional signaling and consider the simplified SDSL up-stream example of Section 3.4. Since 3 information bits are transmitted per symbol, an 8-ary ASK signal set is used.

Figure 5.22 shows a signal segment of length 4000 symbols over time. In the first row, the 8-ary data symbols $a[k]$, assuming equiprobably the values $\pm 1, \pm 3, \pm 5$, and ± 7, are displayed. Given this signal, Tomlinson-Harashima precoding generates channel symbols $x[k]$, which lie in the interval $[-8,\ 8)$; cf. second row in Figure 5.22. The dashed lines mark the boundaries of the support range for $x[k]$. As known from theory (cf. Section 3.2.2), the channel symbols are uniformly distributed over $[-8,\ 8)$. This behavior is very visible when looking at the signal over time. Additionally, numerical estimations of the pdf, which more clearly show the amplitude distribution, are shown on the right-hand side, next to the respective signals. The middle row gives the corresponding effective data sequence $\langle v[k] \rangle$. The symbols are only odd integers, and reducing them modulo 16 into the interval $[-8,\ 8)$ again results in the data symbols $a[k]$. Due to the channel filter, the effective data sequence is low-pass, which can be inferred from the figure. Please note the different scaling of the y-axis.

The signals obtained when applying shaping without scrambling are shown in row 4 ($x[k]$) and row 5 ($v[k]$) of Figure 5.22. The support region for channel symbols $x[k]$ is now extended to the interval $[-16,\ 16)$. A close look at $\langle x[k] \rangle$ gives the impression that symbols with smaller amplitude occur more often than symbols with larger amplitude. Every once in a while, symbols whose amplitude exceeds the limit imposed in Tomlinson-Harashima precoding (dashed lines) are visible. A numerical simulation of the pdf shows that here, as expected, an almost Gaussian distribution is present. The effective data symbols $v[k]$, shown on the bottom of Figure 5.22, are again given as a periodic extension of the data symbols $a[k]$, i.e., $v[k]$ is congruent modulo 16 to $a[k]$. The statistical properties of this sequence do not differ significantly from those when using Tomlinson-Harashima precoding.

For further illustration, Figure 5.23 shows a zoom into the sequences. Now, only 100 consecutive symbols are shown.

At the very beginning, both precoding schemes result in the same symbols. But then, shaping without scrambling (+) sometimes produces symbols $x[k]$ with amplitude larger than 8, in order to be able to choose symbols with small amplitude later on, and hence minimize average energy in the long run. Tomlinson-Harashima precoding ($\circ$) strictly insists on channel symbols from the region $[-8,\ 8)$.

In both schemes, the effective data symbols are congruent modulo 16 to the data symbols $a[k]$. Tomlinson-Harashima precoding and shaping without scrambling only differ in the choice of the actual representative from the set modulo congruent signal levels. The effective data symbols produced by these algorithms only differ in integer multiples of 16.

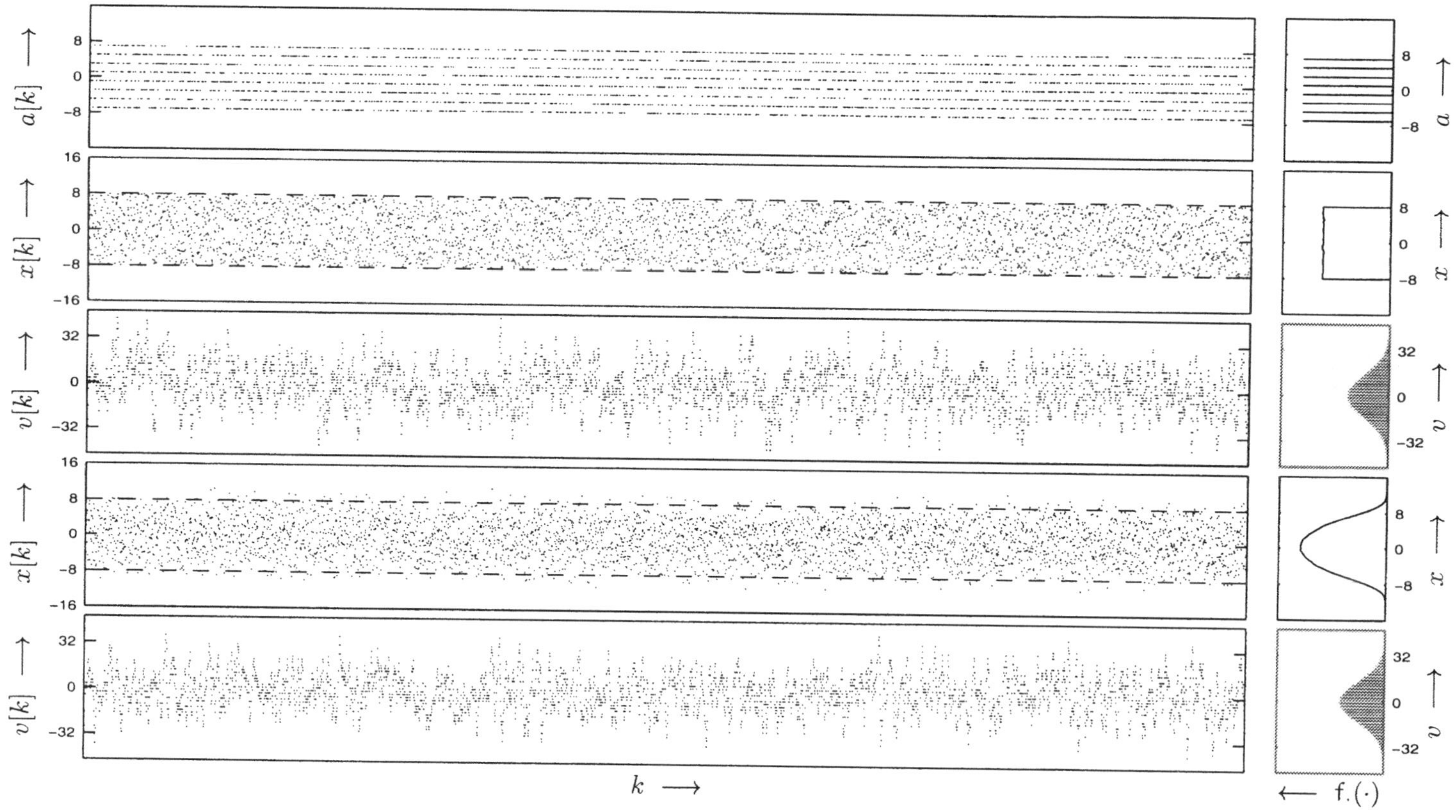

Fig. 5.22 PAM data symbols $a[k]$, channel symbols $x[k]$, and effective data symbols $v[k]$ over time for Tomlinson-Harashima precoding and shaping without scrambling, respectively.

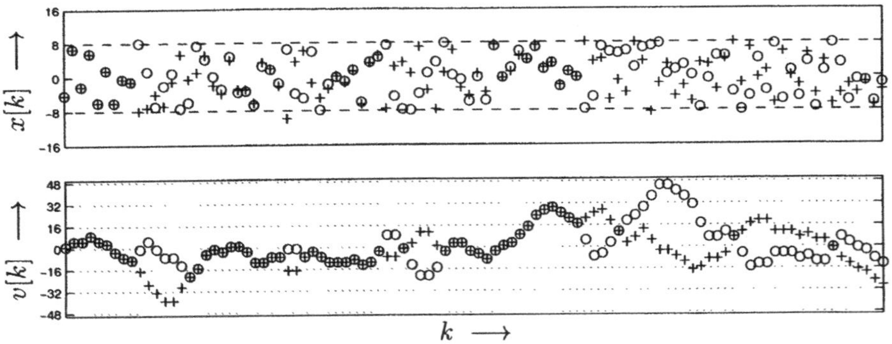

Fig. 5.23 Channel symbols $x[k]$ (top), and effective data symbols $v[k]$ (bottom) over time. o: Tomlinson-Harashima precoding; $+$: shaping without scrambling.

Example 5.7: *Numerical Simulations of Precoding/Shaping Schemes*

We continue Example 3.6 of Chapter 3 on the numerical simulation of precoding schemes. Now, the combined precoding and shaping techniques presented above are compared. Again, the SDSL transmission scenario (baseband signaling with 3 information bits per one-dimensional symbol, cable length 3 km, self-NEXT) and the ADSL scenario (passband signaling with 5 information bits per two-dimensional symbol, cable length 5 km, white noise) are considered. In both cases, the T-spaced discrete-time end-to-end channel model $H(z)$ with monic impulse response of order $p = 10$ is calculated via the Yule-Walker equations.

As in Chapter 3, the results are displayed over the *transmit energy* per information bits E_b at the output of the precoder, divided by the normalized (one-sided) noise power spectral density N_0'. Due to precoding, in effect a unit-gain discrete-time AWGN channel with noise variance $\sigma_n^2 = \frac{N_0'}{T}$ is present.

In Figure 5.24 the symbol error rates (SER) over the signal-to-noise ratio E_b/N_0' in dB for transmission employing shaping without scrambling and trellis precoding, respectively, are displayed. In both precoding/shaping schemes, the shaping decoder operates on a 16-state trellis, hence the complexity is the same. The decoding delays are adjusted such that the same shaping gain is achieved. For reference, the simulation results for Tomlinson-Harashima precoding are given, as well. All transmission schemes do not employ channel coding.

The figure at the top shows the results for the SDSL scenario (baseband transmission). Tomlinson-Harashima precoding and shaping without scrambling use an 8-ary one-dimensional PAM signal constellation, and the channel symbols $x[k]$ are restricted to the intervals $[-8,\ 8)$ and $[-16,\ 16)$, respectively. Trellis precoding is based on a 16-ary constellation, and restricts the channel symbols to the interval $[-16,\ 16)$, too. The decoding delays are 14 symbols in shaping without scrambling and 8 symbols in trellis precoding, which gives the same shaping gain of 0.56 dB.

The error rates of the ADSL scheme are sketched on the bottom of the figure. Here, Tomlinson-Harashima precoding and shaping without scrambling use a rotated square constellation with 32 signal points (cf. Figure 3.10). Due to constellation expansion in shaping,

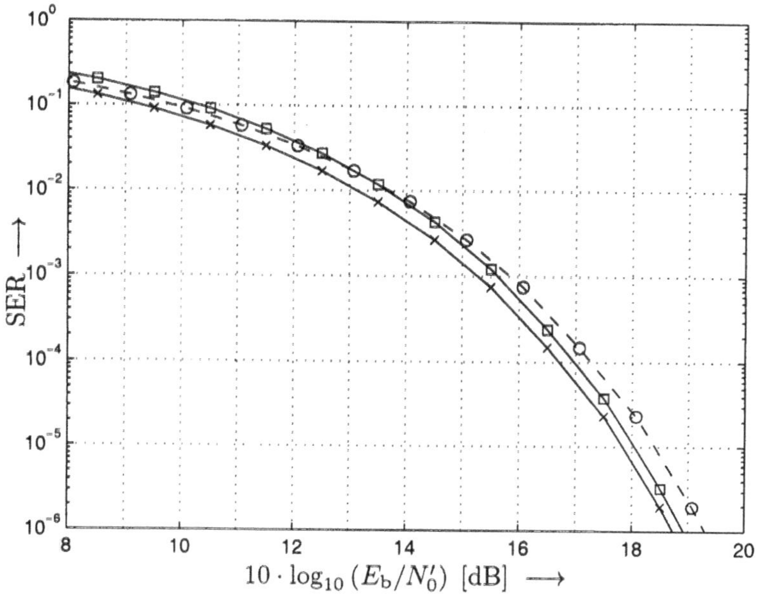

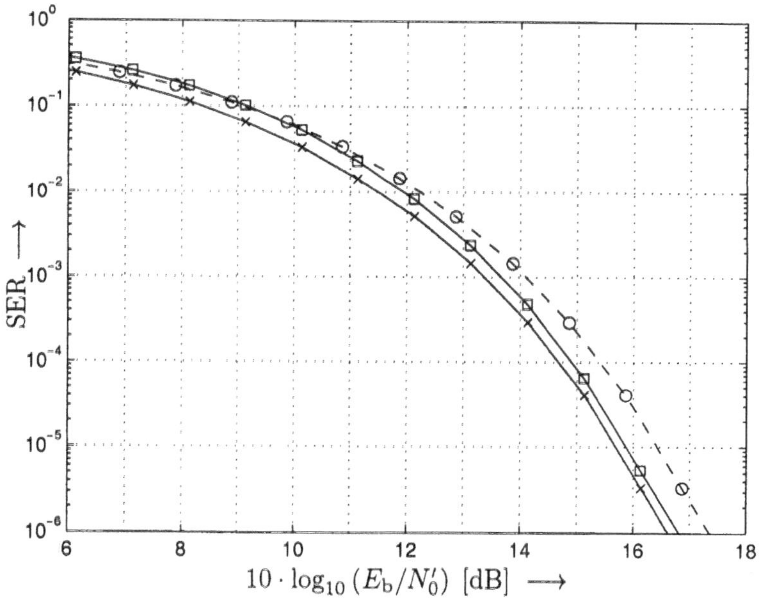

Fig. 5.24 Symbol error rates versus the signal-to-noise ratio. ×: shaping without scrambling; □: trellis precoding; ○: Tomlinson-Harashima precoding. Top: SDSL (baseband) scenario, bottom: ADSL (passband) scenario.

shaping without scrambling and trellis precoding produce channel symbols restricted to the square region $[-8,\ 8)^2$. Conversely, Tomlinson-Harashima precoding restricts the channel symbols to a rotated square, which is circumscribed by the support regions of the shaping schemes. A shaping gain of 0.74 dB is achieved for a decoding delay of 12 symbols in shaping without scrambling and 8 symbols in trellis precoding.

The simulations show a clear superiority of shaping without scrambling over the other transmission schemes. When employing shaping without scrambling, and thus achieving shaping gain, the performance curve is obtained from that of Tomlinson-Harashima precoding by simply shifting it to the left by the shaping gain. However, in trellis precoding the effect of error multiplication in the syndrome former at the receiver side can be recognized. The same gross shaping gain, i.e., reduction in average transmit energy, is achieved, but part of the gain is lost due to the error propagation at the receiver. The whole shaping gain can be utilized as a gain in performance only for very low error rates.

To summarize, shaping without scrambling is preferable over trellis precoding since it is completely compatible with Tomlinson-Harashima precoding and, additionally, can utilize the whole shaping gain.

5.3 PRECODING AND SHAPING UNDER ADDITIONAL CONSTRAINTS

Up to now, the task of combined precoding/shaping techniques has been to generate a transmit signal which (a) results in an equalized signal at the receiver input and (b) has minimum average energy. But using signal shaping, almost any desired property may be achieved. We now discuss precoding and shaping under additional constraints, as well as for generating further signal properties. Simultaneous control of different signals, e.g., transmit and receive signals, is of special interest. In particular, we study the restriction of the dynamic range of the receive signal and peak-to-average energy ratio reduction at the transmitter.

5.3.1 Preliminaries on Receiver-Side Dynamics Restriction

Compared to linear preequalization, Tomlinson-Harashima precoding employs an expanded signal set, where all points congruent modulo some precoding lattice Λ_{p} represent the same information, cf. Section 3.2. These effective data symbols $v[k]$, which are present at the input of the decision device at the receiver, exhibit a wide dynamic range. Moreover, the maximum amplitude of $v[k]$ is proportional to the absolute sum $\sum_{k=0}^{p} |h[k]|$ over the coefficients $h[k]$ of the T-spaced discrete-time end-to-end channel model, see (3.2.5). Hence, the stronger the intersymbol interference, the larger the dynamic range at the receiver. As a consequence, receiver implementation is complicated since the sensitivity to residual equalization errors and jitter of the symbol timing is increased.

Second, in some situations it is desirable to blindly adjust an adaptive filter at the receiver to compensate for the mismatch of precoder and actual overall channel characteristics. In [FGH95] we have shown that, over a remarkably wide range, adjusting the precoder to a compromise application and to compensate for the mismatch by linear equalization causes almost no loss in performance. Here, the necessity of a backward channel is avoided and even point-to-multipoint transmission is possible when using a fixed compromise precoder.

Unfortunately, the effective data symbols $v[k]$ are almost discrete Gaussian distributed; cf. Figure 5.22. This inhibits the use of low-complex blind adaptation algorithms since, in principle, T-spaced blind equalization based on second-order statistics is impossible if the signal to be recovered has a Gaussian distribution [BGR80]. Furthermore, the application of a decision-directed least-mean-square algorithm [Pro01, Hay96] is not satisfying in Tomlinson-Harashima precoding schemes, as the region of convergence is much too small for operation; see [FGH95, Ger98].

In this section we show how the dynamic range of the effective data symbols can be reduced. In particular, we propose two new precoding procedures which we call *dynamics limited precoding* and *dynamics shaping*. Both schemes are straightforward extensions of Tomlinson-Harashima precoding and shaping without scrambling, respectively. Using these techniques, the requirements for receiver implementation can be lowered. Additionally, signals suited for blind adaptive equalization, e.g., using the Sato algorithm [BGR80, Sat75] or its modified version [Ben84], are generated.

5.3.2 Dynamics Limited Precoding

In [FGH95], we have introduced a straightforward extension of Tomlinson-Harashima precoding, called *dynamics limited precoding (DLP)*. As in Tomlinson-Harashima precoding, the initial signal set $\mathcal{A}$ is extended periodically and all symbols modulo-congruent with respect to the precoding lattice $\mathbf{\Lambda}_\mathrm{p}$ represent the same data. This sequence of effective data symbols is then filtered with the inverse of the channel filter $H(z)$. But here, an additional constraint is imposed on the support region for the effective data symbols $v[k]$. Only symbols falling in a predetermined region $\mathcal{R}_\mathcal{V}$ are allowed. Hence, in dynamics limited precoding the expanded signal set reads

$$\mathcal{V} = (\mathcal{A} + \mathbf{\Lambda}_\mathrm{p}) \cap \mathcal{R}_\mathcal{V} \tag{5.3.1a}$$
$$= \{a + d \mid a \in \mathcal{A},\ d \in \mathbf{\Lambda}_\mathrm{p}\} \cap \mathcal{R}_\mathcal{V}\,. \tag{5.3.1b}$$

If no restriction is imposed on $\mathcal{R}_\mathcal{V}$, i.e., $\mathcal{R}_\mathcal{V} = \mathbb{R}^D$ in D-dimensional signaling, the usual Tomlinson-Harashima precoding is present. Conversely, for $\mathcal{R}_\mathcal{V} = \mathcal{R}_\mathrm{V}(\mathbf{\Lambda}_\mathrm{p})$, the Voronoi region of the precoding lattice, the signal sets $\mathcal{A}$ and $\mathcal{V}$ are identical, and linear preequalization results. Hence, dynamics limited precoding offers a trade-off between these two equalization strategies.

As in Tomlinson-Harashima precoding, the actual representative—or equivalently the precoding symbols $d[k]$—is chosen symbol-by-symbol from the set $\mathcal{V}$. It is obvious that the symbol $v[k]$ should be selected, which, after preequalization, results in the channel symbol $x[k]$ with least amplitude. Ties may be resolved arbitrarily. Since the support of the effective data symbols is restricted, in general, it can no longer be guaranteed that the corresponding transmit symbols exclusively fall into the Voronoi region $\mathcal{R}_\mathrm{V}(\mathbf{\Lambda}_\mathrm{p})$ of the precoding lattice.

Finally, at the receiver side, a modulo-$\mathbf{\Lambda}_\mathrm{p}$ operation eliminates the influence of the precoding symbol $d[k]$ and recovers the data symbol $a[k]$. The receiver of Tomlinson-Harashima precoding schemes can thus be used without any modification in dynamics limited precoding.

We now study dynamics limited precoding for one-dimensional M-ary ASK constellations $\mathcal{A} = \{\pm1, \pm3, \ldots, \pm(M-1)\}$ more closely. Restricting the maximum amplitude of the effective data symbols $v[k]$ to V_max, i.e., $|v[k]| \le V_\mathrm{max}$, the expanded signal set reads

$$\mathcal{V} = (\mathcal{A} + 2M\mathbb{Z}) \cap [-V_\mathrm{max},\ V_\mathrm{max}]\ . \tag{5.3.2}$$

Here, $\mathbf{\Lambda}_\mathrm{p} = 2M\mathbb{Z}$ and $\mathcal{R}_\mathcal{V} = [-V_\mathrm{max},\ V_\mathrm{max}]$.

As in Tomlinson-Harashima precoding (cf. Figure 3.4), the selection of the actual representative $v[k]$ can be done implicitly using a nonlinear device. Figure 5.25 illustrates the dynamics limited precoder employing a variable, memoryless nonlinearity. This function $f_M(q, a)$ depends on the number M of signal points and on the current data symbol $a[k]$, hence, in general, there are up to M different functions. Their mathematical definition reads

$$f_M(q, a) \triangleq q + 2M \cdot d\,, \quad \text{with} \quad d = \operatorname*{argmin}_{\delta \in \mathbb{Z},\ |a + 2M\cdot\delta| \le V_\mathrm{max}} |q + 2M \cdot \delta|\ . \tag{5.3.3}$$

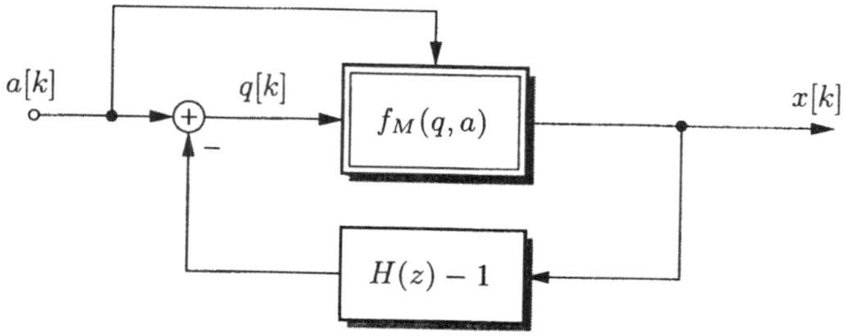

Fig. 5.25 Dynamics limited precoding for one-dimensional signaling using a variable nonlinear device.

If the restriction on the dynamic range is dropped ($V_{\max} \to \infty$), the nonlinear function becomes independent of a and reads $f_M(q) = q + 2M \cdot \operatorname{argmin}_{d \in \mathbf{Z}} |q + 2M \cdot d|$, which is simply a modulo reduction of q into the interval $[-M,\ M)$. Once more, Tomlinson-Harashima precoding is obtained.

Example 5.8: *Nonlinear Functions in Dynamics Limited Precoding*

This example gives the memoryless nonlinear functions used in dynamics limited precoding. For clarity, a quaternary signal set $\mathcal{A} = \{-3,\ -1,\ 1,\ 3\}$ is assumed. The dynamic range of the expanded signal set is limited to $V_{\max} = 10$, i.e., the set reads $\mathcal{V} = \{\pm 1, \pm 3, \pm 5, \pm 7, \pm 9\}$. Points spaced by $2M = 8$ are congruent. Figure 5.26 plots the four different functions $f_4(q, a)$.

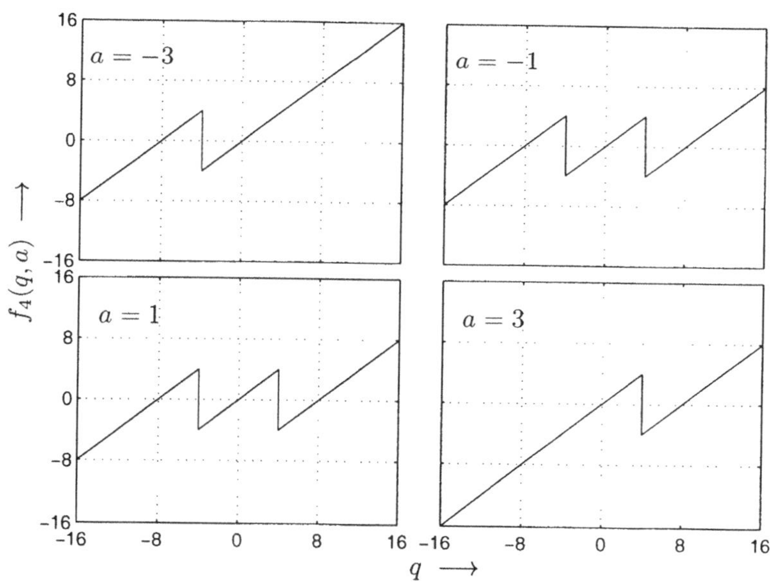

Fig. 5.26 Nonlinear functions $f_4(q, a)$ according to (5.3.3). $M = 4$, $V_{\max} = 10$.

Since the initial signal point -3 is congruent to the point 5, but all other points spaced by 8 exceed the dynamic limitation, the respective nonlinear function $f_4(q,a)$ has sawtooth characteristics with two branches, corresponding to $d=0$ and $d=1$. By changing the signs, the same is true for signal point $+3$.

In contrast, the points -1 and $+1$ may be represented by the points $-9,-1,7$ and $-7,1,9$, respectively, from the expanded signal set. In turn, the respective nonlinear function $f_4(q,a)$ has three branches ($d=0$, $d=1$, and $d=-1$).

In general, a peak limitation for the signal set $\mathcal{V}$ of modulo $2M$ congruent symbols immediately leads to nonlinear functions with sawtooth characteristics in a certain range and "linear branches" for large input values.

If the size of the signal set is at least doubled, i.e., if $V_{\max} \geq 2M$ holds for the dynamics restriction, dynamics limited precoding is guaranteed to produce a stable output, even if the T-spaced discrete-time end-to-end transfer function $H(z)$ has discrete spectral nulls.[2] Here, all sequences $\langle a[k]\rangle$ with spectral components, which would be boosted infinitely by the spectral poles of $1/H(z)$, can be converted into alternative sequences $\langle v[k]\rangle$ with symbols drawn from the expanded signal set $\mathcal{V}$. The power spectral density $\Phi_{vv}(\mathrm{e}^{\mathrm{j}2\pi fT})$ of the effective data sequence $\langle v[k]\rangle$ will have spectral zeros at the same frequency points that $H(\mathrm{e}^{\mathrm{j}2\pi fT})$ has. In turn, the power spectral density $\Phi_{xx}(\mathrm{e}^{\mathrm{j}2\pi fT}) = \Phi_{vv}(\mathrm{e}^{\mathrm{j}2\pi fT})/|H(\mathrm{e}^{\mathrm{j}2\pi fT})|^2$ of the channel symbols $x[k]$ will remain finite for all f.

For example, consider a channel $H(z)$ which blocks DC, i.e., $H(\mathrm{e}^{\mathrm{j}2\pi 0T}) = 0$, and quaternary ($M=4$) signaling. Filtering the DC sequence

$$\langle a[k]\rangle = \langle \ldots, 3, 3, 3, 3, 3, 3, 3, 3, \ldots \rangle \tag{5.3.4a}$$

by $1/H(z)$ leads to an unstable output. When doubling the size of the constellation, the signal point "3" is congruent to the point "-5" and $\langle a[k]\rangle$ can be converted into the DC-free sequence

$$\langle v[k]\rangle = \langle \ldots, 3, 3, -5, 3, -5, 3, -5, 3, \ldots \rangle\,, \tag{5.3.4b}$$

which leads to a bounded output of the preequalization filter.

As already mentioned above, dynamics limited precoding offers a variation between linear preequalization ($V_{\max} = M$) and Tomlinson-Harashima precoding, when choosing $V_{\max} \geq 2 \cdot \mathrm{INT}\left[\left(M \cdot \sum_{k=0}^{p} |h[k]| + 1\right)/2\right] - 1$, cf. (3.2.5). The advantage of dynamics limited precoding compared to Tomlinson-Harashima precoding is a well-prescribed dynamic range of the signal at the threshold decision. But the price to be paid is that the transmit symbols $x[k]$ no longer exclusively fall into the interval $[-M, +M)$. Hence, the dynamic range of $x[k]$ is (slightly) increased in order to limit the dynamic range of $v[k]$. By choosing $V_{\max}$, a trade-off between

[2]For the special, but common, case of a zero at DC, $2M-1$ points, together with a shift of the elements of $\mathcal{V}$ from odd to even integers, are sufficient, because then one symbol is represented by zero.

the dynamic range of these two signals is possible. The extreme cases of dynamics limited precoding offer a minimum dynamic range of $x[k]$ at the expense of a maximum dynamic range of $v[k]$ (Tomlinson-Harashima precoding) and a minimum dynamic range of $v[k]$ at the expense of a maximum dynamic range of $x[k]$ (linear preequalization), respectively.

Moreover, average transmit power of dynamics limited precoding will also vary between these two extreme cases. The minimum is achieved for Tomlinson-Harashima precoding, whereas the average transmit power of linear preequalization is larger by a factor equal to the asymptotic prediction gain; cf. Section 3.1. As long as the dynamics limitation is not too extreme, only a slight increase in transmit power will occur. In the following, we combine dynamics limitation with signal shaping and thus mitigate or even overcompensate this effect.

Example 5.9: *Performance of Dynamics Limited Precoding*

This example first shows the sequence of channel symbols $x[k]$ and that of effective data symbols $v[k]$ over the discrete-time index when using dynamics limited precoding. For clarity of presentation, one-dimensional signaling using an ($M = 8$)-ary ASK signal set is assumed, and the simplified SDSL up-stream example of Section 3.4 is adopted. The discrete-time channel model $H(z)$ is of order $p = 10$.

The upper two rows are valid for $V_{\max} \to \infty$, i.e., no dynamics limitation is active, and Tomlinson-Harashima precoding is present. As already seen in Example 5.6, the channel symbols $x[k]$ are uniformly distributed over the interval $[-8,\ 8)$. The corresponding effective data symbols are almost discrete Gaussian distributed.

Next, dynamics limited precoding with $V_{\max} = 2M = 16$ is assessed (the two rows in the middle). Here, the number of signal points in the effective data sequence is doubled compared to the set of data symbols $a[k]$. It is clearly visible that the effective data symbols $v[k]$ are limited to 16 discrete values (the odd integers from -15 to 15). Due to this limitation, the channel symbols $x[k]$ no longer exclusively lie in the interval $[-8,\ 8)$; sometimes points with larger amplitudes appear.

Finally, the last two rows correspond to a limitation of the dynamic range to $V_{\max} = M = 8$—no expansion of the signal set is allowed. The effective data symbols are equal to the data symbols and linear preequalization results. The eight possible data symbols are uniformly distributed but the channel symbols are almost Gaussian. Since the present channel has no spectral nulls, stable operation is guaranteed.

Comparing the pdfs for Tomlinson-Harashima precoding and linear preequalization, a certain duality can be observed. The distributions for dynamics limited precoding lie between these two extreme cases.

The effect of limiting the dynamic range of the effective data symbols on the pdf of the channel symbols and its power spectral density is shown in Figure 5.28. The left column displays the pdf of the channel symbols $x[k]$ and the column in the middle gives the respective power spectral density. Finally, the right column plots the probabilities of the effective data symbols $v[k]$. The limitation of the dynamic range varies from $V_{\max} = \infty$ (no restriction, Tomlinson-Harashima precoding), $V_{\max} = 32, 24, 16, 12$, down to $V_{\max} = 8$ (linear preequalization). For reference, the theoretical power spectral densities for Tomlinson-Harashima precoding (constant PSD with $\Phi_{xx}(e^{j2\pi fT}) = \sigma_x^2 = 64/3 = 21.33$) and linear preequalization (high-pass spectrum with $\Phi_{xx}(e^{j2\pi fT}) = \sigma_a^2 / \left|H(e^{j2\pi fT})\right|^2$, with $\sigma_a^2 = 21$) are shown (dotted lines).

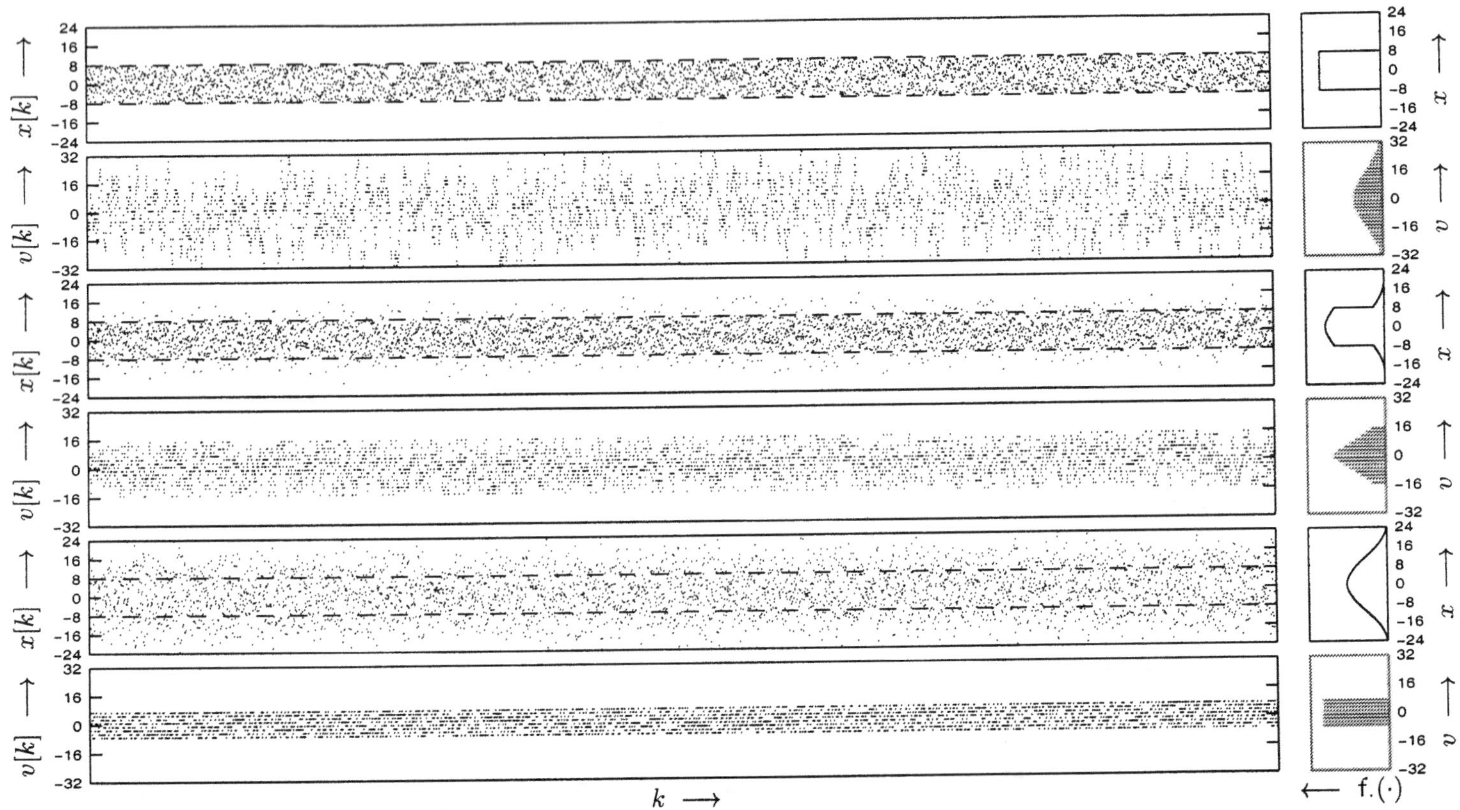

Fig. 5.27 Channel symbols $x[k]$ and effective data symbols $v[k]$ for dynamics limited precoding over time. Dynamics restriction: $V_{\max} \rightarrow \infty$ (Tomlinson-Harashima precoding), 16, and 8 (linear preequalization).

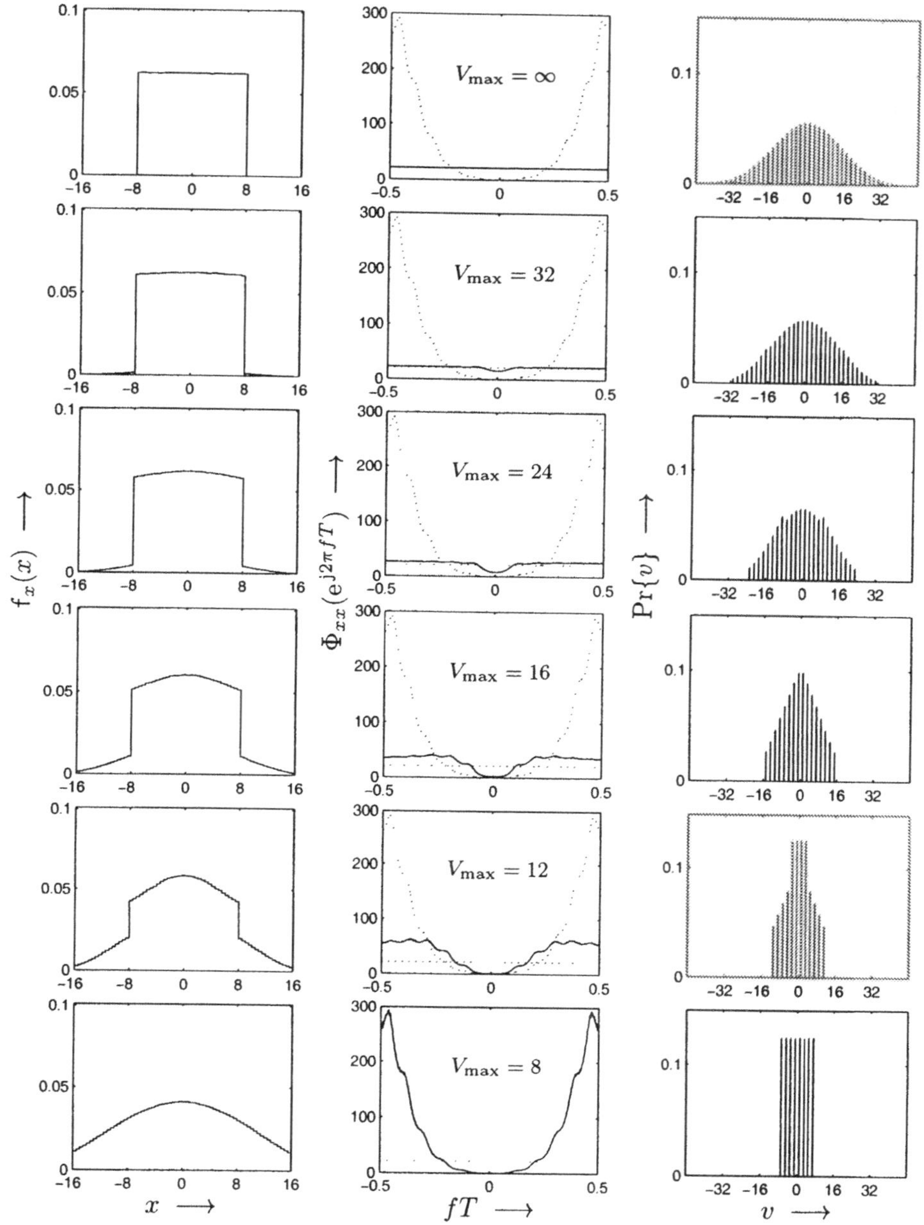

Fig. 5.28 Probability density functions (left) and estimated power spectral density (middle) of the channel symbols $x[k]$, and probability density functions of the effective data symbols $v[k]$ (right). Top to bottom: variation of the dynamics restriction V_{max}. Dotted lines: theoretical PSD for Tomlinson-Harashima precoding and linear preequalization.

The above described phenomena are clearly visible. For $V_{\max} = 32$ almost no difference occurs compared to Tomlinson-Harashima precoding. Note that the maximum magnitude of the effective data symbols when using Tomlinson-Harashima precoding here turns out to be 71 (cf. (3.2.5)). When restricting the dynamic range of $v[k]$ more and more, the pdf of $x[k]$ is broadened. At the same time, the PSD changes from a constant one to a high-pass spectrum. For linear preequalization, a Gaussian distributed channel input results and, as expected, the PSD is proportional to the inverse of the squared magnitude of the channel filter spectrum.

Finally, the trade-off between dynamic range $V_{\max}$ of the effective data symbols and transmit power is plotted in Figure 5.29. Here, the transmit power is given relative to that of

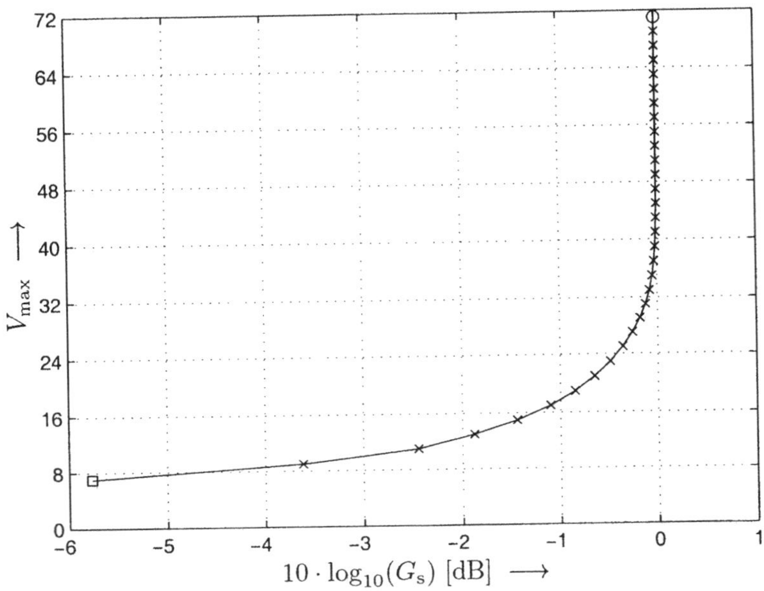

Fig. 5.29 Trade-off between dynamic range $V_{\max}$ of the effective data symbols and shaping gain (in dB) in dynamics limited precoding. ($M = 8$)-ary signaling. o: Tomlinson-Harashima precoding; □: linear preequalization.

Tomlinson-Harashima precoding, i.e., as shaping gain, which in the present case is always a loss.

Without any restrictions, starting from 8-ary data symbols $a[k]$, the effective data symbols $v[k]$ assume values up to ± 71 in Tomlinson-Harashima precoding (marked with o). Without any noticeable loss, the dynamic range can be lowered significantly. The loss becomes sizable only if the dynamics restriction is lower than about $3M = 24$. For $V_{\max} = 8$ no signal expansion is allowed ($\mathcal{A} = \mathcal{V} = \{\pm 1, \pm 3, \pm 5, \pm 7\}$) and linear preequalization results (marked with □). The loss of almost 6 dB is exactly the prediction gain G_p of noise prediction/DFE, or equivalently, precoding, over linear equalization, cf. Figure 2.18 for $p = 10$. This example demonstrates the potential of restricting the dynamic range in precoding.

Finally a remark concerning flexible precoding. In principle, a reduction of the dynamic range of the received signal would be possible, as well. Similar to the proceeding above, the quantization in the feedback part of the flexible precoder has to be suitably adapted. Again, due to the restriction of the dynamic range of the decoder input signal, average transmit power will be increased. A compensation of this effect is only possible by combining signal shaping and precoding. But the separation of the these two operations is exactly the basic aim of flexible precoding. Moreover, when restricting the dynamic range, the additive dither sequence $m[k]$ no longer lies exclusively in the Voronoi region of the underlying signal lattice. Hence, if power shaping is performed prior to precoding, there would no longer be a guarantee that a shaping gain is achieved or that the characteristics of the input distribution are preserved. Additionally, the inverse precoder cannot be implemented by a simple slicer (cf. Figure 3.22), but the symbols $m[k]$ have to be explicitly regenerated and subtracted from the estimate of $x[k]$.

5.3.3 Dynamics Shaping

In the last section we saw that reducing the dynamic range of the effective data symbols results in an increased average and peak transmit energy. On the other hand, by choosing the effective data symbols in the long run, a reduction in average energy is possible, and shaping gain can be achieved. Hence, an obvious approach is to combine the principles of *shaping without scrambling* and *dynamics limited precoding*. The dynamic range of the effective data symbols should be strictly bounded, but the loss in transmit energy should be as small as possible. We denote the resulting scheme, first presented in [FH94, FGH95], as *dynamics shaping*.

It is straightforward to incorporate a limitation of the dynamic range into shaping without scrambling. The operation is as described in Section 5.2 and depicted in Figure 5.15, only the modulo-$\mathbf{\Lambda}_s$ operation has to be replaced by appropriate nonlinear, memoryless functions. Given the data symbols $a[k]$, the shaping decoder determines a binary sequence $\langle b[k] \rangle$, which for each time index selects one of two subsets of the precoding lattice $\mathbf{\Lambda}_p$. That precoding symbol $d[k]$ is implicitly chosen from the current subset, which after preequalization, gives the channel symbol $x[k]$ with the smallest magnitude. This selection is done symbol-by-symbol and under the additional constraint that the amplitude of the effective data symbol $v[k] = a[k]+d[k]$ does not exceed a given bound.

Again, we study dynamics shaping more closely for one-dimensional M-ary signaling using $\mathcal{A} = \{\pm 1, \pm 3, \ldots, \pm(M-1)\}$. The dynamic range of the effective data symbols should be limited to the interval $[-V_{\max}, V_{\max}]$ ($|v[k]| \le V_{\max}$), hence the expanded signal set of effective data symbols reads

$$\mathcal{V} = (\mathcal{A} + 2M\mathbb{Z}) \cap [-V_{\max}, V_{\max}] \; . \tag{5.3.5}$$

For the operation of shaping, this set of effective data symbols is divided into two disjoint subsets $\mathcal{V}_0$ and $\mathcal{V}_1$, i.e., $\mathcal{V}_0 \cup \mathcal{V}_1 = \mathcal{V}$ and $\mathcal{V}_0 \cap \mathcal{V}_1 = \{\}$, namely

$$\mathcal{V}_0 = (\mathcal{A} + 4M\mathbb{Z}) \cap [-V_{\max}, V_{\max}] \tag{5.3.6a}$$

$$\mathcal{V}_1 = (\mathcal{A} + 4M\mathbb{Z} + 2M) \cap [-V_{\max}, V_{\max}] \ . \tag{5.3.6b}$$

Equivalently, the precoding symbols $d[k]$ are either taken from the set $4M\mathbb{Z}$ or from $4M\mathbb{Z} + 2M$. The union of both sets gives the precoding lattice $\mathbf{\Lambda}_\mathrm{p} = 2M\mathbb{Z}$.

Figure 5.30 shows the block diagram of dynamics shaping. The shaping algorithm

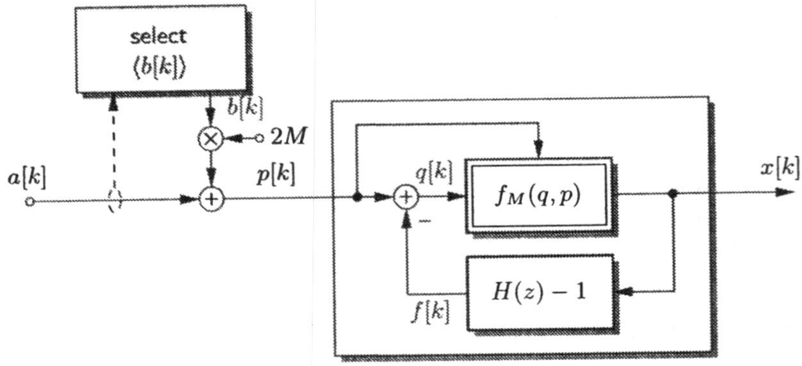

Fig. 5.30 Block diagram of dynamics shaping for one-dimensional signaling using a variable nonlinear device. Decoding delay not shown.

determines a binary sequence $\langle b[k]\rangle$, which, in each time step, selects one of the two subsets $\mathcal{V}_{b[k]}$. This addressing is done by either leaving the data symbols $a[k]$ unchanged or shifting them by $2M$. The subsequent precoder part is identical to dynamics limited precoding. From the current set $\mathcal{V}_{b[k]}$, that symbol $v[k]$ is selected which minimizes the magnitude of the corresponding channel symbol $x[k]$. This selection is implicitly done by a variable nonlinear device $f_M(q,p)$. Since $p[k]$ may assume $2M$ discrete values, in dynamics shaping there are $2M$ possible nonlinear functions $f_M(q,p)$, which are defined as

$$f_M(q,p) \triangleq q + 4M \cdot d \ , \quad \text{with} \quad d = \underset{\delta\in\mathbb{Z},\ |p+4M\cdot\delta|\le V_{\max}}{\operatorname{argmin}} |q + 4M\cdot\delta| \ . \tag{5.3.7}$$

Finally, in order to reduce the average energy of the channel symbols $x[k]$, the shaping decoder employs the branch metric $\lambda[k] = |x[k]|^2$. In detail, given some state S of the decoder with its associated hypothetical past transmit symbols $x^{(S)}[k-\kappa]$, $\kappa = 1, 2, \ldots, p$, and the binary symbols $b[k]$ determining the branch, the metrics reads (cf. Figure 5.30)

$$\begin{aligned}\lambda^{(S,b)}[k] &= |x^{(S,b)}[k]|^2 \\ &= \left| f_M\left(a[k] - 2Mb[k] - \sum\nolimits_{\kappa=1}^{p} h[\kappa]x^{(S)}[k-\kappa], a[k] - 2Mb[k]\right)\right|^2 .\end{aligned} \tag{5.3.8}$$

Compared to the metric (5.2.3) in shaping without scrambling, only the modulo-($\mathbf{\Lambda}_\mathrm{s} = 4M$) operation is replaced by the function $f_M(q,p)$. These functions have a modulo behavior only over a certain range and linear course for large input amplitudes, cf. Example 5.8.

Notice, if $V_{\max} < 2M - 1$, i.e., the size of the expanded signal set $\mathcal{V}$ is less than twice the size of the initial set $\mathcal{A}$, some symbols $a[k]$ have only one representative $v[k]$ in the subset $\mathcal{V}_0$. Here, no valid precoding symbol $d[k]$ can be found for $b[k] = 1$ and the corresponding branch in the shaping trellis has to be blocked. This can be handled, e.g., by setting $f_M(q,p)$ and hence $x[k]$ to an infinite value. Moreover, in order to guarantee the desired restriction of the dynamic range, it is essential that decisions made by the shaping decoder correspond to a contiguous path through the shaping trellis.

In contrast to all previous shaping techniques, dynamics shaping works with regard to *two* signals: the average energy of the *transmit* symbols $x[k]$ should be minimized under the restriction of a limited amplitude of the (noiseless) *receive* symbols $v[k]$. Thus, a mixed criterion on the l_2/l_∞ norm [BS98] of channel *in- and output* signals is applied. But in general, using suitable branch metrics, any desired property of these signals may be controlled by shaping.

Example 5.10: *Performance of Dynamics Shaping*

This example continues Examples 5.6 and 5.9. Figure 5.31 displays the sequences of channel symbols $x[k]$ and of effective data symbols $v[k]$ when using Tomlinson-Harashima precoding, dynamics limited precoding, and dynamics shaping, respectively, over the discrete-time index. Once more, one-dimensional signaling using an $(M = 8)$-ary ASK signal set is assumed, and the simplified SDSL up-stream example (cf. Section 3.4) is considered. The discrete-time channel model $H(z)$ is of order $p = 10$.

The upper two rows are valid for Tomlinson-Harashima precoding. As already noted several times, the channel symbols $x[k]$ are uniformly distributed over the interval $[-8,\ 8)$, and the corresponding effective data symbols $v[k]$ are almost discrete Gaussian.

The two rows in the middle show the signals for dynamics limited precoding with a dynamics restriction to $V_{\max} = 2M = 16$. The effective data symbols $v[k]$ are limited to 16 discrete values, but the channel symbols $x[k]$ no longer exclusively lie in the interval $[-8,\ 8)$. The average and peak energy of the channel symbols are increased compared to Tomlinson-Harashima precoding.

Finally, the last two rows correspond to dynamics shaping. The dynamic range is limited to $V_{\max} = 2M = 16$, too, and the shaping algorithm works on an imaginary scrambler with 16-state trellis. The decoding delay is adjusted to 32 symbols. Again, the effective data symbols $v[k]$ are limited to 16 discrete values. But due to the shaping algorithm, the channel symbols tend to be Gaussian distributed. Here, the average energy is reduced compared to dynamics limited precoding. Moreover, since peaks in the signal $x[k]$ are avoided due to the long-term selection of the effective data symbols, peak energy is reduced, too.

Compared to Tomlinson-Harashima precoding, which is the baseline for performance evaluation, in dynamics limited precoding, the average transmit energy is increased by 1.44 dB (a shaping loss occurs). Conversely, for reducing the dynamics of the effective data symbols from 71 down to 16, average energy in dynamics shaping is only increased by 0.44 dB. Hence, here signal shaping provides a gain of about 1 dB over a symbol-by-symbol selection.

As in Figure 5.28 for dynamics limited precoding, Figure 5.32 shows the effect of dynamics shaping on the pdf of the channel symbols and its power spectral density. In each case a 16-state trellis is used for signal shaping, and the decoding delay is fixed to 32 symbols.

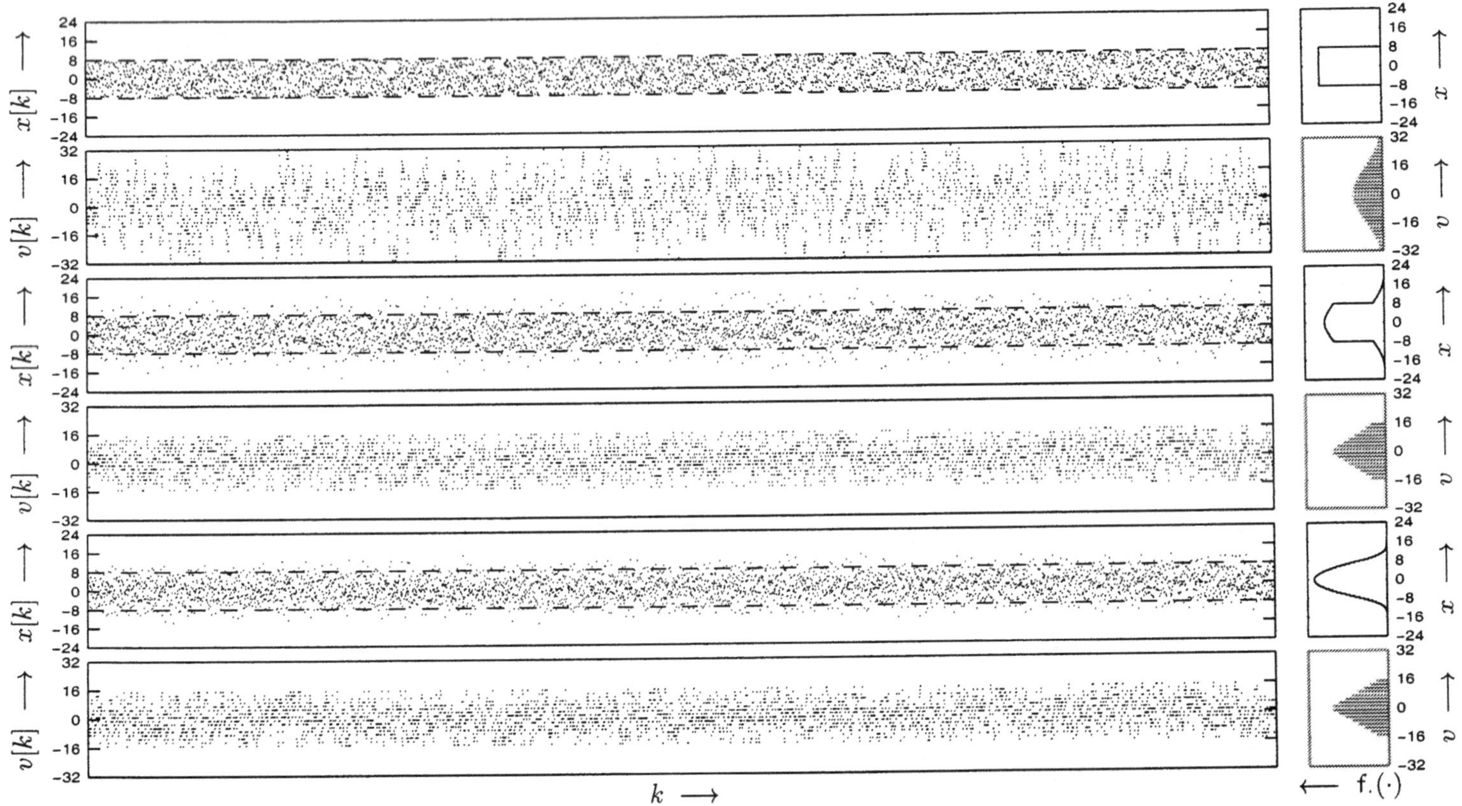

Fig. 5.31 Channel symbols $x[k]$ and effective data symbols $v[k]$ over time. Top: Tomlinson-Harashima precoding; middle: dynamics limited precoding; bottom: dynamics shaping. Dynamics restriction: $V_{\max} = 16$.

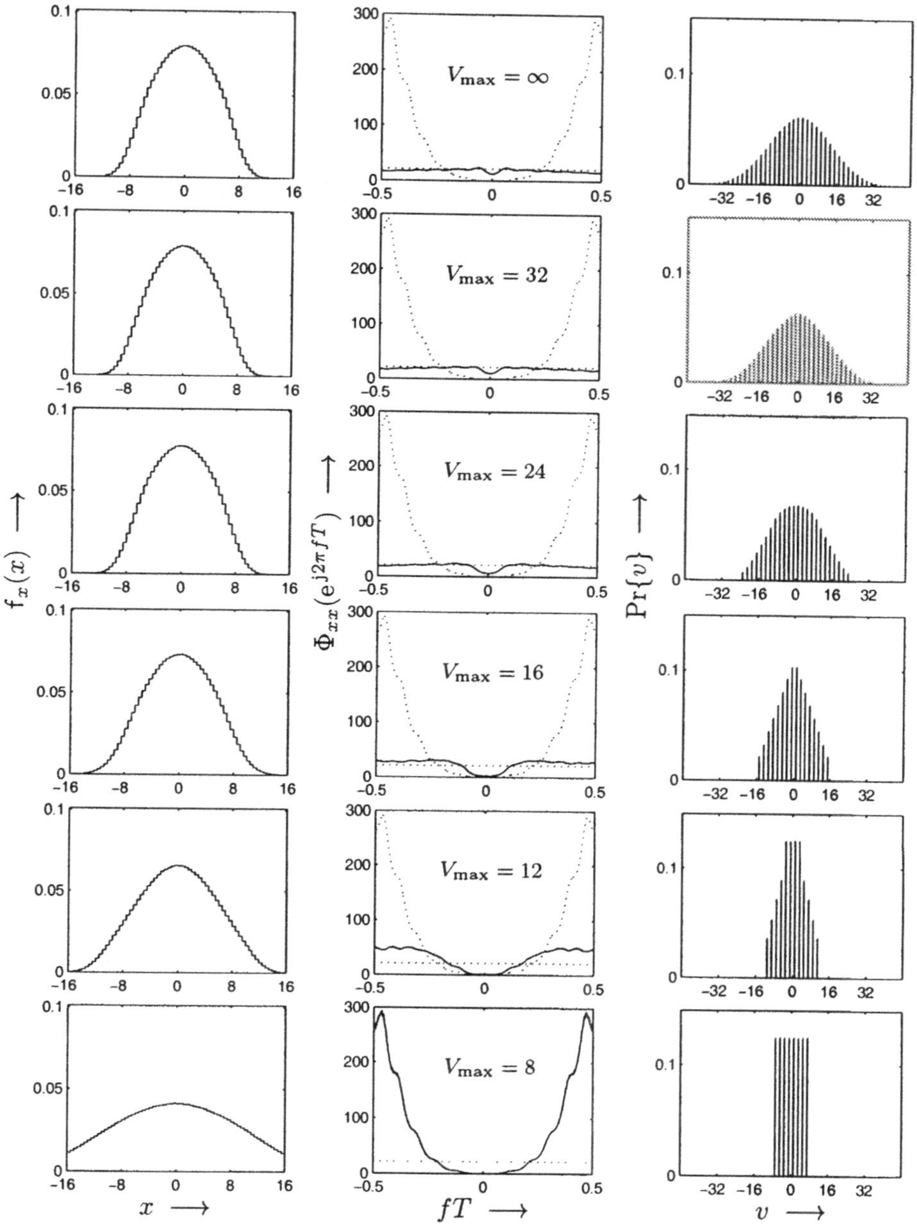

Fig. 5.32 Probability density functions (left) and estimated power spectral density (middle) of the channel symbols $x[k]$, and probability density functions of the effective data symbols $v[k]$ (right). Top to bottom: variation of the dynamics restriction $V_{\max}$. Dotted lines: theoretical PSD for Tomlinson-Harashima precoding and linear preequalization.

The left column displays the pdf of the channel symbols $x[k]$ and the column in the middle gives the respective power spectral density. For the sake of completeness, the right column plots the probabilities of the effective data symbols $v[k]$. The limitation of the dynamic range varies from $V_{\max} = \infty$ (no restriction, shaping without scrambling) $V_{\max} = 32, 24, 16, 12$, down to $V_{\max} = 8$ (linear preequalization). Again, the theoretical power spectral densities for Tomlinson-Harashima precoding and linear preequalization are included (dotted lines).

In each case the pdf of the channel symbols is almost Gaussian, but has different variance for different degrees of dynamics limitation. Shaping without scrambling (no dynamics limitation) produces an almost white transmit signal; compared to Tomlinson-Harashima precoding, the PSD is lowered by the shaping gain. When restricting the dynamic range of $v[k]$ more and more, the PSD of $x[k]$ tends from an almost constant spectrum to a high-pass one, which is proportional to the inverse of the squared magnitude of the channel filter spectrum.

The reader should compare this figure with Figure 5.28, valid for dynamics limited precoding.

Example 5.11: *Comparison of Dynamics Limitation Strategies*

This example compares Tomlinson-Harashima precoding with dynamics limited precoding and dynamics shaping, respectively. Again, the simplified SDSL up-stream example (self-NEXT limited environment) with cable length 3 km and 8-ary one-dimensional signaling is assumed. The discrete-time channel model $H(z)$ is calculated via the Yule-Walker equations (2.3.14).

The reference point for the comparison is linear zero-forcing equalization. Figure 5.33 shows the dynamics of the effective data sequence versus the gain over linear zero-forcing equalization.

The points marked with $\square$, and linearly interpolated with a dashed line, are valid for Tomlinson-Harashima precoding. Here, an exchange of gain and dynamic range of $v[k]$ is only possible via a variation of the order p of the discrete-time noise whitening filter $H(z)$. Increasing p also increases the prediction gain G_{p} over linear zero-forcing equalization ($p = 0$), cf. Figure 2.18. At the same time, the sum $\sum_{k=0}^{p} |h[k]|$ over the magnitudes of the taps of the whitening filter increases, which in turn increases the dynamic range of the effective data symbols, see (3.2.5). As p tends to infinity, the prediction gain approaches the ultimate prediction gain (6.04 dB, in this particular case), which is marked by the vertical dash-dotted line. An extremely high dynamic range of the effective data symbols is required for approaching this ultimate limit.

For dynamics limited precoding and dynamics shaping, the order of the whitening filter is fixed to $p = 10$. Starting for this point with $V_{\max} = 71$, dynamics limited precoding offers a better trade-off between gain and dynamic range. Since here the average transmit energy is increased compared to Tomlinson-Harashima precoding (negative shaping gain in dB), part of the prediction gain of 5.83 dB is lost. In the range relevant in practice ($V_{\max} \approx 2M \ldots 3M = 16 \ldots 32$), dynamics limited precoding offers a gain of about 0.5 dB for the fixed dynamic range.

The trade-off between gain and dynamic range is even better when using dynamics shaping. The order of the whitening filter is chosen to $p = 10$, too. Here, shaping gain G_{s} is achieved, which adds (in dB) to the noise prediction gain G_{p}. As in the previous examples, a 16-state trellis is used for shaping and the decoding delay is adjusted to 32 symbols. For $V_{\max} > 32$ the total gain is approximately 6.43 dB. In the region $V_{\max} \approx 16 \ldots 32$, dynamics shaping is

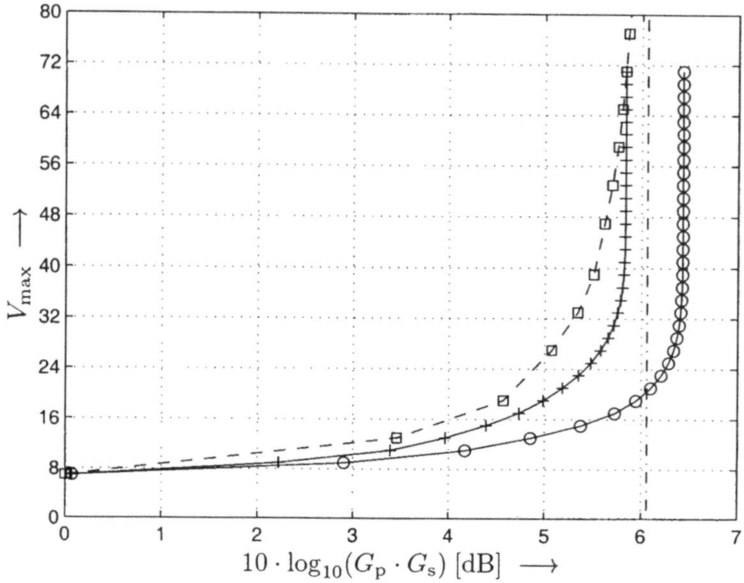

Fig. 5.33 Trade-off between dynamic range V_{max} of the effective data symbols and the gain relative to linear zero-forcing equalization (prediction gain plus shaping gain (in dB)). □: Tomlinson-Harashima precoding (variation of p); $+$: dynamics limited precoding ($p = 10$); ∘: dynamics shaping ($p = 10$). Dash-dotted line: asymptotic prediction gain.

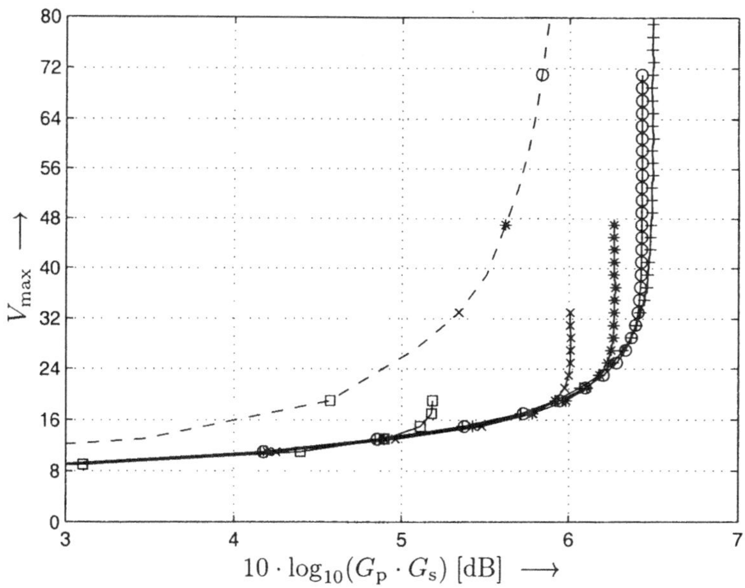

Fig. 5.34 Trade-off between dynamic range V_{max} of the effective data symbols and the gain relative to linear zero-forcing equalization (prediction gain plus shaping gain (in dB)) when using dynamics shaping. Left to right: $p = 2$ (□), 4 ($\times$), 6 ($*$), 10 ($\circ$), and 20 ($+$). Dashed line: Tomlinson-Harashima precoding (variation of p).

superior to Tomlinson-Harashima precoding by more then 1.5 dB. Note that this gain can be larger than than the "ultimate shaping gain" of 1.53 dB (cf. Chapter 4), since here the shaping and prediction gains add up.

The gain of dynamics shaping is much more pronounced when channels with spectral nulls are considered. For example, forcing a null at DC leads to an even larger dynamic range than without a spectral null. Here, gains of up to 6 dB have been reported. For further examples on the performance of dynamics shaping see [FGH95].

To conclude this example, Figure 5.34 shows the trade-off between the dynamic range of the effective data symbols and the total gain for various orders p of the noise whitening filter. Here, dynamics shaping with a 16-state trellis is assumed. For reference, the performance of Tomlinson-Harashima precoding is also plotted, where the same marker types correspond to the same order p. The horizontal distance of the points to the corresponding curves is the achieved shaping gain.

When p is small, the prediction gain is low, but the dynamic range without any restriction is also low. Conversely, using a high prediction order, the curves start at large total gains and large dynamics. Here, a dynamics restriction is particularly effective. With almost no loss, $V_{\max}$ can be lowered significantly. Interestingly, all curves merge and approach a common envelope. Hence, the lesson learned from this figure is to use a noise whitening filter with a degree as high as possible, and then reduce the dynamics by signal shaping. Using this approach, the best possible trade-off between performance and dynamic range is achieved.

5.3.4 Reduction of the Peak-to-Average Power Ratio

Up to now, we have primarily focused on the properties of the sequence $\langle x[k] \rangle$ of discrete-time channel symbols. The shaping algorithms were focused on producing the desired properties of this signal. However, in order to obtain the actual, analog transmit signal $s(t)$, the channel symbols $x[k]$ are passed through the transmit or pulse-shaping filter with transfer function $G_{\mathrm{T}}(f)$; see Figure 5.35. Due to the

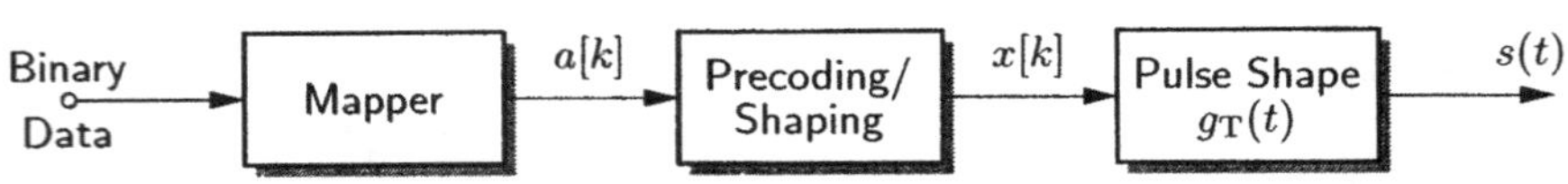

Fig. 5.35 Generation of the PAM transmit signal in precoding/shaping schemes.

dispersive nature of this filter, the pdf of $x[k]$ is smeared and the average[3] pdf of $s(t)$ may look different from that of $x[k]$. As the time span of the transmit pulse gets longer—which is the case as the band-edges get steeper—the pdf tends to be Gaussian. Even if the support region of the symbols $x[k]$ is strictly limited, large

[3]Since the continuous-time signals in PAM schemes are cyclostationary with a period equal to the symbol duration T, all quantities have to be averaged over one symbol period.

peaks may occur in the transmit signal $s(t)$, and hence its peak-to-average power ratio is increased compared to that of $x[k]$.

The line driver/power amplifier which amplifies the analog transmit signal has to cope with the full dynamic range of $s(t)$, and to work linearly over the entire region. Hence, implementation becomes more difficult, and usually the power loss at the amplifier tends to be significant. Power consumption is of major concern, especially if the transceiver at the customer premises or repeaters should be supplied remotely with power, i.e., over the same twisted-pair line that carry the data signal, or the transceiver is run by a battery. Therefore, the loss at the drivers—which is directly related to the PAR—has to be as low as possible.

Additionally, in many applications, (near-end) crosstalk from other systems is the main source of disturbance. For example, in DSL transmission, many systems may be installed within a binder group, and thus operate on lines in close proximity. Here, the transmitted signal power of one user directly translates to noise in the other systems. A reduction of average transmit power without sacrificing performance in the present system is thus highly appreciated.

Finally, as we have discussed recently, in order to become more robust against residual equalization errors and symbol timing jitter at the receiver, the dynamic range of the effective data symbols $v[k]$ should be limited and significantly reduced compared to Tomlinson-Harashima precoding.

In summary, a combined precoding/shaping technique should try to meet the following three aims:

- The average *transmit* power should be as low as possible.
- The peak-to-average power ratio of the analog *transmit* signal should be as low as possible.
- The effective data symbols, present at the *receiver*, should be limited to a small and fixed prescribed range.

Obviously, these three aims contradict each other—only a trade-off between the three demands above is possible. For example, decreasing the dynamic range at the receiver increases the PAR at the transmitter, as do shaping techniques for reducing average transmit power. A proper system design will look for the best exchange of dynamic range at the receiver, peak-to-average power ratio, and average power at the transmitter.

Extension of Dynamics Shaping Dynamics shaping, presented in the last section, provides a trade-off between two of the three demands discussed. We now show how dynamics shaping can be modified in order to additionally control the peak-to-average power ratio of the analog transmit signal $s(t)$.

Since we want to reduce the peaks in the continuous-time *transmit* signal $s(t)$, the associated hypothetical analog transmit signal has to be calculated for each path the shaping decoder processes. In practice, a sampled version with U samples per modulation interval T is sufficient, i.e., the samples $s[l] \triangleq s(t = lT_s)$, $T_s = T/U$, $l \in \mathbb{Z}$. Using a polyphase filter [Sch94, PM96], this can be done very efficiently.

Finally, for signal shaping, given a block of U consecutive samples $s[l]$ an appropriate metric increment

$$\lambda[k] = f(s[kU], s[kU+1], \ldots, s[kU+U-1]) , \quad k \in \mathbb{Z} \tag{5.3.9}$$

has to be calculated. Thereby, the signals $\langle x[k] \rangle$ and $\langle v[k] \rangle$ have to be taken into account, too, as before.

Figure 5.36 depicts the block diagram of dynamics shaping with the modification for peak power control. Again, the unavoidable decoding delay of the shaping algorithm is not shown.

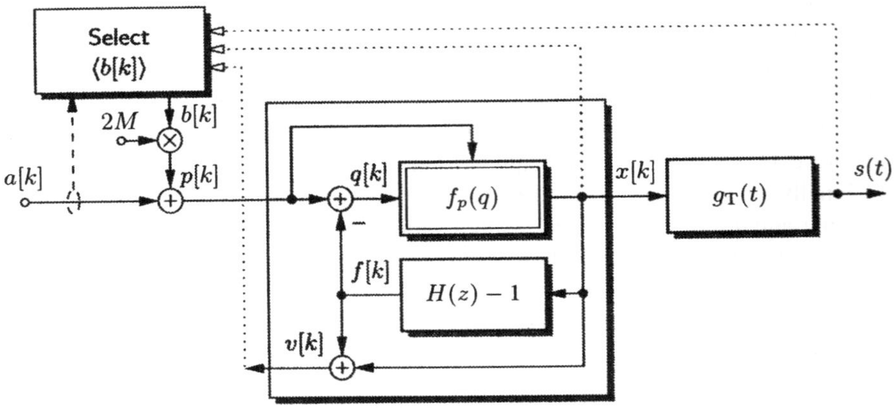

Fig. 5.36 Block diagram of extended dynamics shaping for one-dimensional signaling. Decoding delay not shown. The signals controlled by shaping are identified.

The remaining task is to select a suitable branch metric $\lambda[k]$ for achieving a peak power reduction. We now briefly discuss some possible approaches.

Maximum Amplitude An obvious strategy for peak power reduction is to look only at the maximum values of the transmit signal. Mathematically, the branch metric is chosen as

$$\lambda[k] = \max_{u=0,1,\ldots,U-1} |s[kU+u]| . \tag{5.3.10}$$

Unfortunately, using this approach, the average transmit power is not taken into account, and hence will increase significantly. Consequently, the metric has to incorporate some additional term reflecting average power.

Clipping In [Reu99], during SDSL standards activities, it was proposed that power metric be used combined with clipping of paths which exceed a given threshold or clipping level C. Using this approach, metrics can be written as

$$\lambda[k] = \sum_{u=0}^{U-1} f(s[kU+u]) , \qquad \text{with} \qquad f(s) = \begin{cases} |s|^2, & |s| \leq C \\ \infty, & |s| > C \end{cases} . \tag{5.3.11}$$

The disadvantage of this approach is that the clipping level C has to be selected very carefully. For large clipping levels, signal peaks exceeding the chosen level will be eliminated. But if the threshold is chosen too low, the shaping algorithm may run into dead ends and produce arbitrary signals. As a result, the shaping aim will be missed.

Soft Clipping Instead of assigning infinity to the metric if a threshold is exceeded, a course of the nonlinearity $f(s)$ which penalizes large signal values harder than the quadratic function may be applied. For example, functions

$$f(s) = \begin{cases} |s|^2, & |s| \le C \\ |s|^m, & |s| > C \end{cases}, \quad m > 2\,, \tag{5.3.12}$$

are attractive. Here, proper operation of the shaping algorithm can always be guaranteed. However, threshold c and exponent m still have to be optimized jointly for best performance.

mth-Power Metric A very promising approach is to simply replace the instantaneous power metric $|s|^2$ by the mth power, $m > 2$. The branch metric then simply

$$\lambda[k] = \sum_{u=0}^{U-1} |s[kU + u]|^m\,. \tag{5.3.13}$$

Here, only the exponent m has to be optimized. For increasing m, large signal peaks are penalizes ever harder, whereas the contribution of signals with small amplitudes to the accumulated metric tends to vanish. Setting $m = 2$, pure power shaping is active.

Interestingly, minimizing the mth moment of a distribution $f_s(s)$ for fixed entropy results in a pdf of the form $c_1 \cdot e^{-c_2|s|^m}$, where c_1, c_2 are positive real constants. This can be seen in Section 4.1.2 by simply replacing $|\cdot|^2$ with $|\cdot|^m$. This distribution shows that it decreases faster for increasing m, which is the desired effect. As m goes to infinity, the distribution converges to a uniform one.

Because of its appealing simplicity, in the subsequent examples we focus on the mth-power metric for PAR reduction.

Example 5.12: *Dynamics Shaping for Peak Power Reduction*

In this example, we assess the proposed shaping scheme for peak power reduction by numerical simulations. Here, we study a SDSL transmission scenario that is closer to practical situations.

We regard the up-stream direction and assume a data rate of 2.312 Mbits/s, the maximum data rate for SDSL within European networks [ITU00, Annex B]. The primary line parameters (resistance, impedance, and capacity per length) of the cable are taken from [ETSI96], which reflects a 0.4-mm-gauge, polyethylenian isolated cable (Loop model #2 in [ITU00, Annex B]). The cable length is 3 km. In SDSL, a ($M = 16$)-ary PAM signal set is used where each symbol carries 3 bits of useful information, while the 4th bit is redundancy introduced by trellis-coded modulation. The interference is adjusted according to noise model type "B"

[ITU00, Annex B], which represents a medium penetration scenario. Details can be found in [KPN99].

Pulse shaping is done as specified in ITU-T recommendation G.991.2 [ITU00]. A rectangular pulse equal to the symbol duration $T = 3/(2.312\,\text{Mbit/s}) = 1.3\ \mu\text{s}$ is filtered with a 6th-order Butterworth filter with 3 dB cutoff frequency at half the symbol rate of $1/T = 770.66$ kHz. The nominal transmit power is fixed to 14.5 dBm (28.18 mW).

From the above parameters, a T-spaced discrete-time channel model $H(z)$ of order $p = 32$ is calculated via the Yule-Walker equations. Note that the results are valid for a broad spectrum of scenarios and mainly dependent on the shape of the transmit filter. The discrete-time model $H(z)$ does not vary significantly as the parameters are slightly changed.

In each case, a Viterbi algorithm with 16 states is employed for shaping purposes. The respective trellis is defined by an imaginary scrambler with polynomial $1 \oplus D \oplus D^4$, and the decoding delay is adjusted to 64 symbols. For metric generation, $U = 4$ samples per symbol interval T of the analog transmit signal $s(t)$ are calculated. For shaping it is essential that the pulse-shaping filter is minimum-phase. Otherwise, the impact of the current selection is needlessly delayed. This introduces delay in the control loop, which in turn can lead to instabilities. These effects can only be combatted by increasing the number of states.

The performance of the given shaping technique is assessed by observing the *clipping probability*, i.e., the probability that the instantaneous power $|s(t)|^2$ of the analog transmit signal exceeds a certain threshold. Mathematically, the average (over one symbol interval T) complementary cumulative distribution function of $|s(t)|^2$ is studied. For convenience, the instantaneous signal power is normalized to the average transmit power S_{THP} when using Tomlinson-Harashima precoding, i.e.,

$$S_{\text{norm}} = |s(t)|^2 / S_{\text{THP}} \,. \tag{5.3.14}$$

Again, conventional Tomlinson-Harashima precoding constitutes the base line for performance evaluation. First, we neglect the dynamics restriction at the receiver side, i.e., $V_{\max} \to \infty$ is chosen, then additional constraints are imposed.

Variation of the Branch Metric In Figure 5.37, the complementary cumulative distribution function of $|s(t)|^2$ when using the mth-power metric is plotted for different values of m. For $m = 2$, pure power shaping is active. Since the distribution of $s(t)$ tends to be Gaussian, average power is reduced by about 0.54 dB and shaping gain—measured with respect to the continuous-time signal $s(t)$—is achieved, but the peak value is increased significantly. By going to larger m, the peaks are eliminated at the cost of average power. For a given clipping probability, e.g., 10^{-6}, a gain of up to 1.5 dB in clipping level can be achieved. In practice, clipping probabilities lower than $\approx 10^{-6}$ should be tolerable. Due to channel coding no decision errors will occur, and out of band radiation when using nonlinear amplifiers is still acceptable.

In the following, we restrict ourself to $m = 16$, which seems to offer a good trade-off between complexity (four successive multiplications are required, since $s^{16} = (((s^2)^2)^2)^2$ holds) and achievable performance.

Restriction of the Dynamic Range Now we consider an additional restriction on the dynamic range at the receiver side. Note that for Tomlinson-Harashima precoding the maximum possible amplitude calculates to about $V_{\text{THP}} = 357 \approx 22M$. Figure 5.38 plots the complementary cumulative distribution function of the instantaneous transmit power $|s(t)|^2$ for $m = 16$ and various amplitude restrictions $V_{\max}$ (solid lines). For comparison purposes, the performance

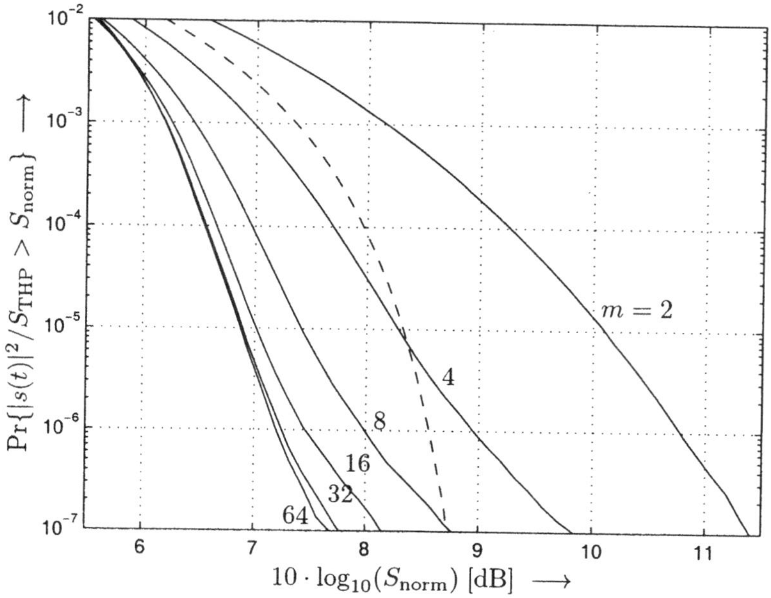

Fig. 5.37 Average complementary cumulative distribution function of $|s(t)|^2$ when applying dynamics shaping with mth-power metrics. $m = 2, 4, 8, 16, 32, 64$. Dashed: Tomlinson-Harashima precoding.

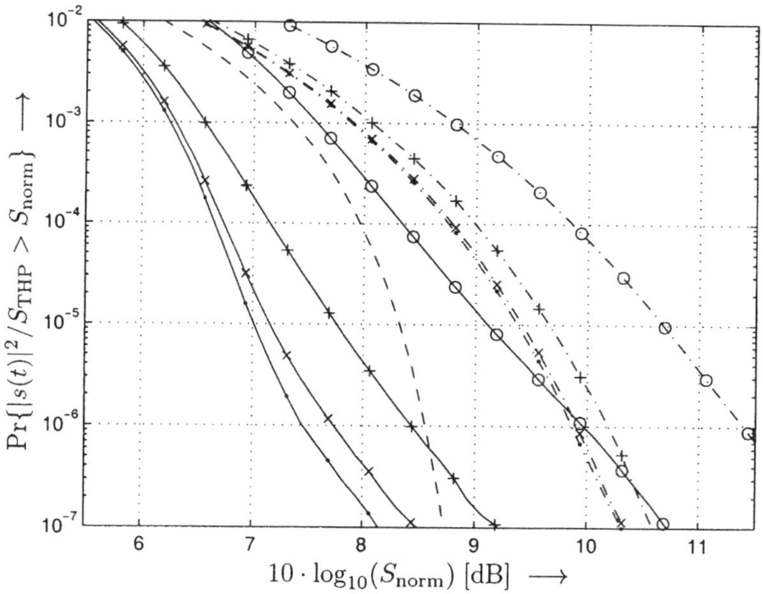

Fig. 5.38 Average complementary cumulative distribution function of $|s(t)|^2$. Dashed: Tomlinson-Harashima precoding. Dash-dotted: dynamics shaping with respect to discrete-time channel input signal. Solid: shaping on the continuous-time signal $s(t)$ using 16th-power metrics. Dynamics restriction at receiver: $V_{\max} = 3M$ (∘), $5M$ (+), $7M$ (×), and ∞ (·).

of Tomlinson-Harashima precoding is shown, as well as that of dynamics shaping based solely on the energy of the discrete-time channel symbols $x[k]$; cf. Section 5.3.3.

Restricting the dynamic range of the *receive* signal clearly leads to an increase in the peak power at the *transmitter*. This effect is increased as the restriction on the dynamic range $V_{\max}$ is tightened. If signal shaping is done based on the channel symbols $x[k]$ prior to pulse shaping, the performance is worse than that for Tomlinson-Harashima precoding. But if the branch metric is calculated from $s(t)$, a gain in peak power at the transmitter is possible, even for a restriction at the receiver down to $V_{\max} = 5M$. Moreover, in this case, the average transmit power is decreased by 0.29 dB compared to Tomlinson-Harashima precoding.

Comparision A summary of the performance is given in Figure 5.39. The curves are parameterized by the dynamics restriction $V_{\max}$ and are normalized to the performance of Tomlinson-Harashima precoding ($V_{\mathrm{THP}} = 22M$), which corresponds to the origin (□). The x-axis shows the clipping level gain when the clipping probability at the transmitter is fixed at 10^{-6}

$$G_{\mathrm{cl}} \triangleq \frac{S_{\mathrm{norm}}\big|_{\Pr\{|s_{\mathrm{THP}}(t)|^2/S_{\mathrm{THP}} > S_{\mathrm{norm}}\} = 10^{-6}}}{S_{\mathrm{norm}}\big|_{\Pr\{|s(t)|^2/S_{\mathrm{THP}} > S_{\mathrm{norm}}\} = 10^{-6}}}\,, \tag{5.3.15}$$

when the clipping probability at the transmitter is fixed at 10^{-6}. The y-axis gives the respective shaping gain $G_{\mathrm{s}} \triangleq \mathrm{E}\{|s_{\mathrm{THP}}(t)|^2\}/\,\mathrm{E}\{|s(t)|^2\}$, i.e., the reduction in average transmit power.

Restricting $V_{\max}$ as low as $5M$ in the proposed signal shaping method causes no loss. Shaping gain, as well as gain in dynamics reduction, is possible using the proposed algorithm. Specifically, a shaping gain of 0.29 dB and a clipping level gain of 0.15 dB are achievable

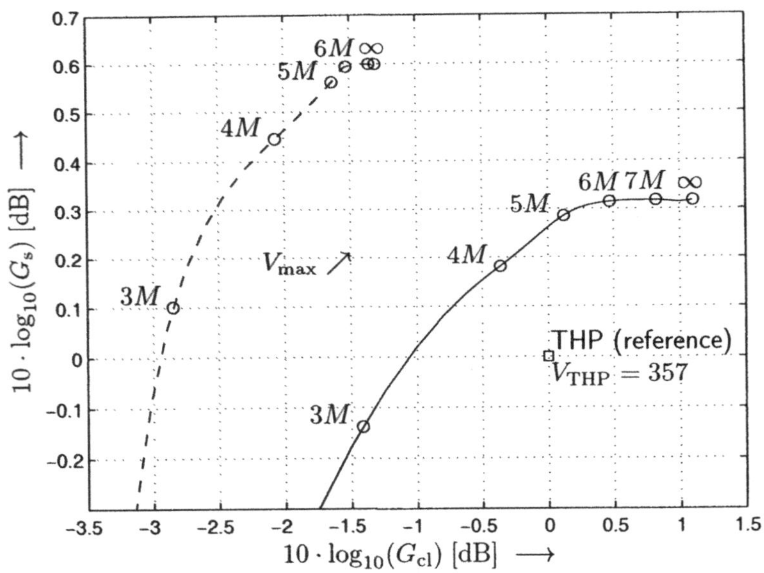

Fig. 5.39 Shaping gain G_{s} (in dB) versus clipping level gain G_{cl} (in dB); clipping probability 10^{-6}. □: Tomlinson-Harashima precoding (reference). Dashed line: dynamics shaping with respect to the discrete-time channel input signal $x[k]$. Solid line: shaping on the continuous-time transmit signal $s(t)$ using 16th-power metrics. Variation of the dynamics restriction $V_{\max}$ at the receiver side.

with $V_{\max} = 5M$. Signal power as well as clipping level are increased only if the maximum amplitude at the receiver is limited below $4M$. For dynamics shaping as described in Section 5.3.3, higher shaping gains are possible at the price of increased dynamics at the transmitter.

In summary, employing the new algorithm, the three demands—(a) low average transmit power, (b) low peak-to-average power ratio, and (c) limited dynamic range of the effective data symbols—can be fulfilled simultaneously over a wide range.

Influence on the Power Spectral Density Finally, the influence of shaping on the power spectral density is visualized in Figure 5.40. Additionally, the spectral mask given in [ITU00, Appendix B] are drawn. When comparing these spectra with that of Tomlinson-Harashima

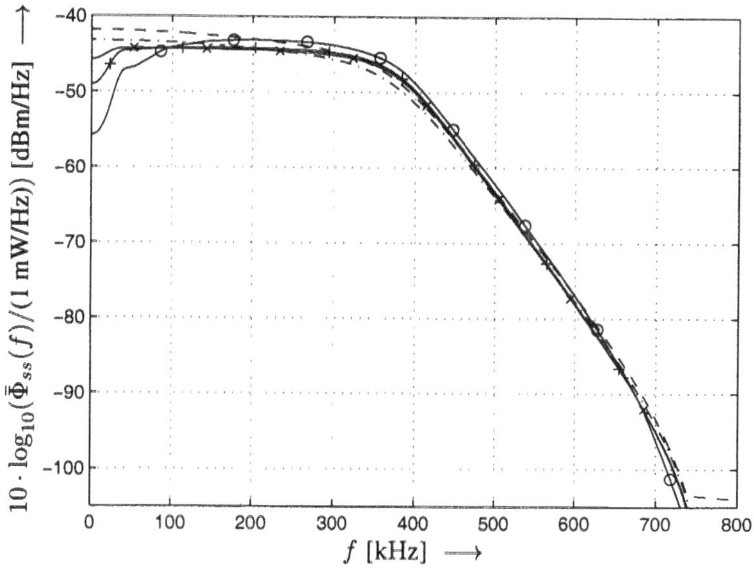

Fig. 5.40 Power spectral density of the continuous-time transmit signal $s(t)$ when applying shaping using 16th-power metrics. Dynamics restriction at the receiver: $V_{\max} = 3M$ (o), $5M$ (+), $7M$ (×). Dash-dotted line: Tomlinson-Harashima precoding; dashed line: spectral mask for SDSL according to [ITU00, Appendix B].

precoding, some accentuation is visible around Nyquist frequency ($\approx$ 380 kHz). Since the spectral mask is not exhausted in this region when using Tomlinson-Harashima precoding, no violation of the regulations occur. The power spectral density exceeds the spectral mask only if the dynamic range at the receiver is restricted down to $3M$.

Note that, the same investigations have also been performed for the OPTIS power spectrum, defined for HDSL2 [ANSI99], and also used in [ITU00, Appendix A]. Compared to the PSD used above, it has a much steeper roll-off region. This implies that, even for Tomlinson-Harashima precoding, an almost Gaussian distributed transmit signal $s(t)$ is obtained. In this case, even pure power shaping is rewarding, since reducing the average power of a Gaussian distribution also reduces the clipping probability. It can be observed that the proposed shaping algorithm can achieve even better results for transmit spectra with sharper band-edges [Tzs00].

5.4 GEOMETRICAL INTERPRETATION OF PRECODING AND SHAPING

The operation of precoding or combined precoding/signal shaping can be best understood if it is geometrically illustrated. For that purpose we recall that in digital transmission, sequences of symbols are preferably visualized in *signal space*, a concept dating back to Shannon [Sha49]. Each discrete-time index k is associated with one dimension, and the PAM symbols at the respective time indices give the coordinates. The fact that in each time step the data symbols $a[k]$ may be chosen independently of the preceding or subsequent symbols is depicted by using a Cartesian coordinate system, i.e., the axes are mutually perpendicular. If a sequence of N symbols, each drawn from a D-dimensional PAM signal set, is considered, the corresponding geometrical representation is a point in the ND-dimensional space; cf. [Sha49, FGL$^+$84]. For illustration it is best to think of baseband PAM transmission using a one-dimensional signal set $\mathcal{A}$.

When sending a sequence of PAM symbols through a linear filter channel with transfer function $H(z) = 1 + \sum_{k\geq 1} h[k]\, z^{-k}$, intersymbol interference occurs. The dimensions are no longer independent or, equivalently, orthogonality is lost [Mes73]. Mathematically, a semi-infinite sequence of symbols $a[k]$, $k = 0, 1, \ldots$, written as vector $\boldsymbol{a} = [a[0],\, a[1],\, \ldots]$ is transformed into the sequence of (noise-free) receive symbols $v[k]$, written as vector $\boldsymbol{v} = [v[0],\, v[1],\, \ldots]$, by multiplication of $\boldsymbol{a}$ by a channel matrix $\boldsymbol{H}$, i.e.,

$$\boldsymbol{v} = \boldsymbol{a} \cdot \boldsymbol{H}\,, \quad \text{with} \quad \boldsymbol{H} = \begin{bmatrix} 1 & h[1] & h[2] & h[3] & \cdots \\ 0 & 1 & h[1] & h[2] & \cdots \\ 0 & 0 & 1 & h[1] & \cdots \\ \vdots & \ddots & \ddots & \ddots & \ddots \end{bmatrix}; \tag{5.4.1}$$

cf. [FU98]. Geometrically, $\boldsymbol{H}$ specifies a *coordinate transformation*. Since the diagonal elements of $\boldsymbol{H}$ are all 1, its determinant equals 1, hence the transformation from $\boldsymbol{a}$ to $\boldsymbol{v}$ is at least volume-preserving [FU98].

5.4.1 Combined Precoding and Signal Shaping

The intersymbol interference introduced by $H(z)$ may be equalized by linear preequalization at the transmitter side. Geometrically, the sequence of PAM data symbols is first transformed into the sequence of channel symbols $x[k]$ through multiplication by $\boldsymbol{H}^{-1}$, the inverse of the channel matrix. This matrix is also upperdiagonal and Toeplitz as $\boldsymbol{H}$ is, and the entries are the coefficients of the impulse response corresponding to $1/H(z)$. This procedure is shown in Figure 5.41 (top) for a two-dimensional signal space, i.e., two consecutive one-dimensional data symbols. Orthogonality is achieved at the receiver, but the peak as well as the average power of the channel symbols $x[k]$ are substantially increased by linear predistortion.

These disadvantages are avoided by application of Tomlinson-Harashima precoding, cf. the middle row in Figure 5.41. Now, the initial PAM signal set $\mathcal{A}$ is extended periodically, and is predistorted by the inverse of the channel filter $H(z)$.

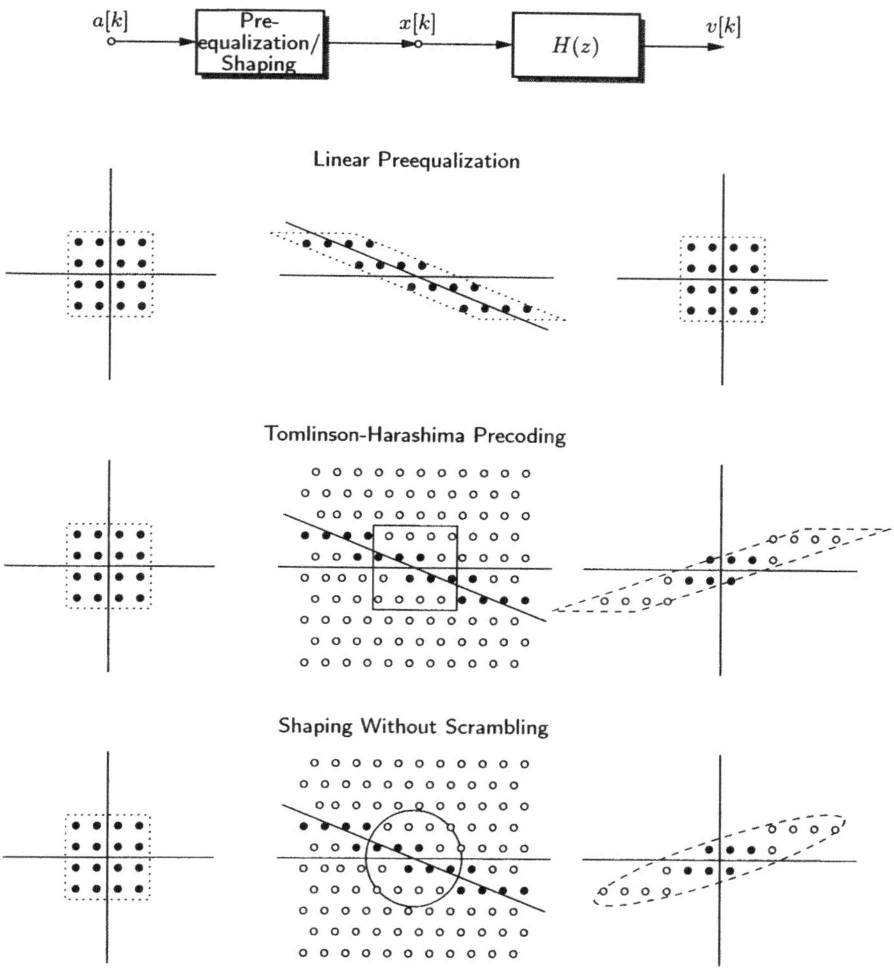

Fig. 5.41 Geometrical illustration of linear preequalization (top), Tomlinson-Harashima precoding (middle), and shaping without scrambling (bottom).

Geometrically, in Tomlinson-Harashima precoding, from the set of possible channel symbol sequences $\langle x[k] \rangle$ those are cut out [Ung91], which lie in a *hypercube*; see solid boundary in Figure 5.41. Since the predistortion filter is a causal system, the selection of the actual representative can be done symbol-by-symbol, i.e., consecutive for each dimension, but taking the preceding dimensions into account. The pdf of the channel symbols can be obtained as the projection of the high-dimensional formation onto an axis. Since the signal points no longer lie exclusively on the integer grid, a continuous distribution results, which was also covered in Chapter 3.

At the channel output, the effective data sequences then lie within a parallelepiped, cf. Figure 5.41. Again, the coordinates are orthogonal and the signal points form a regular grid. The projection onto an axis is no longer uniform, but tends to be (discrete) Gaussian as the dimensionality of the signal space increases. Symbol-by-symbol, i.e., in each dimension separately, a decision can be made and data are recovered. Moreover, by a modulo reduction in each dimension, the initial signal point sequences $\langle a[k] \rangle$ are recovered.

The situation is very similar if Tomlinson-Harashima precoding is replaced by shaping without scrambling (see Section 5.2). This is shown on the bottom of Figure 5.41. Instead of restricting the channel symbols to a *hypercube*, they are now drawn from a *hypersphere* with the same volume (number of signal points). Here, the average transmit power is reduced and shaping gain achieved. The projection of a hypersphere onto an axis results in a Gaussian distribution (Section 4.2.3) for the channel symbols. Conversely, the distribution of the effective data symbols is nearly unchanged compared to Tomlinson-Harashima precoding and is still discrete Gaussian.

5.4.2 Limitation of the Dynamic Range

We now turn to the situation when, as an additional constraint, the dynamic range of the effective data symbols is limited to the interval $[-V_{\max}, V_{\max}]$ (Section 5.3). The data symbols are again periodically extended, but under the restriction that $|v[k]| \le V_{\max}, \forall k$. The support region for sequences of initial data symbols $a[k]$ (dotted line) and the support region for sequences of effective data symbols $v[k]$ (dashed line) in signal space are shown at the top of Figure 5.42. For clarity, the individual signal points are not shown here.

All accessible channel symbol sequences $\langle x[k] \rangle$ after preequalization now lie in a parallelepiped, which is obtained by transformation of the hypercube of side length $2V_{\max}$. In dynamics limited precoding, a symbol-by-symbol selection of the actual representative is performed. Due to the restriction of the dynamic range of $v[k]$, the channel symbol sequences cannot be exclusively selected from a hypercube (solid line). In high-dimensional signal space, a formation with strange vertices and swellings results. These vertices immediately correspond to the linear branches of the nonlinear function $f_a(\cdot)$ for large input values. Hence, compared to Tomlinson-Harashima precoding, an increase in average transmit power occurs, as it also does in peak transmit power.

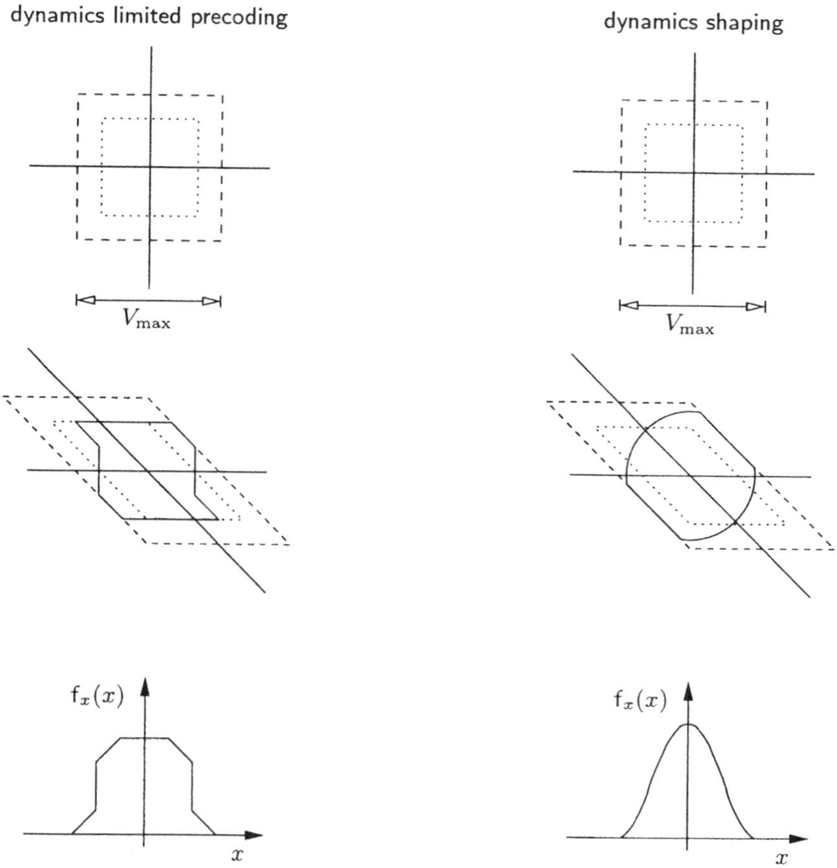

Fig. 5.42 Geometrical illustration of dynamics limited precoding and dynamics shaping. Top: support regions for sequences of data symbols $a[k]$ (dotted line) and that for sequences of effective data symbols $v[k]$ (dashed line). Middle: regions for sequences of channel symbols $x[k]$ (solid line). Bottom: densities induced on one dimension.

The probability density induced onto an axis is plotted at the bottom of Figure 5.42. Instead of a uniform pdf, as in Tomlinson-Harashima precoding, a broadened pdf with increased peak and average power is present. Note that this simple illustration matches perfectly with the pdfs obtained from numerical simulations; cf., e.g., Figures 5.27 and 5.28.

Starting from dynamics limited precoding, based on a long-term selection of the effective data symbols, dynamics shaping additionally minimizes the transmit power. Obviously, the optimum region for sequences of channel symbols is the intersection of the parallelepiped due to dynamics limitation and a *hypersphere* (see Figure 5.42, right column). The radius of this hypersphere is adjusted so that the number of sequences contained in the intersection is the same as in Tomlinson-Harashima precoding and dynamics limited precoding. In contrast to dynamics limited precoding, the high-dimensional formation will exhibit no strange vertices and swellings. In turn, peak power as well as the average power of the channel symbols will be lowered. The projection onto one dimension is nearly Gaussian, as revealed by numerical simulations; cf. Figure 5.31 and 5.32.

In addition, it becomes obvious that the gain in average power of dynamics shaping compared to dynamics limited precoding is not bounded by $\frac{\pi e}{6}$, the ultimate shaping gain of a hypersphere over a hypercube. Moreover, if the restriction of the dynamic range is moderate, no notable loss in shaping gain will occur, since the signal space region deviates only slightly from a hypersphere, cf. Figure 5.33. This statement agrees with the results on shaping under a transmit peak constraint, as discussed in Section 4.2.4.

5.5 CONNECTION TO QUANTIZATION AND PREDICTION

Even though this book is devoted to digital transmission over linear distorting channels, at some points we have pointed out the connection to source coding. In Chapter 4 it was shown that signal shaping is an operation dual to source coding. To be precise, coding of memoryless sources has to be considered. We now briefly show that coding for sources with memory, in particular Markov sources [Pap91], is dual to combined precoding and signal shaping. Based on this insight, source-coding schemes are derived in [LL00] from the combination of shell mapping and flexible precoding.

Geometrically, precoding/shaping techniques try to bound the transmit sequence in a hypersphere and to achieve good distance properties of the receive signal; cf. the last section. The signals to be controlled are related by filtering with the end-to-end channel model $H(z)$. In N-space, the boundary of the set of sequences at the transmitter is of concern. We have seen that choosing the points in signal space from a hypersphere rather than from a hypercube results in an ultimate shaping gain of 1.53 dB. Conversely, at the receiver the internal arrangement, the packing, of the signal points is decisive. Channel coding controls this packing in Euclidean space and typically achieves gains on the order of 4 to 6 dB.

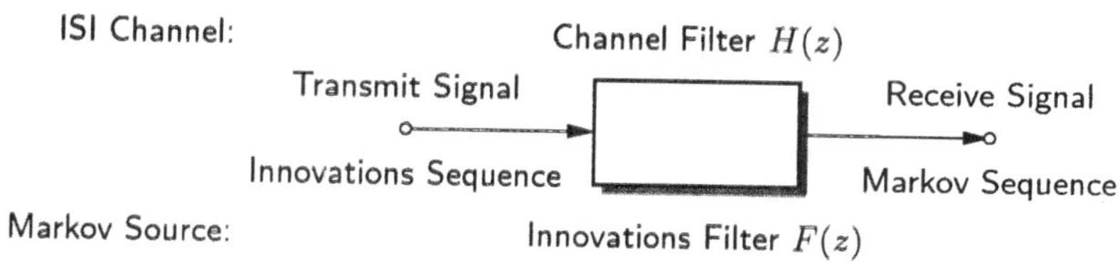

Fig. 5.43 Illustration of the duality of precoding/shaping for ISI channels and vector quantization/prediction for Markov sources.

We now turn to source coding, in particular vector quantization of Gaussian Markov sources. To show the duality to precoding/shaping, Figure 5.43 depicts the correspondences. It is well-known that we can model Markov sequences to be generated from a memoryless *innovations sequence* by filtering with an *innovations filter* with some transfer function $F(z)$, e.g., [Pap91]. Hence, we can find a correspondence of the channel filter $H(z)$ to the innovations filter $F(z)$. Moreover, the innovations sequence is the counterpart to the transmit sequence, and the Markov sequence plays the role of the receive sequence.

Given the actual Markov sequence, a straightforward approach would be to apply *prediction*, i.e., to filter the source sequence with the *prediction-error filter* or *whitening filter* with transfer function $1/H(z)$. Then, this memoryless signal is quantized (cf. *differential pulse code modulation (DPCM)* systems). Usually, this will be done using a vector quantizer which offers a high *granular gain*, i.e., the Voronoi cells around the representatives have a low second moment, which is directly related to distortion measured as the mean-squared error (MSE). The granular gain is dual to the shaping gain, and is also limited to $\pi e/6$, corresponding to 0.255 bits per dimension [EF93, ZF96, GN98]. It is achieved when the Voronoi regions are "quasi-spherical." Since the innovations sequence is i.i.d. and Gaussian, with high probability, in N-space the points representing all these sequences lie within a hypersphere, the so-called *source sphere* [EF93, LL00]. Hence, the innovations codebook—the codebook used for vector quantization of the innovations sequence—should only comprise representatives which lie inside this sphere. Choosing a suitable boundary of the codebook results in a *boundary gain*. In source coding, the boundary gain takes over the role of the coding gain in channel coding.

The source decoder has to reconstruct the initial Markov sequence. Therefore the quantized version of the innovations sequence has to be filtered with the innovations filter $F(z)$. Considering N-space, the representatives then lie within a *hyperellipsoid* [JN84], see also Figure 5.41. This boundary is optimum, but the quantization cells are linearly distorted. They are no longer quasi-spherical, but also have some ellipsoidal shape. In turn, the reconstruction error is increased. This phenomenon is the counterpart to a reduced minimum Euclidean distance of linearly distorted PAM sequences.

Precoding aims to generate transmit sequences which result in an equalized channel output, i.e., the distance properties of the initial data sequence are preserved. Moreover, combined precoding/shaping additionally minimizes average transmit power. Transforming this to source coding means that the innovations codebook should be "predistorted," so that the quantization cells of the source codebook—the codebook for vector quantization of the source sequence—are quasi-spherical. At the same time, the spherical boundary of the innovations codebook has to be preserved.

In summary, combined precoding and signal shaping for transmission over intersymbol-interference channels has a counterpart in source coding, namely, vector quantization of sources with memory. In [LL00], a precoded trellis-based scalar vector quantizer is developed by applying this duality to flexible precoding combined with shell mapping. Shell mapping has a source-coding counterpart in structured fixed-rate vector quantization [LF93a, LF93b]. Precoder and inverse precoder (cf. Section 3.3) interchange their role and are adapted to perform (nonlinear) prediction and reconstruction, respectively. A detailed discussion of this source-coding scheme is far beyond the scope of this book; the reader is referred to the original literature [LL00].

Finally, Table 5.1 summarizes the most important correspondences between precoding/shaping for transmission over intersymbol-interference channels and vector quantization of sources with memory.

Table 5.1 Dualities between combined precoding/shaping for transmission over intersymbol-interference channels and vector quantization of sources with memory.

Transmission over intersymbol-interference channels	Vector quantization of sources with memory
Sequence of transmit symbols	Innovations sequence
Sequence of (noisy) receive symbols	Markov source sequence
Linear ISI channel $H(z)$	Innovations filter $F(z)$
Precoder	Prediction
Voronoi cells of signal lattice	Boundary of quantization codebook
Boundary of signal constellation	Shape of the quantization cells
Receive point lies in a noise sphere around transmit point (AWGN channel)	Source sequences lie within a sphere (memoryless Gaussian source)
Minimum average energy of transmit signal	Minimum distortion (squared error) in source domain
Maximum Euclidean distance of channel output sequences	Maximum boundary gain of innovations codebook
Gain in signal-to-noise ratio (in dB) at given error rate	Gain in reduction of average rate (in bits) at given distortion
Coding gain	Boundary gain
Shaping gain ($\leq \pi e/6$)	Granular gain ($\leq \frac{1}{2}\log_2(\pi e/6) = 0.255$ bit/dim.)
Constellation design: (a) max. min. Euclidean distance (b) minimize average energy	Quantizer design: (a) max. boundary gain (b) min. second moment of quantizer cells

REFERENCES

[ANSI99] American National Standards Institute (ANSI). Draft for HDSL2 Standard. Draft American National Standard for Telecommunications, March 1999.

[Ben84] A. Benveniste. Blind Equalizers. *IEEE Transactions on Communications*, COM-32, pp. 871–883, August 1984.

[BGR80] A. Benveniste, M. Goursat, and G. Rouget. Robust Identification of a Nonminimum Phase System: Blind Adjustment of a Linear Equalizer in Data Communications. *IEEE Transactions on Automatic Control*, AC-25, pp. 385–399, June 1980.

[BS98] I. N. Bronstein and K. A. Semendjajew. *Handbook of Mathematics*. Springer Verlag, Berlin, Heidelberg, Reprint of the third edition, 1998.

[CE89] P. R. Chevillat and E. Eleftheriou. Decoding of Trellis-Encoded Signals in the Presence of Intersymbol Interference and Noise. *IEEE Transactions on Communications*, COM-37, pp. 669–676, July 1989.

[DH89] A. Duel-Hallen and C. Heegard. Delayed Decision-Feedback Sequence Estimation. *IEEE Transactions on Communications*, COM-37, pp. 428–436, May 1989.

[EF92] M. V. Eyuboğlu and G. D. Forney. Trellis Precoding: Combined Coding, Precoding and Shaping for Intersymbol Interference Channels. *IEEE Transactions on Information Theory*, IT-38, pp. 301–314, March 1992.

[EF93] M. V. Eyuboğlu and G. D. Forney. Lattice and Trellis Quantization with Lattice- and Trellis-Bounded Codebooks—High-Rate Theory for Memoryless Sources. *IEEE Transactions on Information Theory*, IT-39, pp. 46–59, January 1993.

[EFDL93] M. V. Eyuboğlu, G. D. Forney, P. Dong, and G. Long. Advanced Modulation Techniques for V.fast. *European Transactions on Telecommunications*, ETT-4, pp. 243–256, May/June 1993.

[EQ88] M. V. Eyuboğlu and S. U. H. Qureshi. Reduced-State Sequence Estimation with Set Partitioning and Decision Feedback. *IEEE Transactions on Communications*, COM-36, pp. 13–20, January 1988.

[EQ89] M. V. Eyuboğlu and S. U. H. Qureshi. Reduced-State Sequence Estimation for Coded Modulation on Intersymbol Interference Channels. *IEEE Journal on Selected Areas in Communications*, JSAC-7, pp. 989–995, August 1989.

[ETSI96] ETSI ETR 152. *High Bit-rate Digital Subscriber Line (HDSL) Transmission System on Metallic Local Lines; HDSL Core Specification and Applications for 2048 kbit/s Based Access Digital Sections.* European Telecommunications Standards Institute (ETSI), Sophia Antipolis, France, 3. edition, December 1996.

[FE91] G. D. Forney and M. V. Eyuboğlu. Combined Equalization and Coding using Precoding. *IEEE Communications Magazine*, 29, pp. 25–34, December 1991.

[FGH95] R. Fischer, W. Gerstacker, and J. Huber. Dynamics Limited Precoding, Shaping, and Blind Equalization for Fast Digital Transmission over Twisted Pair Lines. *IEEE Journal on Selected Areas in Communications*, JSAC-13, pp. 1622–1633, December 1995.

[FGL+84] G. D. Forney, R. G. Gallager, G. R. Lang, F. M. Longstaff, and S. U. H. Qureshi. Efficient Modulation for Band-Limited Channels. *IEEE Journal on Selected Areas in Communications*, JSAC-2, pp. 632–647, September 1984.

[FH94] R. Fischer and J. Huber. Signalformung zur Begrenzung der Dynamik bei der Tomlinson-Harashima-Vorcodierung. In *ITG-Fachtagung "Codierung für Quelle und Kanal" (Fachbericht 130)*, pp. 457–466, München, October 1994. (In German.)

[Fis92] R. Fischer. *Vergleich von Vorcodierungsverfahren für digitale PAM Übertragungsverfahren.* Diploma Thesis, Lehrstuhl für Nachrichtentechnik, Universität Erlangen–Nürnberg, Erlangen, Germany, 1992. (In German.)

[For73] G. D. Forney. The Viterbi Algorithm. *Proceedings of the IEEE*, 61, pp. 268–278, March 1973.

[For92] G. D. Forney. Trellis Shaping. *IEEE Transactions on Information Theory*, IT-38, pp. 281–300, March 1992.

[FU98] G. D. Forney and G. Ungerböck. Modulation and Coding for Linear Gaussian Channels. *IEEE Transactions on Information Theory*, IT-44, pp. 2384–2415, October 1998.

[Ger98] W. Gerstacker. Entzerrverfahren für die schnelle digitale Übertragung über symmetrische Leitungen. PhD Thesis, Technische Fakultät der Universität Erlangen–Nürnberg, Erlangen, Germany, December 1998. (In German.)

[GN98] R. M. Gray and D. L. Neuhoff. Quantization. *IEEE Transactions on Information Theory*, IT-44, pp. 2325–2383, October 1998.

[Hay96] S. Haykin. *Adaptive Filter Theory*. Prentice-Hall, Inc., Englewood Cliffs, NJ, 3ed edition, 1996.

[Hub92] J. Huber. *Trelliscodierung*. Springer Verlag, Berlin, Heidelberg, 1992. (In German.)

[ITU00] International Telecommunication Union (ITU). Draft Recommendation G.991.2: Single-Pair High-Speed Digital Subscriber Line (SHDSL) Transceivers — Draft G.991.2, April 2000.

[JN84] N. S. Jayant and P. Noll. *Digital Coding of Waveforms*. Prentice-Hall, Inc., Englewood Cliffs, NJ, 1984.

[JZ99] R. Johannesson and K. Sh. Zigangirov. *Fundamentals of Convolutional Coding*. IEEE Press, Piscataway, NJ, 1999.

[KPN99] R. van den Brink and B. van den Heuvel. Update of SDSL Noise Models, as Requested by ETSI-TM6. ETSI TM6, Edinburgh, UK, September 1999.

[LF93a] R. Laroia and N. Farvardin. A Structured Fixed-Rate Vector Quantizer Derived from a Variable-Length Scalar Quantizer: Part I—Memoryless Sources. *IEEE Transactions on Information Theory*, IT-39, pp. 851–867, May 1993.

[LF93b] R. Laroia and N. Farvardin. A Structured Fixed-Rate Vector Quantizer Derived from a Variable-Length Scalar Quantizer: Part II—Vector Sources. *IEEE Transactions on Information Theory*, IT-39, pp. 868–876, May 1993.

[LFT94] R. Laroia, N. Farvardin, and S. A. Tretter. On Optimal Shaping of Multidimensional Constellations. *IEEE Transactions on Information Theory*, IT-40, pp. 1044–1056, July 1994.

[LL00] C.-C. Lee and R. Laroia. Trellis-Based Scalar Vector Quantization of Sources with Memory. *IEEE Transactions on Information Theory*, IT-46, pp. 153–170, January 2000.

[Mes73] D. G. Messerschmitt. A Geometrical Theory of Intersymbol Interference—Part I and II. *Bell System Technical Journal*, 52, pp. 1483–1539, November 1973.

[Pap91] A. Papoulis. *Probability, Random Variables, and Stochastic Processes*. McGraw-Hill, New York, 3rd edition, 1991.

[PM96] J. G. Proakis and D. G. Manolakis. *Digital Signal Processing: Principles, Alogrithms, and Applications*. Prentice-Hall, Inc., Upper Saddle River, NJ, 3rd edition, 1996.

[Pro01] J. G. Proakis. *Digital Communications*. McGraw-Hill, New York, 4th edition, 2001.

[Reu99] I. Reuven. Combined Constellation Shaping and Reduction of Peak-to-Average Ratio. ETSI TM6, Edinburgh, UK, September 1999.

[Sat75] Y. Sato. A Method of Self-Recovering Equalization for Multilevel Amplitude-Modulation Systems. *IEEE Transactions on Communications*, COM-23, pp. 679–682, June 1975.

[Sch94] H. W. Schüßler. *Digitale Signalverarbeitung, Band I*. Springer Verlag, Berlin, Heidelberg, 4th edition, 1994. (In German.)

[Sha49] C. E. Shannon. Communications in the Presence of Noise. *Proceedings of the Institute of Radio Engineers*, 37, pp. 10–21, 1949.

[Tom71] M. Tomlinson. New Automatic Equaliser Employing Modulo Arithmetic. *Electronics Letters*, 7, pp. 138–139, March 1971.

[Tzs00] R. Tzschoppe. *Evolution of Signal Shaping Algorithms for Fast Digital Transmission over Twisted-Pair Lines*. Student Research Project, Lehrstuhl für Nachrichtentechnik II, Universität Erlangen–Nürnberg, Erlangen, Germany, 2000.

[Ung91] G. Ungerböck. Personal Communications. Zurich, 1991.

[Wes87] K. Wesolowski. Efficient Digitial Receiver Structure for Trellis-Coded Signals Transmitted through Channels with Intersymbol Interference. *Electronics Letters*, 23, pp. 1265–1267, November 1987.

[ZF96] R. Zamir and M. Feder. On Lattice Quantization Noise. *IEEE Transactions on Information Theory*, IT-42, pp. 1152–1159, July 1996.

Appendix A
Wirtinger Calculus

The optimization of system parameters is a very common problem in communications and engineering. For example, the optimal tap weights of an equalizer should be adapted for minimum deviation (e.g., measured by the mean-squared error or the peak distortion) of the output signal from the desired (reference) signal. For an analytical solution, a cost function is set up and the partial derivatives with respect to the adjustable parameters are set to zero. Solving this set of equations results in the desired optimal solution.

Often, however, the problem is formulated using complex-valued parameters. In digital communications, signals and systems are preferably treated in the *equivalent complex baseband* [Fra69, Tre71, Pro01]. For solving such optimization problems, derivation with respect to a complex variable is required. Starting from well-known principles, this Appendix derives a smart and easily remembered calculus, sometimes known as the *Wirtinger Calculus* [FL88, Rem89].

A.1 REAL AND COMPLEX DERIVATIVES

First, we consider a real-valued function of a real variable:

$$f : \mathbb{R} \ni x \mapsto y = f(x) \in \mathbb{R} . \tag{A.1.1}$$

The point x_{opt}, for which $f(x)$ is maximum[1] is obtained by taking the derivative of f with respect to x and setting it to zero. For x_{opt} the following equation has to be valid:

$$\left. \frac{\mathrm{d}f}{\mathrm{d}x} \right|_{x_{\text{opt}}} \stackrel{!}{=} 0 . \tag{A.1.2}$$

Here we assume $f(x)$ to be continuous in some region $\mathcal{R}$, and the derivative to exist. Whether the solution of the above equation actually gives a minimum or maximum point has to be checked via additional considerations or by inspecting higher-order derivatives.

Analogous to real functions, a derivative can be defined for complex functions of a complex variable

$$f : \mathbb{C} \ni z \mapsto w = f(z) \in \mathbb{C} \tag{A.1.3}$$

as well:

$$f'(z_0) \triangleq \left. \frac{\mathrm{d}f}{\mathrm{d}z} \right|_{z_0} = \lim_{z \to z_0} \frac{f(z) - f(z_0)}{z - z_0} . \tag{A.1.4}$$

The above limit has to exist for the infinitely many series $\{z_n\}$ which approach z_0, i.e., $\lim_{n\to\infty} z_n = z_0$. If $f'(z)$ exists in a region $\mathcal{R} \subset \mathbb{C}$, the function $f(z)$ is called *analytic*, *holomorphic*, or *regular* in $\mathcal{R}$. In the following, the relations between real and complex derivatives are discussed.

A complex function can be decomposed into two real functions, each depending on two real variables x and y, the real and imaginary parts of z:

$$f(z) = f(x + \mathrm{j}\, y) \triangleq u(x, y) + \mathrm{j}\, v(x, y), \qquad z = x + \mathrm{j}\, y . \tag{A.1.5}$$

It can be shown that in order for $f(z)$ to be holomorphic, the component functions $u(x, y)$ and $v(x, y)$ have to meet the *Cauchy-Riemann differential equations*, which read (e.g., [FL88, Rem89]):

$$\frac{\partial u(x,y)}{\partial x} = \frac{\partial v(x,y)}{\partial y} , \tag{A.1.6a}$$

$$\frac{\partial v(x,y)}{\partial x} = -\frac{\partial u(x,y)}{\partial y} . \tag{A.1.6b}$$

[1] The same considerations are valid for minimization.

The complex derivative of a holomorphic function $f(z)$ can then be expressed by the partial derivatives of the real functions $u(x, y)$ and $v(x, y)$:

$$\frac{\mathrm{d}f(z)}{\mathrm{d}z} = \frac{\partial u(x,y)}{\partial x} + \mathrm{j}\,\frac{\partial v(x,y)}{\partial x} \,. \tag{A.1.7}$$

The complex derivative of a complex function plays an important role in complex analysis—in communications it has almost no significance. In fact, a more common problem is the optimization of *real* functions, depending on *complex* parameters. Complex cost functions are of no interest, because in the field of complex numbers no ordering (relations $<$ and $>$) is defined and thus minimization or maximization makes no sense.

A.2 WIRTINGER CALCULUS

As already stated, we have to treat real functions of one or more complex variables. Thus, let us now consider functions

$$f : \mathbb{C} \ni z = x + \mathrm{j}\,y \mapsto w = f(z) = u(x,y) \in \mathbb{R} \,. \tag{A.2.1}$$

Since $v(x, y) \equiv 0$ holds (cf. (A.1.5)), $f(z)$ generally is not holomorphic. A real function would only be regular if, according to (A.1.6), $\frac{\partial u(x,y)}{\partial x} \equiv 0$ and $\frac{\partial u(x,y)}{\partial y} \equiv 0$ are valid. But this only holds for a real constant, and hence can be disregarded.

The straightforward solution to the optimization of the above function is as follows: Instead of regarding $f(z)$ as a real function of *one complex* variable, we view $f(z) = u(x, y)$ as a function of *two real* variables. Thus optimization can be done as for multidimensional real functions. We want to find

$$f(z) \to \text{opt.} \quad \equiv \quad u(x,y) \to \text{opt.} \,,$$

which requires

$$\frac{\partial u(x,y)}{\partial x} \stackrel{!}{=} 0 \quad \text{and} \quad \frac{\partial u(x,y)}{\partial y} \stackrel{!}{=} 0 \,. \tag{A.2.2}$$

In order to obtain a more compact representation, both of the above real-valued equations for the optimal components x_{opt} and y_{opt} can be linearly combined into one complex-valued equation:

$$a_1 \frac{\partial u(x,y)}{\partial x} + \mathrm{j} \cdot a_2 \frac{\partial u(x,y)}{\partial y} \stackrel{!}{=} \text{“}0 + \mathrm{j}\,0\text{”} = 0 \,, \tag{A.2.3}$$

where, for the moment, a_1 and a_2 are arbitrary real and nonzero constants. Equations (A.2.2) and (A.2.3) are equivalent (and hence, of course, result in the same solution) because real and imaginary part are orthogonal. As already stated, this procedure is mainly intended to get a compact representation.

Writing real part and imaginary part of $z = x + \mathrm{j}\, y$ as the tuple (x, y), we can define the following differential operator:

$$\frac{\partial}{\partial z} \triangleq a_1 \frac{\partial}{\partial x} + \mathrm{j} \cdot a_2 \frac{\partial}{\partial y} \,. \qquad \text{(A.2.4)}$$

This operator can, of course, also be applied to complex functions (A.1.5). This is reasonable, because real cost functions are often composed of complex components, e.g., $f(z) = |z|^2 = z \cdot z^* \triangleq f_1(z) \cdot f_2(z)$, with an obvious definition of $f_{1,2}(z) \in \mathbb{C}$. Note, $z^* \triangleq x - \mathrm{j}\, y$ denotes the complex conjugate of $z = x + \mathrm{j}\, y$.

The remaining task is to chose suitable constants a_1 and a_2. The main aim is to obtain a calculus that is easily remembered and easy to apply. As will be shown later, the choice $a_1 = \frac{1}{2}$ and $a_2 = -\frac{1}{2}$ meets all requirements. To honor the work of the Austrian mathematician *Wilhelm Wirtinger* (1865–1945) who established this differential calculus, we call it *Wirtinger Calculus*.

Definition A.1: *Wirtinger Calculus*

The partial derivatives of a (complex) function $f(z)$ of a complex variable $z = x + \mathrm{j}\, y \in \mathbb{C}$, $x, y \in \mathbb{R}$, with respect to z and z^*, respectively, are defined as:

$$\frac{\partial f}{\partial z} \triangleq \frac{1}{2} \cdot \left(\frac{\partial f}{\partial x} - \mathrm{j} \frac{\partial f}{\partial y} \right) \qquad \text{(A.2.5)}$$

and

$$\frac{\partial f}{\partial z^*} \triangleq \frac{1}{2} \cdot \left(\frac{\partial f}{\partial x} + \mathrm{j} \frac{\partial f}{\partial y} \right) \,. \qquad \text{(A.2.6)}$$

A.2.1 Examples

We now study some important examples. First, let $f(z) = cz$, where $c \in \mathbb{C}$ is a constant. Derivation of $f(z)$ yields

$$\frac{\partial f}{\partial z} = \frac{1}{2} \left(\frac{\partial c(x + \mathrm{j}\, y)}{\partial x} - \mathrm{j} \frac{\partial c(x + \mathrm{j}\, y)}{\partial y} \right) = \frac{1}{2} \left(c - \mathrm{j} \left(\mathrm{j}\, c \right) \right) = c \,, \qquad \text{(A.2.7)}$$

and

$$\frac{\partial f}{\partial z^*} = \frac{1}{2} \left(\frac{\partial c(x + \mathrm{j}\, y)}{\partial x} + \mathrm{j} \frac{\partial c(x + \mathrm{j}\, y)}{\partial y} \right) = \frac{1}{2} \left(c + \mathrm{j} \left(\mathrm{j}\, c \right) \right) = 0 \,. \qquad \text{(A.2.8)}$$

Similarly, for $f(z) = cz^*$, we arrive at:

$$\frac{\partial f}{\partial z} = \frac{1}{2} \left(\frac{\partial c(x - \mathrm{j}\, y)}{\partial x} - \mathrm{j} \frac{\partial c(x - \mathrm{j}\, y)}{\partial y} \right) = \frac{1}{2} \left(c - \mathrm{j} \left(-\mathrm{j}\, c \right) \right) = 0 \,, \qquad \text{(A.2.9)}$$

and

$$\frac{\partial f}{\partial z^*} = \frac{1}{2}\left(\frac{\partial c(x-\mathrm{j}\,y)}{\partial x} + \mathrm{j}\,\frac{\partial c(x-\mathrm{j}\,y)}{\partial y}\right) = \frac{1}{2}\left(c + \mathrm{j}\,(-\mathrm{j}\,c)\right) = c\,. \tag{A.2.10}$$

Next, we consider the function $f = zz^* = |z|^2 = x^2 + y^2$. Here the derivatives read:

$$\frac{\partial}{\partial z}\,zz^* = \frac{1}{2}\left(\frac{\partial(x^2+y^2)}{\partial x} - \mathrm{j}\,\frac{\partial(x^2+y^2)}{\partial y}\right) = \frac{1}{2}\,(2x - \mathrm{j}\,2y) = z^*\,, \tag{A.2.11}$$

and

$$\frac{\partial}{\partial z^*}\,zz^* = \frac{1}{2}\left(\frac{\partial(x^2+y^2)}{\partial x} + \mathrm{j}\,\frac{\partial(x^2+y^2)}{\partial y}\right) = \frac{1}{2}\,(2x + \mathrm{j}\,2y) = z\,. \tag{A.2.12}$$

To summarize, the correspondences in Table A.1 are valid.

Table A.1 Wirtinger derivatives of some important functions.

$f(z)$	$\frac{\partial f(z)}{\partial z}$	$\frac{\partial f(z)}{\partial z^*}$
cz	c	0
cz^*	0	c
zz^*	z^*	z

Note that using the Wirtinger Calculus differentiation is formally done as with real functions. Moreover, and somewhat unexpected, z^* is formally considered as a constant when derivating with respect to z and vice versa. It is also easy to show that the sum, product, and quotient rules still hold. For example, given $f(z) = f_1(z) \cdot f_2(z)$, we obtain

$$\begin{aligned}
\frac{\partial}{\partial z} f_1(z)\cdot f_2(z) &= \frac{1}{2}\left(\frac{\partial f_1(z)\cdot f_2(z)}{\partial x} - \mathrm{j}\,\frac{\partial f_1(z)\cdot f_2(z)}{\partial y}\right) \\
&= \frac{1}{2}\left(\frac{\partial f_1(z)}{\partial x} f_2(z) + f_1(z)\frac{\partial f_2(z)}{\partial x}\right. \\
&\qquad \left. - \mathrm{j}\,\frac{\partial f_1(z)}{\partial y} f_2(z) - \mathrm{j}\,f_1(z)\frac{\partial f_2(z)}{\partial y}\right) \\
&= \frac{1}{2}\left(\frac{\partial f_1(z)}{\partial x} f_2(z) - \mathrm{j}\,\frac{\partial f_1(z)}{\partial y} f_2(z)\right) \\
&\qquad + \frac{1}{2}\left(f_1(z)\frac{\partial f_2(z)}{\partial x} - \mathrm{j}\,f_1(z)\frac{\partial f_2(z)}{\partial y}\right) \\
&= \frac{\partial f_1(z)}{\partial z} f_2(z) + f_1(z)\frac{\partial f_2(z)}{\partial z}\,.
\end{aligned} \tag{A.2.13}$$

Finally, for $f(z) = h(g(z)) \triangleq h(w)$, $g : \mathbb{C} \mapsto \mathbb{C}$, the following chain rules hold [FL88, Rem89]:

$$\frac{\partial f(z)}{\partial z} = \left.\frac{\partial h(w)}{\partial w}\right|_{w=g(z)} \cdot \frac{\partial g(z)}{\partial z} + \left.\frac{\partial h(w)}{\partial w^*}\right|_{w=g(z)} \cdot \frac{\partial g^*(z)}{\partial z} \tag{A.2.14a}$$

$$\frac{\partial f(z)}{\partial z^*} = \left.\frac{\partial h(w)}{\partial w}\right|_{w=g(z)} \cdot \frac{\partial g(z)}{\partial z^*} + \left.\frac{\partial h(w)}{\partial w^*}\right|_{w=g(z)} \cdot \frac{\partial g^*(z)}{\partial z^*} . \tag{A.2.14b}$$

A.2.2 Discussion

The Wirtinger derivative can be considered to lie inbetween the real derivative of a real function and the complex derivative of a complex function. Rewriting (A.2.5) and (A.2.6), we arrive at:

$$\begin{aligned} \frac{\partial f}{\partial z} &= \frac{1}{2} \cdot \left(\frac{\partial\,(u(x,y) + \mathrm{j}\, v(x,y))}{\partial x} - \mathrm{j}\, \frac{\partial\,(u(x,y) + \mathrm{j}\, v(x,y))}{\partial y} \right) \\ &= \frac{1}{2} \cdot \left(\frac{\partial u(x,y)}{\partial x} + \frac{\partial v(x,y)}{\partial y} + \mathrm{j} \cdot \left[\frac{\partial v(x,y)}{\partial x} - \frac{\partial u(x,y)}{\partial y} \right] \right) , \end{aligned}$$

$$\begin{aligned} \frac{\partial f}{\partial z^*} &= \frac{1}{2} \cdot \left(\frac{\partial\,(u(x,y) + \mathrm{j}\, v(x,y))}{\partial x} + \mathrm{j}\, \frac{\partial\,(u(x,y) + \mathrm{j}\, v(x,y))}{\partial y} \right) \\ &= \frac{1}{2} \cdot \left(\frac{\partial u(x,y)}{\partial x} - \frac{\partial v(x,y)}{\partial y} + \mathrm{j} \cdot \left[\frac{\partial v(x,y)}{\partial x} + \frac{\partial u(x,y)}{\partial y} \right] \right) . \end{aligned}$$

In the special case, when $f(z)$ is holomorphic, the Cauchy-Riemann equations (A.1.6) hold, and we obtain:

$$\frac{\partial f}{\partial z} = \frac{\partial u(x,y)}{\partial x} + \mathrm{j}\, \frac{\partial v(x,y)}{\partial x} \tag{A.2.15a}$$

$$\frac{\partial f}{\partial z^*} \equiv 0 \tag{A.2.15b}$$

On the one hand, equation (A.2.15a) states that for holomorphic functions the Wirtinger derivative with respect to z agrees with the ordinary derivative of a complex function (cf. (A.1.7)). On the other hand, (A.2.15b) can be interpreted in the way that holomorphic functions do not formally depend on z^*.

Contrary to the usual complex derivative, the Wirtinger derivative exists for all functions, in particular nonholomorphic ones, such as real functions. Since both operators $\frac{\partial}{\partial z}$ and $\frac{\partial}{\partial z^*}$ are merely a compact notation incorporating two real differential quotients, they can be applied to arbitrary functions of complex variables. For nonholomorphic functions, $\frac{\partial f}{\partial z^*} \neq 0$ usually holds, and thus either the derivative with respect to z or z^* can be used for optimization. The actual cost functions determines which one is more advantageous; if quadratic forms are considered, the operator $\frac{\partial f}{\partial z^*}$ is preferable.

To summarize, it should again be emphasized, that, because of its compact notation, Wirtinger Calculus is very well suited for optimization in engineering. It circumvents a separate inspection of real part and imaginary part of the cost function. Because of the simple arithmetic rules—mostly it can be calculated as known from real functions—the Wirtinger Calculus is very clear.

A.3 GRADIENTS

For the majority of applications the cost function does not only depend on one, but on many variables, e.g., we have $f : \mathbb{C}^n \ni z = [z_1, z_2, \ldots, z_n]^{\mathsf{T}} \mapsto w = f(z) \in \mathbb{R}$. For optimization, all n partial derivatives with respect to the complex variables $z_1 = x_1 + \mathrm{j}\, y_1$ through $z_n = x_n + \mathrm{j}\, y_n$ have to be calculated. Usually, these derivatives are again combined into a vector, [2] the so called *gradient*:

$$\frac{\partial f}{\partial z} \triangleq \begin{bmatrix} \frac{\partial f}{\partial z_1} \\ \frac{\partial f}{\partial z_2} \\ \vdots \\ \frac{\partial f}{\partial z_n} \end{bmatrix} \quad \text{and} \quad \frac{\partial f}{\partial z^*} \triangleq \begin{bmatrix} \frac{\partial f}{\partial z_1^*} \\ \frac{\partial f}{\partial z_2^*} \\ \vdots \\ \frac{\partial f}{\partial z_n^*} \end{bmatrix}, \tag{A.3.1}$$

which, in the optimum, has to equal the zero vector $\mathbf{0} \triangleq [0, 0, \ldots, 0]^{\mathsf{T}}$. Wirtinger Calculus is especially well suited for such multidimensional functions, because here only with a great effort can the real part and the imaginary part be separated and inspected independently. Using the above definitions of the partial derivatives ((A.2.5) and (A.2.6)), we arrive at simple arithmetic rules, now expressed using vectors and matrices.

A.3.1 Examples

We now again study some important examples. First, let $f(z) = c^{\mathsf{T}} z = \sum_{i=1}^{n} c_i z_i$ or $f(z) = c^{\mathsf{T}} z^* = \sum_{i=1}^{n} c_i z_i^*$, respectively, with $c = [c_1, c_2, \ldots, c_n]^{\mathsf{T}}$ and c_i constant. It is easy to prove the following properties for gradients:

$$\begin{aligned} \frac{\partial}{\partial z}\, c^{\mathsf{T}} z &= c\,, & \frac{\partial}{\partial z}\, c^{\mathsf{T}} z^* &= \mathbf{0}\,, \\ \frac{\partial}{\partial z^*}\, c^{\mathsf{T}} z &= \mathbf{0}\,, & \frac{\partial}{\partial z^*}\, c^{\mathsf{T}} z^* &= c\,. \end{aligned} \tag{A.3.2}$$

[2] Here we use column vectors, but the same considerations also apply to row vectors.

Finally, considering the quadratic form $f(\boldsymbol{z}) = \boldsymbol{z}^{\mathsf{H}}\mathbf{M}\boldsymbol{z}$, where $\mathbf{M}$ is a constant $n \times n$ matrix, derivation results in:

$$\frac{\partial}{\partial \boldsymbol{z}}\, \boldsymbol{z}^{\mathsf{H}}\mathbf{M}\boldsymbol{z} = \left(\boldsymbol{z}^{\mathsf{H}}\mathbf{M}\right)^{\mathsf{T}}, \qquad \frac{\partial}{\partial \boldsymbol{z}^*}\, \boldsymbol{z}^{\mathsf{H}}\mathbf{M}\boldsymbol{z} = \mathbf{M}\boldsymbol{z}, \tag{A.3.3}$$

and for $f(\boldsymbol{z}) = \boldsymbol{z}^{\mathsf{H}}\boldsymbol{z}$ we arrive at:

$$\frac{\partial}{\partial \boldsymbol{z}}\, \boldsymbol{z}^{\mathsf{H}}\boldsymbol{z} = \boldsymbol{z}^*, \qquad \frac{\partial}{\partial \boldsymbol{z}^*}\, \boldsymbol{z}^{\mathsf{H}}\boldsymbol{z} = \boldsymbol{z}. \tag{A.3.4}$$

To summarize, the correspondences in Table A.2 are valid.

Table A.2 Wirtinger derivatives (gradients) of some important functions.

$f(\boldsymbol{z})$	$\dfrac{\partial f(\boldsymbol{z})}{\partial \boldsymbol{z}}$	$\dfrac{\partial f(\boldsymbol{z})}{\partial \boldsymbol{z}^*}$
$\mathbf{c}^{\mathsf{T}}\boldsymbol{z} = \boldsymbol{z}^{\mathsf{T}}\mathbf{c}$	$\mathbf{c}$	$\mathbf{0}$
$\mathbf{c}^{\mathsf{T}}\boldsymbol{z}^* = \boldsymbol{z}^{\mathsf{H}}\mathbf{c}$	$\mathbf{0}$	$\mathbf{c}$
$\boldsymbol{z}^{\mathsf{H}}\boldsymbol{z} = \boldsymbol{z}^{\mathsf{T}}\boldsymbol{z}^*$	$\boldsymbol{z}^*$	$\boldsymbol{z}$
$\boldsymbol{z}^{\mathsf{H}}\mathbf{M}\boldsymbol{z} = \boldsymbol{z}^{\mathsf{T}}\mathbf{M}^{\mathsf{T}}\boldsymbol{z}^*$	$\mathbf{M}^{\mathsf{T}}\boldsymbol{z}^*$	$\mathbf{M}\boldsymbol{z}$

A.3.2 Discussion

The gradient with respect to the Wirtinger derivatives, is related to the gradient

$$\nabla = \begin{bmatrix} \frac{\partial}{\partial x_1} + \mathrm{j}\frac{\partial}{\partial y_1} \\ \frac{\partial}{\partial x_2} + \mathrm{j}\frac{\partial}{\partial y_2} \\ \vdots \\ \frac{\partial}{\partial x_n} + \mathrm{j}\frac{\partial}{\partial y_n} \end{bmatrix}, \tag{A.3.5}$$

which is frequently used, e.g., in [Hay96] by

$$\nabla f(\boldsymbol{z}) = 2\frac{\partial f(\boldsymbol{z})}{\partial \boldsymbol{z}^*}, \tag{A.3.6}$$

or since $f(\boldsymbol{z})$ is real-valued,

$$(\nabla f(\boldsymbol{z}))^* = 2\frac{\partial f(\boldsymbol{z})}{\partial \boldsymbol{z}}. \tag{A.3.7}$$

The first disadvantage of definition (A.3.6) compared to Wirtinger Calculus is that an undesired factor of 2 occurs. In particular, if the chain rule is applied l times, the result is artificially multiplied by 2^l. Second, the calculus is not very elegant, because the gradient of, e.g., $f(\boldsymbol{z}) = \mathbf{c}^{\mathsf{T}}\boldsymbol{z}$ is $\nabla(\mathbf{c}^{\mathsf{T}}\boldsymbol{z}) = 0$, but that of $f(\boldsymbol{z}) = \mathbf{c}^{\mathsf{T}}\boldsymbol{z}^*$ calculates to $\nabla(\mathbf{c}^{\mathsf{T}}\boldsymbol{z}^*) = 2\boldsymbol{z}$. Hence, because of its much clearer arithmetic rules, we exclusively apply the Wirtinger Calculus.

REFERENCES

[FL88] W. Fischer and I. Lieb. *Funktionentheorie.* Vieweg-Verlag, Braunschweig, Germany, 1988. (In German.)

[Fra69] L. E. Franks. *Signal Theory.* Prentice-Hall, Inc., Englewood Cliffs, NJ, 1969.

[Hay96] S. Haykin. *Adaptive Filter Theory.* Prentice-Hall, Inc., Englewood Cliffs, NJ, 3rd edition, 1996.

[Pro01] J. G. Proakis. *Digital Communications.* McGraw-Hill, New York, 4th edition, 2001.

[Rem89] R. Remmert. *Funktionentheorie I.* Springer Verlag, Berlin, Heidelberg, 1989. (In German.)

[Tre71] H. L. van Trees. *Detection, Estimation, and Modulation Theory—Part III: Radar-Sonar Signal Processing and Gaussian Signals in Noise.* John Wiley & Sons, Inc., New York, 1971.

Appendix B Parameters of the Numerical Examples

For the numerical examples in this book, we consider typical digital subscriber line scenarios. The background of most of the examples is the emerging *Eropean Telecommunications Standards Institute (ETSI)* standard for fast digital transmission over a single twisted-pair line, called *Single-Pair Digital Subscriber Lines (SDSL)*. As a second example, we consider the single-carrier variant of *Asymmetric Digital Subscriber Lines (ADSL)*.

B.1 FUNDAMENTALS OF DIGITAL SUBSCRIBER LINES

In order to obtain insight into the basic principles and phenomena, we use two basic and simplified transmission models. Figure B.1 shows the scenario which is studied.

Transmission takes place between a *central office (CO)*, where the *local loops* of a number of subscribers terminate and are connected to switching equipment and the customer premises. The DSL modems installed at the CO are called *line termination (LT)*, and this end of the loop is denominated *LT side*. At the CO a

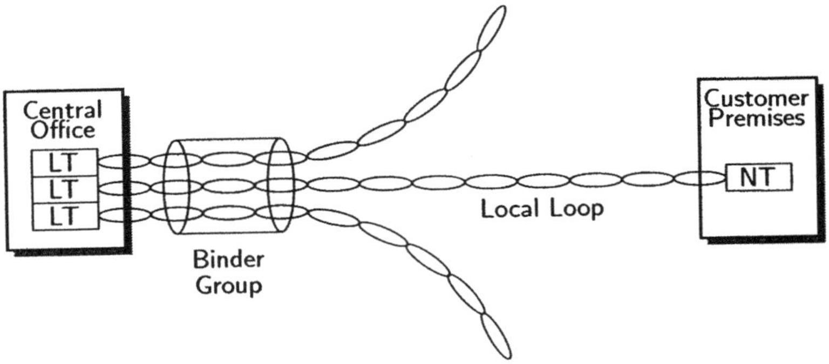

Fig. B.1 Transmission scenario for the numerical examples.

lot of pairs are installed in parallel, typically even in the same binder group of a multipair cable. Hence, at the LT side, *near-end crosstalk (NEXT)* of other users is the dominating impairment. In order to simplify, we assume only the NEXT of other, identical systems (self NEXT). This assumption is justified if the penetration of the considered system is high compared to other services. In practice, a "technology mix" which reflects the actual installation of various DSL systems (e.g., ADSL, HDSL, ISDN) has to be used.

The other end of the local loop terminates at the customer premises. Here, the modem is called *network termination (NT)*; the denomination *NT side* is often used. Typically, each subscriber has only a single pair, and the pairs of the binder groups are dispersed over the field length. Hence, we can consider thermal white noise to be the dominating impairment at the NT side. For a short loop length, *far-end crosstalk (FEXT)* has to be taken into account as well. Here we omit this source of noise for simplicity.

In summary, we deal with two situations: the down-stream (LT $\rightarrow$ NT) direction is assumed to be corrupted by white noise, whereas NEXT of identical systems is considered for the up-stream (NT $\rightarrow$ LT).

For numerical evaluation, we use twisted-pair lines of Deutsche Telekom AG with line diameter 0.4 mm (close to 26 AWG). The analytic expressions for attenuation and phase distortion have been determined through a broad measurement campaign and are documented in [PW95]. Measuring the field length ℓ in kilometers [km] and the frequency in Hertz [Hz], the line transfer function reads:

$$H_{\mathrm{C}}(f,\,\ell) = 10^{-\ell\cdot\alpha(f)/20}\cdot \mathrm{e}^{-\mathrm{j}\,\ell\cdot\beta(f)} \tag{B.1.1}$$

for $f \geq 0$ and $H_{\mathrm{C}}(-f,\,\ell) = H_{\mathrm{C}}^{*}(f,\,\ell)$. Line attenuation $\alpha(f)$ in $\left[\frac{\mathrm{dB}}{\mathrm{km}}\right]$ and phase $\beta(f)$ in $\left[\frac{\mathrm{rad}}{\mathrm{km}}\right]$ are given as

$$\alpha(f) = \begin{cases} 6.9 + 13.4\left(\frac{f}{1\,\mathrm{MHz}}\right)^{0.99}, & 0\,\mathrm{Hz} \leq f \leq 500\,\mathrm{kHz} \\ 0.3 + 18.9\left(\frac{f}{1\,\mathrm{MHz}}\right)^{0.50}, & 500\,\mathrm{kHz} < f \leq 5\,\mathrm{MHz} \\ 10.4 + 11.5\left(\frac{f}{1\,\mathrm{MHz}}\right)^{0.64}, & 5\,\mathrm{MHz} < f \leq 30\,\mathrm{MHz} \end{cases}, \tag{B.1.2a}$$

$$\beta(f) = 32.9\left(\frac{f}{1\ \text{MHz}}\right) + 2.26\sqrt{\frac{f}{1\ \text{MHz}}}\ . \tag{B.1.2b}$$

For sufficiently long lines, the NEXT transfer function $H_X(f)$ is independent of the cable length ℓ, and we use the commonly accepted approximation [Wel89], [ANSI98, Annex B]

$$|H_X(f)|^2 = 0.8536 \cdot 10^{-14} \cdot \left(\frac{|f|}{1\ \text{Hz}}\right)^{1.5} . \tag{B.1.3}$$

The given constant is valid for a single NEXT disturber. If more NEXT systems are present, their sum PSD has to be calculated appropriately. Because of the different distances between the pairs within the binder group, the PSD of a single system is approximately scaled by $N^{0.6}$ when N identical NEXT disturbers are active. If, e.g., the whole binder group (50 pairs) is used for DSL transmission, the NEXT PSD has to be increased by a factor of $49^{0.6} = 10.33 \mathrel{\hat{=}} 10.14$ dB. This situation is assumed for the numerical results.

B.2 SINGLE-PAIR DIGITAL SUBSCRIBER LINES

Single-pair digital subscriber lines (SDSL) are intended to provide symmetric data rates from the central office to the customer and back over only a single wire pair. Symmetric transmission rates up to 2.312 Mbits/s shall be supported.

For SDSL it is agreed (cf. also the ITU-T recommendation G.991.2 [ITU00]) to use baseband pulse amplitude modulation (one-dimensional ASK signal set) with 3 information bits per symbol, i.e., 8-ary ASK for uncoded transmission and 16-ary ASK when trellis-coded modulation with one redundant bit per symbol is applied. Hence, assuming the maximum data rate, the symbol duration calculates to

$$T = \frac{3\ \text{bits}}{2.312\ \text{Mbits/s}} = 1.298\ \mu\text{s}\ , \tag{B.2.1}$$

or the symbol rate is $1/T = 770.66$ kHz. This rate implies that the Nyquist frequency in the examples is 385.33 kHz.

Pulse shaping is done as specified in ITU-T recommendation G.991.2 [ITU00]. A rectangular pulse equal to the symbol duration $T = 3/(2.312\ \text{Mbit/s}) = 1.3\ \mu$s is filtered with a 6th-order Butterworth filter with 3 dB cutoff frequency at half the symbol rate of $1/T = 770.66$ kHz. The nominal transmit power is fixed to 14.5 dBm (28.18 mW).

For the transmit filter we resort to the one specified in Annex B.4 of ITU-T recommendation G.991.2 [ITU00]. The pulse shape is obtained by filtering a rectangular pulse of the same duration as the symbol period T with a 6th-order Butterworth low-pass filter with 3-dB cutoff frequency at Nyquist frequency $1/(2T)$ (half the symbol rate of $1/T = 770.66$ kHz). Thus, mathematically the squared magnitude (two-sided spectrum) of the transmit filter is given by

$$|H_T(f)|^2 = 0.0366 \cdot \frac{\sin(\pi f T)}{\pi f T} \cdot \frac{1}{1 + (2fT)^{12}} . \tag{B.2.2}$$

Using this transfer function, the average transmit PSD can be calculated as

$$\bar{\Phi}_{ss}(f) = \sigma_a^2 \cdot T \cdot |H_{\mathrm{T}}(f)|^2 \,. \tag{B.2.3}$$

The constant in (B.2.2) is chosen such that for a white data sequence with variance $\sigma_a^2 = 1$s the transmit power into 135 Ω would equal 14.5 dBm (28.18 mW) [ITU00].

Note that when dealing with baseband transmission, the noise power spectral density N_0 has to be replaced by $N_0/2$ in all formulas, since only one quadrature component is present.

B.3 ASYMMETRIC DIGITAL SUBSCRIBER LINES

Contrary to SDSL, asymmetric digital subscriber lines (ADSL) support high data rates from the central office to the customer, but only a low-rate channel is offered in reverse direction. Here we concentrate on the high-rate down-stream direction, assuming the E1 data rate of 2.048 Mbits/s.

The *American National Standards Institute (ANSI)* and the *European Telecommunications Standards Institute (ETSI)* have selected *discrete multitone (DMT)*, an efficient multicarrier modulation technique, to be standard. As an alternative to DMT, *carrierless AM/PM (CAP)* [Wer93, IW95] is under discussion for ADSL (cf. the ANSI draft on rate-adaptive ADSL [ANSI97]). CAP is a variant of QAM, which omits an explicit mixing with the carrier. Basically, CAP and QAM are equivalent in performance, and their descriptions in the equivalent complex baseband [Fra80, Tre71, Pro01] coincide.

For the numerical examples, we assume 5 information bits to be mapped onto the two-dimensional (QAM) signal constellation. In uncoded transmission either a 32-ary cross constellation, or a rotated square constellation is used, whereas when trellis-coded modulation with one redundant bit per complex symbol is applied, a 64-ary square QAM signal set is chosen. Here, the symbol duration calculates to $T = \frac{5\ \text{bits}}{2.048\ \text{Mbits/s}} = 2.441\ \mu\text{s}$, and the symbol rate is $\frac{1}{T} = 409.6$ kHz.

Since ADSL has to coexist on the same line with already installed telephone service, only the frequency band above approximately 150 kHz can be used. Assuming a bandwidth-excess factor α, and fixing the start frequency to 150 kHz, the carrier (center) frequency f_c reads

$$f_c = 150\ \text{kHz} + \frac{1+\alpha}{2T} \,. \tag{B.3.1}$$

If we choose the excess factor to be $\alpha = 0.3$ in our examples, the carrier frequency calculates to 416.24 kHz. Because of the low-pass characteristics of the line, the lowest possible carrier frequency is preferable.

Finally, the transmit pulse is assumed to be a square-root raised cosine filter. In the equivalent complex baseband, the squared magnitude of the transmit filter then reads (e.g., [Pro01])

$$|H_{\mathrm{T}}(f)|^2 = \mathrm{const} \cdot \begin{cases} 1\,, & |f| < \frac{1-\alpha}{2T} \\ \frac{1}{2}\left(1 - \sin\left(\frac{\pi T}{\alpha}(|f| - \frac{1}{2T})\right)\right)\,, & \frac{1-\alpha}{2T} \le |f| \le \frac{1+\alpha}{2T} \\ 0\,, & \text{else} \end{cases} . \tag{B.3.2}$$

The transmit power can be adjusted appropriately via the constant.

REFERENCES

[ANSI97] American National Standards Institute (ANSI). Rate Adaptive Asymmetric Digital Subscriber Line (RADSL) Metallic Interface. Draft American National Standard for Telecommunications, September 1997.

[ANSI98] American National Standards Institute (ANSI). Network and Customer Installation Interfaces — Asymmetric Digital Subscriber Line (ADSL) Metallic Interface. Draft American National Standard for Telecommunications, December 1998.

[Fra80] L. E. Franks. Carrier and Bit Synchronization in Data Communication — A Tutorial Review. *IEEE Transactions on Communications*, COM-28, pp. 1107–1121, August 1980.

[ITU00] International Telecommunication Union (ITU). Draft Recommendation G.991.2: Single-Pair High-Speed Digital Subscriber Line (SHDSL) Transceivers — Draft G.991.2, April 2000.

[IW95] G.-H. Im and J. J. Werner. Bandwidth-Efficient Digital Transmission over Unshielded Twisted-Pair Wiring. *IEEE Journal on Selected Areas in Communications*, JSAC-13, pp. 1643–1655, December 1995.

[Pro01] J. G. Proakis. *Digital Communications*. McGraw-Hill, New York, 4th edition, 2001.

[PW95] M. Pollakowski and H.-W. Wellhausen. Eigenschaften symmetrischer Ortsanschlußkabel im Frequenzbereich bis 30 MHz. *Der Fernmelde–Ingenieur*, 49. Jahrgang, pp. 1–58, September/October 1995. (In German.)

[Tre71] H. L. van Trees. *Detection, Estimation, and Modulation Theory—Part III: Radar-Sonar Signal Processing and Gaussian Signals in Noise*. John Wiley & Sons, Inc., New York, 1971.

[Wel89] H.-W. Wellhausen. Eigenschaften symmetrischer Kabel der Ortsnetze und generelle Übertragungsmöglichkeiten. *Der Fernmelde–Ingenieur*, 43. Jahrgang, pp. 1–51, October/November 1989. (In German.)

[Wer93] J. J. Werner. Tutorial on Carrierless AM/PM — Part I: Fundamentals and Digital CAP Transmitter; Part II: Performance of Bandwith-Efficient Line Codes. AT&T Bell Laboratories, Middletown, NJ, 1992/1993.

Appendix C
Introduction to Lattices

In practice, signal constellations form a regular grid. For analysis of the transmission properties of (large) signal sets, the boundary effects are often disregarded and an infinite set of points is assumed. A powerful tool for that is the theory of *lattices*, regular arrays of points in N-dimensional space. In this Appendix, the basics of lattices are discussed and the fundamental concepts are introduced. Much more detailed discussion can be found, e.g., in [CS88, For88a, For88b, FW89, For89]. Here we focus mainly on the application of lattices to digital transmission schemes.

C.1 DEFINITION OF LATTICES

Consider an infinite, discrete set of vectors (points) $\boldsymbol{\lambda}$ in Euclidean space $\mathbb{R}^N$, i.e.,

$$\Lambda = \{\, \boldsymbol{\lambda} \,\} \subset \mathbb{R}^N \quad \text{with} \quad \boldsymbol{\lambda} = \begin{bmatrix} l_1 \\ l_2 \\ \vdots \\ l_N \end{bmatrix} \in \mathbb{R}^N \,. \tag{C.1.1}$$

Geometrically, $\boldsymbol{\Lambda}$ forms an *N-dimensional lattice* if it constitutes a regular arrangement of an infinite number of points in N dimensions.[1] Because of the underlying Euclidean space, we can endow a lattice with attributes such as distances, volumes or shapes.

Regarding its algebraic properties, a lattice is a (*Abelian*) *group* under ordinary vector addition in $\mathbb{R}^N$. Mathematically, for $\boldsymbol{\Lambda}$ beeing a *lattice*, the following requirements have to be met:

G1:	Closure:	$\boldsymbol{\lambda}_i + \boldsymbol{\lambda}_j \in \boldsymbol{\Lambda}, \ \forall \boldsymbol{\lambda}_i, \boldsymbol{\lambda}_j \in \boldsymbol{\Lambda}$
G2:	Associativity:	$(\boldsymbol{\lambda}_i + \boldsymbol{\lambda}_j) + \boldsymbol{\lambda}_k = \boldsymbol{\lambda}_i + (\boldsymbol{\lambda}_j + \boldsymbol{\lambda}_k)$
G3:	Identity:	$\exists\, \mathbf{0} \in \boldsymbol{\Lambda}$ with $\boldsymbol{\lambda}_i + \mathbf{0} = \mathbf{0} + \boldsymbol{\lambda}_i = \boldsymbol{\lambda}_i, \ \forall \boldsymbol{\lambda}_i \in \boldsymbol{\Lambda}$
G4:	Inverse:	$\exists -\boldsymbol{\lambda}_i \in \boldsymbol{\Lambda}$ with $\boldsymbol{\lambda}_i + (-\boldsymbol{\lambda}_i) = \mathbf{0}, \ \forall \boldsymbol{\lambda}_i \in \boldsymbol{\Lambda}$
G5:	Commutativity:	$\boldsymbol{\lambda}_i + \boldsymbol{\lambda}_j = \boldsymbol{\lambda}_j + \boldsymbol{\lambda}_i, \ \forall \boldsymbol{\lambda}_i, \boldsymbol{\lambda}_j \in \boldsymbol{\Lambda}$

Given a set of N linear independent vectors $\boldsymbol{b}_i$, because of the group structure, each lattice point $\boldsymbol{\lambda} \in \boldsymbol{\Lambda}$ can be expressed as a *linear combination* of integer multiples of these *basis vectors* $\boldsymbol{b}_i$

$$\boldsymbol{\lambda} = \sum_{i=1}^{N} k_i \boldsymbol{b}_i \,, \qquad k_i \in \mathbb{Z} \,. \tag{C.1.2}$$

The basis vectors have to be linearly independent and to span the N-space.[2] Note that the choice of the basis is not unique. It is common to combine the basis vectors $\boldsymbol{b}_i$ as columns into the *generator matrix*

$$\boldsymbol{B} = [\,\boldsymbol{b}_1,\, \boldsymbol{b}_2,\, \cdots,\, \boldsymbol{b}_N\,] \,. \tag{C.1.3}$$

With it, the lattice can be specified as

$$\boldsymbol{\Lambda} = \left\{ \boldsymbol{B} \begin{bmatrix} k_1 \\ \vdots \\ k_N \end{bmatrix} \;\middle|\; k_i \in \mathbb{Z},\, i = 1, 2, \ldots, N \right\} \triangleq \boldsymbol{B}\mathbb{Z}^N \,. \tag{C.1.4}$$

Note, the generator matrix is not unique; any integer linear combination of the basis vectors, which preserves the full rank of $\boldsymbol{B}$, can be used as well. The right hand side expression of (C.1.4)—its meaning should be intuitively clear—is defined for convenience. In addition, for the combined *scaling* and *translation* operation, we write briefly, in terms of sets

$$a\boldsymbol{\Lambda} + \boldsymbol{b} \triangleq \left\{ a\boldsymbol{\lambda} + \boldsymbol{b} \;\middle|\; \boldsymbol{\lambda} \in \boldsymbol{\Lambda} \right\}, \qquad \text{with} \quad a \in \mathbb{R},\, \boldsymbol{b} \in \mathbb{R}^N \,. \tag{C.1.5}$$

[1]Here, we only treat *real lattices*, i.e., points taken from $\mathbb{R}^N$. Generalizations, e.g., by using complex vectors instead of real vectors, are not considered, since we do not require them for our applications.

[2]We assume that an N-dimensional lattice spans N dimensions and does not degenerate to a smaller number of dimensions, i.e., a subspace.

A *fundamental region* $\mathcal{R}(\boldsymbol{\Lambda}) \subseteq \mathbb{R}^N$ of a lattice $\boldsymbol{\Lambda}$ (i) includes one and only one point of $\boldsymbol{\Lambda}$, and (ii) when shifting it to any lattice point, i.e., considering $\mathcal{R}(\boldsymbol{\Lambda}) + \boldsymbol{\lambda}$, and when $\boldsymbol{\lambda}$ ranges through all points in $\boldsymbol{\Lambda}$, the whole real N-space $\mathbb{R}^N$ is tiled. Thus, a fundamental region is a building block for the entire N-space, and in the notation of sets, we may write

$$\mathbb{R}^N = \mathcal{R}(\boldsymbol{\Lambda}) + \boldsymbol{\Lambda} . \tag{C.1.6}$$

Here we define an "addition" of sets as follows

$$\mathcal{A} + \mathcal{B} \triangleq \left\{ \boldsymbol{a} + \boldsymbol{b} \mid \boldsymbol{a} \in \mathcal{A},\, \boldsymbol{b} \in \mathcal{B} \right\} . \tag{C.1.7}$$

Note that the basis vectors $\boldsymbol{b}_i$ span the *fundamental parallelepiped*

$$\mathcal{P} = \left\{ \boldsymbol{B} \begin{bmatrix} r_1 \\ \vdots \\ r_N \end{bmatrix} \mid r_i \in \mathbb{R},\, 0 \leq r_i < 1,\, i = 1, 2, \ldots, N \right\} \triangleq \boldsymbol{B}\,[0,\, 1)^N , \tag{C.1.8}$$

which is a particular fundamental region of the lattice.

The concept of *arithmetic modulo* $\boldsymbol{\Lambda}$ is closely connected to the fundamental region. Two points $\boldsymbol{\lambda}_1$ and $\boldsymbol{\lambda}_2$ are called *equivalent modulo* $\boldsymbol{\Lambda}$, if their difference is a lattice point, i.e., $\boldsymbol{\lambda}_1 - \boldsymbol{\lambda}_2 \in \boldsymbol{\Lambda}$. We define a *modulo reduction* with respect to $\boldsymbol{\Lambda}$ as that point out of all equivalent points which lies in the fundamental region containing the origin

$$\boldsymbol{x} \bmod \boldsymbol{\Lambda} \triangleq \boldsymbol{x} + \boldsymbol{\lambda} , \quad \text{with} \quad \boldsymbol{\lambda} \in \boldsymbol{\Lambda} \quad \text{so that} \quad \boldsymbol{x} + \boldsymbol{\lambda} \in \mathcal{R}(\boldsymbol{\Lambda}) . \tag{C.1.9}$$

The *Voronoi region* $\mathcal{R}_{\mathrm{V}}(\boldsymbol{\Lambda})$ of a lattice $\boldsymbol{\Lambda}$ is that fundamental region in which every point is closer to the origin than to any other lattice point. Because we deal with Euclidean N-space, distances are measured as the Euclidean norm of the difference vector. Hence, the Voronoi region is a special fundamental region and contains only the minimum weight (= distance from the origin) point of equivalent points (ties are resolved arbitrarily). In the literature, the Voronoi region is sometimes defined with respect to a particular point (e.g., [CS88]). Because of the regular arrangement of the lattice points, all these regions are congruent and we only look at that region—and call it Voronoi region—containing the origin.

Note that the set $\mathcal{R}_{\mathrm{V}}(\boldsymbol{\Lambda}) + \boldsymbol{\Lambda}$ of translates of the Voronoi region constitute the decision regions for a minimum-distance decoder.

Example C.1: *Lattice, Basis, Fundamental Region, Voronoi Region*

Lattice $\mathbb{Z}$ In one dimension, the set $\mathbb{Z}$ of integers forms a lattice. The generator matrix is simply $\boldsymbol{B} = 1$, and the fundamental parallelepiped is the interval $[0,\, 1)$. Calculation modulo $\mathbb{Z}$ gives the "fractional part." The Voronoi region is the interval $\mathcal{R}_{\mathrm{V}}(\mathbb{Z}) = (-0.5,\, 0.5]$, where we have arbitrarily chosen the point 0.5 to be included. Figure C.1 depicts a portion of the lattice. Note that in all subsequent figures, only the relevant part of the lattice is shown, as a lattice always contains an infinite number of points.

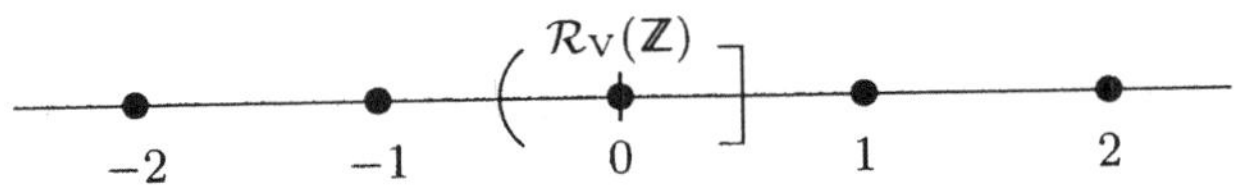

Fig. C.1 The lattice $\mathbb{Z}$ and its Voronoi region $\mathcal{R}_V(\mathbb{Z})$.

Lattice $\mathbb{Z}^N$ Taking the N-fold Cartesian product of the lattice $\mathbb{Z}$ results in the lattice $\mathbb{Z}^N$. The generator matrix is the N-dimensional identity matrix $\boldsymbol{I}$

$$\boldsymbol{B} = \boldsymbol{I} = \begin{bmatrix} 1 & & 0 \\ & \ddots & \\ 0 & & 1 \end{bmatrix},$$

and the fundamental parallelepiped is the hypercube $[0,\ 1)^N$. There are many more choices for a fundamental region in N-dimensions compared to the one-dimensional case, where it is unique. Figure C.2 shows the lattice $\mathbb{Z}^2$, a possible fundamental region, and the Voronoi region $(-0.5,\ 0.5]^2$.

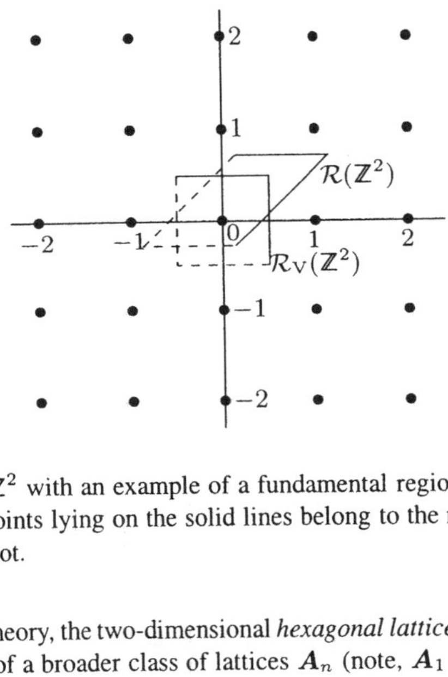

Fig. C.2 The lattice $\mathbb{Z}^2$ with an example of a fundamental region $\mathcal{R}(\mathbb{Z}^2)$ and the Voronoi region $\mathcal{R}_V(\mathbb{Z}^2)$. The points lying on the solid lines belong to the region, whereas the points on the dashed lines do not.

Lattice A_2 In lattice theory, the two-dimensional *hexagonal lattice* is denoted as $\boldsymbol{A}_2$, because it is one representative of a broader class of lattices $\boldsymbol{A}_n$ (note, $\boldsymbol{A}_1 = \mathbb{Z}$) [CS88]. A possible generator matrix is

$$\boldsymbol{B} = \begin{bmatrix} 1 & 1/2 \\ 0 & \sqrt{3}/2 \end{bmatrix}, \tag{C.1.10}$$

i.e., the basis vectors are chosen to be $\boldsymbol{b}_1 = [1, 0]^\mathsf{T}$ and $\boldsymbol{b}_2 = [1/2, \sqrt{3}/2]^\mathsf{T}$. The lattice $\boldsymbol{A}_2$ is shown in Figure C.3 and the relevant quantities are displayed. Note, the Voronoi region is a hexagon, which names the lattice.

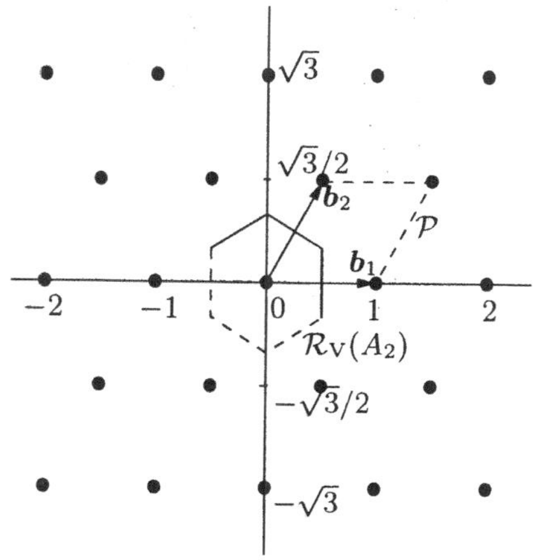

Fig. C.3 The lattice A_2 with its fundamental parallelepiped and Voronoi region $\mathcal{R}_V(A_2)$, and the basis vectors. The points lying on the solid lines belong to the region, whereas the points on the dashed lines do not.

C.2 SOME IMPORTANT PARAMETERS OF LATTICES

Now, we will discuss some important parameters for dealing with lattices.

One of the primary parameters of a lattice is the *minimum squared distance* $d^2_{\min}(\Lambda)$ between any two points from Λ. Mathematically, because of the group properties of the lattice, $d^2_{\min}(\Lambda)$ is identical to the minimum, nonzero weight of the lattice points ($|\lambda|$: Euclidean norm of λ)

$$d^2_{\min}(\Lambda) \triangleq \min_{\substack{\lambda_i,\lambda_j\in\Lambda \\ \lambda_i\neq\lambda_j}} |\lambda_i - \lambda_j|^2 = \min_{\lambda_i\in\Lambda\setminus\{0\}} |\lambda_i|^2 \,. \tag{C.2.1}$$

The ***channel coding problem*** [CS88] of lattice theory is to find lattices which, under certain additional constraints, maximize the minimum squared distance.

A *volume (fundamental volume)* $V(\Lambda)$, which is defined as the volume of a fundamental region, is associated with each lattice. Note that, in contrast to its shape, the volume of a fundamental region is unique, and hence is also equal to the volume of the Voronoi region. Using the generator matrix $\boldsymbol{B}$, from basic geometry, the volume $V(\Lambda)$ is given as

$$V(\Lambda) \triangleq |\det(\boldsymbol{B})| = \sqrt{\det(\boldsymbol{B}^{\mathsf{T}}\boldsymbol{B})} \,. \tag{C.2.2}$$

Sometimes, an equivalent quantity, the *determinant of* $\mathbf{\Lambda}$ is used

$$\det\mathbf{\Lambda} \triangleq \det(\boldsymbol{B}^{\mathsf{T}}\boldsymbol{B}) = V^2(\mathbf{\Lambda}) \,. \tag{C.2.3}$$

Using the minimum squared distance, the *packing radius* $\rho(\mathbf{\Lambda})$ of the lattice $\mathbf{\Lambda}$ is defined as

$$\rho(\mathbf{\Lambda}) \triangleq \frac{1}{2}\sqrt{d^2_{\min}(\mathbf{\Lambda})} \,. \tag{C.2.4}$$

The interpretation of this quantity is as follows: spheres centered at the lattice points can have a maximum radius $\rho(\mathbf{\Lambda})$ without intersecting each other. Regarding the Voronoi region, $\rho(\mathbf{\Lambda})$ is the *inradius*, i.e., the radius of the largest sphere, which can be inscribed into the Voronoi region.

The *density* $\Delta(\mathbf{\Lambda})$ of a lattice $\mathbf{\Lambda}$ is the portion of N-space that is occupied by the above-mentioned spheres

$$\Delta(\mathbf{\Lambda}) \triangleq \frac{\text{Volume of one } N\text{-dim. sphere with radius } \rho(\mathbf{\Lambda})}{V(\mathbf{\Lambda})} \,. \tag{C.2.5}$$

The ***packing problem*** [CS88] is to find that lattice which has maximum density.

Counting the number of lattice points which are at minimum distance from a fixed lattice point gives the *kissing number* $\tau(\mathbf{\Lambda})$ (the terminology is borrowed from the billiards). Because the lattice is well structured, the kissing number can be calculated as

$$\tau(\mathbf{\Lambda}) \triangleq \left|\{\boldsymbol{\lambda} \in \mathbf{\Lambda} \mid |\boldsymbol{\lambda}|^2 = d^2_{\min}(\mathbf{\Lambda})\}\right| \,, \tag{C.2.6}$$

and $|\cdot|$ denotes cardinality if sets are considered. In channel coding, the kissing number is usually known as *number of nearest neighbors*.

Closely related to the packing problem is the question known as ***kissing number problem*** [CS88], dating back (at least in low dimensions) some hundred years [CS88, page 21]: How many equal-sized spheres can be arranged around another sphere, so that they all touch it. Here, the spheres are assumed to be rigid bodies and may not intersect.

Contrary to the packing radius, the *covering radius* $\rho_c(\mathbf{\Lambda})$ is the largest distance a point in $\mathbb{R}^N$ can have to the closest lattice point. Again keeping the group properties of $\mathbf{\Lambda}$ in mind, we have the definition

$$\begin{aligned}\rho_c(\mathbf{\Lambda}) &= \sup_{x\in\mathbb{R}^N}\ \inf_{\boldsymbol{\lambda}\in\mathbf{\Lambda}} |x - \boldsymbol{\lambda}| \\ &= \sup_{x\in\mathcal{R}_V(\mathbf{\Lambda})} |x| \,. \end{aligned} \tag{C.2.7}$$

Geometrically, the covering radius is the *circumradius*, i.e., $\rho_c(\mathbf{\Lambda})$ is the radius of the smallest sphere, which circumscribes the Voronoi region.

With this in mind, the ***covering problem*** [CS88] seeks for the least dense way of completely covering space with equal-sized and overlapping N-spheres.

The last parameter we discuss in this section is the *normalized second moment* $\mathsf{G}(\mathcal{R})$ (or dimensionless second moment) of a region $\mathcal{R}$. It is defined as the ratio of the variance per dimension (the Nth part of the moment of inertia) and the volume normalized to two-dimensions

$$\mathsf{G}(\mathcal{R}) \triangleq \frac{\frac{1}{N}\int_{\mathcal{R}} \frac{1}{V(\mathcal{R})}|\boldsymbol{r}|^2 \, \mathrm{d}\boldsymbol{r}}{V(\mathcal{R})^{(2/N)}} = \frac{\int_{\mathcal{R}} |\boldsymbol{r}|^2 \, \mathrm{d}\boldsymbol{r}}{N\,V(\mathcal{R})^{(1+2/N)}} . \tag{C.2.8}$$

For lattices, we define the normalized second moment to be the respective quantity of the Voronoi region, i.e.

$$\mathsf{G}(\boldsymbol{\Lambda}) \triangleq \mathsf{G}(\mathcal{R}_{\mathrm{V}}(\boldsymbol{\Lambda})) . \tag{C.2.9}$$

This parameter is useful, e.g., when designing an appropriate boundary for the signal constellation (signal shaping).

Moreover, the ***quantization problem*** [CS88] is to find an N-dimensional lattice, for which the normalized second moment is minimum, as (assuming an uniformly distributed input in N dimensions) $\mathsf{G}(\boldsymbol{\Lambda})$ is proportional to the mean-squared quantization error.

Example C.2: *Packing and Covering*

Lattice $\mathbb{Z}^2$ Figure C.4 depicts the packing and covering for the $\mathbb{Z}^2$ lattice. The packing radius is $\rho(\mathbb{Z}^2) = 1/2$, and the covering radius $\rho_c(\mathbb{Z}^2) = 1/\sqrt{2}$. Since the squared Euclidean distance is $d^2_{\min}(\mathbb{Z}^2) = 1$, and the volume equals $V(\mathbb{Z}^2) = 1$, the density of this lattice $\mathbb{Z}^2$ is $\Delta(\mathbb{Z}^2) = \pi/4$. By inspection, the kissing number is 4 and the normalized second moment calculates to $1/12 \approx 0.0833$.

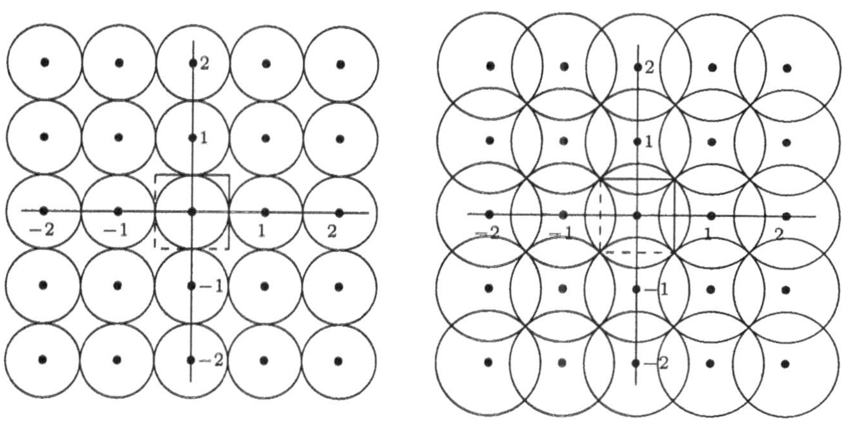

Fig. C.4 Packing and covering for the lattice $\mathbb{Z}^2$.

Lattice $\boldsymbol{A}_2$ Packing and covering are depicted for the hexagonal lattice $\boldsymbol{A}_2$ in Figure C.5. Here, for a squared Euclidean distance of $d^2_{\min}(\boldsymbol{A}_2) = 1$, the packing radius is $\rho(\boldsymbol{A}_2) = 1/2$ and the covering radius equals $\rho_c(\boldsymbol{A}_2) = 1/\sqrt{3}$. From basic geometry, the volume calculates to $V(\boldsymbol{A}_2) = \sqrt{3}/2$; the density of $\boldsymbol{A}_2$ is $\Delta(\boldsymbol{A}_2) = \pi/(2\sqrt{3})$, and the kissing number is 6. The normalized second moment reads $5/(36\sqrt{3}) \approx 0.0802$. As one can see, with respect to all parameters, the $\boldsymbol{A}_2$ lattice can be judged to be better than the $\mathbb{Z}^2$ lattice. It has been proved

[CS88] that the two-dimensional hexagonal lattice is the best with respect to all questions of lattice theory. Comparing Figures C.4 and C.5, we see that the lattice $\boldsymbol{A}_2$ has a denser packing, as well as a thinner covering—the areas of double covering are much smaller.

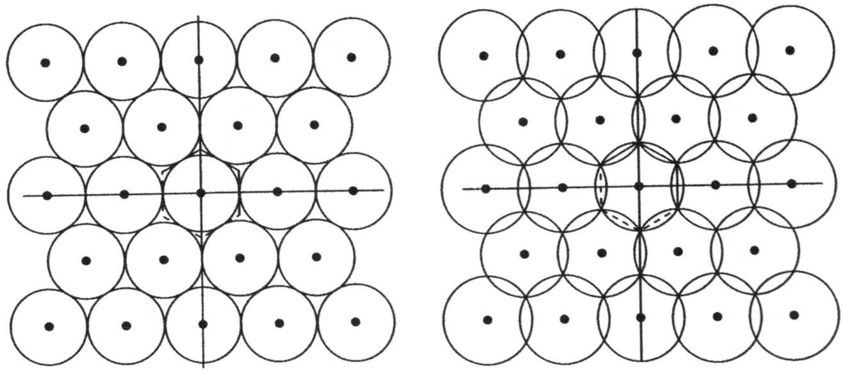

Fig. C.5 Packing and covering for the lattice $\boldsymbol{A}_2$.

C.3 MODIFICATIONS OF LATTICES

Given a lattice $\boldsymbol{\Lambda}$, we now discuss some modifications, which preserve lattice property. The proofs are obvious and hence omitted here.

Scaling Let $\boldsymbol{\Lambda}$ be an N-dimensional lattice and $r \in \mathbb{R}$ a real scalar. Then, $r\boldsymbol{\Lambda} \triangleq \{r\boldsymbol{\lambda} \mid \boldsymbol{\lambda} \in \boldsymbol{\Lambda}\}$ is a lattice, too, consisting of all scaled vectors $r\boldsymbol{\lambda}$ of the initial lattice $\boldsymbol{\Lambda}$. If $\boldsymbol{B}$ is the generator matrix of $\boldsymbol{\Lambda}$, the respective generator matrix of $r\boldsymbol{\Lambda}$ reads $r\boldsymbol{B}$. Scaling changes the parameters according to $d^2_{\min}(r\boldsymbol{\Lambda}) = r^2 d^2_{\min}(\boldsymbol{\Lambda})$, $V(r\boldsymbol{\Lambda}) = r^N V(\boldsymbol{\Lambda})$, $\rho(r\boldsymbol{\Lambda}) = r\rho(\boldsymbol{\Lambda})$, $\rho_c(r\boldsymbol{\Lambda}) = r\rho_c(\boldsymbol{\Lambda})$, and density, kissing number, and normalized second moment are scaling-invariant.

Orthonormal Transformation Let $\boldsymbol{R}$ be a real-valued orthonormal matrix, i.e., $\boldsymbol{R}^{\mathsf{T}}\boldsymbol{R} = \boldsymbol{I}$. The transformed set of points $\boldsymbol{R\Lambda}$ is again a lattice, which is generated by $\boldsymbol{RB}$. Since $\boldsymbol{R}$ is orthonormal, we have

$$V(\boldsymbol{R\Lambda}) = \sqrt{\det(\boldsymbol{B}^{\mathsf{T}}\boldsymbol{R}^{\mathsf{T}}\boldsymbol{R}\boldsymbol{B})} = \sqrt{\det(\boldsymbol{B}^{\mathsf{T}}\boldsymbol{B})} = V(\boldsymbol{\Lambda}) ,$$

i.e., the volume and all other parameters are preserved. Geometrically, this transformation is a rotation in N dimensions.

Reflection The lattice property is also preserved, if Λ is reflected at an $N-1$ dimensional plane. Mathematically, this operation is described by a matrix $\boldsymbol{U}$ with integer entries and $\det(\boldsymbol{U}) = \pm 1$. Here, all parameters of the lattice are unaffected.

If one lattice can be transformed into another lattice by combined rotation, reflection, and scaling, the lattices are called *equivalent*.

Cartesian Product Let Λ be an N-dimensional lattice. The M-fold Cartesian product of Λ, denoted by Λ^M, is a lattice, too. It is defined as the set of MN-dimensional vectors of the form $\left[\boldsymbol{\lambda}_1^{\mathsf{T}}, \boldsymbol{\lambda}_2^{\mathsf{T}}, \ldots, \boldsymbol{\lambda}_M^{\mathsf{T}}\right]^{\mathsf{T}}$, with $\boldsymbol{\lambda}_i \in \Lambda, i = 1, 2, \ldots, M$.

Note that **translation** of a lattice, i.e., taking $\Lambda + \boldsymbol{t}$, $\boldsymbol{t} \in \mathbb{R}^N$, does not give a lattice in general! Only if the translation vector $\boldsymbol{t}$ is itself a lattice point are the lattice characteristics preserved. From the group properties we have $\Lambda + \boldsymbol{t} = \Lambda$, if and only if $\boldsymbol{t} \in \Lambda$.

With respect to digital transmission over the AWGN channel, translation of a lattice has an effect on the error probability. But note, translation of the boundary region of a constellation may cause an increase in average transmit power.

Example C.3: *Modification of Lattices*

Lattice $\mathbb{Z}$:

Scaling The lattice $\mathbb{Z}$ can be scaled by an arbitrary scalar r. For example, the set of even integers $2\mathbb{Z}$ is a lattice.

Cartesian Product By taking the 2-fold Cartesian product of the integer lattice $\mathbb{Z}$, we get the "square lattice" $\mathbb{Z}^2$

$$\mathbb{Z}^2 = \left\{ \begin{bmatrix} x_1 \\ x_2 \end{bmatrix} \middle| x_1, x_2 \in \mathbb{Z} \right\} .$$

This procedure can of course be generalized to N dimensions.

Lattice $\mathbb{Z}^N$:

Scaling The lattice $\mathbb{Z}^N$ scaled by the scalar r is again a lattice, which can also be interpreted as the N-fold Cartesian product of the one-dimensional lattice $r\mathbb{Z}$

$$r\mathbb{Z}^N = (r\mathbb{Z})^N .$$

Rotation Let the combined scaling/rotation matrix be

$$\boldsymbol{R} = \begin{bmatrix} 1 & -1 \\ 1 & 1 \end{bmatrix} ,$$

which performs rotation by 45° counterclockwise and scaling by $\sqrt{2}$. $\boldsymbol{R}\mathbb{Z}^2$ is the rotated and scaled version of $\mathbb{Z}^2$, which can also be described by

$$\boldsymbol{R}\mathbb{Z}^2 = \boldsymbol{D}_2 = \left\{ \begin{bmatrix} x_1 \\ x_2 \end{bmatrix} \middle| x_1, x_2 \in \mathbb{Z}, \text{ and } x_1 + x_2 \in 2\mathbb{Z} \right\} .$$

$\boldsymbol{D}_2$ is also referred to as the two-dimensional checkerboard lattice.

Because $\boldsymbol{R}^2 = \begin{bmatrix} 0 & -2 \\ 2 & 0 \end{bmatrix}$ performs a rotation by 90° and a scaling by 2, and because $\mathbb{Z}^2$ rotated by 90° coincides with itself, the following holds

$$\boldsymbol{R}\left(\boldsymbol{R}\mathbb{Z}^2\right) = 2\mathbb{Z}^2 = (2\mathbb{Z})^2 .$$

Figure C.6 depicts the lattices $\mathbb{Z}^2$, $\boldsymbol{R}\mathbb{Z}^2$, and $2\mathbb{Z}^2$ and the respective Voronoi regions.

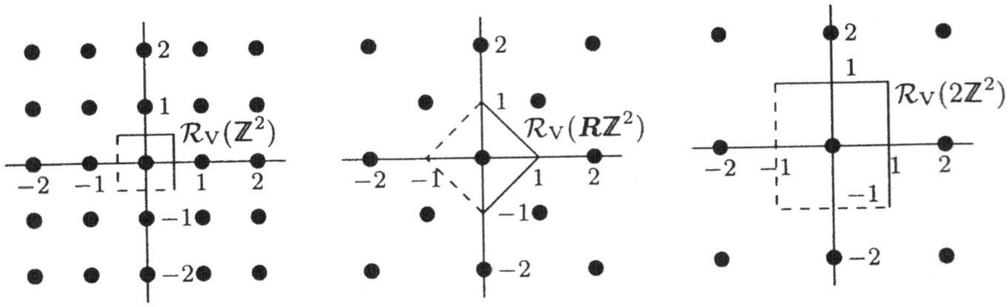

Fig. C.6 Lattices $\mathbb{Z}^2$, $\boldsymbol{R}\mathbb{Z}^2$, and $2\mathbb{Z}^2$.

C.4 SUBLATTICES, COSETS, AND PARTITIONS

From the last example, it can be seen that lattices $\boldsymbol{R}\mathbb{Z}^2$ and $2\mathbb{Z}^2$ are subsets of the lattice $\mathbb{Z}^2$. This leads us to subgroups: A subset $\mathcal{S}$ of a group $\mathcal{G}$, which by itself forms a group, i.e., meets the above group requirements **G1** through **G5**, is called a *subgroup* of the group $\mathcal{G}$. A subset of a lattice, which itself exhibits lattice properties, is called a *sublattice*.

Since real Euclidean space $\mathbb{R}^N$ forms a group (and thus can also be interpreted as a nondiscrete lattice), any real lattice ($\boldsymbol{\lambda}_i \in \mathbb{R}^N$) is a sublattice of $\mathbb{R}^N$.

Given a group $\mathcal{G}$, any subgroup $\mathcal{S}$ induces a *coset decomposition*. The group can be partitioned into $|\mathcal{G}|/|\mathcal{S}|$ disjoint *cosets*. One of them is equal to $\mathcal{S}$, the others are given as translates of the subgroup $\mathcal{S}$, i.e., to each element of $\mathcal{S}$ an element $g \in \mathcal{G}$, but $g \notin \mathcal{S}$ is "added."

Speaking in geometric terms, we have *lattice partitions*. Let $\boldsymbol{\Lambda}$ be a lattice and $\boldsymbol{\Lambda}' \subset \boldsymbol{\Lambda}$ a sublattice thereof. The lattice partition is the set of cosets, i.e., a set the elements of which are sets, and is usually denoted by $\boldsymbol{\Lambda}/\boldsymbol{\Lambda}'$. The (finite) cardinality of $\boldsymbol{\Lambda}/\boldsymbol{\Lambda}'$, i.e., the number m of cosets including the sublattice itself, is called the *order* or *depth* of the partition and is written $m = |\boldsymbol{\Lambda}/\boldsymbol{\Lambda}'|$.

The cosets can be specified by a set $\mathcal{A} = [\boldsymbol{\Lambda}/\boldsymbol{\Lambda}'] \triangleq \{\boldsymbol{a}\}$, comprising m *coset leaders* or *coset representatives* $\boldsymbol{a}$. In the context of lattices, the denomination $[\cdot]$ symbolizes the extraction of the m coset leaders from the lattice partition; one particular representative is chosen from each element of $\boldsymbol{\Lambda}/\boldsymbol{\Lambda}'$ (which are sets). The cosets can then be written as

$$\Lambda' + a, \quad \text{with} \quad a \in \mathcal{A} . \tag{C.4.1}$$

The sublattice itself is the *zero coset* with coset leader $a = 0$. It follows from the definition of cosets that all points in one coset are equivalent modulo Λ', i.e., their differences lie in Λ' (cf. Section C.1). In order for $\mathcal{A}$ to specify m disjoint subsets, no pair of coset leaders must not be equivalent modulo Λ'. Finally, in order to unambiguously define $\mathcal{A}$, here we force the coset leaders to lie within the Voronoi region $\mathcal{R}_V(\Lambda')$ of the sublattice. Note, for the fundamental volumes $V(\Lambda') = m \cdot V(\Lambda)$ holds. Unfortunately, the other parameters of Λ', e.g., the minimum distance, cannot be calculated as easily as the volume.

Based on the cosets, the lattice Λ can be expressed as

$$\Lambda = \bigcup_{a \in \mathcal{A}} (\Lambda' + a) , \tag{C.4.2a}$$

or for short [For88a]

$$\Lambda = \Lambda' + [\Lambda/\Lambda'] . \tag{C.4.2b}$$

Note that algebraically the partition Λ/Λ' forms the *quotient group* or *factor group* [BB91], where cosets are the group elements. Defining the sum of two cosets as that coset which is given by the sum of the respective coset leaders (and a possible modulo reduction to $\mathcal{R}_V(\Lambda')$), the set Λ/Λ' has group properties (e.g., the identity element is the sublattice Λ').

In the design of coded modulation schemes, lattice partitions and the indexing of the cosets by their coset leaders, plays an important role. Since the partition Λ/Λ' is a group, codes based on it are linear codes [For88a]. Only when the constellation is restricted to a finite region, due to the boundary effects, the code becomes nonlinear, i.e., here the sum of any two codewords is no longer a valid codeword.

Returning to Euclidean space, any translate $\Lambda + t$, $t \in \mathbb{R}^N$, of a lattice Λ is simply a coset of Λ in $\mathbb{R}^N$. Here, the set of coset representatives is identical to the Voronoi region $\mathcal{R}_V(\Lambda)$.

Of course, the partitioning procedure can be applied more than once. Given lattices Λ, Λ', and Λ'', with $\Lambda'' \subset \Lambda' \subset \Lambda$, a *lattice partition chain* is induced. Considering the definition of a lattice partition, a lattice partition chain is the set of sets which again are sets of sets (i.e., nested sets). A lattice partition chain is concisely written as

$$\Lambda/\Lambda'/\Lambda'' . \tag{C.4.3}$$

Now, Λ can be uniquely expressed as

$$\Lambda = \bigcup_{a' \in \mathcal{A}'} \bigcup_{a \in \mathcal{A}} (\Lambda'' + a' + a) = \Lambda'' + [\Lambda'/\Lambda''] + [\Lambda/\Lambda'] , \tag{C.4.4}$$

where $\mathcal{A}$ and $\mathcal{A}'$ are the set of coset representatives of the partition steps Λ/Λ' and Λ'/Λ'', respectively. If the partitions Λ/Λ' and Λ'/Λ'' are of order $m = |\Lambda/\Lambda'|$

and $m' = |\Lambda'/\Lambda''|$, respectively, then the order of Λ/Λ'' is $m \cdot m'$. Repeating the partitioning, partition chains of arbitrary depth can be generated.

In practice, *binary lattices* [FW89] are the most common. Such a lattice Λ is a sublattice of $\mathbb{Z}^N$, and has $2^n\mathbb{Z}^N$ as a sublattice for some integer n. Hence, for binary lattices, $\mathbb{Z}^N/\Lambda/2^n\mathbb{Z}^N$ is a partition chain. This fact guarantees that the addressing of lattice points can be done by iterative binary partitioning, and thus the labels of the points can be written as binary numbers.

Example C.4: *Coset Decomposition and Partitions*

Lattice $\mathbb{Z}$ The integer lattice $\mathbb{Z}$ can be repeatedly two-way partitioned, which results in the partition chain

$$\mathbb{Z}/2\mathbb{Z}/4\mathbb{Z}/8\mathbb{Z}/16\mathbb{Z}/\cdots$$

The coset representatives for the ith partition step $(2^{(i-1)}\mathbb{Z}/2^i\mathbb{Z},\ i \in \mathbb{N})$ are

$$\mathcal{A} = [\mathbb{Z}/2^i\mathbb{Z}] = \left\{0, 2^{(i-1)}\right\} .$$

Lattice $\mathbb{Z}^2$ The lattice $\boldsymbol{R}\mathbb{Z}^2$, with the rotation operator $\boldsymbol{R}$ given in the last example, is a sublattice of $\mathbb{Z}^2$. The order of $\mathbb{Z}^2/\boldsymbol{R}\mathbb{Z}^2$ is 2, and the set of coset representatives is

$$\mathcal{A} = [\mathbb{Z}^2/\boldsymbol{R}\mathbb{Z}^2] = \left\{ \begin{bmatrix} 0 \\ 0 \end{bmatrix}, \begin{bmatrix} 1 \\ 0 \end{bmatrix} \right\} .$$

Figure C.7 illustrates the lattices $\mathbb{Z}^2$, $\boldsymbol{R}\mathbb{Z}^2$, and its coset and shows the set $\mathcal{A}$ of coset leaders.

Partitioning the lattice $\boldsymbol{R}\mathbb{Z}^2$ once again results in the partition $\boldsymbol{R}\mathbb{Z}^2/2\mathbb{Z}^2$ or, starting with $\mathbb{Z}^2$, we have the four-way partition $\mathbb{Z}^2/2\mathbb{Z}^2$. Figure C.8 shows the four cosets and the four

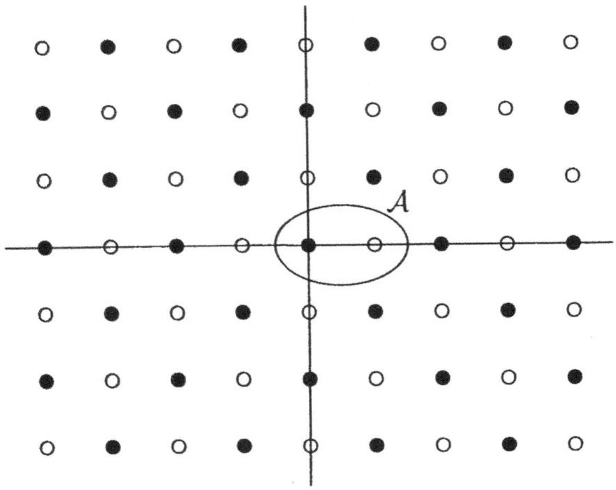

Fig. C.7 Two-way partitioning of the lattice $\mathbb{Z}^2$. The solid dots represent the sublattice $\boldsymbol{R}\mathbb{Z}^2$, the open dots the coset $\boldsymbol{R}\mathbb{Z}^2 + \begin{bmatrix} 1 \\ 0 \end{bmatrix}$.

coset representatives

$$\mathcal{A} = [\mathbb{Z}^2/2\mathbb{Z}^2] = \left\{ \begin{bmatrix} 0 \\ 0 \end{bmatrix}, \begin{bmatrix} 1 \\ 0 \end{bmatrix}, \begin{bmatrix} 0 \\ 1 \end{bmatrix}, \begin{bmatrix} 1 \\ 1 \end{bmatrix} \right\} .$$

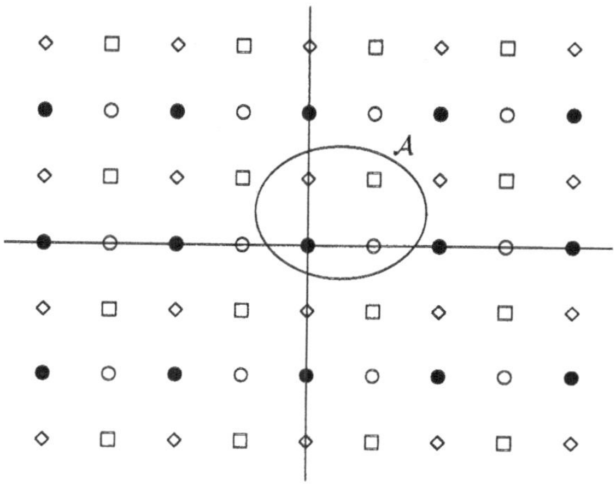

Fig. C.8 Four-way partitioning of the lattice $\mathbb{Z}^2$. The solid dots represent the sublattice $2\mathbb{Z}^2$, the others mark the three further cosets of $2\mathbb{Z}^2$.

Lattice A_2 Rotating the hexagonal lattice A_2 by 90° and scaling it by $\sqrt{3}$ gives a sublattice of A_2, which induces a three-way partitioning. Rotation and scaling can be described by the *ternary rotation operator* [For89]

$$R' = \begin{bmatrix} 0 & -\sqrt{3} \\ \sqrt{3} & 0 \end{bmatrix} .$$

A set of coset representatives is

$$\mathcal{A} = [A_2/R'A_2] = \left\{ \begin{bmatrix} 0 \\ 0 \end{bmatrix}, \begin{bmatrix} 1 \\ 0 \end{bmatrix}, \begin{bmatrix} 1/2 \\ \sqrt{3}/2 \end{bmatrix} \right\} .$$

Since $R'^2 = -3I$, we have the infinite three-way partitioning chain

$$A_2/R'A_2/3A_2/3R'A_2/9A_2/\cdots$$

On the other hand, scaling the A_2 lattice by a factor of 2 results in the sublattice $2A_2$, which leads to a four-way partition. Here, the coset representatives are

$$\mathcal{A} = [A_2/2A_2] = \left\{ \begin{bmatrix} 0 \\ 0 \end{bmatrix}, \begin{bmatrix} 1 \\ 0 \end{bmatrix}, \begin{bmatrix} 1/2 \\ \sqrt{3}/2 \end{bmatrix}, \begin{bmatrix} 3/2 \\ \sqrt{3}/2 \end{bmatrix} \right\} .$$

Repeated partitioning gives the partition chain

$$A_2/2A_2/4A_2/8A_2/16A_2/\cdots$$

Last, we consider a seven-way partition. Here, performing a rotation by $\tan^{-1}(\sqrt{3}/2)$ degree and scaling by $\sqrt{7}$, i.e., applying

$$\boldsymbol{R}'' = \begin{bmatrix} 2 & -\sqrt{3} \\ \sqrt{3} & 2 \end{bmatrix}$$

gives a sublattice. The set of seven coset representatives is

$$\begin{aligned} \mathcal{A} &= [\boldsymbol{A}_2/\boldsymbol{R}''\boldsymbol{A}_2] \\ &= \left\{ \begin{bmatrix} 0 \\ 0 \end{bmatrix}, \begin{bmatrix} 1 \\ 0 \end{bmatrix}, \begin{bmatrix} 1/2 \\ \sqrt{3}/2 \end{bmatrix}, \begin{bmatrix} -1/2 \\ \sqrt{3}/2 \end{bmatrix}, \begin{bmatrix} -1 \\ 0 \end{bmatrix}, \begin{bmatrix} -1/2 \\ -\sqrt{3}/2 \end{bmatrix}, \begin{bmatrix} 1/2 \\ -\sqrt{3}/2 \end{bmatrix} \right\} . \end{aligned}$$

The reader is encouraged to sketch the lattices, sublattices, and cosets as an exercise.

C.5 SOME IMPORTANT LATTICES AND THEIR PARAMETERS

Table C.1 collocates the parameters of some important examples for lattices [CS88, For88a, FW89]. These include the integer lattices $\mathbb{Z}^N$, the two-dimensional hexagonal lattice $\boldsymbol{A}_2$, the four-dimensional checkerboard lattice $\boldsymbol{D}_4$ (also called Schläfli lattice), the Gosset lattice $\boldsymbol{E}_8$. In this section, a short description of these lattices is given. For higher-dimensional lattices, such as the Barnes-Wall lattice $\boldsymbol{\Lambda}_{16}$ or the Leech lattice $\boldsymbol{\Lambda}_{24}$, and for an in-depth discussion of important lattices and their properties, the reader is referred to the literature (e.g., [CS88, Chapter 4]).

Table C.1 Parameters of some important lattices.

	$d^2_{\min}$	V	ρ	Δ	τ	ρ_c	G
$\mathbb{Z}$	1	1	$\frac{1}{2}$	1	2	1	$\frac{1}{12}$
$\mathbb{Z}^2$	1	1	$\frac{1}{2}$	$\frac{\pi}{4}$	4	$\frac{1}{\sqrt{2}}$	$\frac{1}{12}$
$\mathbb{Z}^N$	1	1	$\frac{1}{2}$	$\frac{\pi^{N/2}}{2^N(N/2)!}$	$2N$	$\frac{\sqrt{N}}{2}$	$\frac{1}{12}$
$\boldsymbol{A}_2$	1	$\frac{\sqrt{3}}{2}$	$\frac{1}{2}$	$\frac{\pi}{2\sqrt{3}}$	6	$\frac{1}{\sqrt{3}}$	$\frac{5}{36\sqrt{3}}$
$\boldsymbol{D}_4$	2	2	$\frac{1}{\sqrt{2}}$	$\frac{\pi^2}{16}$	24	1	≈ 0.0766
$\boldsymbol{E}_8$	2	1	$\frac{1}{\sqrt{2}}$	$\frac{\pi^4}{348}$	240	1	≈ 0.0717

Lattice $\mathbb{Z}^N$ The N-dimensional integer (or cubic) lattice ("square lattice" in two dimensions) is defined as

$$\mathbb{Z}^N \triangleq \left\{ \begin{bmatrix} \lambda_1 \\ \vdots \\ \lambda_N \end{bmatrix} \mid \lambda_i \in \mathbb{Z},\ i = 1, 2, \ldots, N \right\} . \tag{C.5.1}$$

The generator matrix reads

$$\boldsymbol{B}_{\mathbb{Z}^N} = \begin{bmatrix} 1 & & 0 \\ & \ddots & \\ 0 & & 1 \end{bmatrix} .$$

The lattice $\mathbb{Z}$ is the unique lattice in one dimension.

Lattice $\boldsymbol{A}_N$ The family of N-dimensional lattices $\boldsymbol{A}_N$ is favorably described in $N+1$ dimensions as

$$\boldsymbol{A}_N \triangleq \left\{ \sqrt{2}\,[\lambda_0, \ldots, \lambda_N]^{\mathsf{T}} \in \mathbb{Z}^{N+1} \mid \textstyle\sum_{i=0}^{N} \lambda_i = 0 \right\} , \tag{C.5.2}$$

which describes an N-dimensional hyperplane in $N+1$-dimensional space. The normalization ($\sqrt{2}$) is chosen to make $d^2_{\min} = 1$. This family includes $\boldsymbol{A}_1 = \mathbb{Z}$ and the two-dimensional hexagonal lattice $\boldsymbol{A}_2$ (cf. the above examples), which is the optimum lattice in two dimensions. Here, a generator matrix is given by

$$\boldsymbol{B}_{\boldsymbol{A}_2} = \begin{bmatrix} 1 & 1/2 \\ 0 & \sqrt{3}/2 \end{bmatrix} .$$

Lattice $\boldsymbol{D}_N$ The two-dimensional checkerboard lattice $\boldsymbol{D}_2$ and the Schläfli lattice $\boldsymbol{D}_4$ are representatives for lattices $\boldsymbol{D}_N$, defined as

$$\boldsymbol{D}_N \triangleq \left\{ [\lambda_1, \ldots, \lambda_N]^{\mathsf{T}} \in \mathbb{Z}^N \mid \textstyle\sum_{i=1}^{N} \lambda_i \text{ even} \right\} . \tag{C.5.3}$$

Here, $d^2_{\min} = 2$ is valid. $\boldsymbol{D}_4$ is the densest lattice in 4 dimensions; it can be described by the generator matrix

$$\boldsymbol{B}_{\boldsymbol{D}_4} = \begin{bmatrix} 2 & 1 & 1 & 1 \\ 0 & 1 & 0 & 0 \\ 0 & 0 & 1 & 0 \\ 0 & 0 & 0 & 1 \end{bmatrix} .$$

Lattice $\boldsymbol{E}_8$ The eight-dimensional lattice $\boldsymbol{E}_8$ is called Gosset lattice, and for $d^2_{\min} = 2$ it can be defined as follows:

$$\boldsymbol{E}_8 \triangleq \left\{ [\lambda_1, \ldots, \lambda_8]^{\mathsf{T}} \mid \text{all } \lambda_i \in \mathbb{Z}, \text{ or all } \lambda_i \in \mathbb{Z} + 1/2 \text{ and } \textstyle\sum_{i=1}^{8} \lambda_i \text{ even} \right\} . \tag{C.5.4}$$

It can be shown that the Gosset lattice, which is the densest lattice in eight dimensions, can be composed as the union of the lattice $\boldsymbol{D}_8$ and a translated version thereof [CS88]

$$\boldsymbol{E}_8 = \boldsymbol{D}_8 \cup \left(\boldsymbol{D}_8 + [1/2, \ldots, 1/2]^{\mathsf{T}} \right) . \tag{C.5.5}$$

Alternatively, the lattice can be defined by its generator matrix

$$B_{E_8} = \begin{bmatrix} 2 & -1 & 0 & 0 & 0 & 0 & 0 & 1/2 \\ 0 & 1 & -1 & 0 & 0 & 0 & 0 & 1/2 \\ 0 & 0 & 1 & -1 & 0 & 0 & 0 & 1/2 \\ 0 & 0 & 0 & 1 & -1 & 0 & 0 & 1/2 \\ 0 & 0 & 0 & 0 & 1 & -1 & 0 & 1/2 \\ 0 & 0 & 0 & 0 & 0 & 1 & -1 & 1/2 \\ 0 & 0 & 0 & 0 & 0 & 0 & 1 & 1/2 \\ 0 & 0 & 0 & 0 & 0 & 0 & 0 & 1/2 \end{bmatrix}.$$

Because fast quantization algorithms exist for E_8, this lattice is widely used in lattice quantizers [CS82, CS83].

Some Useful Partition Chains We give some selected lattice partition chains and the corresponding squared minimum distances for reference and without proof.

In one dimension, we have the partition chain and its corresponding squared distances

$$\mathbb{Z}/2\mathbb{Z}/4\mathbb{Z}/8\mathbb{Z}/16\mathbb{Z}/\cdots \qquad 1/4/16/64/256/\cdots$$

For $N = 2$, and $R = \begin{bmatrix} 1 & -1 \\ 1 & 1 \end{bmatrix}$, a two-way partition chain for $\mathbb{Z}^2$ exists, where at each step the squared distance increases by a factor of 2:

$$\mathbb{Z}^2/R\mathbb{Z}^2/2\mathbb{Z}^2/2R\mathbb{Z}^2/4\mathbb{Z}^2/\cdots \qquad 1/2/4/8/16/\cdots$$

The hexagonal lattice A_2 has a three-way partition chain based on the rotation operator $R' = \begin{bmatrix} 0 & -\sqrt{3} \\ \sqrt{3} & 0 \end{bmatrix}$. Here, the distances increase by factors of 3:

$$A_2/R'A_2/3A_2/3R'A_2/9A_2/\cdots \qquad 1/3/9/27/81/\cdots$$

Another partition (four-way) is based on scaling A_2 by a factor of 2. Correspondingly, the distances increase by factors of 4

$$A_2/2A_2/4A_2/8A_2/16A_2/\cdots \qquad 1/4/16/64/256/\cdots$$

Last, we regard the seven-way partition induced by the rotation operator $R'' = \begin{bmatrix} 2 & -\sqrt{3} \\ \sqrt{3} & 2 \end{bmatrix}$, where the distances increase by factors of 7. Since no multiple of 2π is divisible by the angle of the rotation (which is an irrational number), a scaled version of the initial A_2 lattice never occurs and no partition step is a scaled version of any prior partition.

A standard binary partition chain for coded modulation in 4 dimensions is the following. Using the four-dimensional rotation operator $R_4 = \begin{bmatrix} R & 0 \\ 0 & R \end{bmatrix}$, where R is the two-dimensional scaling/rotation matrix, a partition chain is obtained, where the minimum distance does not increase at every partition step [For88a]

$$\mathbb{Z}^4/D_4/R_4\mathbb{Z}^4/R_4D_4/2\mathbb{Z}^4/2D_4/2R_4\mathbb{Z}^4/\cdots \qquad 1/2/2/4/4/8/8/16/\cdots$$

REFERENCES

[BB91] E. J. Borowski and J. M. Borwein. *The HarperCollins Dictionary of Mathematics*. HarperPerennial, New York, 1991.

[CS82] J. H. Conway and N. J. A. Sloane. Fast Quantizing and Decoding Algorithms for Lattice Quantizers and Codes. *IEEE Transactions on Information Theory*, IT-28, pp. 227–232, March 1982.

[CS83] J. H. Conway and N. J. A. Sloane. A Fast Encoding Method for Lattice Codes and Quantizers. *IEEE Transactions on Information Theory*, IT-29, pp. 820–824, November 1983.

[CS88] J. H. Conway and N. J. A. Sloane. *Sphere Packings, Lattices and Groups*. Springer Verlag, New York, Berlin, 1988.

[For88a] G. D. Forney. Coset Codes — Part I: Introduction and Geometrical Classification. *IEEE Transactions on Information Theory*, IT-34, pp. 1123–1151, September 1988.

[For88b] G. D. Forney. Coset Codes — Part II: Binary Lattices and Related Codes. *IEEE Transactions on Information Theory*, IT-34, pp. 1152–1187, September 1988.

[For89] G. D. Forney. Multidimensional Constellations — Part II: Vornonoi Constellations. *IEEE Journal on Selected Areas in Communications*, JSAC-7, pp. 941–958, August 1989.

[FW89] G. D. Forney and L.-F. Wei. Multidimensional Constellations — Part I: Introduction, Figures of Merit, and Generalized Cross Constellations. *IEEE Journal on Selected Areas in Communications*, JSAC-7, pp. 877–892, August 1989.

Appendix D Calculation of Shell Frequency Distribution

Shell mapping was described in Section 4.3. The algorithm maps integer numbers (data) to blocks of shell indices. This was, shells with lower cost are selected more often than shells with larger cost. In this Appendix, an algorithm for calculating the frequency distribution of the shell indices is described.

As shown in Section 4.3.6, the key point for the assessment of shell mapping schemes is the calculation of the histograms $H(s, i)$. In principle, $H(s, i)$ could be obtained by tabulating all possible shell combinations and counting the occurrences of the shells. However, if a large number K of bits are mapped to the shell indices, this straightforward procedure is impractical or even impossible.

In this Appendix, we show that using *partial histograms*, which are easily calculated from the generating functions $G^{(n)}(x)$, the histogram $H(s, i)$ can be obtained very easily by running the shell mapping encoder once.

This Appendix is based on *Calculation of Shell Frequency Distributions Obtained with Shell-Mapping Schemes* by R. F. H. Fischer, which appeared in *IEEE Transactions on Information Theory*, pp. 1631–1639, July 1999.

D.1 PARTIAL HISTOGRAMS

The shell mapping encoder and decoder are based on generating functions that give the cost spectrum of shells. Moreover, we have seen (Section 4.2.5) that for concentric circular regions the simple linear cost function $C(s) = s$, $s = 0, 1, \ldots, M-1$, is a good choice. Hence, the generating functions are simply $G^{(n)}(x) = \left(1 + x + x^2 + \cdots + x^{M-1}\right)^n$, $n = 1, 2, \ldots, N$.

With the notation for *multinomial coefficients* [BS98, p. 94] ($\mathbb{N}$: set of natural numbers including zero)

$$\binom{n}{k_0, k_1, \ldots, k_{M-1}} = \frac{n!}{k_0! k_1! \cdots k_{M-1}!}, \qquad \text{with} \quad \sum_{s=0}^{M-1} k_s = n, \quad k_s \in \mathbb{N}, \tag{D.1.1}$$

the coefficient $g^{(n)}(c)$, which is the number of shell n-tuples with total cost $c = \sum_{i=1}^{n} C(s_i)$, is

$$g^{(n)}(c) = \sum_{\substack{\sum_{s=0}^{M-1} k_s = n \\ \sum_{s=0}^{M-1} s k_s = c}} \binom{n}{k_0, k_1, \ldots, k_{M-1}}, \qquad 0 \le c \le (M-1)n, \tag{D.1.2}$$

where the summation runs over all vectors $[k_0, k_1, \ldots, k_{M-1}]$ of nonnegative integers with the given constraints.

For a given M-tuple $[k_0, k_1, \ldots, k_{M-1}]$ with $\sum_{s=0}^{M-1} k_s = n$, the coefficient k_s tells how often shell s occurs in the n-tuples of shell indices. Over all n-tuples of shell indices for which k_j elements are equal to j, $j = 0, 1, \ldots, M-1$, shell s occurs $\binom{n}{k_1, k_2, \ldots, k_M} k_s$ times. Since all permutations of the n elements of a shell index vector occur, shell s occurs equally often in each of the n positions, i.e., $\binom{n}{k_1, k_2, \ldots, k_M} \frac{k_s}{n}$ times. Thus, for a fixed position, over all possible combinations of n shells with total cost c, shell s occurs

$$S_s^{(n)}(c) \triangleq \sum_{\substack{\sum_{i=0}^{M-1} k_i = n \\ \sum_{i=0}^{M-1} i k_i = c}} \binom{n}{k_0, k_1, \ldots, k_{M-1}} \frac{k_s}{n}, \qquad s = 1, 2, \ldots, M-1, \tag{D.1.3}$$

times. $S_s^{(n)}(c)$ can be viewed as a partial histogram of shells giving the frequencies of shells in all n-tuples of shells with total cost c. Table D.1 shows a sample of the possible combinations for $n = 8$.

The sum $S_s^{(n)}(c)$ in each cell of Table D.1 is given by (D.1.3). Using (D.1.2) and $\sum_{s=0}^{M-1} k_s = n$, the sum over one *row* is

$$\sum_{s=0}^{M-1} S_s^{(n)}(c) = g^{(n)}(c). \tag{D.1.4}$$

Table D.1 Number of occurrences of shell s in all possible combinations of 8-tuples with total cost c. Trailing zeros k_i in the multinomial coefficients are omitted, e.g., $\binom{8}{7,0,1} \triangleq \binom{8}{7,0,1,0,0,0,0,0}$.

Total cost c	Shell number s: 0	1	2	3	$\cdots$	$M-1$	$g^{(8)}(c)$
0	$\binom{8}{8}\cdot\frac{8}{8}$ $=1$	—	—	—		—	1
1	$\binom{8}{7,1}\cdot\frac{7}{8}$ $=7$	$\binom{8}{7,1}\cdot\frac{1}{8}$ $=1$	—	—		—	8
2	$\binom{8}{7,0,1}\cdot\frac{7}{8}$ $+\binom{8}{6,2}\cdot\frac{6}{8}$ $=28$	$\binom{8}{6,2}\cdot\frac{2}{8}$ $=7$	$\binom{8}{7,0,1}\cdot\frac{1}{8}$ $=1$	—		—	36
3	$\binom{8}{7,0,0,1}\cdot\frac{7}{8}$ $+\binom{8}{6,1,1}\cdot\frac{6}{8}$ $+\binom{8}{5,3}\cdot\frac{5}{8}$ $=84$	$\binom{8}{6,1,1}\cdot\frac{1}{8}$ $+\binom{8}{5,3}\cdot\frac{3}{8}$ $=28$	$\binom{8}{6,1,1}\cdot\frac{1}{8}$ $=7$	$\binom{8}{7,0,0,1}\cdot\frac{1}{8}$ $=1$		—	120
$\vdots$	$\vdots$	$\vdots$	$\vdots$	$\vdots$	$\cdots$	$\vdots$	$\vdots$

From the definition (D.1.1) of the multinomial coefficient, the following is true:

$$\binom{n}{k_1,\ldots,k_j,\ldots,k_l,\ldots,k_M}\cdot\frac{k_j}{n} \qquad \text{with } {\textstyle\sum_{i=0}^{M-1} k_i=n,\ \sum_{i=0}^{M-1} ik_i=c}$$

$$=\binom{n}{k_1,\ldots,k_j-1,\ldots,k_l+1,\ldots,k_M}\cdot\frac{k_l+1}{n}$$

$$=\binom{n}{k_1,\ldots,k_j-1,\ldots,k_l+1,\ldots,k_M}\cdot\frac{k_l+1}{n}$$

$$\triangleq\binom{n}{k'_1,\ldots,k'_j,\ldots,k'_l,\ldots,k'_M}\cdot\frac{k'_l}{n} \qquad \text{with } {\textstyle\sum_{i=0}^{M-1} k'_i=n,\ \sum_{i=0}^{M-1} ik'_i=c-j+l}\ . \tag{D.1.5}$$

Hence, setting $l=j+m$, $m\in\mathbb{Z}$, we have

$$S_s^{(n)}(c)=S_{s+m}^{(n)}(c+m)\ , \tag{D.1.6}$$

i.e., the terms on the diagonals of the above table are identical—the matrix $\left[S_s^{(n)}(c)\right]$ is Toeplitz. In view of the above relationship, the definition of $S_s^{(n)}(c)$ can be formally extended to all indices $s\le c+1$. For $s>c+1$, it is convenient to define $S_s^{(n)}(c)=0$.

Next, let $C^{(n)}$ be a given integer. There are

$$z^{(n)}(C^{(n)}) = \sum_{c=0}^{C^{(n)}-1} g^{(n)}(c) \tag{D.1.7}$$

combinations of n shells with a total cost *less* than $C^{(n)}$. Among these combinations, the number of occurrences of shell s in each position is (summing up the *columns* of the above table)

$$\begin{aligned}
H_s^{(n)}(C^{(n)}) &\triangleq \sum_{c=s}^{C^{(n)}-1} S_s^{(n)}(c) \\
&= \sum_{s'=0}^{C^{(n)}-s} S_s^{(n)}(C^{(n)} - s'), \qquad \text{with } s' = C^{(n)} - c \\
&= \sum_{s'=1}^{C_n-s} S_{s'}^{(n)}(C^{(n)} - s) \\
&= \sum_{s'=0}^{M-1} S_{s'}^{(n)}(C^{(n)} - s - 1) + \sum_{s'=M}^{2M-1} S_{s'}^{(n)}(C^{(n)} - s - 1) + \cdots \\
&= \sum_{m=0}^{\infty} \sum_{s'=0}^{M-1} S_{s'-mM}^{(n)}(C^{(n)} - s - 1) \\
&= \sum_{m=0}^{\infty} \sum_{s'=0}^{M-1} S_{s'}^{(n)}(C^{(n)} - s - 1 - mM) \\
&= \sum_{m=0}^{\infty} g^{(n)}(C^{(n)} - s - 1 - mM), \qquad s = 0, 1, \ldots, M-1\,.
\end{aligned} \tag{D.1.8}$$

In other words, in order to calculate $H_s^{(n)}(C^{(n)})$, the coefficients $g^{(n)}(c)$ have to be aliased modulo M. Since

$$\begin{aligned}
\sum_{s=0}^{M-1} H_s^{(n)}(C^{(n)}) &= \sum_{s=0}^{M-1} \sum_{m=0}^{\infty} g^{(n)}(C^{(n)} - s - 1 - mM) \\
&= \sum_{c'=0}^{C^{(n)}-1} g^{(n)}(c') \\
&= z^{(n)}(C^{(n)}),
\end{aligned} \tag{D.1.9}$$

the histogram $H_s^{(n)}(C^{(n)})$ comprises $z^{(n)}(C^{(n)})$ n-tuples of shell indices with a total cost *less* than $C^{(n)}$.

In order to find the number of occurrences of shell s within all possible combinations of n shells with a total cost *equal* to c, we have to calculate $S_s^{(n)}(c)$, which may be written as

$$\begin{aligned} S_s^{(n)}(c) &= H_s^{(n)}(c+1) - H_s^{(n)}(c) \\ &= \sum_{m=0}^{\infty} g^{(n)}(c+1-s-mM) - \sum_{m=0}^{\infty} g^{(n)}(c-s-mM) \\ &= \sum_{m=0}^{\infty} \bar{g}^{(n)}(c-s-mM)\,, \end{aligned} \tag{D.1.10}$$

with the definition $\bar{g}^{(n)}(c) = g^{(n)}(c+1) - g^{(n)}(c)$.

Example D.1: *Histograms $H_s^{(n)}(c)$ and $S_s^{(n)}(c)$*

We continue Example 4.10 on the V.34 shell mapper with $M = 3$. Here, the generating function for shell 8-tuples reads

$$\begin{aligned} G^{(8)}(x) &= 1 + 8x + 36x^2 + 112x^3 + 266x^4 + 504x^5 \\ &\quad + 784x^6 + 1016x^7 + 1107x^8 + 1016x^9 + 784x^{10} + 504x^{11} \\ &\quad + 266x^{12} + 112x^{13} + 36x^{14} + 8x^{15} + x^{16}\,. \end{aligned} \tag{D.1.11}$$

Equation (D.1.8) specializes to $H_s^{(8)}(c) = \sum_{m=0}^{\infty} g^{(8)}(c-s-1-3m)$, $s = 0, 1, 2$.

Table D.2 summarizes $H_s^{(8)}(c)$ and $S_s^{(8)}(c)$ for total costs up to 8. Compare these tables with Table D.1 and Example 4.10.

Table D.2 Partial histograms $H_s^{(8)}(c)$ and $S_s^{(8)}(c)$. V.34 shell mapper with $M = 3$.

$H_s^{(8)}(c)$	Shell s		
	0	1	2
$c = 0$	0	0	0
1	1	0	0
2	8	1	0
3	36	8	1
4	113	36	8
5	274	113	36
6	540	274	113
7	897	540	274
8	1290	897	540

$S_s^{(8)}(c)$	Shell s		
	0	1	2
$c = 0$	1	0	0
1	7	1	0
2	28	7	1
3	77	28	7
4	161	77	28
5	266	161	77
6	357	266	161
7	393	357	266
8	357	393	357

D.2 PARTIAL HISTOGRAMS FOR GENERAL COST FUNCTIONS

The above derivations do not apply for general cost functions (e.g., for one-dimensional constellations). In this case it is more appropriate to first calculate the number $S_s^{(n)}(c)$ of the occurrences of shell s in a given position and all possible n-tuples with total cost c. Again (D.1.4) holds, but now the matrix $\left[S_s^{(n)}(c)\right]$ (cf. Table D.1) is no longer Toeplitz. But, following the above arguments, it is easy to see that for a general cost function $C(s)$, the formula

$$S_s^{(n)}(c) = S_{s+m}^{(n)}(c - C(s) + C(s+m)) \tag{D.2.1}$$

is still valid.

From (D.1.4) and (D.2.1), the partial histograms $S_s^{(n)}(c)$ can be calculated iteratively by the following algorithm, which basically does a successive filling of a table which is analogous to Table D.1. This is possible because the value and row of the first nonzero element of each column, and the sum over each row, are known.

1. Let $n = 1$.
   2. Let $c = n \cdot C(0)$.
   3. Calculate

$$S_s^{(n)}(c) = \begin{cases} S_0^{(n)}(c - C(s) + C(0)), & c - C(s) + C(0) \geq 0 \\ 0, & c - C(s) + C(0) < 0 \end{cases},$$

$$\forall s, C(s) > C(0)$$

   and

$$S_s^{(n)}(c) = \Big(g^{(n)}(c) - \sum_{\forall s', C(s') > C(0)} S_{s'}^{(n)}(c)\Big) \Big/ g^{(1)}(C(0)),$$

$$\forall s, C(s) = C(0).$$

   4. Increment c. If $c \leq n \cdot C(M)$ go to Step 3.
5. Increment n. If $n \leq N$ go to Step 2.
6. Finally, calculate

$$H_s^{(n)}(C^{(n)}) = \sum_{c=s}^{C^{(n)}-1} S_s^{(n)}(c), \qquad s = 0, 1, \ldots, M-1, \quad n = 1, 2, \ldots, N.$$

D.3 FREQUENCIES OF SHELLS

The frequencies of the shells can be easily obtained from the histograms defined above. The main idea in calculating the frequencies of shells is to run the shell mapping encoder with the maximum input $I = 2^K - 1$, which yields specific intermediate results and the final shell indices $s_{(1)}$ to $s_{(N)}$, with $s_{(i)} = 0, 1, \ldots, M - 1$. Then, with each step in the encoding procedure a partial histogram based on the quantities $S_s^{(n)}(c)$ can be associated. Summing up these partial histograms gives the final histograms $H(s, i)$.

As an example we consider in detail the shell mapping algorithm used in ITU Recommendation V.34 [ITU94], which has a frame size $N = 8$. However, the methods presented here apply to all kinds of shell mapping schemes using all types of cost functions.

The starting point for the calculation of the histogram $H(s, i)$ is a notional tabulation of all shell N-tuples, as done in Table 4.8. Again, the shell combination associated with the lowest index (zero) is plotted in the top, while the N-tuple corresponding to $2^K - 1$ is shown in the bottom. Due to the specific ordering of the shell N-tuples, such a table can be partitioned into regions, each corresponding to an individual step in the encoding procedure for input $I = 2^K - 1$.

Figure D.1 shows this sorting of all 2^K 8-tuples of shells and the decomposition according to the V.34 shell mapping encoder. Please note that the diagram is vertically not to scale. The corresponding assignment of partial histograms to each step of the encoding procedure is given in Figure D.2.

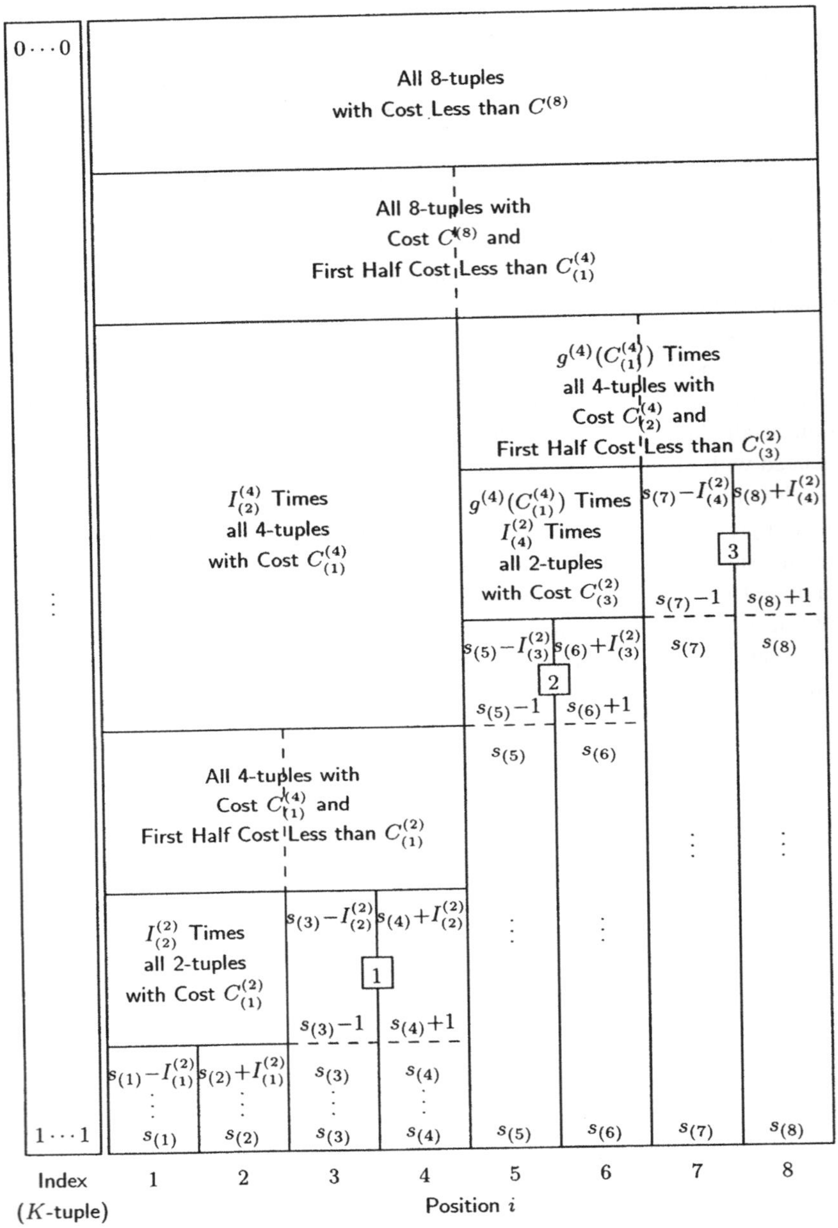

Fig. D.1 Explanation of the sorting and decomposition of all 2^K 8-tuples of shells (not to scale). Repeat each element $\boxed{1}$ $g^{(2)}(C^{(2)}_{(1)})$ times, $\boxed{2}$ $g^{(4)}(C^{(4)}_{(1)})$ times, $\boxed{3}$ $g^{(4)}(C^{(4)}_{(1)}) \cdot g^{(2)}(C^{(2)}_{(3)})$ times.

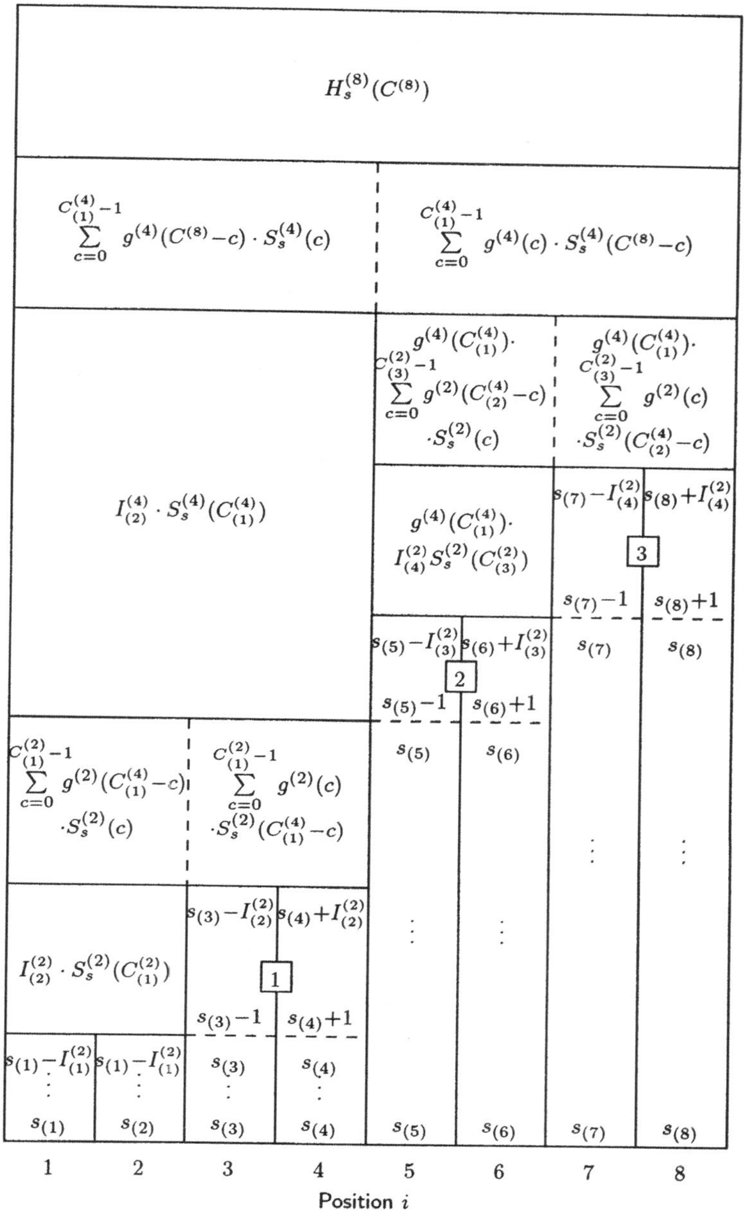

Fig. D.2 Sorting of all 2^K 8-tuples of shells and corresponding partial histograms (not to scale). The sum of column i is $H(s,i)$. Repeat each element [1] $g^{(2)}(C_{(1)}^{(2)})$ times, [2] $g^{(4)}(C_{(1)}^{(4)})$ times, [3] $g^{(4)}(C_{(1)}^{(4)}) \cdot g^{(2)}(C_{(3)}^{(2)})$ times.

For calculating of the frequencies $H(s,i)$ of the shells, the following steps, identical to shell mapping encoding in V.34, are performed. In addition, this example briefly gives the V.34 shell mapping algorithm.

1. *Initialization:*

The encoder input is set to $I = 2^K - 1$, i.e., all K shell mapping bits are set to one.

2. *Calculate total cost $C^{(8)}$:*

The largest integer $C^{(8)}$ is determined for which $z^{(8)}(C^{(8)}) \le I$. $C^{(8)}$ is the total cost of the 8-tuple associated with index I, and $z^{(8)}(C^{(8)})$ is the number of 8-tuples of shells with total cost less than $C^{(8)}$. Let $I^{(8)} = I - z^{(8)}(C^{(8)})$.

Partial Histogram:

Here, for all positions the number of occurrences of shell s is given by $H_s^{(8)}(C^{(8)})$.

3. *Calculate costs $C_{(1)}^{(4)}$, $C_{(2)}^{(4)}$ of first and second half:*

The largest integer $C_{(1)}^{(4)}$ is determined, such that[1]

$$I^{(4)} = I^{(8)} - \sum_{c=0}^{C_{(1)}^{(4)}-1} g^{(4)}(c) \cdot g^{(4)}(C^{(8)} - c)$$

is nonnegative. $C_{(1)}^{(4)}$ is the total cost of the first half of the ring indices, and $C_{(2)}^{(4)} = C^{(8)} - C_{(1)}^{(4)}$ is the total cost of the second half of the ring indices.

Partial Histogram:

The term $\sum_{c=0}^{C_{(1)}^{(4)}-1} g^{(4)}(c) \cdot g^{(4)}(C^{(8)} - c)$ contributes differently to positions 1 to 4 and 5 to 8, respectively. In positions 1 to 4, shell s occurs $\sum_{c=0}^{C_{(1)}^{(4)}-1} g^{(4)}(C^{(8)} - c) \cdot S_s^{(4)}(c)$ times, and in positions 5 to 8, shell s occurs $\sum_{c=0}^{C_{(1)}^{(4)}-1} g^{(4)}(c) \cdot S_s^{(4)}(C^{(8)} - c)$ times.

4. *Calculate index $I_{(1)}^{(4)}$, $I_{(2)}^{(4)}$ of first and second half:*

The integers $I_{(1)}^{(4)}$ and $I_{(2)}^{(4)}$ are determined, such that

$$I^{(4)} = I_{(2)}^{(4)} \cdot g^{(4)}(C_{(1)}^{(4)}) + I_{(1)}^{(4)} , \qquad 0 \le I_{(2)}^{(4)} , \quad 0 \le I_{(1)}^{(4)} < g^{(4)}(C_{(1)}^{(4)}) .$$

Partial Histogram:

The term $I_{(2)}^{(4)} \cdot g^{(4)}(C_{(1)}^{(4)})$ contributes $I_{(2)}^{(4)} \cdot S_s^{(4)}(C_{(1)}^{(4)})$ to the number of occurrences in positions 1 to 4. From now on, in positions 5 to 8 all partial histograms will be multiplied by $g^{(4)}(C_{(1)}^{(4)})$.

[1] $\sum_{c=0}^{-1}(\cdot)$ is defined as 0.

5.1. Calculate costs $C^{(2)}_{(1)}$, $C^{(2)}_{(2)}$ of the first and second quarter:

The largest integer $C_{2,1}$ is determined such that

$$I^{(2)}_{(a)} = I^{(4)}_{(1)} - \sum_{c=0}^{C^{(2)}_{(1)}-1} g^{(2)}(c) \cdot g^{(2)}(C^{(4)}_{(1)} - c)$$

is nonnegative. $C^{(2)}_{(1)}$ is the total cost of the first two ring indices, and $C^{(2)}_{(2)} = C^{(4)}_{(1)} - C^{(2)}_{(1)}$ is the total cost of ring indices 3 and 4.

Partial Histogram:

The term $\sum_{c=0}^{C^{(2)}_{(1)}-1} g^{(2)}(c) \cdot g^{(2)}(C^{(4)}_{(1)} - c)$ contributes differently to positions 1, 2 and 3, 4, respectively. In positions 1 and 2 shell s occurs $\sum_{c=0}^{C^{(2)}_{(1)}-1} g^{(2)}(C^{(4)}_{(1)} - c) \cdot S^{(2)}_s(c)$ times, and in positions 3 and 4, shell s occurs $\sum_{c=0}^{C^{(2)}_{(1)}-1} g^{(2)}(c) \cdot S^{(2)}_s(C^{(4)}_{(1)} - c)$ times.

5.2. Calculate costs $C^{(2)}_{(3)}$, $C^{(2)}_{(4)}$ of the third and fourth quarter:

The largest integer $C^{(2)}_{(3)}$ is determined, such that

$$I^{(2)}_{(b)} = I^{(4)}_{(2)} - \sum_{c=0}^{C^{(2)}_{(3)}-1} g^{(2)}(c) \cdot g^{(2)}(C^{(4)}_{(2)} - c)$$

is nonnegative. $C^{(2)}_{(3)}$ is the total cost of the ring indices 5 and 6, and $C^{(2)}_{(4)} = C^{(4)}_{(2)} - C^{(2)}_{(3)}$ is the total cost of the ring indices 7 and 8.

Partial Histogram:

The term $\sum_{c=0}^{C^{(2)}_{(3)}-1} g^{(2)}(c) \cdot g^{(2)}(C^{(4)}_{(2)} - c)$ contributes differently to positions 5, 6 and 7, 8, respectively. In positions 5 and 6 shell s occurs $g^{(4)}(C^{(4)}_{(1)}) \cdot \sum_{c=0}^{C^{(2)}_{(3)}-1} g^{(2)}(C^{(4)}_{(2)} - c) \cdot S^{(2)}_s(c)$ times, and in positions 7 and 8 shell s occurs $g^{(4)}(C^{(4)}_{(1)}) \cdot \sum_{c=0}^{C^{(2)}_{(3)}-1} g^{(2)}(c) \cdot S^{(2)}_s(C^{(4)}_{(2)} - c)$ times.

6.1. Calculate index $I^{(2)}_{(1)}$, $I^{(2)}_{(2)}$ of the first and second quarter:

The integers $I^{(2)}_{(1)}$ and $I^{(2)}_{(2)}$ are determined such that

$$I^{(2)}_{(a)} = I^{(2)}_{(2)} \cdot g^{(2)}(C^{(2)}_{(1)}) + I^{(2)}_{(1)}\,, \qquad 0 \le I^{(2)}_{(2)}\,, \quad 0 \le I^{(2)}_{(1)} < g^{(2)}(C^{(2)}_{(1)})\,.$$

Partial Histogram:

The term $I^{(2)}_{(2)} \cdot g^{(2)}(C^{(2)}_{(1)})$ contributes $I^{(2)}_{(2)} \cdot S^{(2)}_s(C^{(2)}_{(1)})$ to the number of occurrences in positions 1 and 2. From now on, in positions 3 and 4 all partial histograms will be multiplied by $g^{(2)}(C^{(2)}_{(1)})$.

6.2. Calculate index $I^{(2)}_{(3)}$, $I^{(2)}_{(4)}$ of the third and fourth quarter:

The integers $I^{(2)}_{(3)}$ and $I^{(2)}_{(4)}$ are determined such that

$$I^{(2)}_{(b)} = I^{(2)}_{(4)} \cdot g^{(2)}(C^{(2)}_{(3)}) + I^{(2)}_{(3)} , \qquad 0 \le I^{(2)}_{(4)} , \quad 0 \le I^{(2)}_{(3)} < g^{(2)}(C^{(2)}_{(3)}) .$$

Partial Histogram:

The term $I^{(2)}_{(4)} \cdot g^{(2)}(C^{(2)}_{(3)})$ contributes $g^{(4)}(C^{(4)}_{(1)}) \cdot I^{(2)}_{(4)} \cdot S^{(2)}_s(C^{(2)}_{(3)})$ to the number of occurrences in positions 5 and 6. From now on, in positions 7 and 8 all partial histograms will be multiplied additionally by $g^{(2)}(C^{(2)}_{(3)})$.

7. Calculate final ring indices $s_{(1)}, \ldots, s_{(8)}$:

The final ring indices $s_{(i)}$ and $s_{(i+1)}$ are calculated from the index $J = I^{(2)}_{((i+1)/2)}$ and the cost $K = C^{(2)}_{((i+1)/2)}$, $i = 1, 2, 3, 4$, according to

$$K < M : \quad \begin{aligned} s_{(i)} &= J \\ s_{(i+1)} &= K - J \end{aligned} , \qquad K \ge M : \quad \begin{aligned} s_{(i)} &= K - M + J + 1 \\ s_{(i+1)} &= M - J \end{aligned} .$$

Partial Histogram:

In positions 1 and 2, $I^{(2)}_{(1)} + 1$ pairs of shells are still lacking. Because of the specific sorting in shell mapping, and because the last tuple (belonging to the input $2^K - 1$) is known to be $[s_{(1)}, s_{(2)}]$, in shells $s_{(1)} - I^{(2)}_{(1)}$ through $s_{(1)}$ occur once position 1. Shells $s_{(2)}$ through $s_{(2)} + I^{(2)}_{(1)}$ occur once in position 2. This completes the derivations for the first two positions.

With regard to Figures D.1 and D.2, $I^{(2)}_{(2)} \cdot g^{(2)}(C^{(2)}_{(1)}) + I^{(2)}_{(1)} + 1$ pairs of shells are lacking for positions 3 and 4. In position 3, shells $s_{(2)} - I^{(2)}_{(2)}$ through $s_{(3)} - 1$ occur $g^{(2)}(C^{(2)}_{(1)})$ times and shell $s_{(3)}$ occurs $I^{(2)}_{(1)} + 1$ times. In position 4, shells $s_{(4)} + 1$ through $s_{(4)} + I^{(2)}_{(2)}$ occur $g^{(2)}(C^{(2)}_{(1)})$ times and shell $s_{(4)}$ occurs $I^{(2)}_{(1)} + 1$ times.

The remaining 2-tuples in positions 5, 6 and 7, 8, respectively, can be determined similarly to positions 3 and 4. $I^{(2)}_{(1)}$ has to be replaced by $I^{(4)}_{(1)}$ or $I^{(2)}_{(3)} \cdot g^{(4)}(C^{(4)}_{(1)}) + I^{(4)}_{(1)}$; $I^{(2)}_{(2)}$ by $I^{(2)}_{(3)}$ or $I^{(2)}_{(4)}$; $C^{(2)}_{(2)} = C^{(4)}_{(1)} - C^{(2)}_{(1)}$ by $C^{(2)}_{(3)}$ or $C^{(2)}_{(4)} = C^{(8)} - C^{(4)}_{(1)} - C^{(2)}_{(3)}$; and $g^{(2)}(C^{(2)}_{(1)})$ by $g^{(4)}(C^{(4)}_{(1)})$ or $g^{(4)}(C^{(4)}_{(1)}) \cdot g^{(2)}(C^{(2)}_{(3)})$, respectively.

That is, in position 5, shells $s_{(5)} - I^{(2)}_{(3)}$ through $s_{(5)} - 1$ occur $g^{(4)}(C^{(4)}_{(1)})$ times and shell $s_{(5)}$ occurs $I^{(4)}_{(1)} + 1$ times. In position 6, shells $s_{(6)} + 1$ through $s_{(6)} + I^{(2)}_{(3)}$ occur $g^{(4)}(C^{(4)}_{(1)})$ times and shell $s_{(6)}$ occurs $I^{(4)}_{(1)} + 1$ times.

In position 7, shells $s_{(7)} - I^{(2)}_{(4)}$ through $s_{(7)} - 1$ occur $g^{(4)}(C^{(4)}_{(1)}) \cdot g^{(2)}(C^{(2)}_{(3)})$ times and shell $s_{(7)}$ occurs $I^{(2)}_{(3)} \cdot g^{(4)}(C^{(4)}_{(1)}) + I^{(4)}_{(1)} + 1$ times. In position 8, shells $s_{(8)} + 1$ through $s_{(8)} + I^{(2)}_{(4)}$ occur $g^{(4)}(C^{(4)}_{(1)}) \cdot g^{(2)}(C^{(2)}_{(3)})$ times and shell $s_{(8)}$ occurs $I^{(2)}_{(3)} \cdot g^{(4)}(C^{(4)}_{(1)}) + I^{(4)}_{(1)} + 1$ times.

Example D.2: *Calculation of Histogram $H(s,i)$ for V.34 Shell Mapper*

Once again we continue Example 4.10 on the V.34 shell mapper with $M = 3$ and $K = 5$. Figure D.3 shows the sorting of all 32 shell 8-tuples and the corresponding partial histograms.

Index	Shell 8-tuple							
00000	0	0	0	0	0	0	0	0
00001	0	0	0	0	0	0	0	1
00010	0	0	0	0	0	0	1	0
00011	0	0	0	0	0	1	0	0
00100	0	0	0	0	1	0	0	0
00101	0	0	0	1	0	0	0	0
00110	0	0	1	0	0	0	0	0
00111	0	1	0	0	0	0	0	0
01000	1	0	0	0	0	0	0	0
01001	0	0	0	0	0	0	0	2
01010	0	0	0	0	0	0	1	1
01011	0	0	0	0	0	0	2	0
01100	0	0	0	0	0	1	0	1
01101	0	0	0	0	1	0	0	1
01110	0	0	0	0	0	1	1	0
01111	0	0	0	0	1	0	1	0
10000	0	0	0	0	0	2	0	0
10001	0	0	0	0	1	1	0	0
10010	0	0	0	0	2	0	0	0
10011	0	0	0	1	0	0	0	1
10100	0	0	1	0	0	0	0	1
10101	0	1	0	0	0	0	0	1
10110	1	0	0	0	0	0	0	1
10111	0	0	0	1	0	0	1	0
11000	0	0	1	0	0	0	1	0
11001	0	1	0	0	0	0	1	0
11010	1	0	0	0	0	0	1	0
11011	0	0	0	1	0	1	0	0
11100	0	0	1	0	0	1	0	0
11101	0	1	0	0	0	1	0	0
11110	1	0	0	0	0	1	0	0
11111	0	0	0	1	1	0	0	0

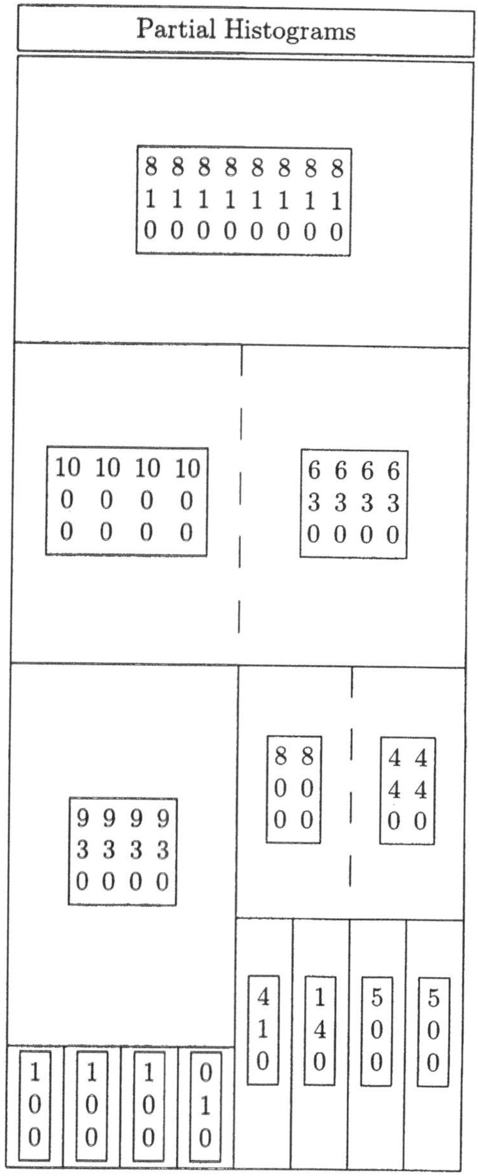

Fig. D.3 Sorting of all 2^K shell 8-tuples and corresponding partial histograms. $M = 3$, $K = 5$.

The partial histograms are given as a matrix, where the columns correspond to the position within the mapping frame, and the rows correspond to shell number $s = 0$ to 2 (cf. the arrangement in Table 4.9). Summing up the partial histograms results in $H(s, i)$, as was already given in Table 4.9.

For reference, as in Figure 4.21, Figure D.4 gives the costs and indices which occur in the shell mapping encoder for input $2^5 - 1 = 31$. Note, since some quantities are zero, not all partial histograms described above occur in the present example. For example, since $I_{(2)}^{(2)} = 0$, the term $I_{(2)}^{(2)} \cdot S_s^{(2)}(C_{(1)}^{(2)})$ does not appear.

<table>
<tr><td colspan="8">$C^{(8)} = 2,\ I^{(8)} = 12$</td></tr>
<tr><td colspan="4">$C_{(1)}^{(4)} = 1,\ I_{(1)}^{(4)} = 0$</td><td colspan="4">$C_{(2)}^{(4)} = 1,\ I_{(2)}^{(4)} = 3$</td></tr>
<tr><td colspan="2">$C_{(1)}^{(2)} = 0,\ I_{(1)}^{(2)} = 0$</td><td colspan="2">$C_{(2)}^{(2)} = 1,\ I_{(2)}^{(2)} = 0$</td><td colspan="2">$C_{(3)}^{(2)} = 1,\ I_{(3)}^{(2)} = 1$</td><td colspan="2">$C_{(4)}^{(2)} = 0,\ I_{(4)}^{(2)} = 0$</td></tr>
<tr><td>$s_{(1)} = 0$</td><td>$s_{(2)} = 0$</td><td>$s_{(3)} = 0$</td><td>$s_{(4)} = 1$</td><td>$s_{(5)} = 1$</td><td>$s_{(6)} = 0$</td><td>$s_{(7)} = 0$</td><td>$s_{(8)} = 0$</td></tr>
</table>

Fig. D.4 Costs and indices which occur in V.34 shell mapper for $M = 3$ and input $2^5 - 1$.

REFERENCES

[BS98] I. N. Bronstein and K. A. Semendjajew. *Handbook of Mathematics*. Springer Verlag, Berlin, Heidelberg, Reprint of the third edition, 1998.

[ITU94] ITU-T Recommendation V.34. *A Modem Operating at Data Signalling Rates of up to 28800 bit/s for Use on the General Switched Telephone Network and on Leased Point-to-Point 2-Wire Telephone-Type Circuits*. International Telecommunication Union (ITU), Geneva, Switzerland, September 1994.

Appendix E
Precoding for MIMO Channels

Consider a transmission system, where N_T transmitters (e.g., antennas) are transmitting in parallel and N_R receivers are receiving simultaneously. Such a scheme is compactly described by a *multiple-input/multiple-output (MIMO)* system. Since all users' signals are superimposed at the receivers, multiuser interference occur. The same situation is present in *direct-sequence code-division multiple access (DS-CDMA)*, when the individual signals interfere due to nonorthogonal (effective) spreading sequences, e.g., when orthogonality is lost on intersymbol-interference channels.

In general, MIMO transmission can be described by the basic relation

$$\boldsymbol{y} = \boldsymbol{H}\boldsymbol{x} + \boldsymbol{n} \,, \tag{E.0.1}$$

where $\boldsymbol{x}$ designates the transmit vector which comprises the transmit symbols of N_T parallel data streams. The vectors $\boldsymbol{y}$ and $\boldsymbol{n}$ of dimension N_R denote the vector of received symbols, and the vector of disturbances, respectively. The MIMO channel is characterized by its $N_\mathrm{R} \times N_\mathrm{T}$ channel matrix $\boldsymbol{H}$. The above input/output relation is intriguingly similar to that of intersymbol-interference channels, which reads in frequency domain

$$Y(z) = H(z)X(z) + N(z) . \tag{E.0.2}$$

Starting from this analogy, in this Appendix, we show how the concept of precoding for equalizing intersymbol interference (temporal equalization) on *single-input/single-output (SISO)* channels can be extended to spatial or combined spatial/temporal equalization of MIMO channels. First, we study the scenario where a single transmitter with multiple antennas is communicating with a single receiver with multiple antennas, as it is the case in BLAST-like (Bell Laboratories layered space-time) approaches. Then, a DS-CDMA downlink scenario is considered, where a central transmitter communicates with a number of distributed receivers.

E.1 CENTRALIZED RECEIVER

First, we consider equalization of multiuser interference when *one* transmitter and *one* receiver with multiple antennas, respectively, are present. Since a common or central receiver is considered, we denote this approach as *centralized.*

E.1.1 Multiple-Input/Multiple-Output Channel

Here, we focus on transmission using N_T transmit antennas and N_R receive antennas. Here, for simplicity, we follow the literature, e.g., [Tel99, FG98], and assume that between each transmit and receive antenna a flat, i.e., nondispersive fading channel is present. The discrete-time system model in the equivalent complex baseband is depicted in Figure E.1.

For the moment, the channel inputs $x_k[\nu]$, $k = 1, \ldots, N_\mathrm{T}$, are assumed to be equal to the data-carrying symbols, which are drawn from some signal constellation $\mathcal{A}$, i.e., $x_k = a_k \in \mathcal{A}$. These symbols are presumed to be (spatially and temporally) uncorrelated, each with the same variance, i.e.,

$$\mathrm{E}\left\{a_k[\nu+\mu]a_\kappa^*[\nu]\right\} = \begin{cases} \sigma_a^2 \,, & \text{if } k = \kappa \text{ and } \mu = 0 \\ 0 \,, & \text{else} \end{cases} . \tag{E.1.1}$$

Conveniently, these symbols, transmitted simultaneously at symbol interval ν over the N_T antennas, are combined into the vector $\boldsymbol{x}[\nu] \triangleq [x_1[\nu], \ldots, x_{N_\mathrm{T}}[\nu]]^\mathsf{T}$. Arranging the corresponding samples at the N_R receive antennas into the vector $\boldsymbol{y}[\nu] \triangleq [y_1[\nu], \ldots, y_{N_\mathrm{R}}[\nu]]^\mathsf{T}$, the input/output relation of the MIMO channel at discrete-time index ν is given by

$$\boldsymbol{y}[\nu] = \boldsymbol{H}[\nu]\boldsymbol{x}[\nu] + \boldsymbol{n}[\nu] . \tag{E.1.2}$$

The $N_\mathrm{R} \times N_\mathrm{T}$ channel matrix $\boldsymbol{H}[\nu] = [h_{k\kappa}[\nu]]$ contains the fading gains $h_{k\kappa}[\nu]$ from transmit antenna κ to receive antenna k at discrete-time index ν. Depending on the actual situation, the fading gains $h_{k\kappa}[\nu]$ may be spatially (indices k and κ) and/or temporally (index ν) correlated.

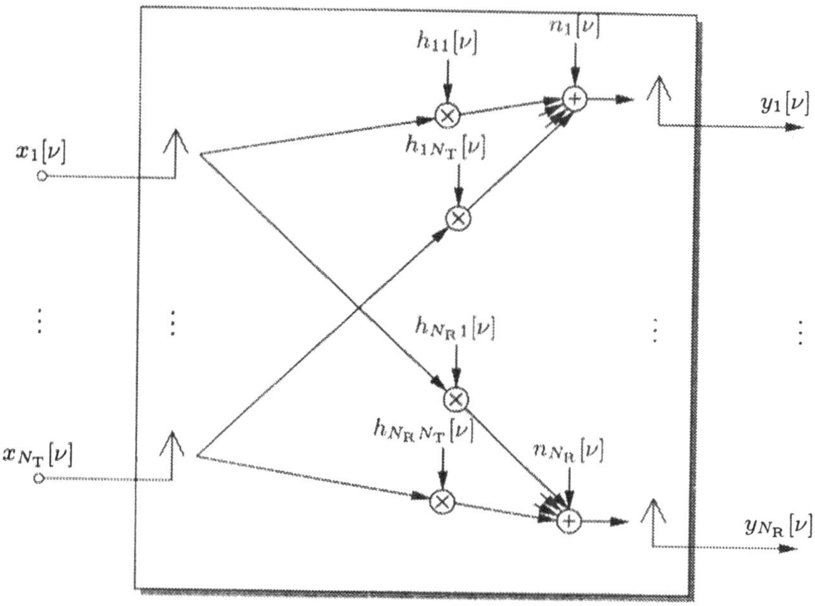

Fig. E.1 System model of a multiple-input/multiple-output channel when using N_T transmit and N_R receive antennas.

The additive channel noise samples at each receive antenna are combined into the vector $\boldsymbol{n}[\nu] \triangleq [n_1[\nu], \ldots, n_{N_R}[\nu]]^T$. Here, the noise samples $n_k[\nu]$, $k = 1, \ldots, N_R$, are expected to be zero mean Gaussian random variables and mutually uncorrelated, i.e.,

$$\mathrm{E}\left\{n_k[\nu+\mu]n_\kappa^*[\nu]\right\} = \begin{cases} \sigma_n^2 \,, & \text{if } k = \kappa \text{ and } \mu = 0 \\ 0 \,, & \text{else} \end{cases} \,. \tag{E.1.3}$$

For the sake of clarity, we consider flat fading channels, hence only *spatial* equalization within one time slot ν has to be performed—each time interval can be processed on its own. Moreover, we assume that the channel is constant over a transmission burst. Hence, for convenience, the discrete time index ν is dropped in the following. The extension to frequency-selective channels is discussed later. Finally, for brevity, we assume $\boldsymbol{H}$ to be a $D \times D$ square matrix (i.e., $D = N_T = N_R$). The extension to the general case is straightforward.

E.1.2 Equalization Strategies for MIMO Channels

The main task when transmitting over MIMO channels is the *separation* or *equalization* of the parallel data streams, i.e., the recovery of the components of the transmitted vector $\boldsymbol{x}$ which interfere at the receiver side.

Spatial equalization in MIMO systems (channel matrix $\boldsymbol{H}$) is tightly related to *temporal* equalization for single-input/single-output transmission over intersymbol-

Table E.1 Corresponding equalization strategies for ISI channels and MIMO channels.

		ISI channel $H(z)$ (temporal equalization)	MIMO channel $\boldsymbol{H}$ (spatial equalization)
Linear	At receiver	Linear equalization via $1/H(z)$	Linear equalization via $\boldsymbol{H}_{(l)}^{-1}$
	At transmitter	Linear preequalization via $1/H(z)$	Linear preequalization via $\boldsymbol{H}_{(r)}^{-1}$
	At transmitter and receiver	OFDM/DMT vector coding	SVD
Nonlinear	At receiver	DFE	Matrix DFE
	At transmitter (and receiver)	Tomlinson-Harashima Precoding	**precoding for MIMO channels**

interference channels (channel transfer function $H(z)$). Each equalization strategy has its direct counterpart in the other domain. The analogies are depicted in Table E.1.

The correspondences for linear equalization at the receiver or transmitter are immediate. The most obvious strategy for separating the data streams is linear equalization at the receiver side. Here, the decision vector is generated by $\boldsymbol{r} = \boldsymbol{H}_{(l)}^{-1}\boldsymbol{y}$, where $\boldsymbol{H}_{(l)}^{-1}$ denotes the left (pseudo) inverse of the channel matrix $\boldsymbol{H}$. It is well-known (cf. Section 2.2) that linear equalization suffers from noise enhancement, and hence has poor power efficiency.

If channel state information is available at the transmitter, the users can be separated by means of linear preequalization. Interference of the users at the receiver side can be completely avoided by multiplying the data vector $\boldsymbol{a}$ with the right (pseudo) inverse $\boldsymbol{H}_{(r)}^{-1}$ of the channel matrix $\boldsymbol{H}$ at the transmitter. Instead of transmitting the data symbols $\boldsymbol{a}$ directly over the channel ($\boldsymbol{x} = \boldsymbol{a}$), the predistorted version $\boldsymbol{x} = \boldsymbol{H}_{(r)}^{-1}\boldsymbol{a}$ is fed into the channel. However, this zero-forcing linear preequalization suffers from the same loss in power efficiency as linear equalization at the receiver side. Here, instead of enhancing the noise, average transmit power is boosted by the same factor.

This disadvantage of linear equalization can be overcome by decision-feedback equalization (DFE), which is a nonlinear equalization strategy at the receiver side (cf. Section 2.3). Its counterpart for MIMO channels is a *matrix DFE* or *multidimensional DFE* for spatial equalization, e.g., [YR94, Due95, CF96]. Unfortunately, error propagation may occur in DFE. Moreover, since immediate decisions are required, the application of channel coding requires some clever interleaving, which in turn introduces significant delay.

In SISO transmission, the feedback part of the DFE can be transferred to the transmitter, leading to *Tomlinson-Harashima precoding (THP)*. Neglecting a very small increase in average transmit power, the performance of DFE and Tomlinson-Harashima precoding is the same (cf. Chapter 3). Since Tomlinson-Harashima precoding is a transmitter technique, error propagation at the receiver is avoided. Moreover, channel coding schemes can be applied in the same way as for the ideal AWGN or flat fading channel. In the remaining part of this Appendix, we discuss nonlinear precoding scheme for MIMO channels in detail.

Finally, for the sake of completeness, the task of channel equalization can be split among transmitter and receiver. A popular strategy is based on the *singular value decomposition (SVD)* of the channel matrix $\boldsymbol{H}$. Writing it as $\boldsymbol{H} = \boldsymbol{U\Sigma V}$, where $\boldsymbol{U}$ and $\boldsymbol{V}$ are unitary matrices and $\boldsymbol{\Sigma}$ is diagonal, and applying $\boldsymbol{V}^{\mathsf{H}}$ at the transmitter and $\boldsymbol{U}^{\mathsf{H}}$ at the receiver, independent, parallel channels are generated. Here, in contrast to linear (pre)equalization, neither transmit power is increased nor is channel noise enhanced. Singular value decomposition for MIMO channels can be identified with *orthogonal frequency-division multiplexing (OFDM)* or *discrete multitone (DMT)* transmission over ISI channels. To be precise, SVD corresponds to a strategy called *vector coding* [KAC90], where blocks of consecutive symbols are processed at the transmitter and receiver based on the eigenvectors of some Toeplitz channel matrix. In both cases, a partitioning of the underlying channel into parallel independent subchannels is performed, which is also the theoretical concept when calculating the channel capacity. Multicarrier transmission (OFDM/DMT), which is a linear but time-variant equalization strategy, performs the same as decision-feedback equalization, which is a nonlinear but time-invariant technique [ZK89, FH97].

E.1.3 Matrix DFE

Assume that the channel symbols x_k are each drawn from some signal constellation $\mathcal{A}$ and equal to the data-carrying symbols, i.e., $x_k = a_k \in \mathcal{A}$, $k = 1, \ldots, D$.

A detection algorithm for uncoded transmission over the above described MIMO channel has been introduced in [GFVW99]. The main point of the algorithm is the successive detection of the parallel data symbols. In the literature, this approach is known as Bell Laboratories Layered Space-Time.

Studying the BLAST algorithm, it turns out that it is equivalent to the matrix DFE, depicted in Figure E.2 (cf. [GC01a]). As in decision-feedback equalization for SISO channels, a feedforward matrix $\boldsymbol{F}$ is present. Additionally, for convenience, a scaling or gain diagonal matrix[1] $\boldsymbol{G} = \mathbf{diag}(g_1, \ldots, g_D)$ is introduced. The feedforward filter in SISO DFE guarantees a causal minimum-phase end-to-end impulse response and whitens the channel noise. Here, the feedforward matrix has to guarantee *spatial whiteness* and *spatial causality*: The decision symbols with index l should only be disturbed by symbols with index $k < l$. Hence, the end-to-end matrix[2]

[1] $\mathbf{diag}(g_1, \ldots, g_D)$ denotes a diagonal matrix with entries g_1 through g_D on the main diagonal.
[2] For the moment we assume a zero-forcing approach.

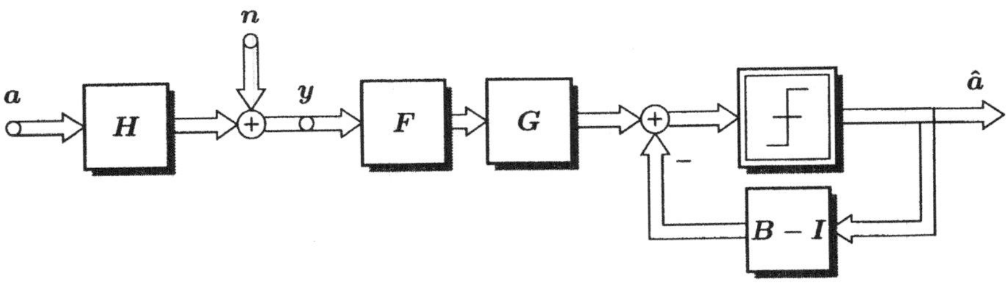

Fig. E.2 Equivalent matrix DFE receiver for BLAST.

$\boldsymbol{B} = [b_{k\kappa}] \triangleq \boldsymbol{GFH}$ has to be lower triangular, i.e., $b_{k\kappa} = 0,\ k \le \kappa$. The scaling matrix $\boldsymbol{G}$ is chosen such that the main diagonal elements of $\boldsymbol{B}$ are one ($b_{kk} = 1$), i.e., for unit-gain signal transmission. This property corresponds to the monic end-to-end impulse response in SISO transmission.

Assuming that decisions $\hat{a}_k$ on the data symbols a_k, $k = 1, 2, \ldots, t-1$, are already available, their portion of the interference can be subtracted out by the feedback matrix $\boldsymbol{B} - \boldsymbol{I}$ ($\boldsymbol{I}$: identity matrix). By construction, the feedback matrix is strictly lower triangular. This ensures a causal loop for successive decisions and interference cancellation starting from the symbol with index 1 and proceeding to symbol number D [Due95].

For achieving optimum performance, the symbols have to be detected according to a specific ordering [GFVW99]. When considering matrix DEF, we always suppose that the transmit antennas were relabeled (permutation of the columns of the channel matrix $\boldsymbol{H}$) according to the optimum ordering.

E.1.4 Tomlinson-Harashima Precoding

Basic Concept In the same way that decision-feedback equalization can be replaced by Tomlinson-Harashima precoding in SISO transmission, this can be done for MIMO channels [HF94, FHK94], see Figure E.3. While in "classical" Tomlinson-Harashima precoding, a single channel is equalized with respect to time, spatial equalization is required for MIMO channels. If, additionally, the channel matrix introduces intersymbol interference, combined temporal and spatial equalization has to be performed. Again, we first restrict ourselves to pure spatial equalization.

If the modulo device at the transmitter were ignored, i.e., the modulo device were replaced by a short circuit/identity matrix, *linear preequalization* of the cascade $\boldsymbol{B} \triangleq \boldsymbol{GFH}$ would be present. Here, because of the triangular structure of the feedback matrix $\boldsymbol{B} - \boldsymbol{I}$, the channel symbols x_k, $k = 1, \ldots, D$, are successively generated from the data symbols $a_k \in \mathcal{A}$:

$$x_1 = a_1\,, \qquad x_k = a_k - \sum_{\kappa=1}^{k-1} b_{k\kappa} x_\kappa\,, \quad k = 2, \ldots, D\,. \tag{E.1.4}$$

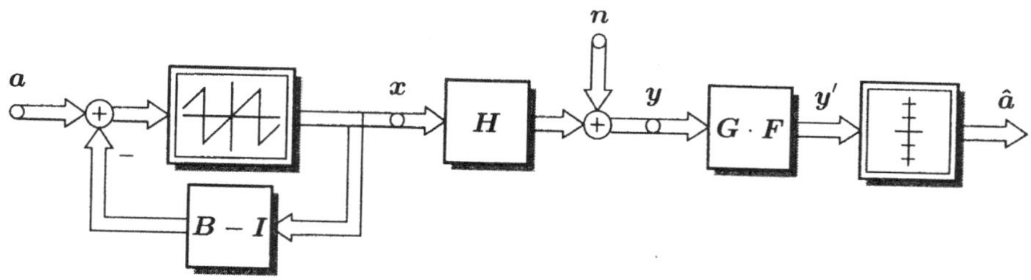

Fig. E.3 Tomlinson-Harashima precoding for MIMO channels.

Average transmit power is boosted compared to a direct transmission of the data vector $\boldsymbol{a}$. Using that $\boldsymbol{x} = \boldsymbol{B}^{-1}\boldsymbol{a}$, and denoting the entries of the lower-triangular, unit-diagonal matrix $\boldsymbol{B}^{-1}$ as $b_{ij}^{(-1)}$, total transmit power for linear preequalization calculates to

$$\mathrm{E}\{\boldsymbol{x}^{\mathsf{H}}\boldsymbol{x}\} = \sigma_a^2 \cdot \mathrm{trace}\left(\boldsymbol{B}^{-\mathsf{H}}\boldsymbol{B}^{-1}\right) = \sigma_a^2\Big(D + \sum_{i,j:\, i>j} |b_{ij}^{(-1)}|^2\Big) > D\sigma_a^2 = E\{\boldsymbol{a}^{\mathsf{H}}\boldsymbol{a}\} . \tag{E.1.5}$$

When the off-diagonal entries of $\boldsymbol{B}^{-1}$ become large, a nonnegligible increase in transmit power occurs.

This increase in transmit power is avoided by *modulo reducing* the channel symbols x_k into the boundary region of $\mathcal{A}$. Assuming the same constellation in all D parallel streams, and that $\mathcal{A}$ is the intersection of a regular grid (signal-point lattice) and the Voronoi region $\mathcal{R}(\boldsymbol{\Lambda}_{\mathrm{p}})$ of the *precoding lattice* $\boldsymbol{\Lambda}_{\mathrm{p}}$, the channel symbols are successively calculated as

$$x_1 = a_1 , \qquad x_k = a_k + d_k - \sum_{\kappa=1}^{k-1} b_{k\kappa} x_\kappa , \quad k = 2, \ldots, D , \tag{E.1.6}$$

where $d_k \in \boldsymbol{\Lambda}_{\mathrm{p}}$. In other words, instead of feeding the data symbols a_k into the linear predistortion, the *effective data symbols* $v_k \triangleq a_k + d_k$ are passed into $\boldsymbol{B}^{-1}$, which is implemented by the feedback structure. That is, the initial signal constellation is extended periodically. Since the *precoding symbols* d_k are matched to the boundary region of the initial signal constellation, the points in the expanded signal are also taken from a regular grid.

All points which are congruent modulo $\boldsymbol{\Lambda}_{\mathrm{p}}$ represent the same data. From these equivalent points, that point is selected symbol-by-symbol for transmission, which results in a channel symbol falling into the boundary region of $\mathcal{A}$.

Since the linear predistortion via $\boldsymbol{B}^{-1}$ equalizes the cascade $\boldsymbol{B} = \boldsymbol{GFH}$, after prefiltering and scaling, the effective data symbols v_k, corrupted by additive noise, are visible at the receiver, i.e., $\boldsymbol{y}' = \boldsymbol{v} + \boldsymbol{n}'$. Here, $\boldsymbol{n}'$ denotes the filtered channel noise and $\boldsymbol{v} = [v_1, \ldots, v_D]^{\mathsf{T}}$. Using a slicer which takes the periodic extension into account, an estimate for the data symbols (vector $\boldsymbol{a}$) can be generated. Alternatively, the received symbols y'_k are first modulo reduced into the boundary region of the signal constellation $\mathcal{A}$. Then, a conventional slicer suffices.

As one can see, the operation of Tomlinson-Harashima precoding for MIMO channel is exactly the same as for SISO channels, cf. Chapter 3. The only difference is that in spatial precoding each symbol interval is processed separately. As a consequence, the channel symbols are not distributed uniformly over the boundary region, but take on more and more discrete levels when going from component x_1 to x_D. Since a continuous uniform distribution is never achieved, the precoding loss in MIMO precoding is slightly lower than that given in Section 3.2.7. Using the same arguments as in Section 3.2.2, the channel symbols x_k can be expected to be mutually uncorrelated, i.e., $\mathsf{E}\{\boldsymbol{x}\boldsymbol{x}^{\mathsf{H}}\} = \sigma_x^2 \boldsymbol{I}$.

Example E.1: *Signals in MIMO Precoding*

For illustration, Figure E.4 shows scatter plots of the channel symbols x_k and the noisy received symbols y'_k for a MIMO channel with $D = 4$ inputs and $D = 4$ outputs. A 16-ary QAM constellation is used in each of the parallel channels.

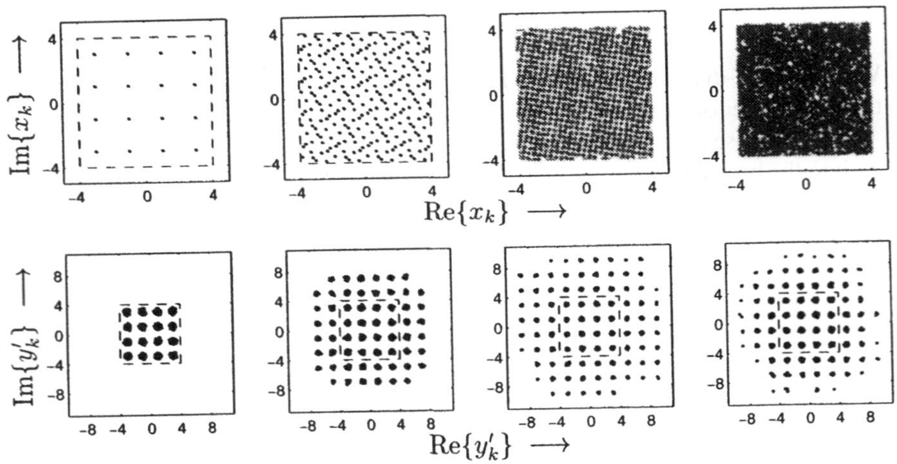

Fig. E.4 Scatter plots of channel symbols x_k and received symbols y'_k when using MIMO precoding. $D = 4$ in- and outputs. 16-QAM constellation. Left to right: Components $k = 1$ through $k = 4$.

From component 1 through 4, the channel symbols tend from the initial 16-QAM constellation to an almost uniform distribution over the boundary region. Simultaneously, the effective data symbols are taken from an increasingly expanded signal set. The nonuniform distribution of the effective data symbols v_k can be seen. In addition, the different noise variances which are effective for the different components, are visible.

Calculation of the Matrix Filters The matrices required for matrix DFE or MIMO precoding can be calculated by performing a *QR-type factorization* of the channel matrix $\boldsymbol{H}$. In what follows, we assume that a relabeling of the transmit antennas for guaranteeing the optimum detection ordering is already included in the channel matrix by suitably permuting its columns. Then, the factorization reads

$$\boldsymbol{H} = \boldsymbol{F}^{\mathsf{H}}\boldsymbol{R}\,, \tag{E.1.7}$$

where $\boldsymbol{F}$ is the unitary (i.e., $\boldsymbol{F}\boldsymbol{F}^{\mathsf{H}} = \boldsymbol{I}$) feedforward matrix and $\boldsymbol{R} = [r_{ij}]$ is a lower triangular matrix ($r_{ij} = 0, i < j$).

For convenience we define $\boldsymbol{B} \triangleq \boldsymbol{G}\boldsymbol{R}$, with $\boldsymbol{G} = \mathbf{diag}(r_{11}^{-1}, \ldots, r_{DD}^{-1})$. The matrix $\boldsymbol{B}$ is thus unit-diagonal lower triangular. The feedback matrix of the precoder is then given as $\boldsymbol{B} - \boldsymbol{I}$. Since $\boldsymbol{H} = \boldsymbol{F}^{\mathsf{H}}\boldsymbol{R}$ and $\boldsymbol{F}$ is a unitary matrix, we have

$$\boldsymbol{H}^{\mathsf{H}}\boldsymbol{H} = \boldsymbol{R}^{\mathsf{H}}\boldsymbol{F}\boldsymbol{F}^{\mathsf{H}}\boldsymbol{R} = \boldsymbol{R}^{\mathsf{H}}\boldsymbol{R}\,. \tag{E.1.8}$$

Hence, the lower triangular matrix $\boldsymbol{R}$ can be obtained by a *Cholesky factorization*[3] [BB91, GL96] of $\boldsymbol{H}^{\mathsf{H}}\boldsymbol{H}$.

The above approach results in filters adjusted according to the *zero-forcing* criterion. For deriving a solution, which optimizes the matrices according to the *minimum mean-squared error (MMSE)* criteria, we consider the error signal at the slicer

$$\boldsymbol{e} = \boldsymbol{G}\boldsymbol{F}\cdot\boldsymbol{y} - \boldsymbol{v} = \boldsymbol{G}\boldsymbol{F}\cdot\boldsymbol{y} - \boldsymbol{B}\cdot\boldsymbol{x}\,. \tag{E.1.9}$$

Regarding the *orthogonality principle* (cf. Section 2.2.3), we require $\boldsymbol{e} \perp \boldsymbol{y}$, which leads to

$$\mathsf{E}\left\{\boldsymbol{e}\boldsymbol{y}^{\mathsf{H}}\right\} = \mathsf{E}\left\{\boldsymbol{G}\boldsymbol{F}\cdot\boldsymbol{y}\boldsymbol{y}^{\mathsf{H}} - \boldsymbol{B}\cdot\boldsymbol{x}\boldsymbol{y}^{\mathsf{H}}\right\} \stackrel{!}{=} \boldsymbol{0} \quad \Rightarrow \quad \boldsymbol{G}\boldsymbol{F}\boldsymbol{\Phi}_{yy} = \boldsymbol{B}\boldsymbol{\Phi}_{xy}\,. \tag{E.1.10}$$

Since $\boldsymbol{y} = \boldsymbol{H}\boldsymbol{x} + \boldsymbol{n}$, $\mathsf{E}\{\boldsymbol{x}\boldsymbol{x}^{\mathsf{H}}\} = \sigma_x^2\boldsymbol{I}$, and $\mathsf{E}\{\boldsymbol{x}\boldsymbol{n}^{\mathsf{H}}\} = \boldsymbol{0}$, the correlation matrices are given by

$$\boldsymbol{\Phi}_{yy} = \sigma_x^2\boldsymbol{H}\boldsymbol{H}^{\mathsf{H}} + \sigma_n^2\boldsymbol{I} \tag{E.1.11a}$$

$$\boldsymbol{\Phi}_{xy} = \sigma_x^2\boldsymbol{H}^{\mathsf{H}} \tag{E.1.11b}$$

and we have

$$\boldsymbol{G}\boldsymbol{F}\left(\sigma_x^2\boldsymbol{H}\boldsymbol{H}^{\mathsf{H}} + \sigma_n^2\boldsymbol{I}\right) = \sigma_x^2\boldsymbol{B}\boldsymbol{H}^{\mathsf{H}}\,. \tag{E.1.12}$$

Using $\zeta = \frac{\sigma_n^2}{\sigma_x^2}$, the error thus can be expressed by

$$\begin{aligned}\boldsymbol{e} &= \boldsymbol{B}\boldsymbol{H}^{\mathsf{H}}\left(\boldsymbol{H}\boldsymbol{H}^{\mathsf{H}} + \zeta\boldsymbol{I}\right)^{-1}\boldsymbol{y} - \boldsymbol{B}\boldsymbol{x} \\ &= \boldsymbol{B}\left(\boldsymbol{H}^{\mathsf{H}}\left(\boldsymbol{H}\boldsymbol{H}^{\mathsf{H}} + \zeta\boldsymbol{I}\right)^{-1}\boldsymbol{y} - \boldsymbol{x}\right) \\ &\triangleq \boldsymbol{B}\tilde{\boldsymbol{e}}\,.\end{aligned} \tag{E.1.13}$$

[3]Here, in contrast to the literature, $\boldsymbol{R}$ is lower triangular. This, however, does not change the main intention of the Cholesky factorization, and is guaranteed to exist, too.

It is easy to prove that the correlation matrix of the newly defined error vector $\tilde{e}$ calculates to

$$\boldsymbol{\Phi}_{\tilde{e}\tilde{e}} = \mathsf{E}\left\{\tilde{e}\tilde{e}^{\mathsf{H}}\right\} = \sigma_x^2\left(\boldsymbol{I} - \boldsymbol{H}^{\mathsf{H}}\left(\boldsymbol{H}\boldsymbol{H}^{\mathsf{H}} + \zeta\boldsymbol{I}\right)^{-1}\boldsymbol{H}\right) . \tag{E.1.14}$$

With the help of the *matrix inversion lemma* (Sherman-Morrison formula) [GL96, PTVF92], the following general statement can be shown

$$\boldsymbol{H}^{\mathsf{H}}\left(\boldsymbol{H}\boldsymbol{H}^{\mathsf{H}} + \zeta\boldsymbol{I}\right)^{-1} = \left(\boldsymbol{H}^{\mathsf{H}}\boldsymbol{H} + \zeta\boldsymbol{I}\right)^{-1}\boldsymbol{H}^{\mathsf{H}} , \tag{E.1.15}$$

and the correlation matrix can be written as

$$\begin{aligned}
\boldsymbol{\Phi}_{\tilde{e}\tilde{e}} &= \sigma_x^2\left(\boldsymbol{I} - \left(\boldsymbol{H}^{\mathsf{H}}\boldsymbol{H} + \zeta\boldsymbol{I}\right)^{-1}\boldsymbol{H}^{\mathsf{H}}\boldsymbol{H}\right) \\
&= \sigma_x^2\left(\boldsymbol{H}^{\mathsf{H}}\boldsymbol{H} + \zeta\boldsymbol{I}\right)^{-1}\left(\boldsymbol{H}^{\mathsf{H}}\boldsymbol{H} + \zeta\boldsymbol{I} - \boldsymbol{H}^{\mathsf{H}}\boldsymbol{H}\right) \\
&= \sigma_n^2\left(\boldsymbol{H}^{\mathsf{H}}\boldsymbol{H} + \zeta\boldsymbol{I}\right)^{-1} .
\end{aligned} \tag{E.1.16}$$

In the optimum, the error e is "white," i.e., $\boldsymbol{\Phi}_{ee} = \mathbf{diag}(\sigma_{e,1}^2, \ldots, \sigma_{e,D}^2)$. Considering that the correlation matrix of the error reads

$$\boldsymbol{\Phi}_{ee} = \boldsymbol{B} \cdot \boldsymbol{\Phi}_{\tilde{e}\tilde{e}} \cdot \boldsymbol{B}^{\mathsf{H}} , \tag{E.1.17}$$

the matrix $\boldsymbol{B}$ has to be the *whitening filter* for the process with correlation matrix $\boldsymbol{\Phi}_{\tilde{e}\tilde{e}}$. The matrix $\boldsymbol{B}$ and the corresponding gain matrix $\boldsymbol{G}$ can be obtained from the matrix $\boldsymbol{R} = \boldsymbol{G}^{-1}\boldsymbol{B}$, which is the result of a *Cholesky factorization* (cf. above) of

$$\boldsymbol{H}^{\mathsf{H}}\boldsymbol{H} + \zeta\boldsymbol{I} \overset{!}{=} \boldsymbol{R}^{\mathsf{H}}\boldsymbol{R} . \tag{E.1.18}$$

Here, $\boldsymbol{R}$ is again a lower triangular matrix. As expected, the MMSE solution approaches the ZF solution for high SNR ($\zeta \to 0$). The feedforward matrix $\boldsymbol{F}$ is then obtained from Eq. (E.1.12) as

$$\boldsymbol{F} = \boldsymbol{G}^{-1}\boldsymbol{B}\left(\boldsymbol{R}^{\mathsf{H}}\boldsymbol{R}\right)^{-1}\boldsymbol{H}^{\mathsf{H}} = \boldsymbol{R}^{-\mathsf{H}}\boldsymbol{H}^{\mathsf{H}} . \tag{E.1.19}$$

Note that for the MMSE solution, the feedforward matrix $\boldsymbol{F}$ is no longer unitary.

Finally, using equations (E.1.16) and (E.1.18) in (E.1.17), the correlation matrix of the error is

$$\boldsymbol{\Phi}_{ee} = \sigma_n^2 \cdot \mathbf{diag}(1/|r_{11}|^2, \ldots, 1/|r_{DD}|^2) , \tag{E.1.20}$$

or the noise variances of the parallel, independent channel induced by precoding are $\sigma_{n,k}^2 = \sigma_n^2/|r_{kk}|^2$, $k = 1, \ldots, D$.

E.2 DECENTRALIZED RECEIVERS

Now we study equalization of multiuser interference when a *central* transmitter communicates with D *distributed* or *decentralized* receivers (or users). Each receiver is assumed to have limited processing power. Hence they perform only linear filtering of their own received signal while no sophisticated detection algorithm is used.

E.2.1 Channel Model

As a prominent example for transmission from a central transmitter to decentralized receivers we look at the simplified DS-CDMA downlink transmission scenario. The equivalent complex baseband representation is depicted in Figure E.5. A base station where all user signals are present communicates with D receivers scattered over the service area. Communication takes place from a central *transmitter* (base station) to *distributed* receivers (mobile terminals).

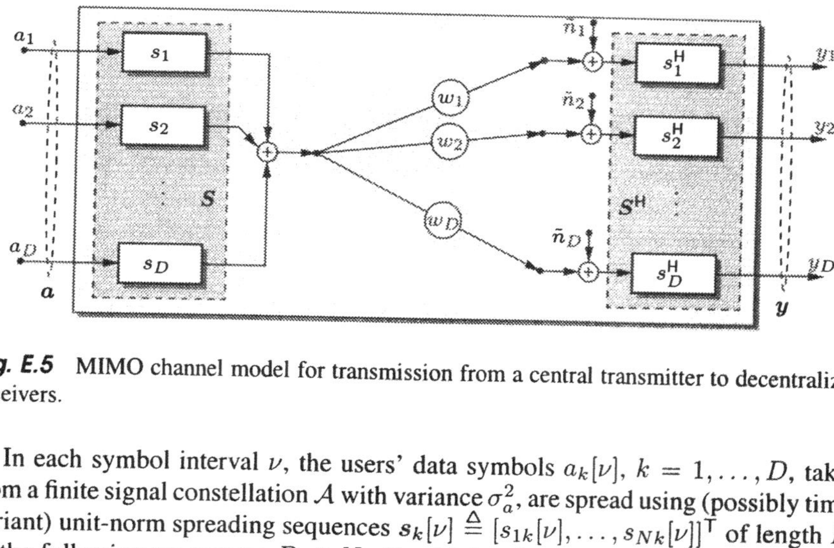

Fig. E.5 MIMO channel model for transmission from a central transmitter to decentralized receivers.

In each symbol interval ν, the users' data symbols $a_k[\nu]$, $k = 1, \ldots, D$, taken from a finite signal constellation $\mathcal{A}$ with variance σ_a^2, are spread using (possibly time-variant) unit-norm spreading sequences $\boldsymbol{s}_k[\nu] \triangleq [s_{1k}[\nu], \ldots, s_{Nk}[\nu]]^{\mathsf{T}}$ of length N. In the following we assume $D \leq N$. Combining the users' signals into the vector $\boldsymbol{a}[\nu] \triangleq [a_1[\nu], \ldots, a_D[\nu]]^{\mathsf{T}}$, and defining an $N \times D$ matrix of spreading sequences $\boldsymbol{S}[\nu] = [\boldsymbol{s}_1[\nu], \ldots, \boldsymbol{s}_D[\nu]]$, the transmit signal in symbol interval ν is given as $\boldsymbol{S}[\nu]\boldsymbol{a}[\nu]$.

The transmit signal is propagated to the D receivers over nondispersive (flat) fading channels with complex-valued path weights $w_1[\nu], \ldots, w_D[\nu]$. These weights are combined into the weight matrix $\boldsymbol{W}[\nu] \triangleq \mathbf{diag}(w_1[\nu], \ldots, w_D[\nu])$.

Each receiver k passes its received signal through the filter matched to its spreading sequences $\boldsymbol{s}_k[\nu]$, which yields the matched-filter output symbols

$$y_k[\nu] = \boldsymbol{s}_k^{\mathsf{H}}[\nu] w_k[\nu] \boldsymbol{S}[\nu] \boldsymbol{a}[\nu] + \boldsymbol{s}_k^{\mathsf{H}}[\nu] \tilde{\boldsymbol{n}}_k[\nu] \,. \tag{E.2.1}$$

Here, $\tilde{n}_k[\nu] = [\tilde{n}_{k1}[\nu], \ldots, \tilde{n}_{kN}[\nu]]^{\mathsf{T}}$ denotes the additive white zero-mean complex Gaussian channel noise at the input of receiver k with variance $\mathsf{E}\{|\tilde{n}_{k\kappa}[\nu]|^2\} = \sigma_n^2$, $\forall k, \kappa$. For decentralized receivers it is natural to assume that the channel noise is independent between the receivers, i.e., $\mathsf{E}\{\tilde{n}_k[\nu]\tilde{n}_\kappa{}^{\mathsf{H}}[\nu]\} = 0, \forall k \neq \kappa$.

Since the flat fading channel introduces no intersymbol interference, and assuming that all signals are wide-sense stationary, we may process the signals in each symbol interval ν separately. Hence, as we did in the last section, we regard one particular time interval and now omit the discrete-time index ν.

It is convenient to combine the matched-filter outputs y_k—although they are present at different locations—into a vector $y \triangleq [y_1, \ldots, y_D]^{\mathsf{T}}$. Then, the end-to-end transmission is given by

$$y = WS^{\mathsf{H}}Sa + n = Ha + n \ . \tag{E.2.2}$$

The overall MIMO channel is hence characterized by the matrix

$$H \triangleq WS^{\mathsf{H}}S \ , \tag{E.2.3}$$

and for the noise vector $n \triangleq [s_1^{\mathsf{H}}\tilde{n}_1, \ldots, s_D^{\mathsf{H}}\tilde{n}_D]^{\mathsf{T}}$ of the MIMO model, $\mathsf{E}\{nn^{\mathsf{H}}\} = \sigma_n^2 I$ holds.

E.2.2 Centralized Receiver and Decision-Feedback Equalization

In order to explain the nonlinear precoding scheme which is suited for the avoidance of multiuser interference at decentralized receivers, it is reasonable to first review the dual problem—the separation of the users' signals at the base station in an uplink scenario. Figure E.6 illustrates the situation together with nonlinear decision-feedback multiuser detection [Ver98]. A comparison with Figure E.2 shows, that this is exactly the matrix DFE structure discussed in the last section.

E.2.3 Decentralized Receivers and Precoding

The desired precoding scheme for a centralized transmitter and decentralized receivers can be derived immediately by taking the dualities between centralized receiver (Figure E.6) and centralized transmitter into consideration (Figure E.5).

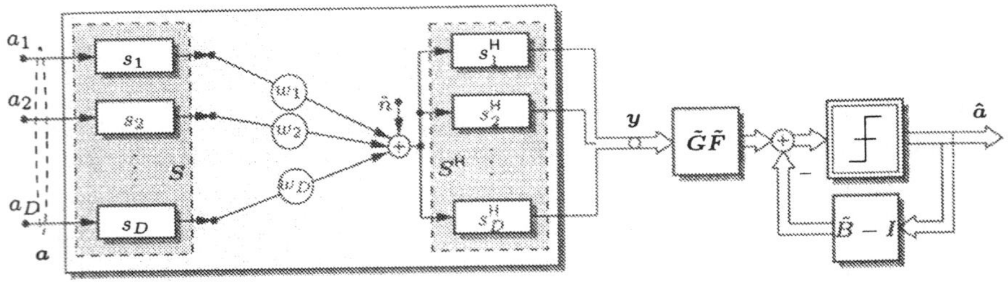

Fig. E.6 Decision-feedback multiuser detection for centralized receiver (cf. Figure E.2).

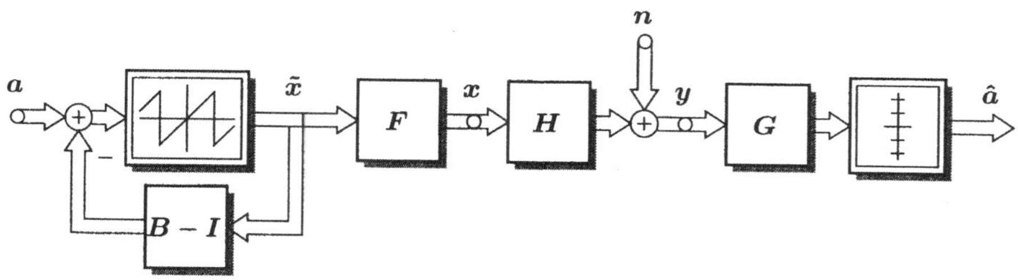

Fig. E.7 Precoding for decentralized receivers.

Basic Concept The counterpart of decision-feedback equalization at the receiver side is again Tomlinson-Harashima precoding at the transmitter side. However, the scheme given in the last section is not applicable here, since it would still require joint processing of the signals at the receiver by applying the feedforward matrix. Hence, the feedforward matrix $\boldsymbol{F}$ has be moved to the transmitter, too. The task of the feedforward matrix is to spatially whiten the channel noise and to force spatial causality. Since the channel noise is assumed to be white, only causality has to be achieved, which—in contrast to noise whitening—is also possible by a matrix at the transmitter. However, the operation of the precoder is still the same as given above. The resulting scheme is depicted in Figure E.7. Note, similar schemes were proposed independently for the multiantenna Gaussian broadcast channel [CS01] (see also [ESZ00]) and for canceling far-end crosstalk in digital subscriber line transmission [GC01b].

Calculation of the Matrix Filters Regarding the situation given above, the required matrices can now be calculated by decomposing the channel matrix according to (cf. equation (E.1.7))

$$\boldsymbol{H} = \boldsymbol{G}^{-1}\boldsymbol{B}\boldsymbol{F}^{\mathsf{H}} , \tag{E.2.4}$$

where $\boldsymbol{F}$ is a unitary matrix, $\boldsymbol{B}$ is a unit-diagonal lower triangular matrix, and $\boldsymbol{G} = \mathbf{diag}(g_1, \ldots, g_D)$ is a diagonal scaling matrix. Again, this is a QR-type decomposition of the channel matrix. The feedback matrix at the precoder is then again given as $\boldsymbol{B} - \boldsymbol{I}$.

Since $\boldsymbol{F}$ is a unitary matrix and defining a lower triangular matrix $\boldsymbol{R} \triangleq \boldsymbol{G}^{-1}\boldsymbol{B}$ as above, (E.2.4) can be rewritten as

$$\boldsymbol{H}\boldsymbol{H}^{\mathsf{H}} = \boldsymbol{R}\boldsymbol{F}^{\mathsf{H}}\boldsymbol{F}\boldsymbol{R}^{\mathsf{H}} = \boldsymbol{R}\boldsymbol{R}^{\mathsf{H}} . \tag{E.2.5}$$

Hence, the required matrices can also be obtained by performing a *Cholesky factorization* [BB91, GL96] of $\boldsymbol{H}\boldsymbol{H}^{\mathsf{H}}$, in contrast to a factorization of $\boldsymbol{H}^{\mathsf{H}}\boldsymbol{H}$, in the case of a central receiver.

For a central transmitter and decentralized receivers, this approach is also optimal with respect to the mean-squared error (MSE). For any choice of feedforward and feedback matrices, the error, present at the decision device, reads

$$\begin{aligned} e &= y - G^{-1} B F^{\mathsf{H}} x \\ &= (H - G^{-1} B F^{\mathsf{H}}) x + n \\ &\triangleq E x + n \, , \end{aligned} \tag{E.2.6}$$

with the obvious definition of E. Since transmit signal x and channel noise n are assumed to be white ($\mathsf{E}\{xx^{\mathsf{H}}\} = \sigma_x^2 I$ and $\mathsf{E}\{nn^{\mathsf{H}}\} = \sigma_n^2 I$) and mutually uncorrelated, the error covariance matrix is given by

$$\mathsf{E}\{ee^{\mathsf{H}}\} = \sigma_x^2 E E^{\mathsf{H}} + \sigma_n^2 I \, . \tag{E.2.7}$$

According to (E.2.4), for the particular choice of F and B $E = H - G^{-1} B F^{\mathsf{H}} = 0$ holds and the error covariance matrix reduces to

$$\mathsf{E}\{ee^{\mathsf{H}}\} = \sigma_n^2 I \, . \tag{E.2.8}$$

Since trace$(EE^{\mathsf{H}}) \geq 0$, in each case the total error power trace$(\mathsf{E}\{ee^{\mathsf{H}}\})$ is lower bounded by $D\sigma_n^2$. Since the ZF solution given above achieves this minimum, it is also optimum with respect to the MSE. That is, in precoding for decentralized receivers, where no joint processing of the received signal is possible, the zero-forcing solution is equal to the (unbiased) MMSE solution.

However, in the case of low-rate transmission, some additional gains are possible due to the properties of underlying *modulo channel* [FTC00]. Moreover, going to higher-dimensional precoding lattices Λ_{p}, the shaping gap can be bridged. In [ESZ00] a scheme denoted by "inflated lattice" precoding is proved to be capacity-achieving. Here, we concentrate on high rates/high SNRs and hence the ZF approach.

E.3 DISCUSSION

In this section, some properties of MIMO Tomlinson-Harashima precoding are discussed and possible extensions are briefly addressed. For implementation issues and performance evaluation of MIMO precoding, please refer to [FWLH02a, FWLH02b].

E.3.1 ISI Channels

Up to now, only flat fading channels have been considered. MIMO Tomlinson-Harashima precoding can be used in a straightforward was for channels which produce intersymbol interference. Then, joint spatial and temporal equalization is performed. Assuming that the channel is (almost) constant over one transmission burst, the elements of the channel matrix will be impulse responses, rather than constant gain factors.

Denoting the matrix of (causal) impulse responses as $\langle H[\nu] \rangle = [\langle h_{k\kappa}[\nu] \rangle]$, $H[\nu] = [h_{kl}[\nu]]$, $\langle h_{k\kappa}[\nu] \rangle = \langle h_{k\kappa}[0] h_{k\kappa}[1] \ldots \rangle$, the received signal in time interval

ν reads

$$y[\nu] = \sum_{\mu=0}^{\infty} H[\mu]x[\nu-\mu] + n[\nu] \,. \qquad \text{(E.3.1)}$$

For calculating the optimum feedforward and feedback matrices, we define the z-transform of the channel matrix as

$$H(z) \triangleq \left[\sum_{\nu=0}^{\infty} h_{k\kappa}[\nu]z^{-\nu}\right] = \sum_{\nu=0}^{\infty} H[\nu]z^{-\nu} \,. \qquad \text{(E.3.2)}$$

Then the Cholesky factorization (E.1.8) for a central receiver has to be replaced by the *spectral factorization problem*

$$H^{\mathsf{H}}(z^{-*})H(z) + \zeta I = R^{\mathsf{H}}(z^{-*})R(z) \,, \qquad \text{(E.3.3a)}$$

or for decentralized receivers by

$$H(z)H^{\mathsf{H}}(z^{-*}) + \zeta I = R(z)R^{\mathsf{H}}(z^{-*}) \,. \qquad \text{(E.3.3b)}$$

Here, the time-domain matrix $R[\nu]$ corresponding to $R(z) = \sum_{\nu=0}^{\infty} R[\nu]z^{-\nu}$ is strictly causal, i.e., $R[\nu] = 0$, $\nu < 0$, and the matrix $R[0] = [r_{k\kappa}[0]]$ is lower triangular. The feedforward matrix is then given as $F(z) = R^{\mathsf{H}}(z^{-*})H^{\mathsf{H}}(z^{-*})$, and the feedback matrix reads $B(z) = (\mathbf{diag}(s_{11}[0], \ldots, s_{DD}[0]))^{-1} S(z)$. For details on the factorization of spectral matrices and a fuller treatment of MIMO channels with ISI, see [You61, Dav63] and [FHK94, Fis96].

E.3.2 Application of Channel Coding

Using MIMO Tomlinson-Harashima precoding, the matrix channel is transformed into D parallel, independent SISO channels. However, these channels show different signal-to-noise ratios. Since in precoding schemes no immediate (zero-delay) decisions at the receiver are required for equalization, channel coding can be applied as on the AWGN channel or on fading channels.

In BLAST, where in effect matrix decision-feedback equalization is performed, the data streams of each antenna are encoded separately. The codewords are first decoded in the temporal (horizontal) direction, and then, using reencoded code symbols, the interference for the other, still undecoded, parallel data streams is cancelled. This approach, of course, involves a large decoding delay for each codeword. Moreover, in order to average over the different fading conditions seen by the parallel data streams, the assignment of (parts of) the codewords to the antennas may be permuted cyclicly. In the D-BLAST scheme [Fos96], the codewords are split up into different groups, where each group of symbols is transmitted over another antenna. At the receiver, processing is done diagonally (hence D-BLAST) over space and time.

For MIMO precoding, a single channel encoder can be used and the data stream may be demultiplexed onto the D antennas. In turn, channel coding in MIMO precoding exhibits a decoding delay which is lower by a factor of D compared to D-BLAST.

E.3.3 Application of Signal Shaping

In the precoding schemes described above, the effective data symbols are (implicitly) selected symbol by symbol. With regard to Section 5.2 on shaping without scrambling, an obvious extension to precoding is to determine the effective data symbols in the long run by a shaping algorithm, i.e., to perform combined precoding/signal shaping. Almost any property of the transmit or receive signal can be influenced with it. Again, the most important property is to create a transmit signal with lowest possible power (power shaping), but other shaping aims, such as controlling the dynamics range of the effective data symbols or reducing the peak-to-average power ratio (cf. Section 5.3), are also imaginable in MIMO precoding/shaping.

For flat MIMO channels, in each time interval the precoder starts from the all-zero state. Hence, the number of possible precoding sequences is finite and an exhaustive search—in contrast to the suboptimum one described in Section 5.2—might be possible. Reduced-state algorithms can be employed in combined spatial and temporal equalization.

E.3.4 Rate and Power Distribution

Since the signal-to-noise ratios of the D parallel channels may differ, the performance on these channels may vary significantly. System optimization inevitably leads to the so-called "loading" problem which initially arose in multicarrier transmission systems, cf., e.g., [FH96]. Here, total rate and total transmit power are distributed (loaded) over the parallel channel such that the system operates at lowest possible error rate. Other optimization criteria, such as, e.g., a maximization of the throughput [CCB95], are also possible.

REFERENCES

[BB91] E. J. Borowski and J. M. Borwein. *The HarperCollins Dictionary of Mathematics*. HarperPerennial, New York, 1991.

[CCB95] P. S. Chow, J. M. Cioffi, and J. A. C. Bingham. A Practical Discrete Multitone Transceiver Loading Algorithm for Data Transmission over Spectrally Shaped Channels. *IEEE Transactions on Communications*, COM-43, pp. 773–775, February/March/April 1995.

[CF96] J. M. Cioffi and G. D. Forney. Canonical Packet Transmission on the ISI Channel with Gaussian Noise. In *Proceedings of the IEEE Global Telecommunications Conference'96*, pp. 1405–1410, London, November 1996.

[CS01] G. Caire and S. Shamai (Shitz). On the Achievable Throughput of a Multi-Antenna Gaussian Broadcast Channel. Submitted for publication, July 2001.

[Dav63] M. C. Davis. Factoring the Spectral Matrix. *IEEE Transactions on Automatic Control*, AC-7, pp. 296-305, October 1963.

[Due95] A. Duel-Hallen. A Family of Multiuser Decision-Feedback Detectors for Asynchronous Code-Division Multiple-Access Channels. *IEEE Transactions on Communications*, COM-43, pp. 421–434, February/March/April 1995.

[ESZ00] U. Erez, S. Shamai, and R. Zamir. Capacity and Lattice-Strategies for Cancelling Known Interference. In *Proceedings of the International Symposium on Information Theory and Its Application (ISITA 2000)*, pp. 681–684, Honolulu, Hawaii, November 2000.

[FG98] G. Foschini and M. Gans. On limits of wireless communication in a fading environment when using multiple antennas. *Wireless Personal Communications*, 6, pp. 311–335, March 1998.

[FH96] R. Fischer and J. Huber. A New Loading Algorithm for Discrete Multitone Transmission. In *Proceedings of the IEEE Global Telecommunications Conference'96*, pp. 724–728, London, November 1996.

[FH97] R. Fischer and J. Huber. Comparison of Precoding Schemes for Digital Subscriber Lines. *IEEE Transactions on Communications*, COM-45, pp. 334–343, March 1997.

[FHK94] R. Fischer, J. Huber, and G. Komp. Coordinated Digital Transmission: Theory and Examples. *Archiv für Elektronik und Übertragungstechnik (International Journal of Electronics and Communications)*, 48, pp. 289–300, November/December 1994.

[Fis96] R. Fischer. *Mehrkanal- und Mehrträgerverfahren für die schnelle digitale Übertragung im Ortsanschlußleitungsnetz.* PhD thesis, Technische Fakultät der Universität Erlangen–Nürnberg, Erlangen, Germany, October 1996. (In German.)

[Fos96] G. Foschini. Layered Space-Time Architecture for Wireless Communication in a Fading Environment when Using Multiple Antennas. *Bell Laboratories Technical Journal*, pp. 41–59, Autumn 1996.

[FTC00] G. D. Forney, M. D. Trott, and S.-Y. Chung. Sphere-Bound-Achieving Coset Codes and Multilevel Coset Codes. *IEEE Transactions on Information Theory*, IT-46, pp. 820–850, May 2000.

[FWLH02a] R. Fischer, C. Windpassinger, A. Lampe, and J. Huber. Space-Time Transmission using Tomlinson-Harashima Precoding. In *Proceedings of the 4th International ITG Conference on Source and Channel Coding*, Berlin, January 2002.

[FWLH02b] R. Fischer, C. Windpassinger, A. Lampe, and J. Huber. Tomlinson-Harashima Precoding in Space-Time Transmission for Low-Rate Backward Channel. In *Proceedings of the 2002 International Zürich Seminar on Broadband Communications*, Zürich, Switzerland, February 2002.

[GC01a] G. Ginis and J. M. Cioffi. On the Relation Between V-BLAST and the GDFE. *IEEE Communications Letters*, COMML-5, pp. 364–366, September 2001.

[GC01b] G. Ginis and J. M. Cioffi. Vectored-DMT: A FEXT Canceling Modulation Scheme for Coordinating Users. In *Proceedings of the IEEE International Conference on Communications (ICC'01)*, Helsinki, Finland, June 2001.

[GFVW99] G. Golden, G. Foschini, R. Valenzuela, and P. Wolniasky. Detection algorithm and initial laboratory results using the V-BLAST space-time communication architecture. *Electronics Letters*, 35, pp. 14–15, January 1999.

[GL96] G. Golub and C. Van Loan. *Matrix Computations*. The Johns Hopkins University Press, Baltimore, MD, 1996.

[HF94] J. Huber and R. Fischer. Dynamically Coordinated Reception of Multiple Signals in Correlated Noise. In *Proceedings of the IEEE International Symposium on Information Theory*, pp. 132, Trondheim, Norway, June 1994.

[KAC90] S. Kasturia, J. T. Aslanis, and J. M. Cioffi. Vector Coding for Partial Response Channels. *IEEE Transactions on Information Theory*, IT-36, pp. 741–762, July 1990.

[PTVF92] W. H. Press, S. A. Teukolsky, W. T. Vetterling, and B. P. Flannery. *Numerical Recipes in C—The Art of Scientific Computing*. Cambridge University Press, Cambridge, 2nd edition, 1992.

[Tel99] E. Telatar. Capacity of multi-antenna Gaussian channels. *European Transactions on Telecommunications*, ETT-10, pp. 585–596, November 1999.

[Ver98] S. Verdú. *Multiuser Detection*. Cambridge University Press, New York, 1998.

[You61] D. C. Youla. On the Factorization of Rational Matrices. *IEEE Transactions on Information Theory*, IT-7, pp. 172–189, July 1961.

[YR94] J. Yang and S. Roy. Joint Transmitter-Receiver Optimization for Multi-Input Multi-Output Systems with Decision Feedback. *IEEE Transactions on Information Theory*, IT-40, pp. 1334–1347, September 1994.

[ZK89] N. Zervosa and I. Kalet. Optimized Decision Feedback Equalization Versus Optimized Orthogonal Frequency Division Multiplexing for High-Speed Data Transmission Over the Local Cable Network. In *Proceedings of the IEEE International Conference on Communications (ICC'89)*, pp. 1080–1085, September 1989.

Appendix F
List of Symbols, Variables, and Acronyms

... the fact that you are able to typeset your formulas with TEX doesn't necessarily mean that you have found the best notation for communicating with the readers of your work. Some notations will be unfortunate even when they are beautifully formatted.

—D. E. Knuth: *The TEXbook*, Addison-Wesley, Reading, MA, 1986.

Having this statement in mind, the present Appendix summarizes the most important notations. Relevant variables, constants, transforms and functions of the present work are tabulated. In addition, the major acronyms are listed.

F.1 IMPORTANT SETS OF NUMBERS AND CONSTANTS

$\mathbb{N}$	set of natural numbers (including 0)
$\mathbb{Z}$	set of integer numbers
$\mathbb{R}$	set of real numbers

$\mathbb{C}$	set of complex numbers
$\mathbb{F}_2$	finite field (Galois field) of cardinality 2
e	Euler number, $\mathrm{e} = 2.718281828\ldots$
j	imaginary number, $\mathrm{j}^2 = -1$
π	$\pi = 3.14159265358979\ldots$
∞	infinity

F.2 TRANSFORMS, OPERATORS, AND SPECIAL FUNCTIONS

Expectation $\mathsf{E}\{\cdot\}$

Fourier transform $X(f) = \mathcal{F}\{x(t)\} \triangleq \int_{-\infty}^{\infty} x(t)\,\mathrm{e}^{-\mathrm{j}2\pi f t}\,\mathrm{d}t$

Inverse Fourier transform

$$x(t) = \mathcal{F}^{-1}\{X(f)\} \triangleq \int_{-\infty}^{\infty} X(f)\,\mathrm{e}^{\mathrm{j}2\pi f t}\,\mathrm{d}f$$

Fourier transform pair

$$x(t) \circ\!\!-\!\!\bullet\, X(f)$$

z-transform $X(z) = \mathcal{Z}\{x[k]\} \triangleq \sum_k x[k]\,z^{-k}$

Inverse z-transform $x[k] = \mathcal{Z}^{-1}\{X(z)\} \triangleq \frac{1}{2\pi\mathrm{j}} \oint X(z)\,z^{k-1}\,\mathrm{d}z$

z-transform pair $x[k] \circ\!\!-\!\!\bullet\, X(z)$

(Linear) convolution $z(t) = x(t) * y(t)$, with $z(t) = \int_{-\infty}^{+\infty} x(\tau)y(t-\tau)\,\mathrm{d}\tau$

Real part and imaginary part of a complex variable

$$a = \mathrm{Re}\{z\},\ b = \mathrm{Im}\{z\}, \text{ with } z = a + \mathrm{j}\,b$$

Minimum/maximum value

$$\min_{x\in\mathcal{X}} f(x),\ \max_{x\in\mathcal{X}} f(x)$$

Argument, for which minimum/maximum value is assumed

$$\operatorname*{argmin}_{x\in\mathcal{X}} f(x),\ \operatorname*{argmax}_{x\in\mathcal{X}} f(x)$$

Probability of the discrete event X (conditioned on Y)

$$\Pr\{X\},\ \Pr\{X \mid Y\}$$

Probability density function (pdf) of random variable x (conditioned on y)

$$\mathsf{f}_x(x),\ \mathsf{f}_x(x \mid y)$$

Discrete-time Dirac impulse

$$\delta[k] \triangleq \begin{cases} 1, & k = 0 \\ 0, & k \neq 0 \end{cases}$$

Dirac delta function $\delta(t)$, with $\int_{-\infty}^{+\infty} x(t)\delta(t - t_0)\,\mathrm{d}t = x(t_0)$

Logarithm of x to base b, natural logarithm of x

$$\log_b(x), \ln(x)$$

Sign function $\mathrm{sgn}(x) \triangleq \begin{cases} 1, & x \geq 0 \\ -1, & x < 0 \end{cases}$

Rect function $\mathrm{rect}(x) \triangleq \begin{cases} 1, & -1/2 \leq x < 1/2 \\ 0, & \text{else} \end{cases}$

Complementary Gaussian error integral

$$\mathrm{Q}(x) \triangleq \frac{1}{\sqrt{2\pi}} \int_x^{\infty} \mathrm{e}^{-t^2/2}\,\mathrm{d}t$$

Absolute value (of a scalar)

$$|\cdot|$$

Conjugation of a complex number

$$z^* = a - jb, \text{ with } z = a + jb$$

Phase angle (argument) of a complex variable

$$\arg\{\cdot\} \in (-\pi, \pi]$$

Conjugation and inversion

$$z^{-*} = (z^*)^{-1}$$

Floor/ceiling function $y = \lfloor x \rfloor$; smallest $y \in \mathbb{Z}$ not less than $x \in \mathbb{R}$

$y = \lceil x \rceil$; largest $y \in \mathbb{Z}$ not exceeding $x \in \mathbb{R}$

Derivation/partial derivation with respect to x

$$\frac{\mathrm{d}}{\mathrm{d}x}, \frac{\partial}{\partial x}$$

Estimate $\hat{\cdot}$

Sequence $\langle \cdot \rangle$

Equal per definition $\triangleq$

$n \times m$ matrix with elements x_{ij}

$$[x_{ij}]_{\substack{i=1,\ldots,n \\ j=1,\ldots,m}}$$

Identity matrix of dimension d

$$\boldsymbol{I}_d$$

Zero matrix of dimension $n \times m$

$$\mathbf{0}_{n\times m}$$

Transpose of matrix/vector $\boldsymbol{X}$

$$\boldsymbol{X}^{\mathsf{T}}$$

Conjugate transpose (Hermitian transpose) of matrix/vector $\boldsymbol{X}$

$$\boldsymbol{X}^{\mathsf{H}} \triangleq \left(\boldsymbol{X}^{\mathsf{T}}\right)^*$$

Inverse of matrix $\boldsymbol{X}$ $\quad \boldsymbol{X}^{-1}$

Determinant of matrix $\boldsymbol{X}$

$$|\boldsymbol{X}|$$

F.3 IMPORTANT VARIABLES

f	frequency
k	discrete time index
t	continuous time
C	capacity
E	energy
E_{s}	energy per symbol
E_{b}	energy per information bit
G_{c}	coding gain
G_{p}	prediction gain
G_{s}	shaping gain
N_0	(one-sided) noise power spectral density
R	rate
S	signal power
SNR	signal-to-noise ratio
T	symbol interval
γ^2	precoding loss
σ^2	variance
ϑ^2	loss factor
$a[k]$	PAM data symbol
$d[k]$	precoding symbols
$e[k]$	error symbols
$f[k]$	sum of postcursors
$n[k]$	noise sample
$x[k]$	channel (input) symbol
$y[k]$, $r[k]$	unprocessed/processed receive symbol
$v[k]$	effective data symbols

$s(t)$	continuous-time transmit signal
$r(t)$	continuous-time receive signal
$n_0(t)$	continuous-time noise signal
$H(f), H(z)$	continuous-time/discrete-time transfer function
$h(t), h[k]$	continuous-time/discrete-time impulse response
$\Phi_{xx}(f), \Phi_{xx}(e^{j2\pi fT})$	continuous-time/discrete-time power spectral density of process x
$\varphi_{xx}(t), \varphi_{xx}[k]$	continuous-time/discrete-time autocorrelation of process x
$\varphi_{xy}(t), \varphi_{xy}[k]$	continuous-time/discrete-time cross-correlation of process x and y
$\mathcal{A}$	set of PAM data symbols $a[k]$ (signal constellation)
$\mathcal{V}$	set of effective data symbols $v[k]$
$\mathbf{\Lambda}$	lattice
$\mathbf{\Lambda}_a$	signal lattice
$\mathbf{\Lambda}_c$	coding or coset lattice lattice
$\mathbf{\Lambda}_p$	precoding lattice
$\mathbf{\Lambda}_s$	shaping lattice
$V(\cdot)$	volume
$\mathcal{R}$	region
$\mathcal{R}_V(\mathbf{\Lambda})$	Voronoi region of lattice $\mathbf{\Lambda}$

F.4 ACRONYMS

acf	autocorrelation function
A/D	analog/digital
ADSL	asymmetric digital subscriber lines
AMI	alternate mark inversion
ANSI	American National Standards Institute
ASIC	application-specific integrated circuit
ASK	amplitude-shift keying
AWGN	additive white Gaussian noise
BER	bit error rate
BIBO	bounded-input/bounded-output
BLAST	Bell Laboratories layered space-time
CAP	carrierless AM/PM
ccdf	complementary cumulative distribution function
CDMA	code-division multiple access
CER	constellation expansion ratio

CFM	constellation figure of merit
CO	central office
DC	direct current
DFE	decision-feedback equalization
DFT	discrete-time Fourier transform
DLP	dynamics limited precoding
DMT	discrete multitone
DPCM	differential pulse code modulation
DSL	digital subscriber lines
DSP	digital signal processor
ECB	equivalent complex baseband
EDS	effective data sequence
ETSI	European Telecommunications Standards Institute
FEXT	far-end crosstalk
FFT	fast Fourier transform
FIR	finite impulse response
FLP	flexible precoding
FPGA	field-programmable gate array
HDSL	high-rate digital subscriber lines
i.i.d.	independent, identically distributed
IIR	infinite impulse response
ISDN	integrated services digital network
ISI	intersymbol interference
LE	linear equalization
LT	line termination
MF	matched filter
MIMO	multiple-input/multiple-output
ML	maximum likelihood
MLC	multilevel coding
MLSE	maximum-likelihood sequence estimation
MMSE	minimum mean-squared error
MSE	mean-squared error
NEXT	near-end crosstalk
NP	noise prediction
NT	network termination
OFDM	orthogonal frequency division multiplexing
ONF	optimum Nyquist filter
PAM	pulse amplitude modulation
PAR	peak-to-average power (energy) ratio
PCM	pulse code modulation
pdf	probability density function
PDFD	parallel decision-feedback decoding
POTS	plain old telephone system
PSD	power spectral density
PSK	phase-shift keying

QAM	quadrature amplitude modulation
RDS	running digital sum
RFS	running filter sum
RSSE	reduced-state sequence estimation
SER	symbol error rate
SDSL	single-pair digital subscriber lines
SISO	single-input/single-output
SNR	signal-to-noise ratio
SVD	singular value decomposition
TCM	trellis-coded modulation
TCQ	trellis-coded quantization
THP	Tomlinson-Harashima precoding
TS	Trellis shaping
VDSL	very high-rate digital subscriber lines
WMF	whitened matched filter
ZF	zero forcing

Index

A

B

C

D

Q

R

W

X

Y

Z

Printed and bound by CPI Group (UK) Ltd, Croydon, CR0 4YY

16/09/2026

14963441-0007